STUDENT'S SOLUTIONS MANUAL
MULTIVARIABLE

WILLIAM ARDIS
Collin County Community College

MARK WOODARD
Furman University

CALCULUS:
EARLY TRANSCENDENTALS
AND
CALCULUS

William Briggs
University of Colorado, Denver

Lyle Cochran
Whitworth University

With the assistance of

Bernard Gillett
University of Colorado, Boulder

Addison-Wesley
is an imprint of

Reproduced by Addison-Wesley from electronic files supplied by the author.

Publishing as Pearson Addison-Wesley, 75 Arlington Street, Boston, MA 02116.

ISBN-13: 978-0-321-66411-2
ISBN-10: 0-321-66411-6

1 2 3 4 5 6 BB 14 13 12 11 10

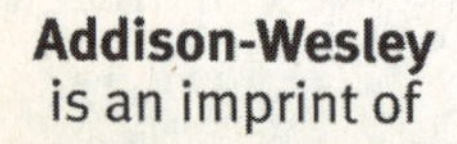

www.pearsonhighered.com

Contents

The chapter numbers in this manual corresonpond to those found in Calculus: Early Transcendentals. *The chapter numbers for* Calculus *are one higher than the numbers listed.*

8 Sequences and Infinite Series 285

8.1 An Overview ... 285
8.2 Sequences ... 289
8.3 Infinite Series ... 294
8.4 The Divergence and Integral Tests ... 298
8.5 The Ratio, Root, and Comparison Tests ... 301
8.6 Alternating Series ... 304
8.7 Chapter Eight Review ... 306

9 Power Series 311

9.1 Approximating Functions With Polynomials ... 311
9.2 Properties of Power Series ... 317
9.3 Taylor Series ... 320
9.4 Working with Taylor Series ... 324
9.5 Chapter Nine Review ... 330

10 Parametric and Polar Curves 335

10.1 Parametric Equations ... 335
10.2 Polar Coordinates ... 344
10.3 Calculus in Polar Coordinates ... 353
10.4 Conic Sections ... 358
10.5 Chapter Ten Review ... 368

11 Vectors and Vector-Valued Functions 377

11.1 Vectors in the Plane ... 377
11.2 Vectors in Three Dimensions ... 381
11.3 Dot Products ... 384
11.4 Cross Products ... 388
11.5 Lines and Curves in Space ... 393
11.6 Calculus of Vector-Valued Functions ... 396
11.7 Motion in Space ... 399
11.8 Length of Curves ... 406
11.9 Curvature and Normal Vectors ... 408
11.10 Chapter Eleven Review ... 413

12 Functions of Several Variables **417**

12.1 Planes and Surfaces 417
12.2 Graphs and Level Curves 426
12.3 Limits and Continuity 431
12.4 Partial Derivatives 434
12.5 The Chain Rule 438
12.6 Directional Derivatives and the Gradient 442
12.7 Tangent Planes and Linear Approximation 449
12.8 Maximum/Minimum Problems 453
12.9 Lagrange Multipliers 458
12.10 Chapter Twelve Review 463

13 Multiple Integration **471**

13.1 Double Integrals over Rectangular Regions 471
13.2 Double Integrals over General Regions 475
13.3 Double Integrals in Polar Coordinates 483
13.4 Triple Integrals 491
13.5 Triple Integrals in Cylindrical and Spherical Coordinates 496
13.6 Integrals for Mass Calculations 502
13.7 Change of Variables in Multiple Integrals 508
13.8 Chapter Thirteen Review 514

14 Vector Calculus **521**

14.1 Vector Fields 521
14.2 Line Integrals 525
14.3 Conservative Vector Fields 530
14.4 Green's Theorem 533
14.5 Divergence and Curl 538
14.6 Surface Integrals 543
14.7 Stokes' Theorem 549
14.8 Divergence Theorem 552
14.9 Chapter Fourteen Review 557

Chapter 8

8.1 An Overview

8.1.1 A *sequence* is a list of numbers $a_1, a_2, a_3, \ldots$, often written $\{a_1, a_2, \ldots\}$ or $\{a_n\}$. For example, the natural numbers $\{1, 2, 3, ...\}$ are a sequence where $a_n = n$ for every n.

8.1.3 $a_1 = 1$ (given); $a_2 = 1 \cdot a_1 = 1$; $a_3 = 2 \cdot a_2 = 2$; $a_4 = 3 \cdot a_3 = 6$; $a_5 = 4 \cdot a_4 = 24$.

8.1.5 An *infinite series* is an infinite sum of numbers. Thus if $\{a_n\}$ is a sequence, then $a_1 + a_2 + \cdots = \sum_{i=1}^{\infty} a_i$, is an infinite series. For example, if $a_i = \frac{1}{i}$, then $\sum_{i=1}^{\infty} a_i = \sum_{i=1}^{\infty} \frac{1}{i}$ is an infinite series.

8.1.7 $S_1 = \sum_{k=1}^{1} k^2 = 1$; $S_2 = \sum_{k=1}^{2} k^2 = 1 + 4 = 5$; $S_3 = \sum_{k=1}^{3} k^2 = 1 + 4 + 9 = 14$; $S_4 = \sum_{k=1}^{4} k^2 = 1 + 4 + 9 + 16 = 30$.

8.1.9 $a_1 = \dfrac{1}{10}; a_2 = \dfrac{1}{100}; a_3 = \dfrac{1}{1000}; a_4 = \dfrac{1}{10000}.$

8.1.11 $a_1 = 1 + \sin(\pi/2) = 2$; $a_2 = 1 + \sin(2\pi/2) = 1 + \sin(\pi) = 1$; $a_3 = 1 + \sin(3\pi/2) = 0$; $a_4 = 1 + \sin(4\pi/2) = 1 + \sin(2\pi) = 1$.

8.1.13 $a_1 = 10$ (given); $a_2 = 3 \cdot a_1 - 12 = 30 - 12 = 18$; $a_3 = 3 \cdot a_2 - 12 = 54 - 12 = 42$; $a_4 = 3 \cdot a_3 - 12 = 126 - 12 = 114$.

8.1.15 $a_1 = 0$ (given); $a_2 = 3 \cdot a_1^2 + 1 + 1 = 2$; $a_3 = 3 \cdot a_2^2 + 2 + 1 = 15$; $a_4 = 3 \cdot a_3^2 + 3 + 1 = 679$.

8.1.17

a. $\frac{1}{32}, \frac{1}{64}$.

b. $a_1 = 1$; $a_{n+1} = \frac{a_n}{2}$.

c. $a_n = \frac{1}{2^{n-1}}$.

8.1.19

a. 32, 64.

b. $a_1 = 1$; $a_{n+1} = 2 \cdot a_n$.

c. $a_n = 2^{n-1}$.

8.1.21

a. 243, 729.

b. $a_1 = 1$; $a_{n+1} = 3 \cdot a_n$.

c. $a_n = 3^{n-1}$.

8.1.23 $a_1 = 9$, $a_2 = 99$, $a_3 = 999$, $a_4 = 9999$. This sequence diverges, since the terms get larger without bound.

8.1.25 $a_1 = -1$, $a_2 = \frac{1}{2}$, $a_3 = -\frac{1}{3}$, $a_4 = \frac{1}{4}$. This sequence converges to 0 since each term is smaller in absolute value than the preceding term and they get arbitrarily close to zero.

8.1.27 Rewrite the recurrence as $a_{n+1} = 10^{-1}a_n^2$. Then $a_0 = 1$, $a_1 = 10^{-1} = \frac{1}{10}$, $a_2 = 10^{-1}(10^{-1})^2 = 10^{-3} = \frac{1}{1000}$, $a_3 = 10^{-1}(10^{-3})^2 = 10^{-7} = \frac{1}{10000000}$, $a_4 = 10^{-1}(10^{-7})^2 = 10^{-15} = \frac{1}{1000000000000000}$. This sequence converges to 0.

8.1.29 $a_0 = 100$, $a_1 = 0.5 \cdot 100 + 50 = 100$, $a_2 = 0.5 \cdot 100 + 50 = 100$, $a_3 = 0.5 \cdot 100 + 50 = 100$, $a_4 = 0.5 \cdot 100 + 50 = 100$. This sequence obviously converges to 100.

8.1.31

a. 1, 2, 3, 4.

b. 1, 2, 3, 4, 5, 6, 7, 8, 9, 10, 11, ... This sequence diverges.

8.1.33

a. 0, 2, 6, 12.

b.

n	1	2	3	4	5	6	7	8	9	10
a_n	0	2	6	12	20	30	42	56	72	90

This sequence appears to diverge.

8.1.35

a. 1/3, 1/2, 3/5, 2/3.

b.

n	2	3	4	5	6	7	8	9	10	11
a_n	1/3	1/2	3/5	2/3	5/7	3/4	7/9	4/5	9/11	5/6

This sequence appears to converge to 1.

8.1.37

a. 5/2, 9/4, 17/8, 33/16.

b. The limit is 2.

8.1.39

a. $a_0 = 3$, $a_1 = 5$,$a_2 = 7$, $a_3 = 9$.

b. $a_n = 2n + 3$.

c.

n	0	1	2	3	4	5	6	7	8	9	10
a_n	3	5	7	9	11	13	15	17	19	21	23

This sequence diverges.

8.1.41

a. $a_0 = 0$, $a_1 = 1$, $a_2 = 3$, $a_3 = 7$, $a_4 = 15$.

b. $a_n = 2^n - 1$.

c.

n	0	1	2	3	4	5	6	7	8	9	10
a_n	0	1	3	7	15	31	63	127	255	511	1023

This sequence diverges.

8.1.43

a. $a_0 = 1$, $a_1 = 3/2$, $a_2 = 7/4$, $a_3 = 15/8$, $a_4 = 31/16$.

b. $a_n = \frac{2^{n+1}-1}{2^n} = 2 - \frac{1}{2^n}$.

c.

n	0	1	2	3	4	5	6	7	8	9	10
a_n	1	3/2	7/4	15/8	31/16	63/32	127/64	255/128	511/256	1023/512	2047/1024

This sequence converges to 2.

8.1.45

a. 20, 10, 5, 2.5.

b. $h_n = 20 \cdot (0.5)^n$.

8.1.47

a. 30, 7.5, 1.875, 0.46875.

b. $h_n = 30 \cdot (0.25)^n$.

8.1.49 $S_1 = 0.3$, $S_2 = 0.33$, $S_3 = 0.333$, $S_4 = 0.3333$. It appears that the infinite series has a value of $0.3333\ldots = \frac{1}{3}$.

8.1.51 $S_1 = 4$, $S_2 = 4.9$, $S_3 = 4.99$, $S_4 = 4.999$. The infinite series has a value of $4.999\cdots = 5$.

8.1.53

a. $S_1 = \frac{2}{3}$, $S_2 = \frac{4}{5}$, $S_3 = \frac{6}{7}$, $S_4 = \frac{8}{9}$.

b. It appears that $S_n = \frac{2n}{2n+1}$.

c. The series has a value of 1 (the partial sums converge to 1).

8.1.55

a. $S_1 = \frac{1}{3}$, $S_2 = \frac{2}{5}$, $S_3 = \frac{3}{7}$, $S_4 = \frac{4}{9}$.

b. $S_n = \frac{n}{2n+1}$.

c. The partial sums converge to $\frac{1}{2}$, which is the value of the series.

8.1.57

a. True. For example, $S_2 = 1 + 2 = 3$, and $S_4 = a_1 + a_2 + a_3 + a_4 = 1 + 2 + 3 + 4 = 10$.

b. False. For example, $\frac{1}{2}, \frac{3}{4}, \frac{7}{8}, \cdots$ where $a_n = 1 - \frac{1}{2^n}$ converges to 1, but each term is greater than the previous one.

c. True. In order for the partial sums to converge, they must get closer and closer together. In order for this to happen, the difference between successive partial sums, which is just the value of a_n, must approach zero.

8.1.59 Using the work from the previous problem:

a. Here $h_0 = 20$, $r = 0.75$, so $S_0 = 40$, $S_1 = 40 + 40 \cdot 0.75 = 70$, $S_2 = S_1 + 40 \cdot (0.75)^2 = 92.5$, $S_3 = S_2 + 40 \cdot (0.75)^3 = 109.375$, $S_4 = S_3 + 40 \cdot (0.75)^4 = 122.03125$

b.

n	0	1	2	3	4	5
a_n	40	70	92.5	109.375	122.0313	131.5234
n	6	7	8	9	10	11
a_n	138.6426	143.9819	147.9865	150.9898	153.2424	154.9318
n	12	13	14	15	16	17
a_n	156.1988	157.1491	157.8618	158.3964	158.7973	159.0980
n	18	19	20	21	22	23
a_n	159.3235	159.4926	159.6195	159.715	159.786	159.839

The sequence converges to 160.

8.1.61

a. 0.5, 0.75, 0.875, .9375.

b. The limit is 1.

8.1.63

a. $\frac{1}{3}, \frac{4}{9}, \frac{13}{27}, \frac{40}{81}$.

b. The limit is 1/2.

8.1.65

a. -1, 0, -1, 0.

b. The limit does not exist.

8.1.67

a. $\frac{3}{10} = 0.3$, $\frac{33}{100} = 0.33$, $\frac{333}{1000} = 0.333$, $\frac{3333}{10000} = 0.3333$.

b. The limit is 1/3.

8.1.69

a. $M_0 = 20$, $M_1 = 20 \cdot 0.5 = 10$, $M_2 = 20 \cdot 0.5^2 = 5$, $M_3 = 20 \cdot 0.5^3 = 2.5$, $M_4 = 20 \cdot 0.5^4 = 1.25$

b. $M_n = 20 \cdot 0.5^n$.

c. The initial mass is M_0. We are given that 50% of the mass is gone after each decade, so that $M_{n+1} = 0.5 \cdot M_n$, $n \geq 0$.

d. The amount of material goes to 0.

8.1.71

a. $d_0 = 200$, $d_1 = 200 \cdot .95 = 190$, $d_2 = 200 \cdot .95^2 = 180.5$, $d_3 = 200 \cdot .95^3 = 171.475$, $d_4 = 200 \cdot .95^4 = 162.90125$.

b. $d_n = 200 \cdot (0.95)^n$, $n \geq 0$.

c. We are given $d_0 = 200$; since 5% of the drug is washed out every hour, that means that 95% of the preceding amount is left every hour, so that $d_{n+1} = 0.95 \cdot d_n$.

d. The sequence converges to 0.

8.1.73

a. $0.333\ldots = \sum_{i=1}^{\infty} 3(0.1)^i$.

b. The limit of the sequence of partial sums is 1/3.

8.1.75

a. $0.111\ldots = \sum_{i=1}^{\infty} (0.1)^i$.

b. The limit of the sequence of partial sums is 1/9.

8.1.77

a. $0.0909\ldots = \sum_{i=1}^{\infty} 9(0.01)^i$.

b. The limit of the sequence of partial sums is 1/11.

8.1.79

a. $0.037037037\ldots = \sum_{i=1}^{\infty} 37(0.001)^i$.

b. The limit of the sequence of partial sums is $37/999 = 1/27$.

8.2 Sequences

8.2.1 There are many examples; one is $a_n = \frac{1}{n}$. This sequence is nonincreasing (in fact, it is strictly decreasing) and has a limit of 0.

8.2.3 There are many examples; one is $a_n = \frac{1}{n}$. This sequence is nonincreasing (in fact, it is strictly decreasing), is bounded above by 1 and below by 0, and has a limit of 0.

8.2.5 $\{r^n\}$ converges for $|r| < 1$ and for $r = 1$. It diverges for all other values of r (see Theorem 8.3).

8.2.7 A sequence a_n converges to l if, given any $\epsilon > 0$, there exists a positive integer N, such that whenever $n > N$, $|a_n - L| < \varepsilon$.

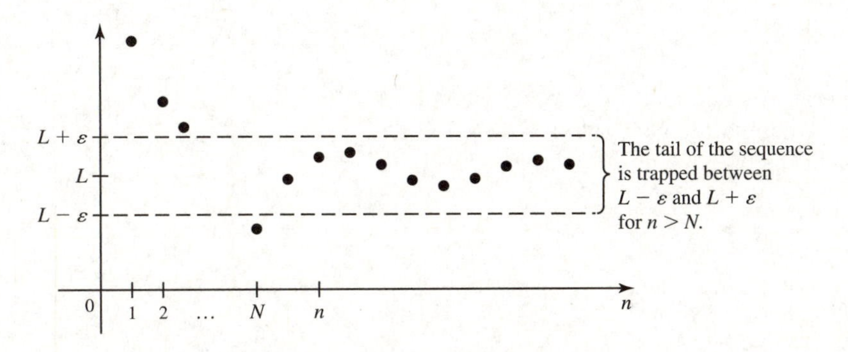

8.2.9 Divide numerator and denominator by n^4 to get $\lim_{n\to\infty} \frac{1/n}{1+\frac{1}{n^4}} = 0$.

8.2.11 Divide numerator and denominator by n^3 to get $\lim_{n\to\infty} \frac{3-n^{-3}}{2+n^{-3}} = \frac{3}{2}$.

8.2.13 As $n \to \infty$, $\tan^{-1} n \to \pi/2$, so $\frac{\tan^{-1} n}{n} \to 0$.

8.2.15 Find the limit of the logarithm of the expression, which is $n \ln\left(1 + \frac{2}{n}\right)$. Using L'Hôpital's rule: $\lim_{n\to\infty} n \ln\left(1 + \frac{2}{n}\right) = \lim_{n\to\infty} \frac{\ln(1+\frac{2}{n})}{1/n} = \lim_{n\to\infty} \frac{\frac{1}{1+(2/n)}\left(\frac{-2}{n^2}\right)}{-1/n^2} = \lim_{n\to\infty} \frac{2}{1+(2/n)} = 2$. Thus the limit of the original expression is e^2.

8.2.17 Take the logarithm of the expression and use L'Hôpital's rule:

$$\lim_{n\to\infty} \frac{n}{2} \ln\left(1 + \frac{1}{2n}\right) = \lim_{n\to\infty} \frac{\ln(1 + (1/2n))}{2/n} = \lim_{n\to\infty} \frac{\frac{1}{1+(1/2n)} \cdot \frac{-1}{2n^2}}{-2/n^2} = \lim_{n\to\infty} \frac{1}{4(1 + (1/2n))} = \frac{1}{4}.$$

Thus the original limit is $e^{1/4}$.

8.2.19 Taking logs, we have $\lim_{n\to\infty} \frac{1}{n} \ln(1/n) = \lim_{n\to\infty} -\frac{\ln n}{n} = \lim_{n\to\infty} \frac{-1}{n} = 0$ by L'Hôpital's rule. Thus the original sequence has limit $e^0 = 1$.

8.2.21 Except for a finite number of terms, this sequence is just $a_n = ne^{-n}$, so it has the same limit as this sequence. Note that $\lim_{n\to\infty} \frac{n}{e^n} = \lim_{n\to\infty} \frac{1}{e^n} = 0$, by L'Hôpital's rule.

8.2.23 $\ln(\sin(1/n)) + \ln n = \ln(n \sin(1/n)) = \ln\left(\frac{\sin(1/n)}{1/n}\right)$. As $n \to \infty$, $\sin(1/n)/(1/n) \to 1$, so the limit of the original sequence is $\ln 1 = 0$.

8.2.25 $\lim_{n\to\infty} n \sin(6/n) = \lim_{n\to\infty} \frac{\sin(6/n)}{1/n} = \lim_{n\to\infty} \frac{\frac{-6\cos(6/n)}{n^2}}{(-1/n^2)} = \lim_{n\to\infty} 6\cos(6/n) = 6$.

8.2.27 When n is an integer, $\sin\left(\frac{n\pi}{2}\right)$ oscillates between the values ± 1 and 0, so this sequence does not converge.

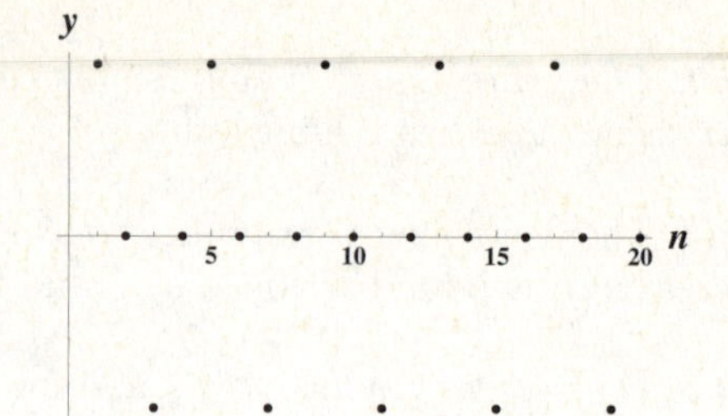

8.2.29 The numerator is bounded in absolute value by 1, while the denominator goes to ∞, so the limit of this sequence is 0.

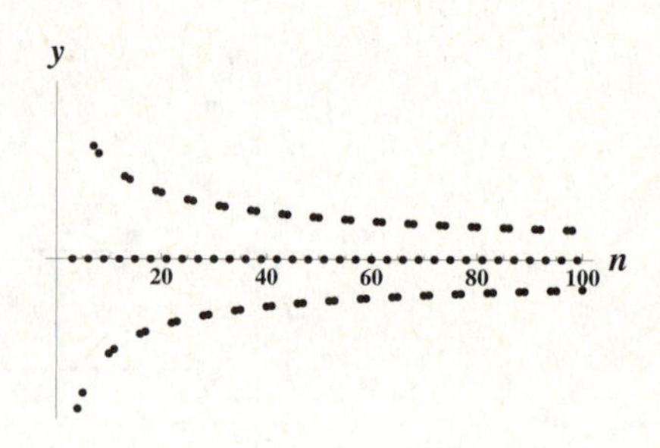

8.2.31 This is the sequence $\frac{\cos n}{e^n}$; the numerator is bounded in absolute value by 1 and the denominator increases without bound, so the limit is zero.

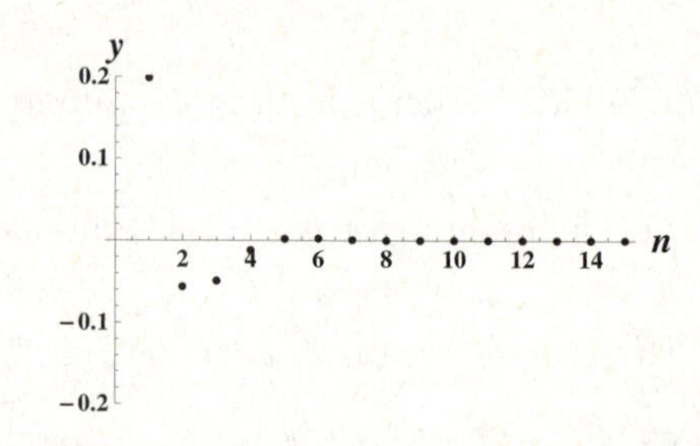

8.2.33 Ignoring the factor of $(-1)^n$ for the moment, we see, taking logs, that $\lim_{n\to\infty} \frac{\ln n}{n} = 0$, so that $\lim_{n\to\infty} \sqrt[n]{n} = e^0 = 1$. Taking the sign into account, the odd terms converge to -1 while the even terms converge to 1. Thus the sequence does not converge.

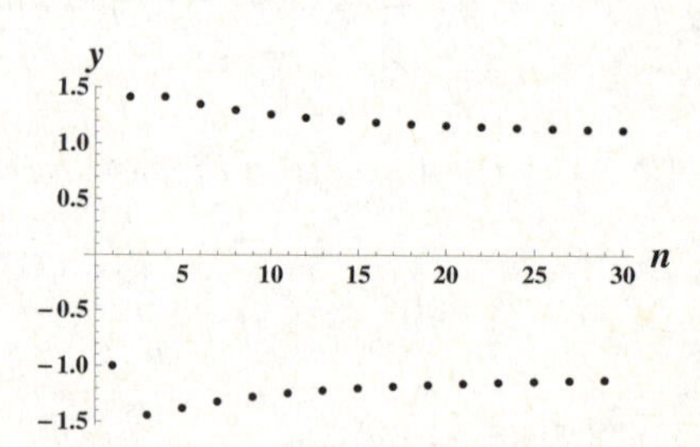

8.2.35 Since $0.2 < 1$, this sequence converges to 0. Since $0.2 > 0$, the convergence is monotone.

8.2.37 Since $|-0.7| < 1$, the sequence converges to 0; since $-0.7 < 0$, it does not do so monotonically.

8.2.39 Since $1.00001 > 1$, the sequence diverges; since $1.00001 > 0$, the divergence is monotone.

8.2.41 Since $|-2.5| > 1$, the sequence diverges; since $-2.5 < 0$, the divergence is not monotone.

8.2.43 Since $-1 \le \sin n \le 1$ for all n, the given sequence satisfies $\frac{-1}{2^n} \le \frac{\sin n}{2^n} \le \frac{1}{2^n}$, and since both $\pm\frac{1}{2^n} \to 0$ as $n \to \infty$, the given sequence converges to zero as well by the Squeeze Theorem.

8.2.45 $\tan^{-1}$ takes values between $-\pi/2$ and $\pi/2$, so the numerator is always between $-\pi$ and π. Thus $\frac{-\pi}{n^3+4} \le \frac{2\tan^{-1} n}{n^3+4} \le \frac{\pi}{n^3+4}$, and by the Squeeze Theorem, the given sequence converges to zero.

8.2.47

a. After the nth dose is given, the amount of drug in the bloodstream is $d_n = 0.5 \cdot d_{n-1} + 80$, since the half-life is one day. The initial condition is $d_1 = 80$.

b. The limit of this sequence is 160 mg.

c. Let $L = \lim_{n\to\infty} d_n$. Then from the recurrence relation, we have $d_n = 0.5 \cdot d_{n-1} + 80$, and thus $\lim_{n\to\infty} d_n = 0.5 \cdot \lim_{n\to\infty} d_{n-1} + 80$, so $L = 0.5 \cdot L + 80$, and therefore $L = 160$.

8.2.49

a.

$$\begin{aligned}
B_0 &= 0\\
B_1 &= 1.0075 \cdot B_0 + \$100 = \$100\\
B_2 &= 1.0075 \cdot B_1 + \$100 = \$200.75\\
B_3 &= 1.0075 \cdot B_2 + \$100 = \$302.26\\
B_4 &= 1.0075 \cdot B_3 + \$100 = \$404.52\\
B_5 &= 1.0075 \cdot B_4 + \$100 = \$507.56
\end{aligned}$$

b. $B_n = 1.0075 \cdot B_{n-1} + \$100$.

c. Using a calculator or computer program, $B_n > \$5,000$ during the 43rd month.

8.2.51 $\{n^2\} \ll \{n^2 \ln n\}$ from the second inequality in Theorem 8.6: $\{n^p\} \ll \{n^p \ln^r n\}$.

8.2.53 $\{100n!\} \ll \{3n^n\}$ since $\{n!\} \ll \{n^n\}$ from Theorem 8.6, and the sequences given are constant multiples of those sequences, so the rankings do not change.

8.2.55 Since $\frac{1}{2} > \frac{1}{10}$, Theorem 8.6 tells us that $\{n^{1/10}\} \ll \{n^{1/2}\}$.

8.2.57 Let $\varepsilon > 0$ be given and let N be an integer with $N > \frac{1}{\varepsilon}$. Then if $n > N$, we have $\left|\frac{1}{n} - 0\right| = \frac{1}{n} < \frac{1}{N} < \varepsilon$.

8.2.59 Let $\varepsilon > 0$ be given. We wish to find N such that for $n > N$, $\left|\frac{3n^2}{4n^2+1} - \frac{3}{4}\right| = \left|\frac{-3}{4(4n^2+1)}\right| = \frac{3}{4(4n^2+1)} < \varepsilon$. But this means that $3 < 4\varepsilon(4n^2+1)$, or $16\varepsilon n^2 + (4\varepsilon - 3) > 0$. Solving the quadratic, we get $n > \frac{1}{4}\sqrt{\frac{3}{\varepsilon} - 4}$, provided $\varepsilon < 3/4$. So let $N = \frac{1}{4}\sqrt{\frac{3}{\varepsilon} - 4}$ if $\epsilon < 3/4$ and let $N = 1$ otherwise.

8.2.61 Let $\varepsilon > 0$ be given. We wish to find N such that for $n > N$, $\left|\frac{cn}{bn+1} - \frac{c}{b}\right| = \left|\frac{-c}{b(bn+1)}\right| = \frac{c}{b(bn+1)} < \varepsilon$. But this means that $\varepsilon b^2 n + (b\varepsilon - c) > 0$, so that $N > \frac{c - b\varepsilon}{b^2\varepsilon}$ will work.

8.2.63

a. True. See Theorem 8.2 part 4.

b. False. For example, if $a_n = e^n$ and $b_n = 1/n$, then $\lim_{n\to\infty} a_n b_n = \infty$.

c. True. The definition of the limit of a sequence involves only the behavior of the nth term of a sequence as n gets large (see the Definition of Limit of a Sequence). Thus suppose a_n, b_n differ in only finitely many terms, and that M is large enough so that $a_n = b_n$ for $n > M$. Suppose a_n has limit L. Then for $\varepsilon > 0$, if N is such that $|a_n - L| < \varepsilon$ for $n > N$, first increase N if required so that $N > M$ as well. Then we also have $|b_n - L| < \varepsilon$ for $n > N$. Thus a_n and b_n have the same limit. A similar argument applies if a_n has no limit.

d. True. Note that a_n converges to zero. Intuitively, the nonzero terms of b_n are those of a_n, which converge to zero. More formally, given ϵ, choose N_1 such that for $n > N_1$, $a_n < \epsilon$. Let $N = 2N_1 + 1$. Then for $n > N$, consider b_n. If n is even, then $b_n = 0$ so certainly $b_n < \epsilon$. If n is odd, then $b_n = a_{(n-1)/2}$, and $(n-1)/2 > ((2N_1+1)-1)/2 = N_1$ so that $a_{(n-1)/2} < \epsilon$. Thus b_n converges to zero as well.

e. False. If $\{a_n\}$ happens to converge to zero, the statement is true. But consider for example $a_n = 2+\frac{1}{n}$. Then $\lim_{n\to\infty} a_n = 2$, but $(-1)^n a_n$ does not converge (it oscillates between positive and negative values increasingly close to ± 2).

f. True. Suppose $\{0.000001a_n\}$ converged to L, and let $\epsilon > 0$ be given. Choose N such that for $n > N$, $|0.000001a_n - L| < \epsilon \cdot 0.000001$. Dividing through by 0.000001, we get that for $n > N$, $|a_n - 1000000L| < \epsilon$, so that a_n converges as well (to $1000000L$).

8.2.65 $\{(n-2)^2 + 6(n-2) - 9\}_{n=3}^{\infty} = \{n^2 + 2n - 17\}_{n=3}^{\infty}$.

8.2.67 Evaluate the limit of each term separately: $\lim_{n\to\infty} \frac{75^{n-1}}{99^n} = \frac{1}{99} \lim_{n\to\infty} \left(\frac{75}{99}\right)^{n-1} = 0$, while $\frac{-5^n}{8^n} \le \frac{5^n \sin n}{8^n} \le \frac{5^n}{8^n}$, so by the Squeeze Theorem, this second term converges to 0 as well. Thus the sum of the terms converges to zero.

8.2.69 Since $\lim_{n\to\infty} 0.99^n = 0$, and since cosine is continuous, the first term converges to $\cos 0 = 1$. The limit of the second term is $\lim_{n\to\infty} \frac{7^n+9^n}{63^n} = \lim_{n\to\infty} \left(\frac{7}{63}\right)^n + \lim_{n\to\infty} \left(\frac{9}{63}\right)^n = 0$. Thus the sum converges to 1.

8.2.71 A graph shows that the sequence appears to converge. Let its supposed limit be L, then $\lim_{n\to\infty} a_{n+1} = \lim_{n\to\infty} (2a_n(1-a_n)) = 2(\lim_{n\to\infty} a_n)(1 - \lim_{n\to\infty} a_n)$, so $L = 2L(1-L) = 2L - 2L^2$, and thus $2L^2 - L = 0$, so $L = 0, \frac{1}{2}$. Thus the limit appears to be either 0 or 1/2; with the given initial condition, doing a few iterations by hand confirms that the sequence converges to 1/2: $a_0 = 0.3$; $a_1 = 2 \cdot 0.3 \cdot 0.7 = .42$; $a_2 = 2 \cdot 0.42 \cdot 0.58 = 0.4872$.

8.2.73 Computing three terms gives $a_0 = 0.5, a_1 = 4 \cdot .5 \cdot 0.5 = 1, a_2 = 4 \cdot 1 \cdot (1-1) = 0$. All successive terms are obviously zero, so the sequence converges to 0.

8.2.75 For $b = 2$, $2^3 > 3!$ but $16 = 2^4 < 4! = 24$. For e, $e^5 \approx 148.41 > 5! = 120$ while $e^6 \approx 403.4 < 6! = 720$. For 10, $24! \approx 6.2 \times 10^{23} < 10^{24}$, while $25! \approx 1.55 \times 10^{25} > 10^{25}$.

8.2.77

a. The profits for each of the first ten days, in dollars are:

n	0	1	2	3	4	5	6	7	8	9	10
h_n	130.00	130.75	131.40	131.95	132.40	132.75	133.00	133.15	133.20	133.15	133.00

b. The profit on an item is revenue minus cost. The total cost of keeping the hippo for n days is $.45n$, and the revenue for selling the hippo on the nth day is $(200+5n) \cdot (.65 - .01n)$, since the hippo gains 5 pounds per day but is worth a penny less per pound each day. Thus the total profit on the nth day is $h_n = (200+5n) \cdot (.65 - .01n) - .45n = 130 + 0.8n - 0.05n^2$. The maximum profit occurs when $-.1n + .8 = 0$, which occurs when $n = 8$. The maximum profit is achieved by selling the hippo on the 8th day.

8.2.79 The approximate first few values of this sequence are:

n	0	1	2	3	4	5	6
c_n	.7071	.6325	.6136	.6088	.6076	.6074	.6073

The value of the constant appears to be around 0.607.

8.2.81

a. If we "cut off" the expression after n square roots, we get a_n from the recurrence given. We can thus *define* the infinite expression to be the limit of a_n as $n \to \infty$.

b. $a_0 = 1$, $a_1 = \sqrt{2}$, $a_2 = \sqrt{1+\sqrt{2}} \approx 1.5538$,, $a_3 \approx 1.598$, $a_4 \approx 1.6118$, and $a_5 \approx 1.6161$.

c. $a_{10} \approx 1.618286$, which differs from $\frac{1+\sqrt{5}}{2} \approx 1.61803394$ by less than .001.

d. Assume $\lim_{n\to\infty} a_n = L$. Then $\lim_{n\to\infty} a_{n+1} = \lim_{n\to\infty} \sqrt{1+a_n} = \sqrt{1+\lim_{n\to\infty} a_n}$, so $L = \sqrt{1+L}$, and thus $L^2 = 1+L$. Therefore we have $L^2 - L - 1 = 0$, so $L = \frac{1\pm\sqrt{5}}{2}$.
Since clearly the limit is positive, it must be the positive square root.

e. Letting $a_{n+1} = \sqrt{p+\sqrt{a_n}}$ with $a_0 = p$ and assuming a limit exists we have $\lim_{n\to\infty} a_{n+1} = \lim_{n\to\infty} \sqrt{p+a_n} = \sqrt{p+\lim_{n\to\infty} a_n}$, so $L = \sqrt{p+L}$, and thus $L^2 = p+L$. Therefore, $L^2 - L - p = 0$, so $L = \frac{1\pm\sqrt{1+4p}}{2}$, and since we know that L is positive, we have $L = \frac{1+\sqrt{4p+1}}{2}$. The limit exists for all positive p.

8.2.83

a. Define a_n as given in the problem statement. Then we can *define* the value of the continued fraction to be $\lim_{n\to\infty} a_n$.

b. $a_0 = 1$, $a_1 = 1 + \frac{1}{a_0} = 2$, $a_2 = 1 + \frac{1}{a_1} = \frac{3}{2} = 1.5$, $a_3 = 1 + \frac{1}{a_2} = \frac{5}{3} \approx 1.67$, $a_4 = 1 + \frac{1}{a_3} = \frac{8}{5} \approx 1.6$, $a_5 = 1 + \frac{1}{a_4} = \frac{13}{8} \approx 1.625$.

c. From the list above, the values of the sequence alternately decrease and increase, so we would expect that the limit is somewhere between 1.6 and 1.625.

d. Assume that the limit is equal to L. Then from $a_{n+1} = 1 + \frac{1}{a_n}$, we have $\lim_{n\to\infty} a_{n+1} = 1 + \frac{1}{\lim_{n\to\infty} a_n}$, so $L = 1 + \frac{1}{L}$, and thus $L^2 - L - 1 = 0$. Therefore, $L = \frac{1\pm\sqrt{5}}{2}$, and since L is clearly positive, it must be equal to $\frac{1+\sqrt{5}}{2}$.

e. Here $a_0 = a$ and $a_{n+1} = a + \frac{b}{a_n}$. Assuming that $\lim_{n\to\infty} a_n = L$ we have $L = a + \frac{b}{L}$, so $L^2 = aL + b$, and thus $L^2 - aL - b = 0$. Therefore, $L = \frac{a\pm\sqrt{a^2+4b}}{2}$, and since $L > 0$ we have $L = \frac{a+\sqrt{a^2+4b}}{2}$.

8.2.85

a. $f_0 = f_1 = 1, f_2 = 2, f_3 = 3, f_4 = 5, f_5 = 8, f_6 = 13, f_7 = 21, f_8 = 34, f_9 = 55, f_{10} = 89$.

b. The sequence is clearly not bounded.

c. $\frac{f_{10}}{f_9} \approx 1.61818$

d. We use induction. Note that $\frac{1}{\sqrt{5}}\left(\varphi + \frac{1}{\varphi}\right) = \frac{1}{\sqrt{5}}\left(\frac{1+\sqrt{5}}{2} + \frac{2}{1+\sqrt{5}}\right) = \frac{1}{\sqrt{5}}\left(\frac{1+2\sqrt{5}+5+4}{2(1+\sqrt{5})}\right) = 1 = f_1$. Also note that $\frac{1}{\sqrt{5}}\left(\varphi^2 - \frac{1}{\varphi^2}\right) = \frac{1}{\sqrt{5}}\left(\frac{3+\sqrt{5}}{2} - \frac{2}{3+\sqrt{5}}\right) = \frac{1}{\sqrt{5}}\left(\frac{9+6\sqrt{5}+5-4}{2(3+\sqrt{5})}\right) = 1 = f_2$. Now note that

$$\begin{aligned} f_{n-1} + f_{n-2} &= \frac{1}{\sqrt{5}}(\varphi^{n-1} - (-1)^{n-1}\varphi^{1-n} + \varphi^{n-2} - (-1)^{n-2}\varphi^{2-n}) \\ &= \frac{1}{\sqrt{5}}((\varphi^{n-1} + \varphi^{n-2}) - (-1)^n(\varphi^{2-n} - \varphi^{1-n})). \end{aligned}$$

Now, note that $\varphi - 1 = \frac{1}{\varphi}$, so that

$$\varphi^{n-1} + \varphi^{n-2} = \varphi^{n-1}\left(1 + \frac{1}{\varphi}\right) = \varphi^{n-1}(\varphi) = \varphi^n$$

and

$$\varphi^{2-n} - \varphi^{1-n} = \varphi^{-n}(\varphi^2 - \varphi) = \varphi^{-n}(\varphi(\varphi - 1)) = \varphi^{-n}$$

Making these substitutions, we get

$$f_{n-1} + f_{n-2} = \frac{1}{\sqrt{5}}(\varphi^n - (-1)^n\varphi^{-n}) = f_n$$

8.2.87

a.

2 :	1
3 :	10, 5, 16, 8, 4, 2, 1
4 :	2, 1
5 :	16, 8, 4, 2, 1
6 :	3, 10, 5, 16, 8, 4, 2, 1
7 :	22, 11, 34, 17, 52, 26, 13, 40, 20, 10, 5, 16, 8, 4, 2, 1
8 :	4, 2, 1
9 :	28, 14, 7, 22, 11, 34, 17, 52, 26, 13, 40, 20, 10, 5, 16, 8, 4, 2, 1
10 :	5, 16, 8, 4, 2, 1

b. From the above, $H_2 = 1, H_3 = 7$, and $H_4 = 2$.

c. This plot is for $1 \le n \le 100$. Like hailstones, the numbers in the sequence a_n rise and fall but eventually crash to the earth. The conjecture appears to be true.

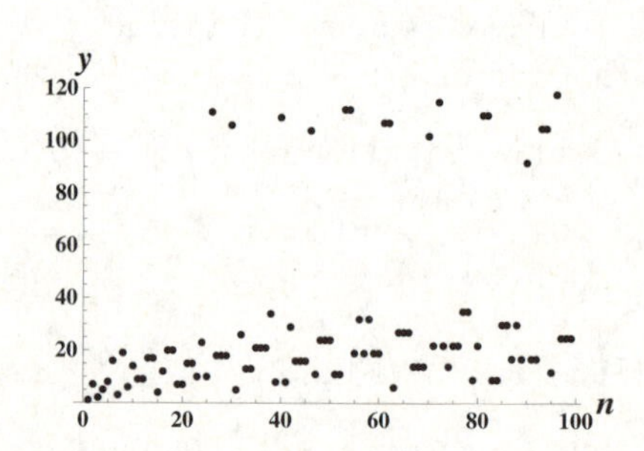

8.3 Infinite Series

8.3.1 A geometric series is a series in which the ratio of successive terms in the underlying sequence is a constant. Thus a geometric series has the form $\sum ar^k$ where r is the constant. One example is $3 + 6 + 12 + 24 + 48 + \cdots$ in which $a = 3$ and $r = 2$.

8.3.3 The ratio is the common ratio between successive terms in the sum.

8.3.5 No. For example, the geometric series with $a_n = 3 \cdot 2^n$ does not have a finite sum.

8.3.7 $S = 1 \cdot \dfrac{1 - 3^9}{1 - 3} = \dfrac{19682}{2} = 9841.$

8.3.9 $S = 1 \cdot \dfrac{1 - (4/25)^{21}}{1 - 4/25} = \dfrac{25^{21} - 4^{21}}{25^{21} - 4 \cdot 25^{20}} \approx 1.1905.$

8.3.11 $S = 1 \cdot \dfrac{1 - (-3/4)^{10}}{1 + 3/4} = \dfrac{4^{10} - 3^{10}}{4^{10} + 3 \cdot 4^9} = \dfrac{141361}{262144} \approx 0.5392.$

8.3.13 $S = 1 \cdot \dfrac{1-\pi^7}{1-\pi} = \dfrac{\pi^7-1}{\pi-1} \approx 1409.84.$

8.3.15 $S = 1 \cdot \dfrac{1-(-1)^{21}}{2} = 1.$

8.3.17 $\dfrac{1093}{2916}.$

8.3.19 $\dfrac{1}{1-1/4} = \dfrac{4}{3}.$

8.3.21 $\dfrac{1}{1-0.9} = 10.$

8.3.23 Divergent, since $r > 1$.

8.3.25 $\dfrac{e^{-2}}{1-e^{-2}} = \dfrac{1}{e^2-1}.$

8.3.27 $\dfrac{2^{-3}}{1-2^{-3}} = \dfrac{1}{7}.$

8.3.29 $\dfrac{1/625}{1-1/5} = \dfrac{1}{500}.$

8.3.31 $\dfrac{1}{1-e/\pi} = \dfrac{\pi}{\pi-e}.$ (Note that $e < \pi$, so $r < 1$ for this series.)

8.3.33 $\displaystyle\sum_{k=0}^{\infty}\left(\frac{1}{4}\right)^k 5^{6-k} = 5^6\sum_{k=0}^{\infty}\left(\frac{1}{20}\right)^k = 5^6 \cdot \frac{1}{1-1/20} = \frac{5^6\cdot 20}{19} = \frac{312500}{19}.$

8.3.35 $\dfrac{1}{1+9/10} = \dfrac{10}{19}.$

8.3.37 $3 \cdot \dfrac{1}{1+1/\pi} = \dfrac{3\pi}{\pi+1}.$

8.3.39 $\dfrac{0.15^2}{1.15} = \dfrac{9}{460} \approx 0.0196.$

8.3.41 $0.121212\ldots = \displaystyle\sum_{i=0}^{\infty} .12 \cdot 10^{-2i} = \frac{.12}{1-1/100} = \frac{12}{99} = \frac{4}{33}.$

8.3.43 $0.456456456\ldots = \displaystyle\sum_{i=0}^{\infty} .456 \cdot 10^{-3i} = \frac{.456}{1-1/1000} = \frac{456}{999} = \frac{152}{333}.$

8.3.45 $0.00952952\ldots = \displaystyle\sum_{i=0}^{\infty} .00952 \cdot 10^{-3i} = \frac{.00952}{1-1/1000} = \frac{9.52}{999} = \frac{952}{99900} = \frac{238}{24975}.$

8.3.47 The second term of each summand cancels with the first term of the succeeding summand, so $S_n = \frac{1}{1+1} - \frac{1}{n+2} = \frac{n}{2n+4}$, and $\lim_{n\to\infty} \frac{n}{2n+4} = \frac{1}{2}$.

8.3.49 $\dfrac{1}{(k+1)(k+2)} = \dfrac{1}{k+1} - \dfrac{1}{k+2}$, so the series given is the same as $\sum_{k=1}^{\infty}\left(\frac{1}{k+1} - \frac{1}{k+2}\right)$. In that series, the second term of each summand cancels with the first term of the succeeding summand, so $S_n = \frac{1}{1+1} - \frac{1}{n+2} = \frac{n}{2n+4}$ and $\lim_{n\to\infty} \frac{n}{2n+4} = \frac{1}{2}$

8.3.51 $\ln\left(\frac{k+1}{k}\right) = \ln(k+1) - \ln k$, so the series given is the same as $\sum_{k=1}^{\infty}(\ln(k+1) - \ln k)$, in which the first term of each summand cancels with the second term of the next summand, so we get $S_{n-1} = \ln n - \ln 1 = \ln n$, and thus the series diverges.

8.3.53 $\frac{1}{(k+p)(k+p+1)} = \frac{1}{k+p} - \frac{1}{k+p+1}$, so that $\sum_{k=1}^{\infty} \frac{1}{(k+p)(k+p+1)} = \sum_{k=1}^{\infty}\left(\frac{1}{k+p} - \frac{1}{k+p+1}\right)$ and this series telescopes to give $S_n = \frac{1}{p+1} - \frac{1}{n+p+1} = \frac{n}{n(p+1)+(p+1)^2}$ so that $\lim_{n\to\infty} S_n = \frac{1}{p+1}$.

8.3.55 Let $a_n = \frac{1}{\sqrt{n+1}} - \frac{1}{\sqrt{n+3}}$. Then the second term of a_n cancels with the first term of a_{n+2}, so the series telescopes and $S_n = \frac{1}{\sqrt{2}} + \frac{1}{\sqrt{3}} - \frac{1}{\sqrt{n-1+3}} - \frac{1}{\sqrt{n+3}}$ and thus the sum of the series is the limit of S_n, which is $\frac{1}{\sqrt{2}} + \frac{1}{\sqrt{3}}$.

8.3.57 $16k^2 + 8k - 3 = (4k+3)(4k-1)$, so $\frac{1}{16k^2+8k-3} = \frac{1}{(4k+3)(4k-1)} = \frac{1}{4}\left(\frac{1}{4k-1} - \frac{1}{4k+3}\right)$. Thus the series given is equal to $\frac{1}{4}\sum_{k=0}^{\infty}\left(\frac{1}{4k-1} - \frac{1}{4k+3}\right)$. This series telescopes, so $S_n = \frac{1}{4}\left(-1 - \frac{1}{4n+3}\right)$ and so the sum of the series is equal to $\lim_{n\to\infty} S_n = -\frac{1}{4}$.

8.3.59

a. True. $\left(\frac{\pi}{e}\right)^{-k} = \left(\frac{e}{\pi}\right)^k$; since $e < \pi$, this is a geometric series with ratio less than 1.

b. True. If $\sum_{k=12}^{\infty} a^k = L$, then $\sum_{k=0}^{\infty} a^k = \left(\sum_{k=0}^{11} a^k\right) + L$.

c. False. For example, let $0 < a < 1$ and $b > 1$.

8.3.61 At the nth stage, there are 2^{n-1} triangles of area $A_n = \frac{1}{8}A_{n-1} = \frac{1}{8^{n-1}}A_1$, so the total area of the triangles formed at the nth stage is $\frac{2^{n-1}}{8^{n-1}}A_1 = \left(\frac{1}{4}\right)^{n-1} A_1$. Thus the total area under the parabola is

$$\sum_{n=1}^{\infty}\left(\frac{1}{4}\right)^{n-1} A_1 = A_1 \sum_{n=1}^{\infty}\left(\frac{1}{4}\right)^{n-1} = A_1 \frac{1}{1-1/4} = \frac{4}{3}A_1.$$

8.3.63 It appears that the loan is paid off after about 470 months. Let B_n be the loan balance after n months. Then $B_0 = 180000$ and $B_n = 1.005 \cdot B_{n-1} - 1000$. Then $B_n = 1.005 \cdot B_{n-1} - 1000 = 1.005(1.005 \cdot B_{n-2} - 1000) - 1000 = (1.005)^2 \cdot B_{n-2} - 1000(1+1.005) = (1.005)^2 \cdot (1.005 \cdot B_{n-3} - 1000) - 1000(1+1.005) = (1.005)^3 \cdot B_{n-3} - 1000(1 + 1.005 + (1.005)^2) = \cdots = (1.005)^n B_0 - 1000(1 + 1.005 + (1.005)^2 + \cdots + (1.005)^{n-1}) = (1.005)^n \cdot 180000 - 1000\left(\frac{(1.005)^n - 1}{1.005 - 1}\right)$. Solving this equation for $B_n = 0$ gives $n \approx 461.66$ months, so the loan is paid off after 462 months.

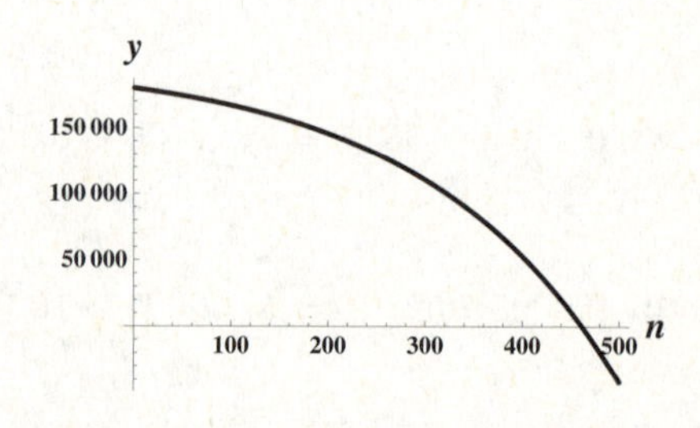

8.3.65 $F_n = 1.015F_{n-1} - 120$ with $F_0 = 4000$. Assume F_n reaches a limit L; then $L = 1.015L - 120$, so $.015L = 120$, and thus $L = 8000$. However, starting with only 4000 fish, the replacement rate of .015% only adds 60 fish in the first month, which is not enough to make up for the harvest of 120 fish. Thus the population in fact decreases to zero. If the initial population were above 8000, it would level off at 8000.

8.3.67 Under the one-child policy, each couple will have one child. Under the one-son policy, we compute the expected number of children as follows: with probability 1/2 the first child will be a son; with probability $(1/2)^2$, the first child will be a daughter and the second child will be a son; in general, with probability $(1/2)^n$, the first $n-1$ children will be girls and the nth a boy. Thus the expected number of children is the sum $\displaystyle\sum_{i=1}^{\infty} i \cdot \left(\frac{1}{2}\right)^i$. To evaluate this series, use the following "trick": Let $f(x) = \displaystyle\sum_{i=1}^{\infty} ix^i$. Then $f(x) + \displaystyle\sum_{i=1}^{\infty} x^i = \sum_{i=1}^{\infty}(i+1)x^i$. Now, let

$$g(x) = \sum_{i=1}^{\infty} x^{i+1} = -1 - x + \sum_{i=0}^{\infty} x^i = -1 - x + \frac{1}{1-x}$$

and

$$g'(x) = f(x) + \sum_{i=1}^{\infty} x^i = f(x) - 1 + \sum_{i=0}^{\infty} x^i = f(x) - 1 + \frac{1}{1-x}.$$

Evaluate $g'(x) = -1 - \frac{1}{(1-x)^2}$; then

$$f(x) = 1 - \frac{1}{1-x} - 1 - \frac{1}{(1-x)^2} = \frac{-1+x+1}{(1-x)^2} = \frac{x}{(1-x)^2}$$

Finally, evaluate at $x = \frac{1}{2}$ to get $f\left(\frac{1}{2}\right) = \sum_{i=1}^{\infty} i \cdot \left(\frac{1}{2}\right)^i = \frac{1/2}{(1-1/2)^2} = 2$. There will thus be twice as many children under the one-son policy as under the one-child policy.

8.3.69 Ignoring the initial drop for the moment, the height after the nth bounce is $10p^n$, so the total time spent in that bounce is $2 \cdot \sqrt{2 \cdot 10p^n/g}$ seconds. The total time before the ball comes to rest (now including the time for the initial drop) is then $\sqrt{20/g} + \sum_{i=1}^{\infty} 2 \cdot \sqrt{2 \cdot 10p^n/g} = \sqrt{\frac{20}{g}} + 2\sqrt{\frac{20}{g}} \sum_{i=1}^{\infty} (\sqrt{p})^n = \sqrt{\frac{20}{g}} + 2\sqrt{\frac{20}{g}} \frac{\sqrt{p}}{1-\sqrt{p}} = \sqrt{\frac{20}{g}}\left(1 + \frac{2\sqrt{p}}{1-\sqrt{p}}\right) = \sqrt{\frac{20}{g}}\left(\frac{1+\sqrt{p}}{1-\sqrt{p}}\right)$ seconds.

8.3.71

a. I_{n+1} is obtained by I_n by dividing each edge into three equal parts, removing the middle part, and adding two parts equal to it. Thus 3 equal parts turn into 4, so $L_{n+1} = \frac{4}{3}L_n$. This is a geometric sequence with a ratio greater than 1, so the nth term grows without bound.

b. As the result of part (a), I_n has $3 \cdot 4^n$ sides of length $\frac{1}{3^n}$; each of those sides turns into an added triangle in I_{n+1} of side length 3^{-n-1}. Thus the added area in I_{n+1} consists of $3 \cdot 4^n$ equilateral triangles with side 3^{-n-1}. The area of an equilateral triangle with side x is $\dfrac{x^2\sqrt{3}}{4}$. Thus $A_{n+1} = A_n + 3 \cdot 4^n \cdot \frac{3^{-2n-2}\sqrt{3}}{4} = A_n + \frac{\sqrt{3}}{12} \cdot \left(\frac{4}{9}\right)^n$, and $A_0 = \frac{\sqrt{3}}{4}$. Thus $A_{n+1} = A_0 + \sum_{i=0}^{n} \frac{\sqrt{3}}{12} \cdot \left(\frac{4}{9}\right)^i$, so that

$$A_\infty = A_0 + \frac{\sqrt{3}}{12}\sum_{i=0}^{\infty}\left(\frac{4}{9}\right)^i = \frac{\sqrt{3}}{4} + \frac{\sqrt{3}}{12}\frac{1}{1-4/9} = \frac{\sqrt{3}}{4}(1+\frac{3}{5}) = \frac{2}{5}\sqrt{3}$$

8.3.73 $|S - S_n| = \left|\displaystyle\sum_{i=n}^{\infty} r^k\right| = \left|\dfrac{r^n}{1-r}\right|$ since the latter sum is simply a geometric series with first term r^n and ratio r.

8.3.75

a. Solve $\left|\frac{(-0.8)^n}{1.8}\right| = \frac{0.8^n}{1.8} < 10^{-6}$ for n to get $n = 60$.

b. Solve $\frac{0.2^n}{0.8} < 10^{-6}$ for n to get $n = 9$.

8.3.77

a. Solve $\frac{1/\pi^n}{1-1/\pi} < 10^{-6}$ for n to get $n = 13$.

b. Solve $\frac{1/e^n}{1-1/e} < 10^{-6}$ for n to get $n = 15$.

8.3.79 $f(x) = \sum_{k=0}^{\infty}(-1)^k x^k = \frac{1}{1+x}$; since f is a geometric series, $f(x)$ exists only when the ratio, $-x$, is such that $|-x| = |x| < 1$. Then $f(0) = 1$, $f(0.2) = \frac{1}{1.2} = \frac{5}{6}$, $f(0.5) = \frac{1}{1+.05} = \frac{2}{3}$. Neither $f(1)$ nor $f(1.5)$ exists. The domain of f is $|x| < 1$.

8.3.81 $f(x)$ is a geometric series with ratio $\frac{1}{1+x}$; thus $f(x)$ converges when $\left|\frac{1}{1+x}\right| < 1$. For $x > -1$, $\left|\frac{1}{1+x}\right| = \frac{1}{1+x}$ and $\frac{1}{1+x} < 1$ when $1 < 1+x$, $x > 0$. For $x < -1$, $\left|\frac{1}{1+x}\right| = \frac{1}{-1-x}$, and this is less than 1 when $1 < -1-x$, i.e. $x < -2$. So $f(x)$ converges for $x > 0$ and for $x < -2$. When $f(x)$ converges, its value is $\frac{1}{1-\frac{1}{1+x}} = \frac{1+x}{x}$, so $f(x) = 3$ when $1 + x = 3x$, $x = \frac{1}{2}$.

8.4 The Divergence and Integral Tests

8.4.1 A series may diverge so slowly that no reasonable number of terms may definitively show that it does so.

8.4.3 Yes. Either the series and the integral both converge, or both diverge.

8.4.5 For the same values of p as in the previous problem – it converges for $p > 1$, and diverges for all other values of p.

8.4.7 The remainder of an infinite series is the error in approximating a convergent infinite series by a finite number of terms.

8.4.9 $\displaystyle\sum_{k=0}^{\infty}\left(3\left(\frac{2}{5}\right)^k - 2\left(\frac{5}{7}\right)^k\right) = 3\sum_{k=0}^{\infty}\left(\frac{2}{5}\right)^k - 2\sum_{k=0}^{\infty}\left(\frac{5}{7}\right)^k = 3\left(\frac{1}{3/5}\right) - 2\left(\frac{1}{2/7}\right) = 5 - 7 = -2.$

8.4.11 $\displaystyle\sum_{k=1}^{\infty}\left(\frac{1}{3}\left(\frac{5}{6}\right)^k + \frac{3}{5}\left(\frac{7}{9}\right)^k\right) = \frac{1}{3}\sum_{k=1}^{\infty}\left(\frac{5}{6}\right)^k + \frac{3}{5}\sum_{k=1}^{\infty}\left(\frac{7}{9}\right)^k = \frac{1}{3}\left(\frac{5/6}{1/6}\right) + \frac{3}{5}\left(\frac{7/9}{2/9}\right) = \frac{5}{3} + \frac{21}{10} = \frac{113}{30}.$

8.4.13 $\displaystyle\sum_{k=1}^{\infty}\left(\left(\frac{1}{6}\right)^k + \left(\frac{1}{3}\right)^{k-1}\right) = \sum_{k=1}^{\infty}\left(\frac{1}{6}\right)^k + \sum_{k=1}^{\infty}\left(\frac{1}{3}\right)^{k-1} = \frac{1/6}{5/6} + \frac{1}{2/3} = \frac{17}{10}.$

8.4.15 $a_k = \frac{k}{2k+1}$ and $\lim_{k\to\infty} a_k = \frac{1}{2}$, so the series diverges.

8.4.17 $a_k = \frac{k}{\ln k}$ and $\lim_{k\to\infty} a_k = \infty$, so the series diverges.

8.4.19 $a_k = \frac{1}{1000+k}$ and $\lim_{k\to\infty} a_k = 0$, so the divergence test is inconclusive.

8.4.21 $a_k = \frac{\sqrt{k}}{\ln^{10} k}$ and $\lim_{k\to\infty} a_k = \infty$, so the series diverges.

8.4.23 Let $f(x) = \frac{1}{x \ln x}$. Then $f(x)$ is continuous and decreasing on $(1, \infty)$, since $x \ln x$ is increasing there. Since $\int_1^\infty f(x)\,dx = \infty$, the series diverges.

8.4.25 Let $f(x) = x \cdot e^{-2x^2}$. This function is continuous for $x \ge 1$. Its derivative is $e^{-2x^2}(1 - 4x^2) < 0$ for $x \ge 1$, so $f(x)$ is decreasing. Since $\int_1^\infty x \cdot e^{-2x^2}\,dx = \frac{1}{4e^2}$, the series converges.

8.4.27 Let $f(x) = \frac{1}{\sqrt{x+8}}$. $f(x)$ is obviously continuous and decreasing for $x \geq 1$. Since $\int_1^\infty \frac{1}{\sqrt{x+8}}\,dx = \infty$, the series diverges.

8.4.29 Let $f(x) = \frac{x}{e^x}$. $f(x)$ is clearly continuous for $x > 1$, and its derivative, $f'(x) = \frac{e^x - xe^x}{e^{2x}} = (1-x)\frac{e^x}{e^{2x}}$, is negative for $x > 1$ so that $f(x)$ is decreasing. Since $\int_1^\infty f(x)\,dx = 2e^{-1}$, the series converges.

8.4.31 This is a p-series with $p = 10$, so this series converges.

8.4.33 $\sum_{k=3}^\infty \frac{1}{(k-2)^4} = \sum_{k=1}^\infty \frac{1}{k^4}$, which is a p-series with $p = 4$, thus convergent.

8.4.35

a. The remainder R_n is bounded by $\int_n^\infty \frac{1}{x^6}\,dx = \frac{1}{5n^5}$.

b. We solve $\frac{1}{5n^5} < 10^{-3}$ to get $n = 3$.

c. $L_n = S_n + \int_{n+1}^\infty \frac{1}{x^6}\,dx = S_n + \frac{1}{5(n+1)^5}$, and $U_n = S_n + \int_n^\infty \frac{1}{x^6}\,dx = S_n + \frac{1}{5n^5}$.

d. $S_{10} \approx 1.017341512$, so $L_{10} \approx 1.017341512 + \frac{1}{5 \cdot 11^5} \approx 1.017342754$, and $U_{10} \approx 1.017341512 + \frac{1}{5 \cdot 10^5} \approx 1.017343512$.

8.4.37

a. The remainder R_n is bounded by $\int_n^\infty \frac{1}{3^x}\,dx = \frac{1}{3^n \ln(3)}$.

b. We solve $\frac{1}{3^n \ln(3)} < 10^{-3}$ to obtain $n = 7$.

c. $L_n = S_n + \int_{n+1}^\infty \frac{1}{3^x}\,dx = S_n + \frac{1}{3^{n+1}\ln(3)}$, and $U_n = S_n + \int_n^\infty \frac{1}{3^x}\,dx = S_n + \frac{1}{3^n \ln(3)}$.

d. $S_{10} \approx 0.4999915325$, so $L_{10} \approx 0.4999915325 + \frac{1}{3^{11}\ln 3} \approx 0.4999966708$, and $U_{10} \approx 0.4999915325 + \frac{1}{3^{10}\ln 3} \approx 0.5000069475$.

8.4.39

a. The remainder R_n is bounded by $\int_n^\infty \frac{1}{x^{3/2}}\,dx = 2n^{-1/2}$.

b. We solve $2n^{-1/2} < 10^{-3}$ to get $n > 4 \times 10^6$, so let $n = 4 \times 10^6 + 1$.

c. $L_n = S_n + \int_{n+1}^\infty \frac{1}{x^{3/2}}\,dx = S_n + 2(n+1)^{-1/2}$, and $U_n = S_n + \int_n^\infty \frac{1}{x^{3/2}}\,dx = S_n + 2n^{-1/2}$.

d. $S_{10} = \sum_{k=1}^{10} \frac{1}{k^{3/2}} \approx 1.995336494$, so $L_{10} \approx 1.995336494 + 2 \cdot 11^{-1/2} \approx 2.598359183$, and $U_{10} \approx 1.995336494 + 2 \cdot 10^{-1/2} \approx 2.627792026$.

8.4.41

a. The remainder R_n is bounded by $\int_n^\infty \frac{1}{x^3}\,dx = \frac{1}{2n^2}$.

b. We solve $\frac{1}{2n^2} < 10^{-3}$ to get $n = 23$.

c. $L_n = S_n + \int_{n+1}^\infty \frac{1}{x^3}\,dx = S_n + \frac{1}{2(n+1)^2}$, and $U_n = S_n + \int_n^\infty \frac{1}{x^3}\,dx = S_n + \frac{1}{2n^2}$.

d. $S_{10} \approx 1.197531986$, so $L_{10} \approx 1.197531986 + \frac{1}{2 \cdot 11^2} \approx 1.201664217$, and $U_{10} \approx 1.197531986 + \frac{1}{2 \cdot 10^2} \approx 1.202531986$.

8.4.43

a. True. The two series differ by a finite amount ($\sum_{k=1}^{9} a_k$), so if one converges, so does the other.

b. True. The same argument applies as in part (a).

c. False. If $\sum a_k$ converges, then $a_k \to 0$ as $k \to \infty$, so that $a_k + 0.0001 \to 0.0001$ as $k \to \infty$, so that $\sum(a_k + 0.0001)$ cannot converge.

d. False. Suppose $p = .9999$. Then $\sum p^k$ converges, but $p + 0.001 = 1.009$ so that $\sum(p+0.001)^k$ diverges.

e. False. Let $p = 1.0005$; then $-p + .001 = -(p - .001) = -.9995$, so that $\sum k^{-p}$ converges (p-series) but $\sum k^{-p+.001}$ diverges.

f. False. Let $a_k = \frac{1}{k}$, the harmonic series.

8.4.45 Converges by the Integral Test since $\displaystyle\int_1^\infty \frac{1}{(3x+1)(3x+4)}\,dx = \int_1^\infty \frac{1}{3(3x+1)} - \frac{1}{3(3x+4)}\,dx =$ $\displaystyle\lim_{b\to\infty}\int_1^b \left(\frac{1}{3(3x+1)} - \frac{1}{3(3x+4)}\right)dx = \lim_{b\to\infty}\frac{1}{9}\left(\ln\left(\frac{3x+1}{3x+4}\right)\right)\Bigg|_1^b = \lim_{b\to\infty} = \frac{-1}{9}\cdot\ln(4/7) \approx 0.06217 < \infty.$

8.4.47 Diverges by the Divergence Test since $\displaystyle\lim_{k\to\infty} a_k = \lim_{k\to\infty}\frac{k}{\sqrt{k^2+4}} = 1 \neq 0.$

8.4.49 Converges by the Integral Test since $\displaystyle\int_2^\infty \frac{4}{x\ln^2 x}\,dx = \lim_{b\to\infty}\left(\frac{-4}{\ln x}\bigg|_2^b\right) = \frac{4}{\ln 2} < \infty.$

8.4.51

a. We must be somewhat careful here, as the function $\dfrac{1}{x\ln x(\ln\ln x)^p}$ has a discontinuity at $x = e$. So instead we evaluate the convergence of the series starting with a lower bound of 3. In this case, $\int \frac{1}{x\ln x(\ln\ln x)^p}\,dx = \frac{1}{1-p}(\ln\ln x)^{1-p}$, and thus the improper integral with bounds n and ∞ exists only if $p > 1$ since $\ln\ln x > 0$ for $x > e$. So this series converges for $p > 1$.

b. For large values of z, clearly $\sqrt{z} > \ln z$, so that $z > (\ln z)^2$. Write $z = \ln x$; then for large x, $\ln x > (\ln\ln x)^2$; multiplying both sides by $x\ln x$ we get that $x\ln^2 x > x\ln x(\ln\ln x)^2$, so that the first series converges faster since the terms get smaller faster.

8.4.53 Let $S_n = \sum_{k=1}^n \frac{1}{\sqrt{k}}$. Then this looks like a left Riemann sum for the function $y = \frac{1}{\sqrt{x}}$ on $[1, n+1]$. Since each rectangle lies above the curve itself, we see that S_n is bounded below by the integral of $\frac{1}{\sqrt{x}}$ on $[1, n+1]$. Now,

$$\int_1^{n+1} \frac{1}{\sqrt{x}}\,dx = \int_1^{n+1} x^{-1/2}\,dx = 2\sqrt{x}\bigg|_1^{n+1} = 2\sqrt{n+1} - 2$$

This integral diverges as $n \to \infty$, so the series does as well by the bound above.

8.4.55 $\sum_{k=1}^\infty ca_k = \lim_{n\to\infty}\sum_{k=1}^n ca_k = \lim_{n\to\infty} c\sum_{k=1}^n a_k = c\lim_{n\to\infty}\sum_{k=1}^n a_k$, so that one sum converges if and only if the other one does.

8.4.57 To approximate the sequence for $\zeta(m)$, note that the remainder R_n after n terms is bounded by

$$\int_n^\infty \frac{1}{x^m}\,dx = \frac{1}{m-1}n^{1-m}.$$

For $m = 3$, if we wish to approximate the value to within 10^{-3}, we must solve $\frac{1}{2}n^{-2} < 10^{-3}$, so that $n = 23$, and $\displaystyle\sum_{i=1}^{23}\frac{1}{i^3} \approx 1.201151926$. The true value is ≈ 1.202056903.

For $m = 5$, if we wish to approximate the value to within 10^{-3}, we must solve $\frac{1}{4}n^{-4} < 10^{-3}$, so that $n = 4$, and $\displaystyle\sum_{i=1}^{4}\frac{1}{i^5} \approx 1.036341789$. The true value is ≈ 1.036927755.

For $m = 7$, if we wish to approximate the value to within 10^{-3}, we must solve $\dfrac{1}{6}n^{-6} < 10^{-3}$, so that $n = 3$, and $\displaystyle\sum_{i=1}^{3} \frac{1}{i^7} \approx 1.008269747$. The true value is ≈ 1.008349277.

8.4.59 $\displaystyle\sum_{k=1}^{\infty} \frac{1}{k^2} = \sum_{k=1}^{\infty} \frac{1}{(2k)^2} + \sum_{k=1}^{\infty} \frac{1}{(2k-1)^2}$, splitting the series into even and odd terms. But $\sum_{k=1}^{\infty} \frac{1}{(2k)^2} = \frac{1}{4}\sum_{k=1}^{\infty} \frac{1}{k^2}$. Thus $\frac{\pi^2}{6} = \frac{1}{4}\frac{\pi^2}{6} + \sum_{k=1}^{\infty} \frac{1}{(2k-1)^2}$, so that the sum in question is $\frac{3\pi^2}{24} = \frac{\pi^2}{8}$.

8.4.61

a. $x_1 = \sum_{k=2}^{2} \frac{1}{k} = \frac{1}{2}$, $x_2 = \sum_{k=3}^{4} \frac{1}{k} = \frac{1}{3} + \frac{1}{4} = \frac{7}{12}$, $x_3 = \sum_{k=4}^{6} \frac{1}{k} = \frac{1}{4} + \frac{1}{5} + \frac{1}{6} = \frac{37}{60}$.

b. x_n has n terms. Each term is bounded below by $\frac{1}{2n}$ and bounded above by $\frac{1}{n+1}$. Thus $x_n \geq n \cdot \frac{1}{2n} = \frac{1}{2}$, and $x_n \leq n \cdot \frac{1}{n+1} < n \cdot \frac{1}{n} = 1$.

c. The right Riemann sum for $\int_1^2 \frac{dx}{x}$ using n subintervals has n rectangles of width $\frac{1}{n}$; the right edges of those rectangles are at $1 + \frac{i}{n} = \frac{n+i}{n}$ for $i = 1, 2, \ldots, n$. The height of such a rectangle is the value of $\frac{1}{x}$ at the right endpoint, which is $\frac{n}{n+i}$. Thus the area of the rectangle is $\frac{1}{n} \cdot \frac{n}{n+i} = \frac{1}{n+i}$. Adding up over all the rectangles gives x_n.

d. The limit $\lim_{n\to\infty} x_n$ is the limit of the right Riemann sum as the width of the rectangles approaches zero. This is precisely $\int_1^2 \frac{dx}{x} = \ln x \Big|_1^2 = \ln 2$.

8.4.63

a. Note that the center of gravity of any stack of dominoes is the average of the locations of their centers. Define the midpoint of the zeroth (top) domino to be $x = 0$, and stack additional dominoes down and to its right (to increasingly positive x-coordinates.) Let $m(n)$ be the x-coordinate of the midpoint of the nth domino. Then in order for the stack not to fall over, the left edge of the nth domino must be placed directly under the center of gravity of dominos 0 through $n-1$, which is $\frac{1}{n}\sum_{i=0}^{n-1} m(i)$, so that $m(n) = 1 + \frac{1}{n}\sum_{i=0}^{n-1} m(i)$. Claim that in fact $m(n) = \sum_{k=1}^{n} \frac{1}{k}$. Use induction. This is certainly true for $n = 1$. Note first that $m(0) = 0$, so we can start the sum at 1 rather than at 0. Now, $m(n) = 1 + \frac{1}{n}\sum_{i=1}^{n-1} m(i) = 1 + \frac{1}{n}\sum_{i=1}^{n-1}\sum_{j=1}^{i} \frac{1}{j}$. Now, 1 appears $n-1$ times in the double sum, 2 appears $n-2$ times, and so forth, so we can rewrite this sum as $m(n) = 1 + \frac{1}{n}\sum_{i=1}^{n-1} \frac{n-i}{i} = 1 + \frac{1}{n}\sum_{i=1}^{n-1} \left(\frac{n}{i} - 1\right) = 1 + \frac{1}{n}\left(n\sum_{i=1}^{n-1} \frac{1}{i} - (n-1)\right) = \sum_{i=1}^{n-1} \frac{1}{i} + 1 - \frac{n-1}{n} = \sum_{i=1}^{n} \frac{1}{i}$, and we are done by induction (noting that the statement is clearly true for $n = 0$, $n = 1$). Thus the maximum overhang is $\sum_{k=2}^{n} \frac{1}{k}$.

b. For an infinite number of dominos, since the overhang is the harmonic series, the distance is potentially infinite.

8.5 The Ratio, Root, and Comparison Tests

8.5.1 Given a series $\sum a_k$ of positive terms, compute $\lim_{k\to\infty} \frac{a_{k+1}}{a_k}$ and call it r. If $0 \leq r < 1$, the given series converges. If $r > 1$ (including $r = \infty$), the given series diverges. If $r = 1$, the test is inconclusive.

8.5.3 Given a series of positive terms $\sum a_k$ that you suspect converges, find a series $\sum b_k$ that you know converges, for which $\lim_{k\to\infty} \frac{a_k}{b_k} = L$ where $L \geq 0$ is a finite number. If you are successful, you will have shown that the series $\sum a_k$ converges.

Given a series of positive terms $\sum a_k$ that you suspect diverges, find a series $\sum b_k$ that you know diverges, for which $\lim_{k\to\infty} \frac{a_k}{b_k} = L$ where $L > 0$ (including the case $L = \infty$). If you are successful, you will have shown that $\sum a_k$ diverges.

8.5.5 The Ratio Test.

8.5.7 The difference between successive partial sums is a term in the sequence. Since the terms are positive, differences between successive partial sums are as well, so the sequence of partial sums is increasing.

8.5.9 The ratio between successive terms is $\frac{a_{k+1}}{a_k} = \frac{1}{(k+1)!} \cdot \frac{(k)!}{1} = \frac{1}{k+1}$, which goes to zero as $k \to \infty$, so the given series converges by the Ratio Test.

8.5.11 The ratio between successive terms is $\frac{a_{k+1}}{a_k} = \frac{(k+1)^2}{4^{(k+1)}} \cdot \frac{4^k}{(k)^2} = \frac{1}{4}\left(\frac{k+1}{k}\right)^2$. The limit is $1/4$ as $k \to \infty$, so the given series converges by the Ratio Test.

8.5.13 The ratio between successive terms is $\frac{a_{k+1}}{a_k} = \frac{(k+1)e^{-(k+1)}}{(k)e^{-(k)}} = \frac{k+1}{(k)e}$. The limit of this ratio as $k \to \infty$ is $1/e < 1$, so the given series converges by the Ratio Test.

8.5.15 The ratio between successive terms is $\frac{2^{k+1}}{(k+1)^{99}} \cdot \frac{(k)^{99}}{2^k} = 2\left(\frac{k}{k+1}\right)^{99}$; the limit as $k \to \infty$ is 2, so the given series diverges by the Ratio Test.

8.5.17 The ratio between successive terms is $\frac{((k+1)!)^2}{(2(k+1))!} \cdot \frac{(2k)!}{((k)!)^2} = \frac{(k+1)^2}{(2k+2)(2k+1)}$; the limit as $k \to \infty$ is $1/4$, so the given series converges by the Ratio Test.

8.5.19 $\lim_{k\to\infty} \sqrt[k]{a_k} = \lim_{k\to\infty} \frac{4k^3+k}{9k^3+k+1} = \frac{4}{9} < 1$, so the given series converges by the Root Test.

8.5.21 $\lim_{k\to\infty} \sqrt[k]{a_k} = \lim_{k\to\infty} \frac{k^{2/k}}{2} = \frac{1}{2} < 1$, so the given series converges by the Root Test.

8.5.23 $\lim_{k\to\infty} \sqrt[k]{a_k} = \lim_{k\to\infty} \left(\frac{k}{k+1}\right)^{2k} = e^{-2} < 1$, so the given series converges by the Root Test.

8.5.25 $\lim_{k\to\infty} \sqrt[k]{a_k} = \lim_{k\to\infty} \sqrt[k]{\left(\frac{1}{k^k}\right)} = \lim_{k\to\infty} \frac{1}{k} = 0$, so the given series converges by the Root Test.

8.5.27 $\frac{1}{k^2+4} < \frac{1}{k^2}$, and $\sum_{k=1}^{\infty} \frac{1}{k^2}$ converges, so $\sum_{k=1}^{\infty} \frac{1}{k^2+4}$ converges as well, by the Comparison Test.

8.5.29 Use the Limit Comparison Test with $\left\{\frac{1}{k}\right\}$. The ratio of the terms of the two series is $\frac{k^3-k}{k^3+4}$ which has limit 1 as $k \to \infty$. Since the comparison series diverges, the given series does as well.

8.5.31 For all k, $\frac{1}{k^{3/2}+1} < \frac{1}{k^{3/2}}$. The series whose terms are $\frac{1}{k^{3/2}}$ is a p-series which converges, so the given series converges as well by the Comparison Test.

8.5.33 $\sin(1/k) > 0$ for $k \geq 1$, so we can apply the Comparison Test with $1/k^2$. $\sin(1/k) < 1$, so $\frac{\sin(1/k)}{k^2} < \frac{1}{k^2}$. Since the comparison series converges, the given series converges as well.

8.5.35 Use the Limit Comparison Test with $\{1/k\}$. The ratio of the terms of the two series is $\frac{k}{2k-\sqrt{k}} = \frac{1}{2-1/\sqrt{k}}$, which has limit $1/2$ as $k \to \infty$. Since the comparison series diverges, the given series does as well.

8.5.37 Use the Limit Comparison Test with $\frac{k^{2/3}}{k^{3/2}}$. The ratio of corresponding terms of the two series is $\frac{\sqrt[3]{k^2+1}}{\sqrt{k^3+1}} \cdot \frac{k^{3/2}}{k^{2/3}} = \frac{\sqrt[3]{k^2+1}}{\sqrt[3]{k^2}} \cdot \frac{\sqrt{k^3}}{\sqrt{k^3+1}}$, which has limit 1 as $k \to \infty$. The comparison series is the series whose terms are $k^{2/3-3/2} = k^{-5/6}$, which is a p-series with $p < 1$, so it, and the given series, both diverge.

8.5.39

a. False. For example, let $\{a_k\}$ be all zeros, and $\{b_k\}$ be all 1's.

b. True. This is a result of the Comparison Test.

c. True. Both of these statements follow from the Comparison Test.

8.5.41 Use the Comparison Test. Each term $\frac{1}{k} + 2^{-k} > \frac{1}{k}$. Since the harmonic series diverges, so does this series.

8.5.43 Use the Ratio Test. $\frac{a_{k+1}}{a_k} = \frac{2^{k+1}(k+1)!}{(k+1)^{k+1}} \cdot \frac{(k)^k}{2^k(k)!} = 2\left(\frac{k}{k+1}\right)^k$, which has limit $\frac{2}{e}$ as $k \to \infty$, so the given series converges.

8.5.45 Use the Limit Comparison Test with $\{1/k^3\}$. The ratio of corresponding terms is $\frac{k^{11}}{k^{11}+3}$, which has limit 1 as $k \to \infty$. Since the comparison series converges, so does the given series.

8.5.47 This is a p-series with exponent greater than 1, so it converges.

8.5.49 $\ln\left(\frac{k+2}{k+1}\right) = \ln(k+2) - \ln(k+1)$, so this series telescopes. We get $\sum_{k=1}^{n} \ln\left(\frac{k+2}{k+1}\right) = \ln(n+2) - \ln 2$. Since $\lim_{n\to\infty} \ln(n+2) - \ln(2) = \infty$, the sequence of partial sums diverges, so the given series is divergent.

8.5.51 For $k > 10$, $\ln k > 2$ so note that $\frac{1}{k^{\ln k}} < \frac{1}{k^2}$. Since $\sum_{k=1}^{\infty} \frac{1}{k^2}$ converges, the given series converges as well.

8.5.53 Use the Limit Comparison Test with the harmonic series. $\frac{\tan(1/k)}{1/k}$ has limit 1 as $k \to \infty$ since $\lim_{x\to 0} \frac{\tan(x)}{x} = 1$. Thus the original series diverges.

8.5.55 Note that $\dfrac{1}{(2k+1)\cdot(2k+3)} = \dfrac{1}{2}\left(\dfrac{1}{2k+1} - \dfrac{1}{2k+3}\right)$. Thus this series telescopes.

$$\sum_{k=0}^{n} \frac{1}{(2k+1)(2k+3)} = \frac{1}{2}\sum_{k=0}^{n}\left(\frac{1}{2k+1} - \frac{1}{2k+3}\right) = \frac{1}{2}\left(-\frac{1}{2n+3} + 1\right),$$

so the given series converges to $1/2$, since that is the limit of the sequence of partial sums.

8.5.57 This series is $\sum_{k=1}^{\infty} \frac{k^2}{k!}$. By the Ratio Test, $\frac{a_{k+1}}{a_k} = \frac{(k+1)^2}{(k+1)!} \cdot \frac{k!}{k^2} = \frac{1}{k+1}\left(\frac{k+1}{k}\right)^2$, which has limit 0 as $k \to \infty$, so the given series converges.

8.5.59 For $p \le 1$ and $k > e$, $\frac{\ln k}{k^p} > \frac{1}{k^p}$. The series $\sum_{k=1}^{\infty} \frac{1}{k^p}$ diverges, so the given series diverges. For $p > 1$, let $q < p - 1$; then for sufficiently large k, $\ln k < k^q$, so that by the Comparison Test, $\frac{\ln k}{k^p} < \frac{k^q}{k^p} = \frac{1}{k^{p-q}}$. But $p - q > 1$, so that $\sum_{k=1}^{\infty} \frac{1}{k^{p-q}}$ is a convergent p-series. Thus the original series is convergent precisely when $p > 1$.

8.5.61 For $p \le 1$, $\frac{(\ln k)^p}{k^p} > \frac{1}{k^p}$, and $\sum_{k=1}^{\infty} \frac{1}{k^p}$ diverges for $p \le 1$, so the original series diverges. For $p > 1$, let $q < p - 1$; then for sufficiently large k, $(\ln k)^p < k^q$. Note that $\frac{(\ln k)^p}{k^p} < \frac{k^q}{k^p} = \frac{1}{k^{p-q}}$. But $p - q > 1$, so $\sum_{k=1}^{\infty} \frac{1}{k^{p-q}}$ converges, so the given series converges. Thus, the given series converges exactly for $p > 1$.

8.5.63 Use the Ratio Test:

$$\frac{a_{k+1}}{a_k} = \frac{1\cdot 3\cdot 5\cdots(2k+1)}{(k+1)p^{k+2}(k+1)!} \cdot \frac{(k)p^{k+1}(k)!}{1\cdot 3\cdot 5\cdots(2k-1)} = \frac{(2k+1)(k)}{(k+1)^2 p}$$

and this expression has limit $\frac{2}{p}$ as $k \to \infty$. Thus the series converges for $p > 2$.

8.5.65 $\lim_{k\to\infty} a_k = \lim_{k\to\infty} \left(1 - \frac{p}{k}\right)^k = e^{-p} \ne 0$, so this sequence diverges for all p by the Divergence Test.

8.5.67 These tests apply only for series with positive terms, so assume $r > 0$. Clearly the series do not converge for $r = 1$, so we assume $r \ne 1$ in what follows. Using the Integral Test, $\sum r^k$ converges if and only if $\int_1^\infty r^x dx$ converges. This improper integral has value $\lim_{b\to\infty} \left.\frac{r^x}{\ln r}\right|_1^b$, which converges only when $\lim_{b\to\infty} r^b$ exists, which occurs only for $r < 1$. Using the Ratio Test, $\frac{a_{k+1}}{a_k} = \frac{r^{k+1}}{r^k} = r$, so by the Ratio Test, the series converges if and only if $r < 1$. Using the Root Test, $\lim_{k\to\infty} \sqrt[k]{a_k} = \lim_{k\to\infty} \sqrt[k]{r^k} = \lim_{k\to\infty} r = r$, so again we have convergence if and only if $r < 1$.

8.5.69 To prove case (2), assume $L = 0$ and that $\sum b_k$ converges. Since $L = 0$, for every $\varepsilon > 0$, there is some N such that for all $n > N$, $|\frac{a_k}{b_k}| < \varepsilon$. Take $\varepsilon = 1$; this then says that there is some N such that for all $n > N$, $0 < a_k < b_k$. By the Comparison Test, since $\sum b_k$ converges, so does $\sum a_k$. To prove case (3), since $L = \infty$, then $\lim_{k\to\infty} \frac{b_k}{a_k} = 0$, so by the argument above, we have $0 < b_k < a_k$ for sufficient large k. But $\sum b_k$ diverges, so by the Comparison Test, $\sum a_k$ does as well.

8.5.71 $\dfrac{a_{k+1}}{a_k} = \dfrac{x^{k+1}}{x^k} = x$. This has limit x as $k \to \infty$, so the series converges for $x < 1$. It clearly does not converge for $x = 1$.

8.5.73 $\dfrac{a_{k+1}}{a_k} = \dfrac{x^{k+1}}{(k+1)^2} \cdot \dfrac{k^2}{x^k} = x\left(\dfrac{k}{k+1}\right)^2$, which has limit x as $k \to \infty$. Thus the series converges for $x < 1$. When $x = 1$, the series is $\frac{1}{k^2}$, which converges. Thus the original series converges for $x \le 1$.

8.5.75 $\dfrac{a_{k+1}}{a_k} = \dfrac{x^{k+1}}{2^{k+1}} \cdot \dfrac{2^k}{x^k} = \dfrac{x}{2}$, which has limit $x/2$ as $k \to \infty$. Thus the series converges for $x < 2$. For $x = 2$, it is obviously divergent.

8.5.77

a. $\ln \prod_{k=0}^{\infty} e^{1/2^k} = \sum_{k=0}^{\infty} \frac{1}{2^k} = 2$, so that the original product converges to e^2.

b. $\ln \prod_{k=2}^{\infty} \left(1 - \frac{1}{k}\right) = \ln \prod_{k=2}^{\infty} \frac{k-1}{k} = \sum_{k=2}^{\infty} \ln \frac{k-1}{k} = \sum_{k=2}^{\infty} (\ln(k-1) - \ln(k))$. This series telescopes to give $S_n = -\ln(n)$, so the original series has limit $\lim_{n\to\infty} P_n = \lim_{n\to\infty} e^{-\ln(n)} = 0$.

8.6 Alternating Series

8.6.1 If $a_n < 0$ (which happens for every other value of n), then $S_n = S_{n-1} + a_n < S_{n-1}$.

8.6.3 Since we are far enough out that the terms are nonincreasing in magnitude, we may assume the series is $\sum_{k=0}^{\infty} (-1)^k a_k$, where $a_k > 0$ for all k, and such that the a_k are nonincreasing in magnitude starting with a_1. Then

$$S = S_{2n+1} + (a_{2n} - a_{2n+1}) + (a_{2n+2} - a_{2n+3}) + \cdots$$

and each term of the form $a_{2k} - a_{2k+1} > 0$, so that $S_{2n+1} < S$. Also

$$S = S_{2n} + (-a_{2n+1} + a_{2n+2}) + (-a_{2n+3} + a_{2n+4}) + \cdots$$

and each term of the form $-a_{2k+1} + a_{2k+2} < 0$, so that $S < S_{2n}$. Thus the sum of the series is trapped between the odd partial sums and the even partial sums.

8.6.5 The remainder is less than the first neglected term since

$$L - S_n = (-1)^{n+1}(a_{n+1} + (-a_{n+2} + a_{n+3}) + \cdots)$$

so that the sum of the series *after* the first disregarded term has the opposite sign from the first disregarded term.

8.6.7 No. If the terms are positive, then the absolute value of each term is the term itself, so convergence and absolute convergence would mean the same thing in this context.

8.6.9 Yes. For example, $\sum \frac{(-1)^k}{k^3}$ converges absolutely and thus not conditionally (see the definition).

8.6.11 The terms of the series decrease in magnitude, and $\lim_{k\to\infty} \frac{1}{k^3} = 0$, so the given series converges.

8.6.13 The terms of the series decrease in magnitude, and

$$\lim_{k\to\infty} \frac{k^2}{k^3+1} = \lim_{k\to\infty} \frac{1}{k+1/k^2} = 0,$$

so the given series converges.

8.6.15 $\lim_{k\to\infty} \frac{k^2-1}{k^2+3} = 1$, so the terms of the series do not tend to zero and thus the given series diverges.

8.6.17 $\lim_{k\to\infty} \left(1+\frac{1}{k}\right) = 1$, so the given series diverges.

8.6.19 The derivative of $f(k) = \frac{k^{10}+2k^5+1}{k(k^{10}+1)}$ is $f'(k) = \frac{-(k^{20}+2k^{10}+12k^{15}-8k^5+1)}{k^2(k^{10}+1)^2}$. The numerator is negative for large enough values of k, and the denominator is always positive, so the derivative is negative for large enough k. Also, $\lim_{k\to\infty} \frac{k^{10}+2k^5+1}{k(k^{10}+1)} = \lim_{k\to\infty} \frac{1+2k^{-5}+k^{-10}}{k+k^{-9}} = 0$. Thus the given series converges.

8.6.21 $\lim_{k\to\infty} k^{1/k} = 1$ (for example, take logs and apply L'Hôpital's rule), so the given series diverges by the Divergence Test.

8.6.23 $\frac{1}{\sqrt{k^2+4}}$ is decreasing and tends to zero as $k \to \infty$, so the given series converges.

8.6.25 We want $\frac{1}{n+1} < 10^{-4}$, or $n+1 > 10^4$, so $n = 10^4$.

8.6.27 The series starts with $k = 0$, so we want $\frac{1}{2n+1} < 10^{-4}$, or $2n+1 > 10^4$, $n = 5000$.

8.6.29 We want $\frac{1}{(n+1)^4} < 10^{-4}$, or $(n+1)^4 > 10^4$, so $n = 10$.

8.6.31 The series starts with $k = 0$, so we want $\frac{1}{3n+1} < 10^{-4}$, or $3n+1 > 10^4$, $n = 3334$.

8.6.33 The series starts with $k = 0$, so we want $\frac{1}{4^n}\left(\frac{2}{4n+1} + \frac{2}{4n+2} + \frac{1}{4n+3}\right) < 10^{-4}$, or $\frac{4^n(4n+1)(4n+2)(4n+3)}{4(20n^2+21n+5)} > 10000$, which occurs first for $n = 6$.

8.6.35 To figure out how many terms we need to sum, we must find n such that $\frac{1}{(n+1)^5} < 10^{-3}$, so that $(n+1)^5 > 1000$; this occurs first for $n = 3$. Thus $\frac{-1}{1} + \frac{1}{2^5} - \frac{1}{3^5} \approx -0.972865$.

8.6.37 To figure how many terms we need to sum, we must find n such that $\frac{1}{(n+1)^{n+1}} < 10^{-3}$, or $(n+1)^{n+1} > 1000$, so $n = 4$ ($5^5 = 3125$). Thus the approximation is $\sum_{k=1}^{4} \frac{(-1)^n}{n^n} \approx -.7831307870$.

8.6.39 The series of absolute values is a p-series with $p = 3/2$, so it converges absolutely.

8.6.41 The series of absolute values is $\sum \frac{|\cos(k)|}{k^3}$, which converges by the Comparison Test since $\frac{|\cos(k)|}{k^3} \le \frac{1}{k^3}$. Thus the series converges absolutely.

8.6.43 The series of absolute values is $\sum \frac{k}{2k+1}$, but $\lim_{k\to\infty} \frac{k}{2k+1} = \frac{1}{2}$, so by the Divergence Test, this series diverges. The original series does not converge conditionally, either, since $\lim_{k\to\infty} a_k = \frac{1}{2} \neq 0$.

8.6.45 The series of absolute values is $\sum \frac{\tan^{-1}(k)}{k^3}$, which converges by the Comparison Test since $\frac{\tan^{-1}(k)}{k^3} < \frac{\pi}{2}\frac{1}{k^3}$, and $\sum \frac{\pi}{2}\frac{1}{k^3}$ converges since it is a constant multiple of a convergent $p-$series. So the original series converges absolutely.

8.6.47

a. False. For example, consider the alternating harmonic series.

b. True. This is part of Theorem 8.21.

c. True. This statement is simply saying that a convergent series converges.

d. True. This is part of Theorem 8.21.

e. False. Let $a_k = \frac{1}{k}$.

f. True. Use the Comparison Test: $\lim_{k\to\infty} \frac{a_k^2}{a_k} = \lim_{k\to\infty} a_k = 0$ since $\sum a_k$ converges, so $\sum a_k^2$ and $\sum a_k$ converge or diverge together. Since the latter converges, so does the former.

g. True, by definition. If $\sum |a_k|$ converged, the original series would converge absolutely, not conditionally.

8.6.49 $\sum_{k=1}^{\infty} \frac{1}{k^2} - \sum_{k=1}^{\infty} \frac{(-1)^{k+1}}{k^2} = 2\sum_{k=1}^{\infty} \frac{1}{(2k)^2} = 2 \cdot \frac{1}{4} \sum_{k=1}^{\infty} \frac{1}{k^2}$, and thus $\sum_{k=1}^{\infty} \frac{(-1)^{k+1}}{k^2} = \frac{\pi^2}{6} - \frac{1}{2} \cdot \frac{\pi^2}{6} = \frac{\pi^2}{12}$.

8.6.51 Write $r = -s$; then $0 < s < 1$ and $\sum r^k = \sum (-1)^k s^k$. Since $|s| < 1$, the terms s^k are nonincreasing and tend to zero, so by the Alternating Series Test, the series $\sum s^k$ converges, so $\sum (-1)^k s^k = \sum r^k$ does too.

8.6.53 Let $S = 1 - \frac{1}{2} + \frac{1}{3} - \cdots$. Then

$$\begin{array}{rcccccccccc} S & = & \left(1 - \frac{1}{2}\right) & + & \left(\frac{1}{3} - \frac{1}{4}\right) & + & \left(\frac{1}{5} - \frac{1}{6}\right) & + & \left(\frac{1}{7} - \frac{1}{8}\right) & + & \cdots \\ \frac{1}{2}S & = & \frac{1}{2} & - & \frac{1}{4} & + & \frac{1}{6} & - & \frac{1}{8} & + & \cdots \end{array}$$

Add these two series together to get

$$\frac{3}{2}S = \frac{3}{2}\ln 2 = 1 + \frac{1}{3} - \frac{1}{2} + \frac{1}{5} + \cdots$$

To see that the results are as desired, consider a collection of four terms:

$$\begin{array}{cccccc} \cdots + & \left(\frac{1}{4k+1} - \frac{1}{4k+2}\right) & + & \left(\frac{1}{4k+3} - \frac{1}{4k+4}\right) & + & \cdots \\ \cdots & + & \frac{1}{4k+2} & - & \frac{1}{4k+4} & + \cdots \end{array}$$

Adding these results in the desired sign pattern. This repeats for each group of four elements.

8.6.55 Both series diverge, so comparisons of their values are not meaningful.

8.7 Chapter Eight Review

8.7.1

a. False. Let $a_n = 1 - \frac{1}{n}$. This sequence has limit 1.

b. False. The terms of a sequence tending to zero is necessary but not sufficient for convergence of the series.

c. True. This is the definition of convergence of a series.

d. False. If a series converges absolutely, the definition says that it does not converge conditionally.

8.7.3 $\lim_{n\to\infty} \frac{8^n}{n!} = 0$ since exponentials grow more slowly than factorials.

8.7.5 Take logs and compute $\lim_{n\to\infty} (1/n) \ln n = \lim_{n\to\infty} (\ln n)/n = \lim_{n\to\infty} \frac{1}{n} = 0$ by L'Hôpital's rule. Thus the original limit is $e^0 = 1$.

8.7.7 Take logs, and then evaluate $\lim_{n\to\infty} \frac{1}{\ln n}\ln(1/n) = \lim_{n\to\infty}(-1) = -1$, so the original limit is e^{-1}.

8.7.9 $a_n = (-1/0.9)^n = (-10/9)^n$. The terms grow without bound so the sequence does not converge.

8.7.11

a. $S_1 = \frac{1}{3}$, $S_2 = \frac{11}{24}$, $S_3 = \frac{21}{40}$, $S_4 = \frac{17}{30}$.

b. $S_n = \dfrac{1}{2}\left(\dfrac{1}{1} + \dfrac{1}{2} - \dfrac{1}{n+1} - \dfrac{1}{n+2}\right)$, since the series telescopes.

c. From part (b), $\lim_{n\to\infty} S_n = \frac{3}{4}$, which is the sum of the series.

8.7.13 $\sum_{k=1}^{\infty} 3(1.001)^k = 3\sum_{k=1}^{\infty}(1.001)^k$. This is a geometric series with ratio greater than 1, so it diverges.

8.7.15 $\frac{1}{k(k+1)} = \frac{1}{k} - \frac{1}{k+1}$, so the series telescopes, and $S_n = 1 - \frac{1}{n+1}$. Thus $\lim_{n\to\infty} S_n = 1$, which is the value of the series.

8.7.17 This series telescopes. $S_n = 3 - \frac{3}{3n+1}$, so that $\lim_{n\to\infty} S_n = 3$, which is the value of the series.

8.7.19 $\displaystyle\sum_{k=1}^{\infty} \frac{2^k}{3^{k+2}} = \frac{1}{9}\sum_{k=1}^{\infty}\left(\frac{2}{3}\right)^k = \frac{1}{9}\cdot\frac{2/3}{1-2/3} = \frac{2}{9}.$

8.7.21

a. It appears that the series converges, since the sequence of partial sums appears to converge to 1.

b. It appears that series B converges to about 4.

c. This series clearly appears to diverge, since the partial sums seem to be growing without bound.

8.7.23 The series can be written $\sum \frac{1}{k^{2/3}}$, which is a p-series with $p = 2/3 < 1$, so this series diverges.

8.7.25 This is a geometric series with ratio $2/e < 1$, so the sum is $\frac{2/e}{1-2/e} = \frac{2}{e-2}$.

8.7.27 Applying the Ratio Test:

$$\lim_{k\to\infty}\frac{a_{k+1}}{a_k} = \lim_{k\to\infty}\frac{2^{k+1}(k+1)!}{(k+1)^{k+1}}\cdot\frac{k^k}{2^k k!} = \lim_{k\to\infty} 2\left(\frac{k}{k+1}\right)^k = \frac{2}{e} < 1,$$

so the given series converges.

8.7.29 Use the Comparison Test: $\frac{3}{2+e^k} < \frac{3}{e^k}$, but $\sum \frac{3}{e^k}$ converges since it is a geometric series with ratio $\frac{1}{e} < 1$. Thus the original series converges as well.

8.7.31 $a_k = \frac{k^{1/k}}{k^3} = \frac{1}{k^{3-1/k}}$. For $k \geq 2$, then, $a_k < \frac{1}{k^2}$. Since $\sum \frac{1}{k^2}$ converges, the given series also converges, by the Comparison Test.

8.7.33 Use the Ratio Test: $\frac{a_{k+1}}{a_k} = \frac{(k+1)^5}{e^{k+1}}\cdot\frac{e^k}{k^5} = \frac{1}{e}\cdot\left(\frac{k+1}{k}\right)^5$, which has limit $1/e < 1$ as $k\to\infty$. Thus the given series converges.

8.7.35 Use the Comparison Test. Since $\displaystyle\lim_{k\to\infty}\frac{\ln k}{k^{1/2}} = 0$, we have that for sufficiently large k, $\ln k < k^{1/2}$, so that $a_k = \frac{2\ln k}{k^2} < \frac{2k^{1/2}}{k^2} = \frac{2}{k^{3/2}}$. Now $\sum \frac{2}{k^{3/2}}$ is convergent, since it is a p-series with $p = 3/2 > 1$. Thus the original series is convergent.

8.7.37 $|a_k| = \frac{1}{k^2-1}$. Use the Limit Comparison Test with the convergent series $\sum \frac{1}{k^2}$. Since $\lim_{k\to\infty} \frac{1}{k^2-1} \Big/ \frac{1}{k^2} = \lim_{k\to\infty} \frac{k^2}{k^2-1} = 1$, the given series converges absolutely.

8.7.39 Use the Ratio Test on the absolute values of the sequence of terms: $\lim_{k\to\infty} \left|\frac{a_{k+1}}{a_k}\right| = \lim_{k\to\infty} \frac{k+1}{e^{k+1}} \cdot \frac{e^k}{k} = \lim_{k\to\infty} \frac{1}{e} \cdot \frac{k+1}{k} = \frac{1}{e} < 1$. Thus, the original series is absolutely convergent.

8.7.41 Use the Ratio Test on the absolute values of the sequence of terms: $\lim_{k\to\infty} \left|\frac{a_{k+1}}{a_k}\right| = \lim_{k\to\infty} \frac{10}{k+1} = 0$, so the series converges absolutely.

8.7.43

a. For $|x| < 1$, $\lim_{k\to\infty} x^k = 0$, so this limit is zero.

b. This is a geometric series with ratio $-4/5$, so the sum is $\frac{1}{1+4/5} = \frac{5}{9}$.

8.7.45 Since the series converges, we must have $\lim_{k\to\infty} a_k = 0$. Since it converges to 8, the partial sums converge to 8, so that $\lim_{k\to\infty} S_k = 8$.

8.7.47 The series converges absolutely for $p > 1$, conditionally for $0 < p \le 1$ in which case $\{k^{-p}\}$ is decreasing to zero.

8.7.49 The sum is 0.2500000000 to ten decimal places. The maximum error is

$$\int_{20}^{\infty} \frac{1}{5^x}\,dx = \lim_{b\to\infty} \frac{-1}{\ln(5)5^x}\bigg|_{20}^{b} = \frac{1}{\ln(5)\cdot 5^{20}} \approx 6.5\times 10^{-15}.$$

8.7.51 The maximum error is a_{n+1}, so we want $a_{n+1} = \frac{1}{(k+1)^4} < 10^{-8}$, or $(k+1)^4 > 10^8$, so $k = 100$.

8.7.53

a. Let T_n be the amount of additional tunnel dug during week n. Then $T_0 = 100$ and $T_n = .95 \cdot T_{n-1} = (.95)^n T_0 = 100(0.95)^n$, so the total distance dug in N weeks is

$$S_N = 100 \sum_{k=0}^{N-1} (0.95)^k = 100 \left(\frac{1-(0.95)^N}{1-0.95}\right) = 2000(1-0.95^N).$$

Then $S_{10} \approx 802.5$ meters and $S_{20} \approx 1283.03$ meters.

b. The longest possible tunnel is $S_\infty = 100\sum_{k=0}^{\infty}(0.95)^k = \frac{100}{1-.95} = 2000$ meters.

8.7.55

a. The area of a circle of radius r is πr^2. For $r = 2^{1-n}$, this is $\pi 2^{2-2n}$. There are 2^{n-1} circles on the nth page, so the total area of circles on the nth page is $2^{n-1}\cdot \pi 2^{2-2n} = 2^{1-n}\pi$.

b. The sum of the areas on all pages is $\sum_{k=1}^{\infty} 2^{1-k}\pi = 2\pi \sum_{k=1}^{\infty} 2^{-k} = 2\pi \cdot \frac{1/2}{1/2} = 2\pi$.

8.7.57

a. $B_n = 1.0025 \cdot B_{n-1} + 100$ and $B_0 = 100$.

b. $B_n = 100 \cdot 1.0025^n + 100 \cdot \frac{1-1.0025^n}{1-1.0025} = 100 \cdot 1.0025^n - 40000(1-1.0025^n)$.

8.7.59

a. $T_1 = \frac{\sqrt{3}}{16}$ and $T_2 = \frac{7\sqrt{3}}{64}$.

b. At stage n, 3^{n-1} triangles of side length $1/2^n$ are removed. Each of those triangles has an area of $\dfrac{\sqrt{3}}{4 \cdot 4^n} = \dfrac{\sqrt{3}}{4^{n+1}}$, so a total of

$$3^{n-1} \cdot \frac{\sqrt{3}}{4^{n+1}} = \frac{\sqrt{3}}{16} \cdot \left(\frac{3}{4}\right)^{n-1}$$

is removed at each stage. Thus

$$T_n = \frac{\sqrt{3}}{16} \sum_{k=1}^{n} \left(\frac{3}{4}\right)^{k-1} = \frac{\sqrt{3}}{16} \sum_{k=0}^{n-1} \left(\frac{3}{4}\right)^{k} = \frac{\sqrt{3}}{4}\left(1 - \left(\frac{3}{4}\right)^{n}\right)$$

c. $\lim_{n\to\infty} T_n = \frac{\sqrt{3}}{4}$ since $\left(\frac{3}{4}\right)^n \to 0$ as $n \to \infty$.

d. The area of the triangle was originally $\frac{\sqrt{3}}{4}$, so none of the original area is left.

Chapter 9

9.1 Approximating Functions With Polynomials

9.1.1 Let the polynomial be $p(x)$. Then $p(0) = f(0)$, $p'(0) = f'(0)$, and $p''(0) = f''(0)$.

9.1.3 The approximations are $p_0(0.1) = 1$, $p_1(0.1) = 1 + \frac{0.1}{2} = 1.05$, and $p_2(0.1) = 1 + \frac{0.1}{2} - \frac{.01}{8} = 1.04875$.

9.1.5 The remainder is the difference between the value of the Taylor polynomial at a point and the true value of the function at that point, $R_n(x) = f(x) - p_n(x)$.

9.1.7

a. $f'(x) = -e^{-x}$, so $p_1(x) = f(0) + f'(0)(x) = 1 - x$.

b. $f''(x) = e^{-x}$, so $p_2(x) = f(0) + f'(0)(x) + \frac{1}{2}f''(0)(x^2) = 1 - x + \frac{1}{2}x^2$.

c. $p_1(0.2) = 0.8$, and $p_2(0.2) = 1 - 0.2 + \frac{1}{2}(0.04) = 0.82$.

9.1.9

a. $f'(x) = -\frac{1}{(x+1)^2}$, so $p_1(x) = f(0) + f'(0)(x) = 1 - x$.

b. $f''(x) = \frac{2}{(x+1)^3}$, so $p_2(x) = f(0) + f'(0)(x) + \frac{1}{2}f''(0)x^2 = 1 - x + x^2$.

c. $p_1(.05) = .95$, and $p_2(.05) = 1 - .05 + .0025 = .9525$.

9.1.11

a. $f'(x) = (1/3)x^{-2/3}$, so $p_1(x) = f(8) + f'(8)(x-8) = 2 + \frac{1}{12}(x-8)$.

b. $f''(x) = (-2/9)x^{-5/3}$, so $p_2(x) = f(8) + f'(8)(x-8) + \frac{1}{2}f''(8)(x-8)^2 = 2 + \frac{1}{12}(x-8) - \frac{1}{288}(x-8)^2$.

c. $p_1(7.5) \approx 1.958333333$, $p_2(7.5) \approx 1.957465278$.

9.1.13 $f(0) = 1, f'(0) = -\sin 0 = 0, f''(0) = -\cos 0 = -1$, so that $p_0(x) = 1$, $p_1(x) = 1$, $p_2(x) = 1 - \frac{1}{2}x^2$.

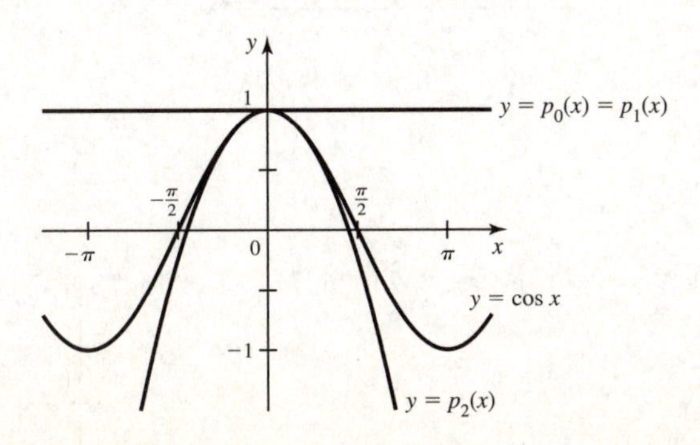

9.1.15 $f(0) = 0$, $f'(0) = \frac{-1}{1-0} = -1, f''(0) = \frac{1}{(1-0)^2} = -1$, so that $p_0(x) = 0$, $p_1(x) = -x$, $p_2(x) = -x - \frac{1}{2}x^2$.

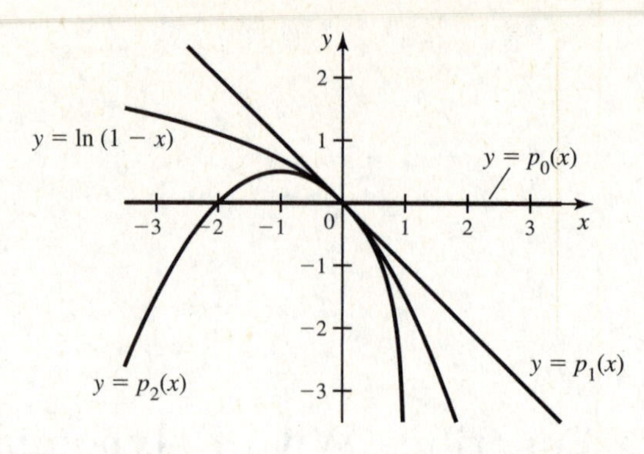

9.1.17 $f(0) = 0$. $f'(x) = 1 + \tan^2 x$, $f''(x) = 2\tan x \sec^2 x$, so that $f'(0) = 1$, $f''(0) = 0$. Thus $p_0(x) = 0$, $p_1(x) = x$, $p_2(x) = x$.

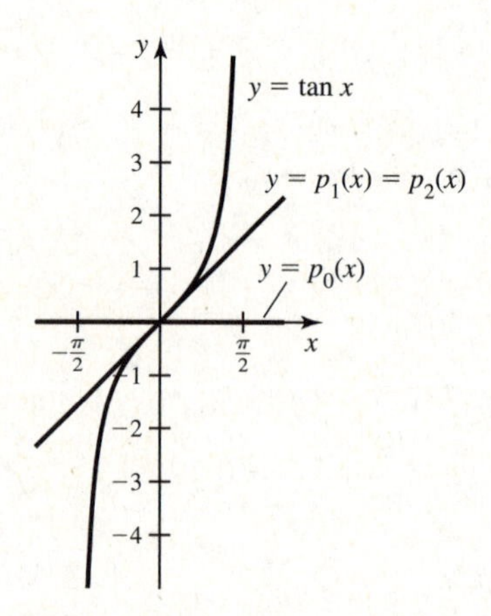

9.1.19 $f(0) = 1$, $f'(0) = -3(1+0)^{-4} = -3$, $f''(0) = 12(1+0)^{-5} = 12$, so that $p_0(x) = 1$, $p_1(x) = 1 - 3x$, $p_2(x) = 1 - 3x + 6x^2$.

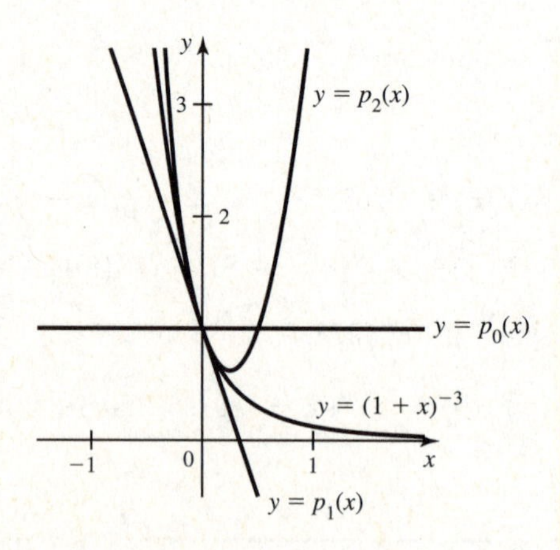

9.1.21

a. $p_2(.05) = 1.0246875$.

b. The absolute error is $\sqrt{1.05} - p_2(1.05) \approx 1.024695077 - 1.0246875 = 7.58 \times 10^{-6}$.

9.1.23

a. $p_2(.08) = 0.9624$.

b. The absolute error is $p_2(.08) - 1/\sqrt{1.08} \approx 0.9624 - 0.9622504482 \approx 1.50 \times 10^{-4}$.

9.1.25

a. $p_2(0.15) = 0.86125$.

b. The absolute error is $p_2(0.15) - e^{-0.15} \approx 0.86125 - 0.8607079764 \approx 5.42 \times 10^{-4}$.

9.1.27 $p_0(x) = \frac{\sqrt{2}}{2}$, $p_1(x) = \frac{\sqrt{2}}{2} + \frac{\sqrt{2}}{2}(x - \frac{\pi}{4})$, $p_2(x) = \frac{\sqrt{2}}{2} + \frac{\sqrt{2}}{2}(x - \frac{\pi}{4}) - \frac{\sqrt{2}}{4}(x - \frac{\pi}{4})^2$.

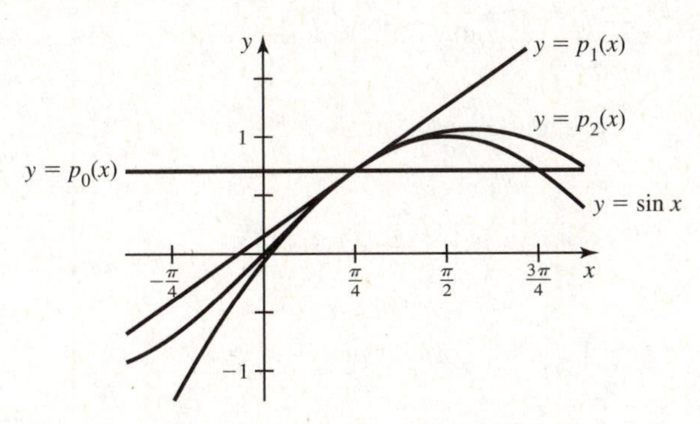

9.1.29 $p_0(x) = 3$, $p_1(x) = 3 + \frac{1}{6}(x - 9)$, $p_2(x) = 3 + \frac{1}{6}(x - 9) - \frac{1}{216}(x - 9)^2$.

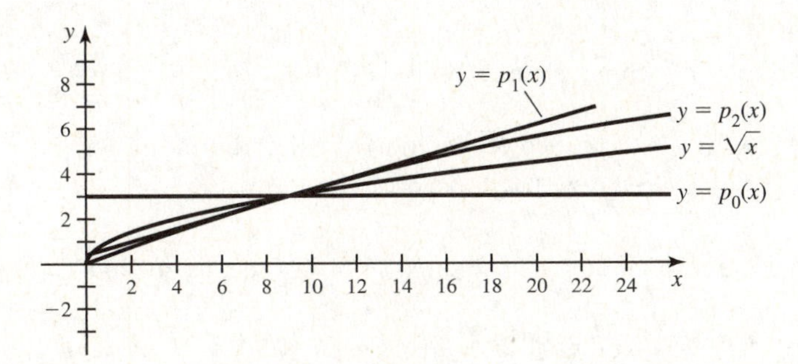

9.1.31 $p_0(x) = 1$, $p_1(x) = 1 + \frac{1}{e}(x - e)$, $p_2(x) = 1 + \frac{1}{e}(x - e) - \frac{1}{2e^2}(x - e)^2$.

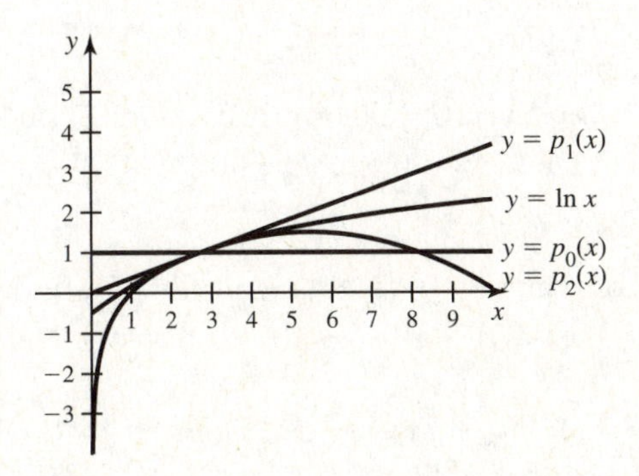

9.1.33

a. $f(x) = e^x$. $p_3(x) = 1 + x + \frac{1}{2}x^2 + \frac{1}{6}x^3$. $p_3(0.12) \approx 1.127488$.

b. $f(0.12) - p_3(0.12) \approx 1.127496852 - 1.127488 = .000008852$.

9.1.35

a. $f(x) = \tan(x)$. $p_3(x) = x + \frac{1}{3}x^3$. $p_3(-0.1) \approx -0.1003333333$.

b. $p_3(-0.1) - f(-0.1) \approx -0.1003333333 + 0.1003346721 = 0.0000013388$.

9.1.37

a. $f(x) = \sqrt{1+x}$. $p_3(x) = 1 + \frac{1}{2}x - \frac{1}{8}x^2 + \frac{1}{16}x^3$. $p_3(0.06) \approx 1.029563500$.

b. $f(0.06) - p_3(0.06) \approx 1.029563014 - 1.029563500 \approx 4.86 \times 10^{-7}$.

9.1.39

a. $f(x) = \sqrt{x}$. $p_3(x) = 10 + \frac{1}{20}(x-100) - \frac{1}{8000}(x-100)^2 + \frac{1}{1600000}(x-100)^3$. $p_3(101) \approx 10.04987563$.

b. $p_3(101) - f(101) \approx 10.04987563 - 10.04987562 \approx 1 \times 10^{-8}$.

9.1.41 $R_n(x) = \dfrac{\sin^{(n+1)}(c)}{(n+1)!}x^{n+1}$ for some c between 0 and x.

9.1.43 $R_n(x) = \dfrac{(-1)^n e^{-c}}{(n+1)!}x^{n+1}$ for some c between 0 and x.

9.1.45 $R_n(x) = \dfrac{\sin^{(n+1)}(c)}{(n+1)!}\left(x - \dfrac{\pi}{2}\right)^{n+1}$ for some c between $\frac{\pi}{2}$ and x.

9.1.47 $f(x) = \sin x$, so $f^{(4)}(x) = \cos x$. Since $\cos x$ is bounded in magnitude by 1, the remainder is bounded by $|R_4(x)| \le \frac{0.3^5}{5!} \approx 2.03 \times 10^{-5}$.

9.1.49 $f(x) = e^x$, so $f^{(5)}(x) = e^x$. Since $e^{0.25}$ is bounded by 2, $|R_4(x)| \le 2 \cdot \frac{0.25^5}{5!} \approx 1.63 \times 10^{-5}$.

9.1.51 $f(x) = e^{-x}$, so $f^{(5)}(x) = -e^{-x}$. Since $f^{(5)}$ achieves its maximum magnitude in the range at $x = -0.5$, which is $\sqrt{e} < 2$, $|R_4(x)| \le 2 \cdot \frac{0.5^5}{5!} \approx 5.2 \times 10^{-4}$.

9.1.53 Here $n = 3$ or 4, so use $n = 4$, and $M = 1$ since $f^{(5)}(x) = \cos x$, so that $R_4(x) \le \frac{(\pi/4)^5}{5!} \approx 2.49 \times 10^{-3}$.

9.1.55 $n = 2$ and $M = e^{1/2} < 2$, so $R_2(x) \le 2 \cdot \frac{(1/2)^3}{3!} \approx 4.17 \times 10^{-2}$.

9.1.57 $n = 2$; $f^{(2)}(x) = \frac{2}{(1+x)^3}$, which achieves its maximum at $x = -0.2$: $|f^{(3)}(x)| = \frac{2}{0.8^3} < 4$. Then $R_2(x) \le 4 \cdot \frac{0.2^3}{3!} \approx 5.3 \times 10^{-3}$.

9.1.59 Use the Taylor series for e^x at $x = 0$. The derivatives of e^x are e^x. On $[-0.5, 0]$, the maximum magnitude of any derivative is thus 1 and $x = 0$, so $|R_n(-0.5)| \le \frac{0.5^{n+1}}{(n+1)!}$, so for $R_n(-0.5) < 10^{-3}$ we need $n = 4$.

9.1.61 Use the Taylor series for $\cos x$ at $x = 0$. The magnitude of any derivative of $\cos x$ is bounded by 1, so $|R_n(-0.25)| \le \frac{0.25^{n+1}}{(n+1)!}$, so for $|R_n(-0.25)| < 10^{-3}$ we need $n = 3$.

9.1.63 Use the Taylor series for $f(x) = \sqrt{x}$ at $x = 1$. Then $|f^{(n+1)}(x)| = \frac{1\cdot 3 \cdot \cdots \cdot (2n-1)}{2^{n+1}}x^{-(2n+1)/2}$, which achieves its maximum on $[1, 1.06]$ at $x = 1$. Then

$$|R_n(1.06)| \le \frac{1 \cdot 3 \cdot \cdots \cdot (2n-1)}{2^{n+1}} \cdot \frac{(1.06-1)^{n+1}}{(n+1)!},$$

and for $|R_n(0.06)| < 10^{-3}$ we need $n = 1$.

9.1.65

a. False. If $f(x) = e^{-2x}$, then $f^{(n)}(x) = (-1)^n 2^n e^{-2x}$, so that $f^{(n)}(0) \ne 0$ and all powers of x are present in the Taylor series.

b. True. The constant term of the Taylor series is $f(0) = 1$. Higher-order terms all involve derivatives of $f(x) = x^5 - 1$ evaluated at $x = 0$; clearly for $n < 5$, $f^{(n)}(0) = 0$, and for $n > 5$, the derivative itself vanishes. Only for $n = 5$, where $f^{(5)}(x) = 5!$, is the derivative nonzero, so the coefficient of x^5 in the Taylor series is $f^{(5)}(0)/5! = 1$ and the Taylor polynomial of order 10 is in fact $x^5 - 1$.

c. True. The odd derivatives of $\sqrt{1+x^2}$ vanish at $x = 0$, while the even ones do not.

9.1.67

a. This matches (C) since for $f(x) = (1+2x)^{1/2}$, $f''(x) = -(1+2x)^{-3/2}$ so $\frac{f''(0)}{2!} = \frac{-1}{2}$.

b. This matches (E) since for $f(x) = (1+2x)^{-1/2}$, $f''(x) = 3(1+2x)^{-5/2}$, so $\frac{f''(0)}{2!} = \frac{3}{2}$.

c. This matches (A) since $f^{(n)}(x) = 2^n e^{2x}$, so that $f^{(n)}(0) = 2^n$, which is (A)'s pattern.

d. This matches (D) since $f''(x) = 8(1+2x)^{-3}$ and $f''(0) = 8$, so that $f''(0)/2! = 4$

e. This matches (B) since $f'(x) = -6(1+2x)^{-4}$ so that $f'(0) = -6$.

f. This matches (F) since $f^{(n)}(x) = (-2)^n e^{-2x}$, so $f^{(n)}(0) = (-2)^n$, which is (F)'s pattern.

9.1.69

a. $p_2(0.1) = 0.1$. The maximum error in the approximation is $1 \cdot \frac{0.1^3}{3!} \approx 1.67 \times 10^{-4}$.

b. $p_2(0.2) = 0.2$. The maximum error in the approximation is $1 \cdot \frac{0.2^3}{3!} \approx 1.33 \times 10^{-3}$.

9.1.71

a. $p_3(0.1) = 1 - .01/2 = 0.995$. The maximum error is $1 \cdot \frac{0.1^4}{4!} \approx 4.17 \times 10^{-6}$.

b. $p_3(0.2) = 1 - .04/2 = 0.98$. The maximum error is $1 \cdot \frac{0.2^4}{4!} \approx 6.67 \times 10^{-5}$.

9.1.73

a. $p_1(0.1) = 1.05$. Since $|f''(x)| = \frac{1}{4}(1+x)^{-3/2}$ has a maximum value of $1/4$ at $x = 0$, the maximum error is $\frac{1}{4} \cdot \frac{0.1^2}{2} = 1.25 \times 10^{-3}$.

b. $p_1(0.2) = 1.1$. The maximum error is $\frac{1}{4} \cdot \frac{0.2^2}{2} = 5 \times 10^{-3}$.

9.1.75

a. $p_1(0.1) = 1.1$. Since $f''(x) = e^x$ is less than 2 on $[0, 0.2]$, the maximum error is less than $2 \cdot \frac{0.1^3}{3!} \approx 3.33 \times 10^{-4}$.

b. $p_1(0.2) = 1.2$. The maximum error is less than $2 \cdot \frac{0.2^3}{3!} \approx 2.67 \times 10^{-3}$.

9.1.77

a.

	$\lvert\sin x - p_3(x)\rvert$	$\lvert\sin x - p_5(x)\rvert$
-0.2	2.66×10^{-6}	2.54×10^{-9}
-0.1	8.33×10^{-8}	1.98×10^{-11}
0.0	0	0
0.1	8.33×10^{-8}	1.98×10^{-11}
0.2	2.66×10^{-6}	2.54×10^{-9}

b. The errors are equal for positive and negative x. This makes sense, since $\sin(-x) = -\sin(x)$ and $p_n(-x) = -p_n(x)$ for $n = 3, 5$. The errors appear to get larger as x gets farther from zero.

9.1.79

a.

	$\lvert e^{-x} - p_1(x)\rvert$	$\lvert e^{-x} - p_2(x)\rvert$
-0.2	2.14×10^{-2}	1.40×10^{-3}
-0.1	5.17×10^{-3}	1.71×10^{-4}
0.0	0	0
0.1	4.84×10^{-3}	1.63×10^{-4}
0.2	1.87×10^{-2}	1.27×10^{-3}

b. The errors are different for positive and negative displacements from zero, and appear to get larger as x gets farther from zero.

9.1.81

a.

	$\lvert\tan(x) - p_1(x)\rvert$	$\lvert\tan(x) - p_3(x)\rvert$
-0.2	2.71×10^{-3}	4.34×10^{-5}
-0.1	3.35×10^{-4}	1.34×10^{-6}
0.0	0	0
0.1	3.35×10^{-4}	1.34×10^{-6}
0.2	2.71×10^{-3}	4.34×10^{-5}

b. The errors are equal for positive and negative x. This makes sense, since $\tan(-x) = -\tan(x)$ and $p_n(-x) = -p_n(x)$ for $n = 1, 3$. The errors appear to get larger as x gets farther from zero.

9.1.83 The true value of $e^{0.35} \approx 1.419067549$. The 6th-order Taylor polynomial for e^x centered at $x = 0$ is

$$p_6(x) = 1 + x + \frac{x^2}{2} + \frac{x^3}{6} + \frac{x^4}{24} + \frac{x^5}{120} + \frac{x^6}{720}.$$

Evaluating the polynomials at $x = 0.35$ produces the following table:

n	$p_n(0.35)$	$\lvert p_n(0.35) - e^{0.35}\rvert$
1	1.350000000	6.91×10^{-2}
2	1.411250000	7.82×10^{-3}
3	1.418395833	6.72×10^{-4}
4	1.419021094	4.65×10^{-5}
5	1.419064862	2.69×10^{-6}
6	1.419067415	1.33×10^{-7}

The 6th-order Taylor polynomial for e^x centered at $x = \ln 2$ is

$$p_6(x) = 2 + 2(x - \ln 2) + (x - \ln 2)^2 + \frac{1}{3}(x - \ln 2)^3 + \frac{1}{12}(x - \ln 2)^4 + \frac{1}{60}(x - \ln 2)^5 + \frac{1}{360}(x - \ln 2)^6.$$

Evaluating the polynomials at $x = 0.35$ produces the following table:

n	$p_n(0.35)$	$\lvert p_n(0.35) - e^{0.35}\rvert$
1	1.313705639	1.05×10^{-1}
2	1.431455626	1.24×10^{-2}
3	1.417987101	1.08×10^{-3}
4	1.419142523	7.50×10^{-5}
5	1.419063227	4.32×10^{-6}
6	1.419067762	2.13×10^{-7}

Comparing the tables shows that using the polynomial centered at $x = 0$ is more accurate for all n. To see why, consider the remainder. Let $f(x) = e^x$. By Theorem 9.2, the magnitude of the remainder when approximating $f(0.35)$ by the polynomial p_n centered at 0 is:

$$|R_n(0.35)| = \frac{|f^{(n+1)}(c)|}{(n+1)!}(0.35)^{n+1} = \frac{e^c}{(n+1)!}(0.35)^{n+1}$$

for some c with $0 < c < 0.35$ while the magnitude of the remainder when approximating $f(0.35)$ by the polynomial p_n centered at $\ln 2$ is:

$$|R_n(0.35)| = \frac{|f^{(n+1)}(c)|}{(n+1)!}|0.35 - \ln 2|^{n+1} = \frac{e^c}{(n+1)!}(\ln 2 - 0.35)^{n+1}$$

for some c with $0.35 < c < \ln 2$. Since $\ln 2 - 0.35 \approx 0.35$, the relative size of the magnitudes of the remainders is determined by e^c in each remainder. Since e^x is an increasing function, the remainder in using the polynomial centered at 0 will be less than the remainder in using the polynomial centered at $\ln 2$, and the former polynomial will be more accurate.

9.1.85

a. The slope of the tangent line to $f(x)$ at $x = a$ is by definition $f'(a)$; by the point-slope form for the equation of a line, we have $y - f(a) = f'(a)(x - a)$, or $y = f(a) + f'(a)(x - a)$.

b. The Taylor polynomial centered at a is $p_1(x) = f(a) + f'(a)(x - a)$, which is the tangent line at a.

9.2 Properties of Power Series

9.2.1 $c_0 + c_1x + c_2x^2 + c_3x^3$.

9.2.3 Generally the Ratio Test or Root Test is used.

9.2.5 The radius of convergence does not change, but the interval may change at the endpoints.

9.2.7 $|x| < \frac{1}{4}$.

9.2.9 Using the Root Test: $\lim_{k\to\infty} \sqrt[k]{a_k} = \lim_{k\to\infty} \frac{x}{3} = \frac{x}{3}$, so the radius of convergence is 3. At -3, the series is $\sum(-1)^k$, which diverges. At 3, the series is $\sum 1$, which diverges. So the interval of convergence is $(-3, 3)$.

9.2.11 Using the Root Test: $\lim_{k\to\infty} \sqrt[k]{a_k} = \lim_{k\to\infty} \frac{x}{k} = 0$, so the radius of convergence is infinite and the interval of convergence is $(-\infty, \infty)$.

9.2.13 Using the Ratio Test: $\lim_{k\to\infty} \left|\frac{(k+1)^2x^{2k+2}}{(k+1)!} \cdot \frac{k!}{k^2x^{2k}}\right| = \lim_{k\to\infty} \frac{k+1}{k^2}x^2 = 0$, so the radius of convergence is infinite, and the interval of convergence is $(-\infty, \infty)$.

9.2.15 Using the Ratio Test: $\lim_{k\to\infty} \left|\frac{a_{k+1}}{a_k}\right| = \left|\frac{x^{2k+3}}{3^k} \cdot \frac{3^{k-1}}{x^{2k+1}}\right| = \frac{x^2}{3}$ so that the radius of convergence is $\sqrt{3}$. At $x = \sqrt{3}$, the series is $\sum 3\sqrt{3}$, which diverges. At $x = -\sqrt{3}$, the series is $\sum(-3\sqrt{3})$, which also diverges, so the interval of convergence is $(-\sqrt{3}, \sqrt{3})$.

9.2.17 Using the Root Test: $\lim_{k\to\infty} \sqrt[k]{|a_k|} = \lim_{k\to\infty} \frac{(|x-1|)k}{k+1} = |x - 1|$, so the series converges when $|x - 1| < 1$, so for $0 < x < 2$. The radius of convergence is 1. At $x = 2$, the series diverges by the Divergence Test. At $x = 0$, the series diverges as well by the Divergence Test. Thus the interval of convergence is $(0, 2)$.

9.2.19 Using the Ratio Test: $\lim_{k\to\infty} \left|\frac{a_{k+1}}{a_k}\right| = \left|\frac{(k+1)^{20}x^{k+1}}{(2k+3)!} \cdot \frac{(2k+1)!}{x^k k^{20}}\right| = \lim_{k\to\infty} \left(\frac{k+1}{k}\right)^{20} \frac{|x|}{(2k+2)(2k+3)} = 0$, so the radius of convergence is infinite, and the interval of convergence is $(-\infty, \infty)$.

9.2.21 $f(3x) = \frac{1}{1-3x} = \sum_{k=0}^{\infty} 3^k x^k$, which converges for $|x| < 1/3$, and diverges at the endpoints.

9.2.23 $h(x) = \frac{2x^3}{1-x} = \sum_{k=0}^{\infty} 2x^{k+3}$, which converges for $|x| < 1$ and is divergent at the endpoints.

9.2.25 $p(x) = \frac{4x^{12}}{1-x} = \sum_{k=0}^{\infty} 4x^{k+12} = 4\sum_{k=0}^{\infty} x^{k+12}$, which converges for $|x| < 1$. It is divergent at the endpoints.

9.2.27 $f(3x) = \ln(1-3x) = -\sum_{k=1}^{\infty} \frac{(3x)^k}{k} = -\sum_{k=1}^{\infty} \frac{3^k}{k} x^k$. Using the Ratio Test: $\lim_{k\to\infty} \left|\frac{a_{k+1}}{a_k}\right| = \lim_{k\to\infty} \frac{3k}{k+1}|x| = 3|x|$, so the radius of convergence is $1/3$. The series diverges at $1/3$ (harmonic series), and converges at $-1/3$ (alternating harmonic series).

9.2.29 $h(x) = x\ln(1-x) = -\sum_{k=1}^{\infty} \frac{x^{k+1}}{k}$. Using the Ratio Test: $\lim_{k\to\infty} \left|\frac{a_{k+1}}{a_k}\right| = \lim_{k\to\infty} \frac{k}{k+1}|x| = |x|$, so the radius of convergence is 1, and the series diverges at 1 (harmonic series) but converges at -1 (alternating harmonic series).

9.2.31 $p(x) = 2x^6\ln(1-x) = -2\sum_{k=1}^{\infty} \frac{x^{k+6}}{k}$. Using the Ratio Test: $\lim_{k\to\infty} \left|\frac{a_{k+1}}{a_k}\right| = \lim_{k\to\infty} \frac{k}{k+1}|x| = |x|$, so the radius of convergence is 1. The series diverges at 1 (harmonic series) but converges at -1 (alternating harmonic series).

9.2.33 The power series for $f(x)$ is $\sum_{k=0}^{\infty} x^k$, convergent for $-1 < x < 1$, so the power series for $g(x) = f'(x)$ is $\sum_{k=1}^{\infty} kx^{k-1} = \sum_{k=0}^{\infty}(k+1)x^k$, also convergent on $|x| < 1$.

9.2.35 The power series for $f(x)$ is $\sum_{k=0}^{\infty} x^k$, convergent for $-1 < x < 1$, so the power series for $g(x) = \frac{1}{6}f'''(x)$ is $\frac{1}{6}\sum_{k=3}^{\infty} k(k-1)(k-2)x^{k-3} = \frac{1}{6}\sum_{k=0}^{\infty}(k+1)(k+2)(k+3)x^k$, also convergent on $|x| < 1$.

9.2.37 The power series for $\frac{1}{1-3x}$ is $\sum_{k=0}^{\infty}(3x)^k$, convergent on $|x| < 1/3$. Since $g(x) = \ln(1-3x) = -3\int \frac{1}{1-3x}\,dx$, the power series for $g(x)$ is $-3\sum_{k=0}^{\infty} 3^k \frac{1}{k+1} x^{k+1} = -\sum_{k=1}^{\infty} \frac{3^k}{k} x^k$, also convergent on $[-1/3, 1/3)$.

9.2.39 Start with $g(x) = \frac{1}{1+x}$. The power series for $g(x)$ is $\sum_{k=0}^{\infty}(-1)^k x^k$. Since $f(x) = g(x^2)$, its power series is $\sum_{k=0}^{\infty}(-1)^k x^{2k}$. The radius of convergence is still 1, and the series is divergent at both endpoints. The interval of convergence is $(-1, 1)$.

9.2.41 Note that $f(x) = \frac{3}{3+x} = \frac{1}{1+(1/3)x}$. Let $g(x) = \frac{1}{1+x}$. The power series for $g(x)$ is $\sum_{k=0}^{\infty}(-1)^k x^k$, so the power series for $f(x) = g((1/3)x)$ is $\sum_{k=0}^{\infty}(-1)^k 3^{-k} x^k = \sum_{k=0}^{\infty}\left(\frac{-x}{3}\right)^k$. Using the Ratio Test: $\lim_{k\to\infty} \left|\frac{a_{k+1}}{a_k}\right| = \lim_{k\to\infty} \left|\frac{3^{-(k+1)}x^{k+1}}{3^{-k}x^k}\right| = \frac{|x|}{3}$, so the radius of convergence is 3. The series diverges at both endpoints. The interval of convergence is $(-3, 3)$.

9.2.43 Note that $f(x) = \ln\sqrt{4-x^2} = \frac{1}{2}\ln(4-x^2) = \frac{1}{2}\left(\ln 4 + \ln\left(1 - \frac{x^2}{4}\right)\right) = \ln 2 + \frac{1}{2}\ln\left(1 - \frac{x^2}{4}\right)$. Now, the power series for $g(x) = \ln(1-x)$ is $-\sum_{k=1}^{\infty} \frac{1}{k}x^k$, so the power series for $f(x)$ is $\ln 2 - \frac{1}{2}\sum_{k=1}^{\infty} \frac{1}{k}\frac{x^{2k}}{4^k} = \ln 2 - \sum_{k=1}^{\infty} \frac{x^{2k}}{k2^{2k+1}}$. Now, $\lim_{k\to\infty} \left|\frac{a_{k+1}}{a_k}\right| = \lim_{k\to\infty} \left|\frac{x^{2k+2}}{(k+1)2^{2k+3}} \cdot \frac{k2^{2k+1}}{x^{2k}}\right| = \lim_{k\to\infty} \frac{k}{4(k+1)}x^2 = \frac{x^2}{4}$, so that the radius of convergence is 2. The series diverges at both endpoints, so its interval of convergence is $(-2, 2)$.

9.2.45

a. True. This power series is centered at $x = 3$, so its interval of convergence will be symmetric about 3.

b. True. Use the Root Test.

c. True. Use the Root Test.

d. True. Since the power series is zero on the interval, all its derivatives are as well, which implies (differentiating the power series) that all the c_k are zero.

9.2.47 $\sum_{k=0}^{\infty}(-1)^k \frac{1}{k+1} x^k$

9.2.49 $\sum_{k=1}^{\infty}(-1)^k \frac{x^{2k}}{k!}$

9.2.51 The power series for $f(x-a)$ is $\sum c_k(x-a)^k$. Then $\sum c_k(x-a)^k$ converges if and only if $|x-a| < R$, which happens if and only if $a - R < x < a + R$, so the radius of convergence is the same.

9.2.53 This is a geometric series with ratio $\sqrt{x} - 2$, so its sum is $\frac{1}{1-(\sqrt{x}-2)} = \frac{1}{3-\sqrt{x}}$. Again using the Root Test, $\lim_{k\to\infty} \sqrt[k]{|a_k|} = \sqrt{x} - 2$, so the interval of convergence is given by $|\sqrt{x} - 2| < 1$, so $1 < \sqrt{x} < 3$ and $1 < x < 9$. The series diverges at both endpoints.

9.2.55 This is a geometric series with ratio e^{-x}, so its sum is $\frac{1}{1-e^{-x}}$. By the Root Test, $\lim_{k\to\infty} \sqrt[k]{|a_k|} = e^{-x}$, so the power series converges for $x > 0$.

9.2.57 This is a geometric series with ratio $(x^2 - 1)/3$, so its sum is $\frac{1}{1-\frac{x^2-1}{3}} = \frac{3}{3-(x^2-1)} = \frac{3}{4-x^2}$. Using the Root Test, the series converges for $\left|x^2 - 1\right| < 3$, so that $-2 < x^2 < 4$ or $-2 < x < 2$. It diverges at both endpoints.

9.2.59 The power series for e^x is $\sum_{k=0}^{\infty} \frac{x^k}{k!}$. Substitute $-x$ for x to get $e^{-x} = \sum_{k=0}^{\infty}(-1)^k \frac{x^k}{k!}$. The series converges for all x.

9.2.61 Substitute $-3x$ for x in the power series for e^x to get $e^{-3x} = \sum_{k=0}^{\infty} \frac{(-3x)^k}{k!} = \sum_{k=0}^{\infty}(-1)^k \frac{3^k}{k!} x^k$. The series converges for all x.

9.2.63 The power series for $x^m f(x)$ is $\sum c_k x^{k+m}$.The ratio used in the Ratio Test is : $\frac{a_{k+1}}{a_k} = \frac{c_{k+1}x^{k+m+1}}{c_k x^{k+m}} = \frac{c_{k+1}}{c_k} x$, which is the same ratio that results when applying the Ratio Test to the original series, so the two power series have the same radius of convergence. Now suppose that the power series for $f(x)$ converges at $x = b$, an endpoint of the interval of convergence. Then $\sum c_k b^k$ converges, and since $\lim_{k\to\infty} \frac{c_k b^{k+m}}{c_k b^k} = b^m < \infty$, the power series for $x^m f(x)$ converges at $x = b$ as well by the Limit Comparison Test.

9.2.65

a. $f(x)g(x) = c_0 d_0 + (c_0 d_1 + c_1 d_0)x + (c_0 d_2 + c_1 d_1 + c_2 d_0)x^2 + \ldots$

b. The coefficient of x^n in $f(x)g(x)$ is $\sum_{i=0}^{n} c_i d_{n-i}$.

9.2.67

a. For both graphs, the difference between the true value and the estimate is greatest at the two ends of the range; the difference at 0.9 is greater than that at -0.9.

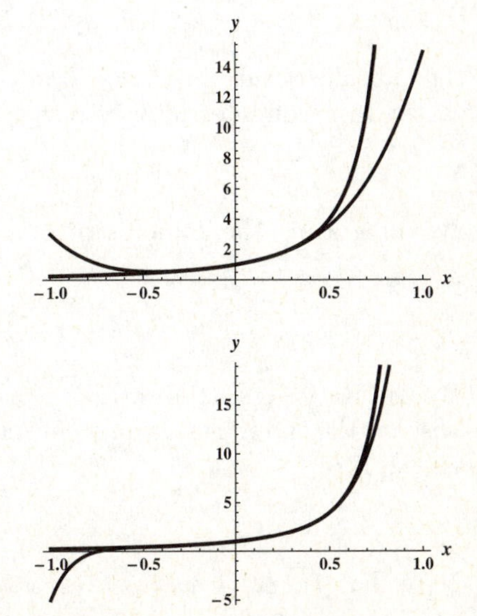

b. The difference between $f(x)$ and $S_n(x)$ is greatest for $x = 0.9$; at that point, $f(x) = \frac{1}{(1-0.9)^2} = 100$, so we want to find n such that $S_n(x)$ is within 0.01 of 100. But $S_{111} \approx 99.98991435$ and $S_{112} \approx 99.99084790$, so $n = 112$.

9.3 Taylor Series

9.3.1 The Taylor series is in a sense the "limit" of the Taylor polynomials. It is the infinite sum of terms of the form appearing in the Taylor polynomials.

9.3.3 The nth coefficient is $\frac{f^{(n)}(a)}{n!}(x-a)^n$.

9.3.5 Substitute x^2 for x in the Taylor series. By theorems proved in the previous section about power series, the interval of convergence does not change except perhaps at the endpoints of the interval.

9.3.7 It means that the limit of the remainder term is zero.

9.3.9

a. Note that $f(0) = 1$, $f'(0) = -1$, $f''(0) = 1$, and $f'''(0) = -1$. So the Maclaurin series is $1 - x + x^2/2 - x^3/6 + \dots$.

b. $\sum_{k=0}^{\infty}(-1)^k \frac{x^k}{k!}$.

c. The series converges on $(-\infty, \infty)$, as can be seen from the Ratio Test.

9.3.11

a. Since the series for $\frac{1}{1+x}$ is $1 - x + x^2 - x^3 + \dots$, the series for $\frac{1}{1+x^2}$ is $1 - x^2 + x^4 - x^6 + \dots$.

b. $\sum_{k=0}^{\infty}(-1)^k x^{2k}$.

c. The absolute value of the ratio of consecutive terms is x^2, so by the Ratio Test, the radius of convergence is 1. The series diverges at the endpoints by the Divergence Test, so the interval of convergence is $(-1, 1)$.

9.3.13

a. Note that $f(0) = 1$, and that $f^{(n)}(0) = 2^n$. Thus, the series is given by $1 + 2x + \frac{4x^2}{2} + \frac{8x^3}{6} + \dots$.

b. $\sum_{k=0}^{\infty} \frac{(2x)^k}{k!}$.

c. The absolute value of the ratio of consecutive terms is $\frac{2|x|}{n}$, which has limit 0 as $n \to \infty$. So by the Ratio Test, the interval of convergence is $(-\infty, \infty)$.

9.3.15

a. By integrating the Taylor series for $\frac{1}{1+x^2}$ (which is the derivative of $\tan^{-1}(x)$), we obtain the series $x - \frac{x^3}{3} + \frac{x^5}{5} - \frac{x^7}{7} + \dots$.

b. $\sum_{k=0}^{\infty}(-1)^k \frac{1}{2k+1} x^{2k+1}$.

c. By the Ratio Test (the ratio of consecutive terms has limit x^2), the radius of convergence is $|x| < 1$. Also, at the endpoints we have convergence by the Alternating Series Test, so the interval of convergence is $[-1, 1]$.

9.3.17

a. Note that $f(\pi/2) = 1$, $f'(\pi/2) = \cos(\pi/2) = 0$, $f''(\pi/2) = -\sin(\pi/2) = -1$, $f'''(\pi/2) = -\cos(\pi/2) = 0$, and so on. Thus the series is given by $1 - \frac{1}{2}\left(x - \frac{\pi}{2}\right)^2 + \frac{1}{24}\left(x - \frac{\pi}{2}\right)^4 - \frac{1}{720}\left(x - \frac{\pi}{2}\right)^6 + \dots$.

b. $\sum_{k=0}^{\infty}(-1)^k \frac{1}{(2k)!}\left(x-\frac{\pi}{2}\right)^{2k}$.

9.3.19

a. Note that $f^{(k)}(1) = (-1)^k \frac{k!}{1^{k+1}} = (-1)^k \cdot k!$. Thus the series is given by $1-(x-1)+(x-1)^2-(x-1)^3+\ldots$.

b. $\sum_{k=0}^{\infty}(-1)^k(x-1)^k$.

9.3.21

a. Note that $f^{(k)}(3) = (-1)^{k-1}\frac{(k-1)!}{3^k}$. Thus the series is given by $\ln(3)+\frac{x-3}{3}-\frac{1}{18}(x-3)^2+\frac{1}{81}(x-3)^3+\ldots$.

b. $\ln(3)+\sum_{k=1}^{\infty}(-1)^{k+1}\frac{1}{k\cdot 3^k}(x-3)^k$.

9.3.23 Since the Taylor series for $\ln(1+x)$ is $x-\frac{x^2}{2}+\frac{x^3}{3}-\frac{x^4}{4}+\cdots$, the first four terms of the Taylor series for $\ln(1+x^2)$ are $x^2-\frac{x^4}{2}+\frac{x^6}{3}-\frac{x^8}{4}$, obtained by substituting x^2 for x.

9.3.25 The Taylor series for e^x-1 is the Taylor series for e^x, less the constant term of 1, so it is $x+\frac{x^2}{2}+\frac{x^3}{3!}+\frac{x^4}{4!}+\cdots$. Thus, the first four terms of the Taylor series for $\frac{e^x-1}{x}$ are $1+\frac{x}{2!}+\frac{x^2}{3!}+\frac{x^3}{4!}$, obtained by dividing the terms of the first series by x.

9.3.27 Since the Taylor series for $(1+x)^{-1}$ is $1-x+x^2-x^3+\cdots$, if we substitute x^4 for x, we obtain $1-x^4+x^8-x^{12}+\cdots$.

9.3.29

a. The binomial coefficients are $\binom{-2}{0}=1$, $\binom{-2}{1}=\frac{-2}{1!}=-2$, $\binom{-2}{2}=\frac{(-2)(-3)}{2!}=3$, $\binom{-2}{3}=\frac{(-2)(-3)(-4)}{3!}=-4$.

Thus the first four terms of the series are $1-2x+3x^2-4x^3$.

b. $1-2\cdot 0.1+3\cdot 0.01-4\cdot 0.001=0.826$

9.3.31

a. The binomial coefficients are $\binom{1/4}{0}=1$, $\binom{1/4}{1}=\frac{1/4}{1}=\frac{1}{4}$, $\binom{1/4}{2}=\frac{(1/4)(-3/4)}{2!}=-\frac{3}{32}$, $\binom{1/4}{3}=\frac{(1/4)(-3/4)(-7/4)}{3!}=\frac{7}{128}$, so the first four terms of the series are $1+\frac{1}{4}x-\frac{3}{32}x^2+\frac{7}{128}x^3$.

b. Substitute $x=0.12$ to get 1.0287445.

9.3.33

a. The binomial coefficients are $\binom{-2/3}{0}=1$, $\binom{-2/3}{1}=-\frac{2}{3}$, $\binom{-2/3}{2}=\frac{(-2/3)(-5/3)}{2!}=\frac{5}{9}$, $\binom{-2/3}{3}=\frac{(-2/3)(-5/3)(-8/3)}{3!}=-\frac{40}{81}$, so the first four terms of the series are $1-\frac{2}{3}x+\frac{5}{9}x^2-\frac{40}{81}x^3$.

b. Substitute $x=0.18$ to get 0.89512.

9.3.35 $\sqrt{1+x^2}=1+\frac{x^2}{2}-\frac{x^4}{8}+\frac{x^6}{16}-\ldots$. By the Ratio Test, the radius of convergence is 1. At the endpoints, the series obtained are convergent by the Alternating Series Test. Thus, the interval of convergence is $[-1,1]$.

9.3.37 $\sqrt{9-9x}=3\sqrt{1-x}=3-\frac{3}{2}x-\frac{3}{8}x^2-\frac{3}{16}x^3-\ldots$. The interval of convergence is $[-1,1)$.

9.3.39 $\sqrt{a^2+x^2}=a\sqrt{1+\frac{x^2}{a^2}}=a+\frac{x^2}{2a}-\frac{x^4}{8a^3}+\frac{x^6}{16a^5}-\ldots$. The series converges when $\frac{x^2}{a^2}$ is less than 1 in magnitude, so the radius of convergence is a. The series given by the endpoints is convergent by the Alternating Series Test, so the interval of convergence is $[-a,a]$.

9.3.41 $(1+4x)^{-2}=1-2(4x)+3(4x)^2-4(4x)^3+\cdots=1-8x+48x^2-256x^3+\ldots$.

9.3.43 $\frac{1}{(4+x^2)^2} = (4+x^2)^{-2} = \frac{1}{16}(1+(x^2/4))^{-2} = \frac{1}{16}\left(1 - 2\cdot\frac{x^2}{4} + 3\cdot\frac{x^4}{16} - 4\cdot\frac{x^6}{64} + \dots\right) = \frac{1}{16} - \frac{1}{32}x^2 + \frac{3}{256}x^4 - \frac{1}{256}x^6 + \dots$

9.3.45 $(3+4x)^{-2} = \frac{1}{9}\left(1+\frac{4x}{3}\right)^{-2} = \frac{1}{9}\left(1 - 2\frac{4}{3}x + 3\frac{16}{9}x^2 - 4\frac{64}{27}x^3 + \dots\right) = \frac{1}{9} - \frac{8}{27}x + \frac{16}{27}x^2 - \frac{256}{243}x^3 + \dots$.

9.3.47 The interval of convergence for the Taylor series for $f(x) = \sin x$ is $(-\infty,\infty)$. The remainder is $R_n(x) = \frac{f^{(n+1)}(c)}{(n+1)!}x^{n+1}$ for some c. Since $f^{(n+1)}(x)$ is $\pm\sin x$ or $\pm\cos x$, we have

$$\lim_{n\to\infty} |R_n(x)| \le \lim_{n\to\infty} \frac{1}{(n+1)!}\left|x^{n+1}\right| = 0$$

for any x.

9.3.49 The interval of convergence for the Taylor series for e^{-x} is $(-\infty,\infty)$. The remainder is $R_n(x) = \frac{(-1)^{n+1}e^{-c}}{(n+1)!}x^{n+1}$ for some c. Thus $\lim_{n\to\infty} |R_n(x)| = 0$ for any x.

9.3.51

a. False. Not all of its derivatives are defined at zero - in fact, none of them are.

b. True. The derivatives of $\csc x$ involve positive powers of $\csc x$ and $\cot x$, both of which are defined at $\pi/2$, so that $\csc x$ has continuous derivatives at $\pi/2$.

c. False. For example, the Taylor series for $f(x^2)$ doesn't converge at $x = 1.9$, since the Taylor series for $f(x)$ doesn't converge at $1.9^2 = 3.61$.

d. False. The Taylor series centered at 1 involves derivatives of f evaluated at 1, not at 0.

e. True. The follows because the Taylor series must itself be an even function.

9.3.53

a. The relevant Taylor series are: $e^x = 1 + x + \frac{x^2}{2!} + \frac{x^3}{3!} + \frac{x^4}{4!} + \frac{x^5}{5!} + \frac{x^6}{6!} + \cdots$ and $e^{-x} = 1 - x + \frac{x^2}{2!} - \frac{x^3}{3!} + \frac{x^4}{4!} - \frac{x^5}{5!} + \frac{x^6}{6!} + \cdots$. Thus the first four terms of the resulting series are $\frac{1}{2}\left(e^x + e^{-x}\right) = 1 + \frac{x^2}{2!} + \frac{x^4}{4!} + \frac{x^6}{6!} + \cdots$.

b. Since each series converges (absolutely) on $(-\infty,\infty)$, so does their sum. The radius of convergence is ∞.

9.3.55

a. Use the binomial theorem. The binomial coefficients are $\binom{-2/3}{0} = 1$, $\binom{-2/3}{1} = -\frac{2}{3}$, $\binom{-2/3}{2} = \frac{(-2/3)(-5/3)}{2!} = \frac{5}{9}$, $\binom{-2/3}{3} = \frac{(-2/3)(-5/3)(-8/3)}{3!} = -\frac{40}{81}$ and then, substituting x^2 for x, we obtain $1 - \frac{2}{3}x^2 + \frac{5}{9}x^4 - \frac{40}{81}x^6 + \dots$.

b. From Theorem 9.6 the radius of convergence is determined from $\left|x^2\right| < 1$, so it is 1.

9.3.57

a. From the binomial formula, the Taylor series for $(1-x)^p$ is $\sum\binom{p}{k}(-1)^k x^k$, so the Taylor series for $(1-x^2)^p$ is $\sum\binom{p}{k}(-1)^k x^{2k}$. Here $p = 1/2$, and the binomial coefficients are $\binom{1/2}{0} = 1$, $\binom{1/2}{1} = \frac{1/2}{1!} = \frac{1}{2}$, $\binom{1/2}{2} = \frac{(1/2)(-1/2)}{2!} = -\frac{1}{8}$, $\binom{1/2}{3} = \frac{(1/2)(-1/2)(-3/2)}{3!} = \frac{1}{16}$ so that $(1-x^2)^{1/2} = 1 - \frac{1}{2}x^2 - \frac{1}{8}x^4 - \frac{1}{16}x^6 + \dots$.

b. From Theorem 9.6 the radius of convergence is determined from $\left|x^2\right| < 1$, so it is 1.

9.3.59

a. $f(x) = (1+x^2)^{-2}$; using the binomial series and substituting x^2 for x we obtain $1 - 2x^2 + 3x^4 - 4x^6 + \dots$.

b. From Theorem 9.6 the radius of convergence is determined from $\left|x^2\right| < 1$, so it is 1.

9.3.61 Since $f(64) = 4$, and $f'(x) = \frac{1}{3}x^{-2/3}$, $f'(64) = \frac{1}{48}$, $f''(x) = -\frac{2}{9}x^{-5/3}$, $f''(64) = -\frac{1}{4608}$, $f'''(x) = \frac{10}{27}x^{8/3}$, and $f'''(64) = \frac{10}{1769472} = \frac{5}{884736}$, the first four terms of the Taylor series are $4+\frac{1}{48}(x-64) - \frac{1}{4608\cdot 2!}(x-64)^2 + \frac{5}{884736\cdot 3!}(x-64)^3$. Evaluating at $x = 60$, we get 3.914870274.

9.3.63 Since $f(16) = 2$, and $f'(x) = \frac{1}{4}x^{-3/4}$, $f'(16) = \frac{1}{32}$, $f''(x) = -\frac{3}{16}x^{-7/4}$, $f''(16) = -\frac{3}{2048}$, $f'''(x) = \frac{21}{64}x^{-11/4}$, and $f'''(16) = \frac{21}{131072}$, the first four terms of the Taylor series are $2+\frac{1}{32}(x-16) - \frac{3}{2048\cdot 2!}(x-16)^2 + \frac{21}{131072\cdot 3!}(x-16)^3$. Evaluating at $x = 13$, we get 1.898937225.

9.3.65 Evaluate the binomial coefficient $\binom{1/2}{k} = \frac{(1/2)(-1/2)(-3/2)\ldots(1/2-k+1)}{k!} = \frac{(1/2)(-1/2)\ldots((3-2k)/2)}{k!} = (-1)^{k-1}2^{-k}\frac{1\cdot 3\cdots(2k-3)}{k!} = (-1)^{k-1}2^{-k}\frac{(2k-2)!}{2^{k-1}\cdot(k-1)!\cdot k!} = (-1)^{k-1}2^{1-2k}\cdot\frac{1}{k}\binom{2k-2}{k-1}$. This is the coefficient of x^k in the Taylor series for $\sqrt{1+x}$. Substituting $4x$ for x, the Taylor series becomes $\sum_{k=0}^{\infty}(-1)^{k-1}2^{1-2k}\cdot\frac{1}{k}\binom{2k-2}{k-1}(4x)^k = \sum_{k=0}^{\infty}(-1)^{k-1}\frac{2}{k}\binom{2k-2}{k-1}x^k$. If we can show that k divides $\binom{2k-2}{k-1}$, we will be done, for then the coefficient of x^k will be an integer. But $\binom{2k-2}{k-1} - \binom{2k-2}{k-2} = \frac{(2k-2)!}{(k-1)!(k-1)!} - \frac{(2k-2)!}{(k-2)!k!} = \frac{(2k-2)!}{(k-1)!(k-1)!} - \frac{(2k-2)!(k-1)}{(k-1)!(k-1)!k} = \frac{k(2k-2)!-(k-1)(2k-2)!}{k(k-1)!(k-1)!} = \frac{1}{k}\frac{(2k-2)!}{(k-1)!(k-1)!} = \frac{1}{k}\binom{2k-2}{k-1}$ and thus we have shown that k divides $\binom{2k-2}{k-1}$.

9.3.67 The Maclaurin series for $\sin x$ is $x - \frac{1}{3!}x^3 + \frac{1}{5!}x^5 - \frac{1}{7!}x^7 + \ldots$. Squaring the first four terms yields

$$\begin{aligned}&(x-\frac{1}{3!}x^3 + \frac{1}{5!}x^5 - \frac{1}{7!}x^7)^2\\ &= x^2 - \frac{2}{3!}x^4 + (\frac{2}{5!} + \frac{1}{3!3!})x^6 + (-2\frac{1}{7!} - 2\frac{1}{3!5!})x^8\\ &= x^2 - \frac{1}{3}x^4 + \frac{2}{45}x^6 - \frac{1}{315}x^8.\end{aligned}$$

The Maclaurin series for $\cos x$ is $1 - \frac{1}{2}x^2 + \frac{1}{4!}x^4 - \frac{1}{6!}x^6 + \frac{1}{8!}x^8 - \ldots$. Substituting $2x$ for x in the Maclaurin series for $\cos x$ and then computing $(1 - \cos 2x)/2$, we obtain

$$\begin{aligned}&(1 - (1-\frac{1}{2}(2x)^2 + \frac{1}{4!}(2x)^4 - \frac{1}{6!}(2x)^6) + \frac{1}{8!}(2x)^8)/2\\ &= (2x^2 - \frac{2}{3}x^4 + \frac{4}{45}x^6 - \frac{2}{315}x^8)/2\\ &= x^2 - \frac{1}{3}x^4 + \frac{2}{45}x^6 - \frac{1}{315}x^8,\end{aligned}$$

and the two are the same.

9.3.69 There are many solutions. For example, first find a series that has $(-1, 1)$ as an interval of convergence, say $\frac{1}{1-x} = \sum_{k=0}^{\infty} x^k$. Then the series $\frac{1}{1-x/2} = \sum_{k=0}^{\infty}\left(\frac{x}{2}\right)^k$ has $(-2, 2)$ as its interval of convergence. Now shift the series up so that it is centered at 4: $\sum_{k=0}^{\infty}\left(\frac{x-4}{2}\right)^k$, and the interval of convergence is $(2, 6)$.

9.3.71 $\frac{1\cdot 3\cdot 5\cdot 7}{2\cdot 4\cdot 6\cdot 8}x^4 - \frac{1\cdot 3\cdot 5\cdot 7\cdot 9}{2\cdot 4\cdot 6\cdot 8\cdot 10}x^5$.

9.3.73 Use the Taylor series for $\cos x$ centered at $\pi/4$: $\frac{\sqrt{2}}{2}(1 - (x-\pi/4) - \frac{1}{2}(x-\pi/4)^2 + \frac{1}{6}(x-\pi/4)^3 + \ldots)$. The remainder after n terms (since the derivatives of $\cos x$ are bounded by 1 in magnitude) is $|R_n(x)| \le \frac{1}{(n+1)!}\cdot\left(\frac{\pi}{4} - \frac{2\pi}{9}\right)^{n+1}$.

Solving for $|R_n(x)| < 10^{-4}$, we obtain $n = 3$. Evaluating the first four terms (through $n = 3$) of the series we get 0.7660427050. The true value is ≈ 0.7660444431.

9.3.75 Use the Taylor series for $f(x) = x^{1/3}$ centered at 64: $4 + \frac{1}{48}(x-64) - \frac{1}{9216}(x-64)^2 + \cdots$. Since we wish to evaluate this series at $x = 83$, $|R_n(x)| = \frac{|f^{(n+1)}(c)|}{(n+1)!}(83-64)^{n+1}$. We compute that $|f^{(n+1)}(c)| = \frac{2\cdot 5\cdot\cdots\cdot(3n-1)}{3^{n+1}c^{(3n+2)/3}}$, which is maximized at $c = 64$. Thus

$$|R_n(x)| \le \frac{2\cdot 5\cdot\cdots\cdot(3n-1)}{3^{n+1}64^{(3n+2)/3}(n+1)!}19^{n+1}$$

Solving for $|R_n(x)| < 10^{-4}$, we obtain $n = 5$. Evaluating the terms of the series through $n = 5$ gives 4.362122553. The true value is ≈ 4.362070671.

9.3.77

a. Use the Taylor series for $(125+x)^{1/3}$ centered at $x=0$. Using the first four terms and evaluating at $x=3$ gives a result (5.03968) accurate to within 10^{-4}.

b. Use the Taylor series for $x^{1/3}$ centered at $x=125$. Note that this gives the identical Taylor series except that the exponential terms are $(x-125)^n$ rather than x^n. Thus we need terms up through $(x-125)^3$, just as before, evaluated at $x=128$, and we obtain the identical result.

c. Since the two Taylor series are the same except for the shifting, the results are equivalent.

9.3.79 Consider the remainder after the first term of the Taylor series. Taylor's Theorem indicates that $R_1(x) = \frac{f''(c)}{2}(x-a)^2$ for some c between x and a, so that $f(x) = f(a) + f'(a)(x-a) + \frac{f''(c)}{2}(x-a)^2$. But $f'(a) = 0$, so that for every x in an interval containing a, there is a c between x and a such that $f(x) = f(a) + \frac{f''(c)}{2}(x-a)^2$.

a. If $f''(x) > 0$ on the interval containing a, then for every x in that interval, we have $f(x) = f(a) + \frac{f''(c)}{2}(x-a)^2$ for some c between x and a. But $f''(c) > 0$ and $(x-a)^2 > 0$, so that $f(x) > f(a)$ and a is a local minimum.

b. If $f''(x) < 0$ on the interval containing a, then for every x in that interval, we have $f(x) = f(a) + \frac{f''(c)}{2}(x-a)^2$ for some c between x and a. But $f''(c) < 0$ and $(x-a)^2 > 0$, so that $f(x) < f(a)$ and a is a local maximum.

9.4 Working with Taylor Series

9.4.1 Replace f and g by their Taylor series centered at a, and evaluate the limit.

9.4.3 Substitute -0.6 for x in the Taylor series for e^x centered at 0. Note that this series is an alternating series, so the error can easily be estimated by looking at the magnitude of the first neglected term.

9.4.5 The series is $f'(x) = \sum_{k=1}^{\infty} kc_k x^{k-1}$, which converges for $|x| < b$.

9.4.7 We compute that

$$\begin{aligned}\frac{e^x - e^{-x}}{x} &= \frac{1}{x}\left(\left(1 + x + \frac{x^2}{2} + \frac{x^3}{6} + \cdots\right) - \left(1 - x + \frac{x^2}{2} - \frac{x^3}{6} + \cdots\right)\right)\\ &= \frac{1}{x}\left(2x + \frac{x^3}{3} + \cdots\right) = 2 + \frac{x^2}{3} + \cdots\end{aligned}$$

so the limit of $\dfrac{e^x - e^{-x}}{x}$ as $x \to 0$ is 2.

9.4.9 We compute that

$$\begin{aligned}\frac{2\cos 2x - 2 + 4x^2}{2x^4} &= \frac{1}{2x^4}\left(2(1 - \frac{(2x)^2}{2} + \frac{(2x)^4}{24} - \frac{(2x)^6}{720} + \cdots) - 2 + 4x^2\right)\\ &= \frac{1}{2x^4}\left(\frac{(2x)^4}{12} - \frac{(2x)^6}{360} + \cdots\right) = \frac{2}{3} - \frac{4x^2}{45} + \cdots\end{aligned}$$

so the limit of $\dfrac{2\cos 2x - 2 + 4x^2}{2x^4}$ as $x \to 0$ is $\dfrac{2}{3}$.

9.4.11 We compute that

$$\begin{aligned}\frac{3\tan x - 3x - x^3}{x^5} &= \frac{1}{x^5}\left(3\left(x+\frac{x^3}{3}+\frac{2x^5}{15}+\frac{17x^7}{315}+\cdots\right)-3x-x^3\right)\\ &= \frac{1}{x^5}\left(\frac{2x^5}{5}+\frac{17x^7}{105}+\cdots\right)=\frac{2}{5}+\frac{17x^2}{105}+\cdots\end{aligned}$$

so the limit of $\dfrac{3\tan x - 3x - x^3}{x^5}$ as $x \to 0$ is $\dfrac{2}{5}$.

9.4.13 We compute that

$$\begin{aligned}\frac{3\tan^{-1} x - 3x + x^3}{x^5} &= \frac{1}{x^5}\left(3\left(x-\frac{x^3}{3}+\frac{x^5}{5}-\frac{x^7}{7}+\cdots\right)-3x+x^3\right)\\ &= \frac{1}{x^5}\left(\frac{3x^5}{5}+\frac{3x^7}{7}+\cdots\right)==\frac{3}{5}+\frac{3x^2}{7}+\cdots\end{aligned}$$

so the limit of $\dfrac{3\tan^{-1} x - 3x + x^3}{x^5}$ as $x \to 0$ is $\dfrac{3}{5}$.

9.4.15 We compute that

$$\begin{aligned}\frac{\sin x - \tan x}{3x^3\cos x} &= \frac{\left(x-\frac{x^3}{6}+\frac{x^5}{120}-\cdots\right)-\left(x+\frac{x^3}{3}+\frac{2x^5}{15}+\cdots\right)}{3x^3\left(1-\frac{x^2}{2}+\cdots\right)}\\ &= \frac{-\frac{x^3}{2}-\frac{x^5}{8}+\cdots}{3x^3-\frac{3x^5}{2}+\cdots}=\frac{-\frac{1}{2}-\frac{x^2}{8}+\cdots}{3-\frac{3x^2}{2}+\cdots}\end{aligned}$$

so the limit of $\dfrac{\sin x - \tan x}{3x^3\cos x}$ as $x \to 0$ is $\dfrac{-1/2}{3} = -\dfrac{1}{6}$.

9.4.17 The Taylor series for $\ln(x-1)$ centered at 2 is

$$\ln(x-1) = (x-2) + \frac{1}{2}(x-2)^2 + \cdots .$$

We compute that

$$\frac{x-2}{\ln(x-1)} = \frac{x-2}{(x-2)+\frac{1}{2}(x-2)^2+\cdots} = \frac{1}{1+\frac{1}{2}(x-2)+\cdots}$$

so the limit of $\dfrac{x-2}{\ln(x-1)}$ as $x \to 2$ is 1.

9.4.19 The Taylor series for $(1+x)^{-2}$ centered at 0 is

$$(1+x)^{-2} = 1 - 2x + 3x^2 - 4x^3 + \cdots .$$

and the Taylor series for $\cos\sqrt{x}$ centered at 0 is

$$\cos\sqrt{x} = 1 - \frac{x}{2} + \frac{x^2}{24} - \frac{x^3}{720} + \cdots$$

We compute that

$$\begin{aligned}&\frac{(1+x)^{-2} - 4\cos\sqrt{x} + 3}{2x^2}\\ &= \frac{1}{2x^2}\left(\left(1-2x+3x^2-4x^3+\cdots\right)-4\left(1-\frac{x}{2}+\frac{x^2}{24}-\frac{x^3}{720}+\cdots\right)+3\right)\\ &= \frac{1}{2x^2}\left(\frac{17x^2}{6}-\frac{719x^3}{180}+\cdots\right)=\frac{17}{12}-\frac{719x}{360}+\cdots\end{aligned}$$

so the limit of $\dfrac{(1+x)^{-2} - 4\cos\sqrt{x} + 3}{2x^2}$ as $x \to 0$ is $\dfrac{17}{12}$.

9.4.21

a. $f'(x) = \frac{d}{dx}(\sum_{k=0}^{\infty} \frac{x^k}{k!}) = \sum_{k=1}^{\infty} k\frac{x^{k-1}}{k!} = \sum_{k=0}^{\infty} \frac{x^k}{k!} = f(x)$.

b. $f'(x) = e^x$ as well.

c. The series converges on $(-\infty, \infty)$.

9.4.23

a. $f'(x) = \frac{d}{dx}(\ln(1+x)) = \frac{d}{dx}(\sum_{k=1}^{\infty}(-1)^{k+1}\frac{1}{k}x^k) = \sum_{k=1}^{\infty}(-1)^{k+1}x^{k-1}$.

b. This is the power series for $\frac{1}{1+x}$.

c. The Taylor series for $\ln(1+x)$ converges on $(-1, 1)$, as does the Taylor series for $\frac{1}{1+x}$.

9.4.25

a. $f'(x) = \frac{d}{dx}(e^{-2x}) = \frac{d}{dx}(\sum_{k=0}^{\infty} \frac{(-2x)^k}{k!}) = (\sum_{k=0}^{\infty}(-2)^k \frac{x^k}{k!})' = -2\sum_{k=1}^{\infty}(-2)^{k-1}\frac{x^{k-1}}{k-1} = -2\sum_{k=0}^{\infty} \frac{(-2x)^k}{k!}$.

b. This is the Taylor series for $-2e^{-2x}$.

c. Since the Taylor series for e^{-2x} converges on $(-\infty, \infty)$, so does this one.

9.4.27

a. Since $y(0) = 2$, we have $0 = y'(0) - y(0) = y'(0) - 2$ so that $y'(0) = 2$. Differentiating the equation gives $y''(0) = y'(0)$, so that $y''(0) = 2$. Successive derivatives also have the value 2 at 0, so the Taylor series is $2\sum_{k=0}^{\infty} \frac{t^k}{k!}$.

b. $2\sum_{k=0}^{\infty} \frac{t^k}{k!} = 2e^t$.

9.4.29

a. $y(0) = 2$, so that $y'(0) = 16$. Differentiating, $y''(t) - 3y'(t) = 0$, so that $y''(0) = 48$, and in general $y^{(k)}(0) = 3y^{(k-1)}(0) = 3^{k-1} \cdot 16$. Thus the power series is $2 + \frac{16}{3}\sum_{k=1}^{\infty} \frac{(3t)^k}{k!}$.

b. $2 + \frac{16}{3}\sum_{k=1}^{\infty} \frac{(3t)^k}{k!} = 2 + \frac{16}{3}(e^{3t} - 1)$.

9.4.31 The Taylor series for e^{-x^2} is $\sum_{k=0}^{\infty}(-1)^k \frac{x^{2k}}{k!}$. Thus, the desired integral is $\int_0^{0.25} \sum_{k=0}^{\infty}(-1)^k \frac{x^{2k}}{k!}\, dx = \sum_{k=0}^{\infty}(-1)^k \frac{x^{2k+1}}{(2k+1)k!}\Big|_0^{0.25} = \sum_{k=0}^{\infty}(-1)^k \frac{1}{(2k+1)k!4^{2k+1}}$. Since this is an alternating series, to approximate it to within 10^{-4}, we must find n such that $a_{n+1} < 10^{-4}$, or $\frac{1}{(2n+3)(n+1)!\cdot 4^{2n+3}} < 10^{-4}$. This occurs for $n = 1$, so $\sum_{k=0}^{1}(-1)^k \frac{1}{(2k+1)\cdot k!\cdot 4^{2k+1}} = \frac{1}{4} - \frac{1}{192} \approx 0.2447916667$.

9.4.33 The Taylor series for $\cos 2x^2$ is $\sum_{k=0}^{\infty}(-1)^k \frac{(2x^2)^{2k}}{(2k)!} = \sum_{k=0}^{\infty}(-1)^k \frac{4^k x^{4k}}{(2k)!}$. Note that $\cos x$ is an even function, so we compute the integral from 0 to 0.35 and double it: $2\int_0^{0.35} \sum_{k=0}^{\infty}(-1)^k \frac{4^k x^{4k}}{(2k)!}\, dx = 2\left(\sum_{k=0}^{\infty}(-1)^k \frac{4^k x^{4k+1}}{(4k+1)(2k)!}\right)\Big|_0^{0.35} = 2\left(\sum_{k=0}^{\infty}(-1)^k \frac{4^k 0.35^{4k+1}}{(4k+1)(2k)!}\right)$. Since this is an alternating series, to approximate it to within $\frac{1}{2}\cdot 10^{-4}$, we must find n such that $a_{n+1} < \frac{1}{2}\cdot 10^{-4}$, or $\frac{4^{n+1}0.35^{4n+3}}{(4n+3)(2n+2)!} < \frac{1}{2}\cdot 10^{-4}$. This occurs first for $n = 1$, and we have $2\left(.35 - \frac{4\cdot 0.35^5}{5\cdot 2!}\right) \approx 0.69579825$.

9.4.35 The Taylor series for $\frac{\sin x}{x}$ is $\sum_{k=0}^{\infty}(-1)^k \frac{x^{2k}}{(2k+1)!}$, so the desired integral is $\int_0^{0.15} \sum_{k=0}^{\infty}(-1)^k \frac{x^{2k}}{(2k+1)!}\, dx = \sum_{k=0}^{\infty}(-1)^k \frac{x^{2k+1}}{(2k+1)(2k+1)!}\Big|_0^{.15} = \sum_{k=0}^{\infty}(-1)^k \frac{0.1)^{2k+1}}{(2k+1)(2k+1)!}$. This is an alternating series, so to approximate it to within 10^{-4}, we must find n such that $a_{n+1} < 10^{-4}$, or $\frac{0.15^{2n+3}}{(2n+3)(2n+3)!} < 10^{-4}$. This occurs first for $n = 1$, and we have $0.15 - \frac{0.15^3}{3\cdot 3!} = 0.1498125$.

9.4.37 The Taylor series for $(1+x^6)^{-1/2}$ is $\sum_{k=0}^{\infty}\binom{-1/2}{k}x^{6k}$, so the desired integral is $\int_0^{0.5}\sum_{k=0}^{\infty}\binom{-1/2}{k}x^{6k}\,dx = \sum_{k=0}^{\infty}\frac{1}{6k+1}\binom{-1/2}{k}x^{6k+1}\Big|_0^{0.5} = \sum_{k=0}^{\infty}\frac{1}{6k+1}\binom{-1/2}{k}0.5^{6k+1}$. This is an alternating series since the binomial coefficients alternate in sign, so to approximate it to within 10^{-4}, we must find n such that $a_{n+1} < 10^{-4}$, or $\left|\frac{1}{6n+7}\binom{-1/2}{n+1}0.5^{6n+7}\right| < 10^{-4}$. This occurs first for $n = 1$, so we have $\binom{-1/2}{0}0.5+\frac{1}{7}\binom{-1/2}{1}(0.5)^7 \approx 0.4994419643$.

9.4.39 Use the Taylor series for e^x at 0: $1+\frac{2}{1!}+\frac{2^2}{2!}+\frac{2^3}{3!}$.

9.4.41 Use the Taylor series for $\cos x$ at 0: $1-\frac{2^2}{2!}+\frac{2^4}{4!}-\frac{2^8}{8!}$.

9.4.43 Use the Taylor series for $\ln(1+x)$ evaluated at $x = 1/2$: $\frac{1}{2}-\frac{1}{2}\cdot\frac{1}{4}+\frac{1}{3}\cdot\frac{1}{8}-\frac{1}{4}\cdot\frac{1}{16}$.

9.4.45 The Taylor series for f centered at 0 is $\frac{-1+\sum_{k=0}^{\infty}\frac{x^k}{k!}}{x} = \frac{\sum_{k=1}^{\infty}\frac{x^k}{k!}}{x} = \sum_{k=1}^{\infty}\frac{x^{k-1}}{k!} = \sum_{k=0}^{\infty}\frac{x^k}{(k+1)!}$. Evaluating both sides at $x = 1$, we have $e-1 = \sum_{k=0}^{\infty}\frac{1}{(k+1)!}$.

9.4.47 The Taylor series for $\ln(1+x)$ centered at 0 is $x-\frac{1}{2}x^2+\frac{1}{3}x^3-\frac{1}{4}x^4+\cdots = \sum_{k=1}^{\infty}(-1)^{k+1}\frac{x^k}{k}$. By the Ratio Test, $\lim_{k\to\infty}\left|\frac{a_{k+1}}{a_k}\right| = \lim_{k\to\infty}\left|\frac{x^{k+1}k}{x^k(k+1)}\right| = |x|$, so the radius of convergence is 1. The series diverges at -1 and converges at 1, so the interval of convergence is $(-1,1]$. Evaluating at 1 gives $\ln 2 = \sum_{k=1}^{\infty}(-1)^{k+1}\frac{1}{k} = 1-\frac{1}{2}+\frac{1}{3}-\frac{1}{4}+\cdots$.

9.4.49 $\sum_{k=0}^{\infty}\frac{x^k}{2^k} = \sum_{k=0}^{\infty}\left(\frac{x}{2}\right)^k = \frac{1}{1-\frac{x}{2}} = \frac{2}{2-x}$.

9.4.51 $\sum_{k=0}^{\infty}(-1)^k\frac{x^{2k}}{4^k} = \sum_{k=0}^{\infty}\left(\frac{-x^2}{4}\right)^k = \frac{1}{1+\frac{x^2}{4}} = \frac{4}{4+x^2}$.

9.4.53 $\ln(1+x) = -\sum_{k=1}^{\infty}(-1)^k\frac{x^k}{k}$, so $\ln(1-x) = -\sum_{k=1}^{\infty}\frac{x^k}{k}$, and finally $-\ln(1-x) = \sum_{k=1}^{\infty}\frac{x^k}{k}$.

9.4.55

$$\begin{aligned}\sum_{k=1}^{\infty}(-1)^k\frac{kx^{k+1}}{3^k} &= \sum_{k=1}^{\infty}(-1)^k\frac{k}{3^k}x^{k+1} = \sum_{k=1}^{\infty}k\left(-\frac{1}{3}\right)^k x^{k+1} \\ &= x^2\sum_{k=1}^{\infty}\left(-\frac{1}{3}\right)^k kx^{k-1} = x^2\sum_{k=1}^{\infty}\left(-\frac{1}{3}\right)^k\frac{d}{dx}(x^k) \\ &= x^2\frac{d}{dx}\left(\sum_{k=1}^{\infty}\left(-\frac{x}{3}\right)^k\right) = x^2\frac{d}{dx}\left(\frac{1}{1+\frac{x}{3}}\right) = -\frac{3x^2}{(x+3)^2}.\end{aligned}$$

9.4.57 $\sum_{k=2}^{\infty}\frac{k(k-1)x^k}{3^k} = x^2\sum_{k=2}^{\infty}\frac{k(k-1)x^{k-2}}{3^k} = x^2\frac{d^2}{dx^2}\left(\sum_{k=2}^{\infty}\frac{x^k}{3^k}\right)$ $= x^2\frac{d^2}{dx^2}\left(\sum_{k=2}^{\infty}\left(\frac{x}{3}\right)^k\right) = x^2\frac{d^2}{dx^2}\left(\frac{x^2}{9}\cdot\frac{1}{1-\frac{x}{3}}\right) = x^2\frac{d^2}{dx^2}\left(\frac{x^2}{9-3x}\right) = x^2\frac{-6}{(x-3)^3} = \frac{-6x^2}{(x-3)^3}$.

9.4.59

a. False, since $\frac{1}{1-x}$ is not continuous at 1, which is in the interval of integration.

b. False, since the Ratio Test shows that the radius of convergence for the Taylor series for $\tan^{-1}(x)$ centered at 0 is 1.

c. True, since $\sum_{k=0}^{\infty}\frac{x^k}{k!} = e^x$. Substitute $x = \ln 2$.

9.4.61 The Taylor series for $\sin x$ centered at 0 is

$$\sin x = x - \frac{x^3}{6} + \frac{x^5}{120} - \cdots.$$

We compute that

$$\frac{\sin ax}{\sin bx} = \frac{ax - \frac{(ax)^3}{6} + \frac{(ax)^5}{120} - \cdots}{bx - \frac{(bx)^3}{6} + \frac{(bx)^5}{120} - \cdots} = \frac{a - \frac{a^3x^2}{6} + \frac{a^5x^4}{120} - \cdots}{b - \frac{b^3x^2}{6} + \frac{b^5x^4}{120} - \cdots}$$

so the limit of $\frac{\sin ax}{\sin bx}$ as $x \to 0$ is $\frac{a}{b}$.

9.4.63 Compute instead the limit of the log of this expression, $\lim_{x\to 0} \frac{\ln(\sin x/x)}{x^2}$. If the Taylor expansion of $\ln(\sin x/x)$ is $\sum_{k=0}^{\infty} c_k x^k$, then $\lim_{x\to 0} \frac{\ln(\sin x/x)}{x^2} = \lim_{x\to 0} \sum_{k=0}^{\infty} c_k x^{k-2} = \lim_{x\to 0} c_0 x^{-2} + c_1 x^{-1} + c_2$, since the higher-order terms have positive powers of x and thus approach zero as x does. So compute the terms of the Taylor series of $\ln\left(\frac{\sin x}{x}\right)$ up through the quadratic term. The relevant Taylor series are: $\frac{\sin x}{x} = 1 - \frac{1}{6}x^2 + \frac{1}{120}x^4 - \cdots$, $\ln(1+x) = x - \frac{1}{2}x^2 + \frac{1}{3}x^3 - \cdots$ and we substitute the Taylor series for $\frac{\sin x}{x} - 1$ for x in the Taylor series for $\ln(1+x)$. Since the lowest power of x in the first Taylor series is 2, it follows that only the linear term in the series for $\ln(1+x)$ will give any powers of x that are at most quadratic. The only term that results is $-\frac{1}{6}x^2$. Thus $c_0 = c_1 = 0$ in the above, and $c_2 = -\frac{1}{6}$, so that $\lim_{x\to 0} \frac{\ln(\sin x/x)}{x^2} = -\frac{1}{6}$ and thus $\lim_{x\to 0} \left(\frac{\sin x}{x}\right)^{1/x^2} = e^{-1/6}$.

9.4.65 The Taylor series we need are $\cos x = 1 - \frac{1}{2}x^2 + \frac{1}{24}x^4 + \ldots$, $e^t = 1 + t + \frac{1}{2!}t^2 + \frac{1}{3!}t^3 + \frac{1}{4!}t^4 + \ldots$. We are looking for powers of x^3 and x^4 that occur when the first series is substituted for t in the second series. Clearly there will be no odd powers of x, since $\cos x$ has only even powers. Thus the coefficient of x^3 is zero, so that $f^{(3)}(0) = 0$. The coefficient of x^4 comes from the expansion of $1 - \frac{1}{2}x^2 + \frac{1}{24}x^4$ in each term of e^t. Higher powers of x clearly cannot contribute to the coefficient of x^4. Thus consider $\left(1 - \frac{1}{2}x^2 + \frac{1}{24}x^4\right)^k$. The term $-\frac{1}{2}x^2$ generates $\binom{k}{2}$ terms of value $\frac{1}{4}x^4$ for $k \geq 2$, while the other term generates k terms of value $\frac{1}{24}x^4$ for $k \geq 1$. These terms all have to be divided by the $k!$ appearing in the series for e^t. So the total coefficient of x^4 is $\frac{1}{24}\sum_{k=1}^{\infty} \frac{k}{k!} + \frac{1}{4}\sum_{k=2}^{\infty} \binom{k}{2}\frac{1}{k!}, = \frac{1}{24}\sum_{k=1}^{\infty} \frac{1}{(k-1)!} + \frac{1}{4}\sum_{k=2}^{\infty} \frac{1}{2\cdot(k-2)!}, = \frac{1}{24}\sum_{k=0}^{\infty} \frac{1}{k!} + \frac{1}{8}\sum_{k=0}^{\infty} \frac{1}{k!},$ $= \frac{1}{24}e + \frac{1}{8}e = \frac{e}{6}$ Thus $f^{(4)}(0) = \frac{e}{6} \cdot 4! = 4e$.

9.4.67 The Taylor series for $\sin t^2$ is $\sin t^2 = t^2 - \frac{1}{3!}t^6 + \frac{1}{5!}t^{10} - \ldots$, so that $\int_0^x \sin t^2\, dt = \frac{1}{3}t^3 - \frac{1}{7\cdot 3!}t^7 + \ldots \Big|_0^x =$ $\frac{1}{3}x^3 - \frac{1}{7\cdot 3!}x^6 + \ldots$. Thus $f^{(3)}(0) = \frac{3!}{3} = 2$ and $f^{(4)}(0) = 0$.

9.4.69 Consider the series $\sum_{k=1}^{\infty} x^k = \frac{x}{1-x}$. Differentiating both sides gives $\frac{1}{(1-x)^2} = \sum_{k=0}^{\infty} kx^{k-1} = \frac{1}{x}\sum_{k=0}^{\infty} kx^k$ so that $\frac{x}{(1-x)^2} = \sum_{k=0}^{\infty} kx^k$. Evaluate both sides at $x = 1/2$ to see that the sum of the series is $\frac{1/2}{(1-1/2)^2} = 2$. Thus the expected number of tosses is 2.

9.4.71

a. We look first for a Taylor series for $(1 - k^2\sin^2(\theta))^{-1/2}$. Since $(1 - k^2x^2)^{-1/2} = (1 - (kx)^2)^{-1/2} = \sum_{i=0}^{\infty} \binom{-1/2}{i}(kx)^{2i}$, and $\sin\theta = \theta - \frac{1}{3!}\theta^3 + \frac{1}{5!}\theta^5 - \ldots$, substituting the second series into the first gives $\frac{1}{\sqrt{1-k^2\sin^2\theta}} = 1 + \frac{1}{2}k^2\theta^2 + \left(-\frac{1}{6}k^2 + \frac{3}{8}k^4\right)\theta^4 + \left(\frac{1}{45}k^2 - \frac{1}{4}k^4 + \frac{5}{16}k^6\right)\theta^6 +$ $\left(\frac{-1}{630}k^2 + \frac{3}{40}k^4 - \frac{5}{16}k^6 + \frac{35}{128}k^8\right)\theta^8 + \ldots$.
Integrating with respect to θ and evaluating at $\pi/2$ (the value of the antiderivative is 0 at 0) gives $\frac{1}{2}\pi + \frac{1}{48}k^2\pi^3 + \frac{1}{160}\left(-\frac{1}{6}k^2 + \frac{3}{8}k^4\right)\pi^5 + \frac{1}{896}\left(\frac{1}{45}k^2 - \frac{1}{4}k^4 + \frac{5}{16}k^6\right)\pi^7 + \frac{1}{4608}\left(-\frac{1}{630}k^2 + \frac{3}{40}k^4 - \frac{5}{16}k^6 + \frac{35}{128}k^8\right)\pi^9$. Evaluating these terms for $k = 0.1$ gives $F(0.1) \approx 1.574749680$. (The true value is approximately 1.574745562.)

b. The terms above, with coefficients of k^n converted to decimal approximations, is $1.5707 + .3918 \cdot k^2 + .3597 \cdot k^4 - .9682 \cdot k^6 + 1.7689 \cdot k^8$. The coefficients are all less than 2 and do not appear to be increasing very much if at all, so if we want the result to be accurate to within 10^{-3} we should probably take n such that $k^n < \frac{1}{2} \times 10^{-3} = .0005$, so $n = 4$ for this value of k.

c. By the above analysis, we would need a larger n since $0.2^n > 0.1^n$ for a given value of n.

9.4.73

a. By the Fundamental Theorem, $S'(x) = \sin x^2$, $C'(x) = \cos x^2$.

b. The relevant Taylor series are $\sin t^2 = t^2 - \frac{1}{3!}t^6 + \frac{1}{5!}t^{10} - \frac{1}{7!}t^{14} + \ldots$, and $\cos t^2 = 1 - \frac{1}{2!}t^4 + \frac{1}{4!}t^8 - \frac{1}{6!}t^{12} + \ldots$. Integrating, we have $S(x) = \frac{1}{3}x^3 - \frac{1}{7\cdot 3!}x^7 + \frac{1}{11\cdot 5!}x^{11} - \frac{1}{15\cdot 7!}x^{15} + \ldots$, and $C(x) = x - \frac{1}{5\cdot 2!}x^5 + \frac{1}{9\cdot 4!}x^9 - \frac{1}{13\cdot 6!}x^{13} + \ldots$.

c. $S(0.05) \approx \frac{1}{3}(0.05)^3 - \frac{1}{42}(0.05)^5 + \frac{1}{1320}(0.05)^{11} - \frac{1}{75600}(0.05)^{15} \approx 4.166664807 \times 10^{-5}$. $C(-0.25) \approx (-0.25) - \frac{1}{10}(-0.25)^5 + \frac{1}{216}(-0.25)^9 - \frac{1}{9360}(-0.25)^{13} \approx -.2499023616$.

d. The series is alternating. Since $a_{n+1} = \frac{1}{(4n+7)(2n+3)!}(0.05)^{4n+7}$, and this is less than 10^{-4} for $n = 0$, only one term is required.

e. The series is alternating. Since $a_{n+1} = \frac{1}{(4n+5)(2n+2)!}(0.25)^{4n+5}$, and this is less than 10^{-6} for $n = 1$, two terms are required.

9.4.75

a. $J_0(x) = 1 - \frac{1}{4}x^2 + \frac{1}{16\cdot 2!^2}x^4 - \frac{1}{2^6\cdot 3!^2}x^6 + \ldots$.

b. Using the Ratio Test: $\left|\frac{a_{k+1}}{a_k}\right| = \frac{x^{2k+2}}{2^{2k+2}((k+1)!)^2} \cdot \frac{2^{2k}(k!)^2}{x^{2k}} = \frac{x^2}{4(k+1)^2}$, which has limit 0 as $k \to \infty$ for any x. Thus the radius of convergence is infinite and the interval of convergence is $(-\infty, \infty)$.

c. Starting only with terms up through x^8, we have $J_0(x) = 1 - \frac{1}{4}x^2 + \frac{1}{64}x^4 - \frac{1}{2304}x^6 + \frac{1}{147456}x^8 + \ldots$, $J_0'(x) = -\frac{1}{2}x + \frac{1}{16}x^3 - \frac{1}{384}x^5 + \frac{1}{18432}x^7 + \ldots$, $J_0''(x) = -\frac{1}{2} + \frac{3}{16}x^2 - \frac{5}{384}x^4 + \frac{7}{18432}x^6 + \ldots$ so that $x^2 J_0(x) = x^2 - \frac{1}{4}x^4 + \frac{1}{64}x^6 - \frac{1}{2304}x^8 + \frac{1}{147456}x^{10} + \ldots$, $xJ_0'(x) = -\frac{1}{2}x^2 + \frac{1}{16}x^4 - \frac{1}{384}x^6 + \frac{1}{18432}x^8 + \ldots$, $x^2 J_0''(x) = -\frac{1}{2}x^2 + \frac{3}{16}x^4 - \frac{5}{384}x^6 + \frac{7}{18432}x^8 + \ldots$, and $x^2 J_0''(x) + xJ_0'(x) + x^2 J_0(x) = 0$.

9.4.77

a. The power series for $\cos x$ has only even powers of x, so that the power series has the same value evaluated at $-x$ as it does at x.

b. The power series for $\sin x$ has only odd powers of x, so that evaluating it at $-x$ gives the opposite of its value at x.

9.4.79

a. Since $f(a) = g(a) = 0$, we use the Taylor series for $f(x)$ and $g(x)$ centered at a to compute that

$$\begin{aligned}
\lim_{x\to a} \frac{f(x)}{g(x)} &= \lim_{x\to a} \frac{f(a) + f'(a)(x-a) + \frac{1}{2}f''(a)(x-a)^2 + \cdots}{g(a) + g'(a)(x-a) + \frac{1}{2}g''(a)(x-a)^2 + \cdots} \\
&= \lim_{x\to a} \frac{f'(a)(x-a) + \frac{1}{2}f''(a)(x-a)^2 + \cdots}{g'(a)(x-a) + \frac{1}{2}g''(a)(x-a)^2 + \cdots} \\
&= \lim_{x\to a} \frac{f'(a) + \frac{1}{2}f''(a)(x-a) + \cdots}{g'(a) + \frac{1}{2}g''(a)(x-a) + \cdots} = \frac{f'(a)}{g'(a)}.
\end{aligned}$$

Since $f'(x)$ and $g'(x)$ are assumed to be continuous at a and $g'(a) \neq 0$,

$$\frac{f'(a)}{g'(a)} = \lim_{x\to a} \frac{f'(x)}{g'(x)}$$

we have that

$$\lim_{x\to a}\frac{f(x)}{g(x)} = \lim_{x\to a}\frac{f'(x)}{g'(x)}$$

which is one form of L'Hôpital's Rule.

b. Since $f(a) = g(a) = f'(a) = g'(a) = 0$, we use the Taylor series for $f(x)$ and $g(x)$ centered at a to compute that

$$\begin{aligned}\lim_{x\to a}\frac{f(x)}{g(x)} &= \lim_{x\to a}\frac{f(a)+f'(a)(x-a)+\frac{1}{2}f''(a)(x-a)^2+\frac{1}{6}f'''(a)(x-a)^3+\cdots}{g(a)+g'(a)(x-a)+\frac{1}{2}g''(a)(x-a)^2+\frac{1}{6}g'''(a)(x-a)^3+\cdots}\\ &= \lim_{x\to a}\frac{\frac{1}{2}f''(a)(x-a)^2+\frac{1}{6}f'''(a)(x-a)^3+\cdots}{\frac{1}{2}g''(a)(x-a)^2+\frac{1}{6}g'''(a)(x-a)^3+\cdots}\\ &= \lim_{x\to a}\frac{\frac{1}{2}f''(a)+\frac{1}{6}f'''(a)(x-a)+\cdots}{\frac{1}{2}g''(a)+\frac{1}{6}g'''(a)(x-a)+\cdots} = \frac{f''(a)}{g''(a)}.\end{aligned}$$

Since $f''(x)$ and $g''(x)$ are assumed to be continuous at a and $g''(a) \neq 0$,

$$\frac{f''(a)}{g''(a)} = \lim_{x\to a}\frac{f''(x)}{g''(x)}$$

we have that

$$\lim_{x\to a}\frac{f(x)}{g(x)} = \lim_{x\to a}\frac{f''(x)}{g''(x)}$$

which is consistent with two applications of L'Hôpital's Rule.

9.5 Chapter Nine Review

9.5.1

a. True. The approximations tend to get better as n increases in size, and also when the value being approximated is closer to the center of the series. Since 2.1 is closer to 2 than 2.2 is, and since $3 > 2$, we should have $|p_3(2.1) - f(2.1)| < |p_2(2.2) - f(2.2)|$.

b. False. The interval of convergence may or may not include the endpoints.

c. True. The interval of convergence is an interval centered at 0, and the endpoints may or may not be included.

d. True. Since $f(x)$ is a polynomial, all its derivatives vanish after a certain point (in this case, $f^{(12)}(x)$ is the last nonzero derivative).

9.5.3 $p_2(x) = 1$.

9.5.5 $p_3(x) = x - \frac{x^2}{2} + \frac{x^3}{3}$.

9.5.7 $p_2(x) = x - 1 - \frac{1}{2}(x-1)^2$.

9.5.9

a. $p_0(x) = 1$, $p_1(x) = 1 + x$, and $p_2(x) = 1 + x + \frac{x^2}{2}$.

b.

n	$p_n(-0.08)$	$\lvert p_n(-0.08) - e^{-0.08}\rvert$
0	1	7.69×10^{-2}
1	0.92	3.12×10^{-3}
2	0.9232	8.37×10^{-5}

9.5.11

a. $p_0(x) = \frac{\sqrt{2}}{2}$, $p_1(x) = \frac{\sqrt{2}}{2}(1 + (x - \pi/4))$, and $p_2(x) = \frac{\sqrt{2}}{2}\left(1 + (x - \pi/4) - \frac{1}{2}(x - \pi/4)^2\right)$.

b.

n	$p_n(\pi/5)$	$\lvert p_n(\pi/5) - \sin(\pi/5)\rvert$
0	0.7071	0.119
1	0.5960	8.25×10^{-3}
2	0.5873	4.74×10^{-4}

9.5.13 The derivatives of $\sin x$ are bounded in magnitude by 1, so $|R_n(x)| \le M\frac{|x|^{n+1}}{(n+1)!} \le \frac{|x|^{n+1}}{(n+1)!}$. But $|x| < \pi$, so $|R_3(x)| \le \frac{\pi^4}{24}$.

9.5.15 Using the Ratio Test, $\lim_{k\to\infty}\left|\frac{a_{k+1}}{a_k}\right| = \lim_{k\to\infty}\left|\frac{(k+1)^2 x^{k+1}}{(k+1)!}\cdot\frac{k!}{k^2 x^k}\right| = \lim_{k\to\infty}\left(\frac{k+1}{k}\right)^2\frac{|x|}{k+1} = 0$, so the interval of convergence is $(-\infty, \infty)$.

9.5.17 Using the Ratio Test, $\lim_{k\to\infty}\frac{a_{k+1}}{a_k} = \lim_{k\to\infty}\left|\frac{(x+1)^{2k+2}}{(k+1)!}\cdot\frac{k!}{(x+1)^{2k}}\right| = \lim_{k\to\infty}\frac{1}{k+1}(x+1)^2 = 0$, so the interval of convergence is $(-\infty, \infty)$.

9.5.19 By the Root Test, $\lim_{k\to\infty}\left(\frac{|x|}{9}\right)^3 = \frac{|x^3|}{729}$, so the series converges for $|x| < 9$. The series given by letting $x = \pm 9$ are both divergent by the Divergence Test. Thus, $(-9, 9)$ is the interval of convergence.

9.5.21 The Maclaurin series for $f(x)$ is $\sum_{k=0}^{\infty} x^{2k}$. By the Root Test, this converges for $\left|x^2\right| < 1$, so $-1 < x < 1$. It diverges at both endpoints, so the interval of convergence is $(-1, 1)$.

9.5.23 The Maclaurin series for $f(x)$ is $\sum_{k=0}^{\infty}(3x)^k = \sum_{k=0}^{\infty} 3^k x^k$. By the Root Test, this has radius of convergence $1/3$. Checking the endpoints, we obtain an interval of convergence of $(-1/3, 1/3)$.

9.5.25 Taking the derivative of $\frac{1}{1-x}$ gives $f(x)$. Thus, the Macluarin series for $f(x)$ is $\sum_{k=1}^{\infty} kx^{k-1}$. Using the Ratio Test, $\lim_{k\to\infty}\left|\frac{a_{k+1}}{a_k}\right| = \lim_{k\to\infty}\left|\frac{(k+1)x^k}{kx^{k-1}}\right| = \lim_{k\to\infty}\frac{k+1}{k}|x| = |x|$, so the radius of convergence is 1. Checking the endpoints, we obtain $(-1, 1)$ for the interval of convergence.

9.5.27 The first three terms are $1 + 3x + \frac{9x^2}{2}$. The series is $\sum_{k=0}^{\infty}\frac{(3x)^k}{k!}$.

9.5.29 The first three terms are $-(x - \pi/2) + \frac{1}{6}(x - \pi/2)^3 - \frac{1}{120}(x - \pi/2)^5$. The series is

$$\sum_{k=0}^{\infty}(-1)^{k+1}\frac{1}{(2k+1)!}\left(x - \frac{\pi}{2}\right)^{2k+1}.$$

9.5.31 The first three terms are $x - \frac{1}{3}x^3 + \frac{1}{5}x^5$. The series is $\sum_{k=0}^{\infty}(-1)^k\frac{x^{2k+1}}{2k+1}$.

9.5.33 $f(x) = \binom{1/3}{0} + \binom{1/3}{1}x + \binom{1/3}{2}x^2 + \cdots = 1 + \frac{1}{3}x - \frac{1}{9}x^2 + \ldots$.

9.5.35 $f(x) = \binom{-3}{0} + \binom{-3}{1}\frac{x}{2} + \binom{-3}{2}\frac{x^2}{4} + \cdots = 1 - \frac{3}{2}x + \frac{3}{2}x^2 + \ldots$.

9.5.37 $R_n(x) = \frac{(-1)^{n+1}e^{-c}}{(n+1)!}x^{n+1}$ for some c between 0 and x, and $\lim_{n\to\infty}|R_n(x)| = e^{-c}\lim_{n\to\infty}\frac{|x|^{n+1}}{(n+1)!} = 0$, since $n!$ grows faster than $|x|^n$ as $n \to \infty$ for all x.

9.5.39 $R_n(x) = \frac{f^{(n+1)}(c)}{(n+1)!}x^{n+1}$ for some c in $(-1/2, 1/2)$. Now, $\left|f^{(n+1)}(c)\right| = \frac{n!}{(1+c)^{n+1}}$, so $\lim_{n\to\infty}|R_n(x)| = \lim_{n\to\infty}\left(\frac{|x|}{1+c}\right)^{n+1}\cdot\frac{1}{n+1} < \lim_{n\to\infty}1^{n+1}\frac{1}{n+1} = 0$.

9.5.41 The Taylor series for $\cos x$ centered at 0 is

$$\cos x = 1 - \frac{x^2}{2} + \frac{x^4}{24} - \frac{x^6}{720} + \cdots .$$

We compute that

$$\begin{aligned}\frac{x^2/2 - 1 + \cos x}{x^4} &= \frac{1}{x^4}\left(x^2/2 - 1 + \left(1 - \frac{x^2}{2} + \frac{x^4}{24} - \frac{x^6}{720} + \cdots\right)\right)\\ &= \frac{1}{x^4}\left(\frac{x^4}{24} - \frac{x^6}{720} + \cdots\right) = \frac{1}{24} - \frac{x^2}{720} + \cdots\end{aligned}$$

so the limit of $\dfrac{x^2/2 - 1 + \cos x}{x^4}$ as $x \to 0$ is $\dfrac{1}{24}$.

9.5.43 The Taylor series for $\ln(x-3)$ centered at 4 is

$$\ln(x-3) = (x-4) - \frac{1}{2}(x-4)^2 + \frac{1}{3}(x-4)^3 - \cdots .$$

We compute that

$$\begin{aligned}\frac{\ln(x-3)}{x^2-16} &= \frac{1}{(x-4)(x+4)}\left((x-4) - \frac{1}{2}(x-4)^2 + \frac{1}{3}(x-4)^3 - \cdots\right)\\ &= \frac{1}{(x-4)(x+4)}\left((x-4)\left(1 - \frac{1}{2}(x-4) + \frac{1}{3}(x-4)^2 - \cdots\right)\right)\\ &= \frac{1}{x+4}\left(1 - \frac{1}{2}(x-4) + \frac{1}{3}(x-4)^2 - \cdots\right)\end{aligned}$$

so the limit of $\dfrac{\ln(x-3)}{x^2-16}$ as $x \to 4$ is $\dfrac{1}{8}$.

9.5.45 The Taylor series for $\sec x$ centered at 0 is

$$\sec x = 1 + \frac{x^2}{2} + \frac{5x^4}{24} + \frac{61x^6}{720} + \cdots$$

and the Taylor series for $\cos x$ centered at 0 is

$$\cos x = 1 - \frac{x^2}{2} + \frac{x^4}{24} - \frac{x^6}{720} + \cdots .$$

We compute that

$$\begin{aligned}&\frac{\sec x - \cos x - x^2}{x^4}\\ &= \frac{1}{x^4}\left(\left(1 + \frac{x^2}{2} + \frac{5x^4}{24} + \frac{61x^6}{720} + \cdots\right) - \left(1 - \frac{x^2}{2} + \frac{x^4}{24} - \frac{x^6}{720} + \cdots\right) - x^2\right)\\ &= \frac{1}{x^4}\left(\frac{x^4}{6} + \frac{31x^6}{360} + \cdots\right) = \frac{1}{6} + \frac{31x^2}{360} + \cdots\end{aligned}$$

so the limit of $\dfrac{\sec x - \cos x - x^2}{x^4}$ as $x \to 0$ is $\dfrac{1}{6}$.

9.5.47 Since $y(0) = 4$, we have $y'(0) - 16 + 12 = 0$, so $y'(0) = 4$. Differentiating the equation $n-1$ times and evaluating at 0 we obtain $y^{(n)}(0) = 4y^{(n-1)}(0)$, so that $y^{(n)}(0) = 4^n$. The Taylor series for $y(x)$ is thus $y(x) = 4 + 4x + \frac{4^2x^2}{2!} + \frac{4^3x^3}{3!} + \ldots$, or $y(x) = 3 + e^{4x}$.

9.5.49

a. The Taylor series for $\ln(1+x)$ is $\sum_{k=1}^{\infty}(-1)^{k+1}\frac{x^k}{k}$. Evaluating at $x=1$ gives $\ln 2 = \sum_{k=1}^{\infty}(-1)^{k+1}\frac{1}{k}$.

b. The Taylor series for $\ln(1-x)$ is $-\sum_{k=1}^{\infty}\frac{x^k}{k}$. Evaluating at $x = 1/2$ gives $\ln(1/2) = -\sum_{k=1}^{\infty}\frac{1}{k2^k}$, so that $\ln 2 = \sum_{k=1}^{\infty}\frac{1}{k2^k}$.

c. $f(x) = \ln\left(\frac{1+x}{1-x}\right) = \ln(1+x) - \ln(1-x)$. Using the two Taylor series above we have $f(x) = \sum_{k=1}^{\infty}(-1)^{k+1}\frac{x^k}{k} - \left(-\sum_{k=1}^{\infty}\frac{x^k}{k}\right) = \sum_{k=1}^{\infty}(1+(-1)^{k+1})\frac{x^k}{k} = 2\sum_{k=0}^{\infty}\frac{x^{2k+1}}{2k+1}$.

d. Since $\frac{1+x}{1-x} = 2$ when $x = \frac{1}{3}$, the resulting infinite series for $\ln 2$ is $2\sum_{k=0}^{\infty}\frac{1}{3^{2k+1}(2k+1)}$.

e. The first four terms of each series are: $1 - \frac{1}{2} + \frac{1}{3} - \frac{1}{4} \approx 0.5833333333$, $\frac{1}{2} + \frac{1}{8} + \frac{1}{24} + \frac{1}{64} \approx 0.6822916667$, $\frac{2}{3} + \frac{2}{81} + \frac{2}{1215} + \frac{2}{15309} \approx 0.6931347573$ The true value is $\ln 2 \approx 0.6931471806$. The third series converges the fastest, since it has 3^{k+1} in the denominator as opposed to 2^k, so its terms get small faster.

Chapter 10

10.1 Parametric Equations

10.1.1 Given an input value of t, the point $(x(t), y(t))$ can be plotted in the xy-plane, generating a curve.

10.1.3 Let $x = R\cos(\pi t/5)$ and $y = R\sin(\pi t/5)$. Note that as t ranges from 0 to 10, $\pi t/5$ ranges from 0 to 2π. Since $x^2 + y^2 = R^2$, this curve represents a circle of radius R.

10.1.5 Let $x = t$ and $y = t^2$ for $t \in (-\infty, \infty)$.

10.1.7

a.

t	x	y
-10	-20	-34
-5	-10	-19
0	0	-4
5	10	11
10	20	26

b.

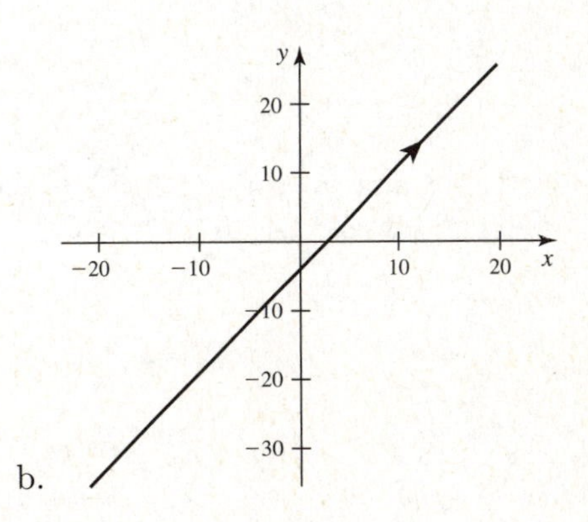

c. Solving $x = 2t$ for t yields $t = x/2$, so $y = 3t - 4 = 3x/2 - 4$.

d. The curve is the line segment from $(-20, -34)$ to $(20, 26)$.

10.1.9

a.

t	x	y
-5	11	-18
-3	9	-12
0	6	-3
3	3	6
5	1	12

b.

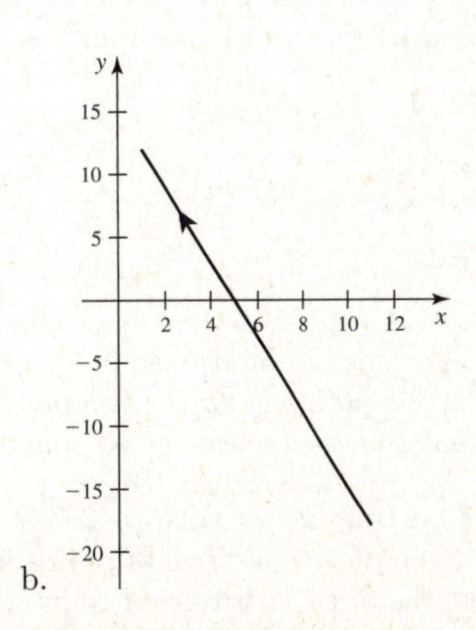

c. Solving $x = -t + 6$ for t yields $t = 6 - x$, so $y = 3t - 3 = 18 - 3x - 3 = 15 - 3x$.

d. The curve is the line segment from $(11, -18)$ to $(1, 12)$.

10.1.11

a. Solving $x = \sqrt{t} + 4$ for t yields $t = (x-4)^2$. Thus, $y = 3\sqrt{t} = 3(x-4)$, where x ranges from 4 to 8. Note that all $t > 0$, $x > 0$, and $y > 0$.

b. The curve is the line segment from $(4, 0)$ to $(8, 12)$.

10.1.13

a. Solving $x = t - 1$ for t yields $t = x + 1$. Thus, $y = t^3 = (x+1)^3$, where $-5 \le x \le 3$.

b. The curve is the part of the standard cubic curve, shifted one unit to the left, from $(-5, -64)$ to $(3, 64)$.

10.1.15 Note that $x^2 + y^2 = 9\cos^2 t + 9\sin^2 t = 9$, so this represents an arc of the circle of radius 3 centered at the origin from $(-3, 0)$ to $(3, 0)$ traversed counterclockwise.

10.1.17 Note that $x^2 + y^2 = 49\cos^2 2t + 49\sin^2 2t = 49$, so this represents an arc of the circle of radius 7 centered at the origin from $(-7, 0)$ to $(-7, 0)$ traversed counterclockwise. (So the whole circle is represented.)

10.1.19

Let $x = 4\cos t$ and $y = 4\sin t$ for $0 \le t \le 2\pi$. Then $x^2 + y^2 = 16\cos^2 t + 16\sin^2 t = 16$.

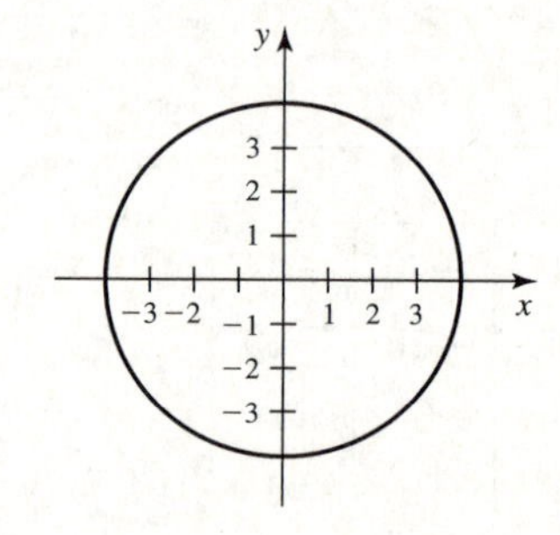

10.1.21

Let $x = -2 + 8\sin t$ and $y = -3 + 8\cos t$ for $0 \le t \le 2\pi$. Then $(x+2)^2 + (y+3)^2 = 64\sin^2 t + 64\cos^2 t = 64$.

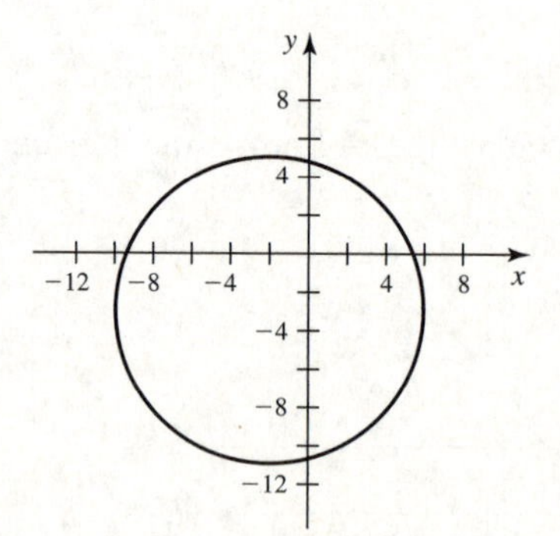

10.1.23 Let t be time in minutes, so $0 \le t \le 1.5$ Let $x = 400\cos(4\pi/3)t$ and $y = 400\sin(4\pi/3)t$. Then since $x^2 + y^2 = 400^2$, the path is a circle of radius 400. Note that the values of x and y are the same at $t = 0$ and $t = 1.5$, and that the circle is traversed counterclockwise.

10.1.25 Let t be time in seconds, so $0 \le t \le 6$ Let $x = 50\cos(\pi/3)t$ and $y = 50\sin(\pi/3)t$. Then since $x^2 + y^2 = 50^2$, the path is a circle of radius 50. Note that the values of x and y are the same at $t = 0$ and $t = 6$, and that the circle is traversed counterclockwise.

10.1.27

Since $t = x - 3$, we have $y = 1 - (x - 3) = 4 - x$, so the line has slope -1. When $t = 0$, we have the point $(3, 1)$.

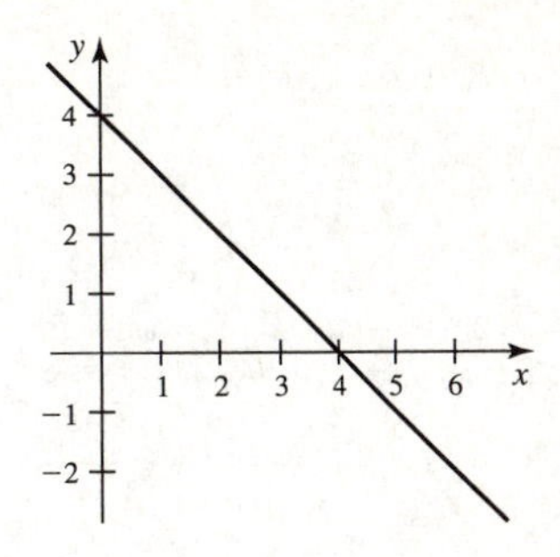

10.1.29

Since $y = 1$, this is a horizontal line with slope 0. When $t = 0$, we have the point $(8, 1)$.

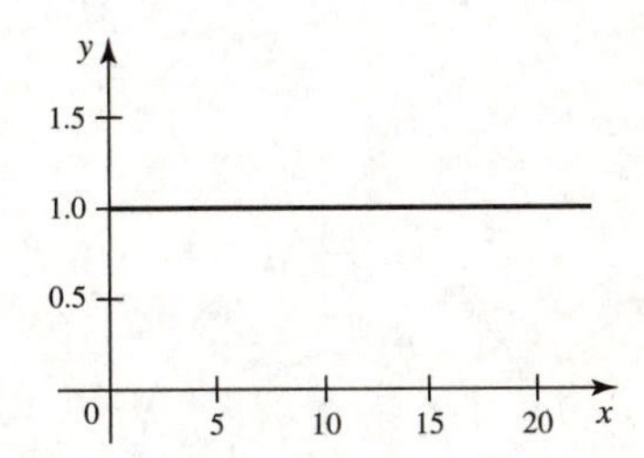

10.1.31 Let $x = x_0 + at$ and $y = y_0 + bt$. Letting $(x_0, y_0) = (0, 0)$, and then finding a and b so that the curve is at the point Q when $t = 1$ yields $x = 2t$, $y = 8t$ for $0 \le t \le 1$.

10.1.33 Let $x = x_0 + at$ and $y = y_0 + bt$. Letting $(x_0, y_0) = (-1, -3)$, and then finding a and b so that the curve is at the point Q when $t = 1$ yields $x = -1 + 7t$, $y = -3 - 13t$ for $0 \le t \le 1$.

10.1.35

Let $x = t$ and $y = 2t^2 - 4$, $-1 \le t \le 5$.

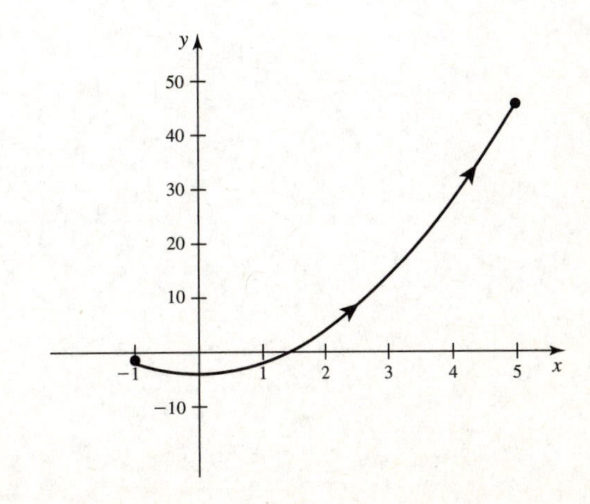

10.1.37

Let $x = -2 + 4t$ and $y = 3 - 6t$, $0 \le t \le 1$, and $x = t + 1$, $y = 8t - 11$ for $1 \le t \le 2$.

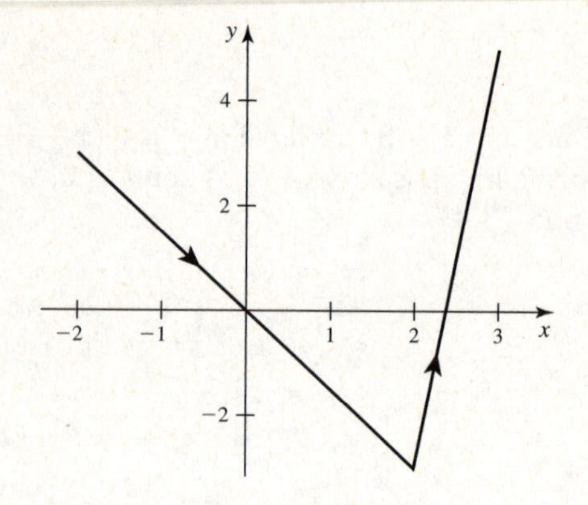

10.1.39

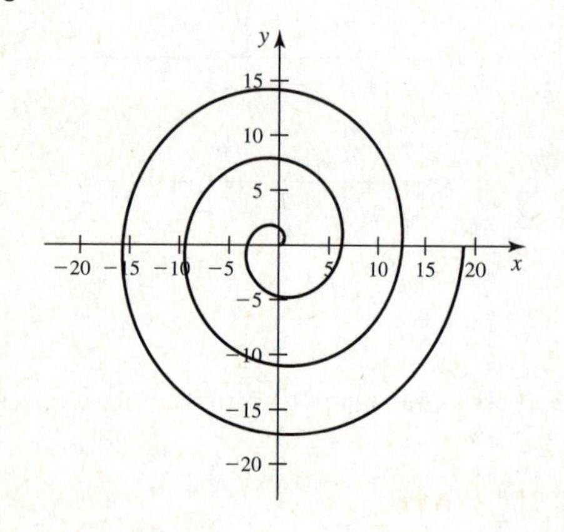

10.1.41

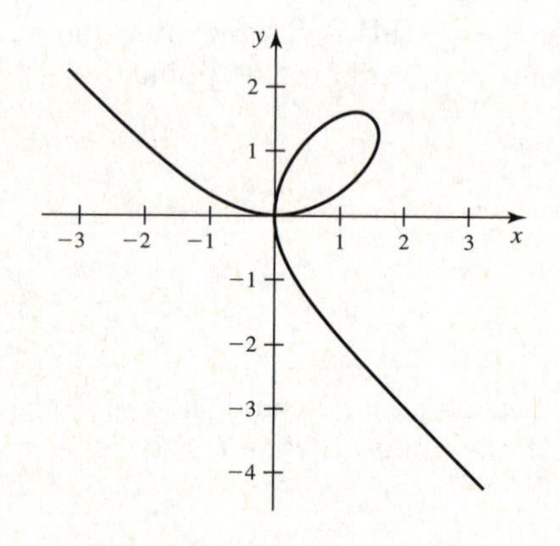

10.1.43

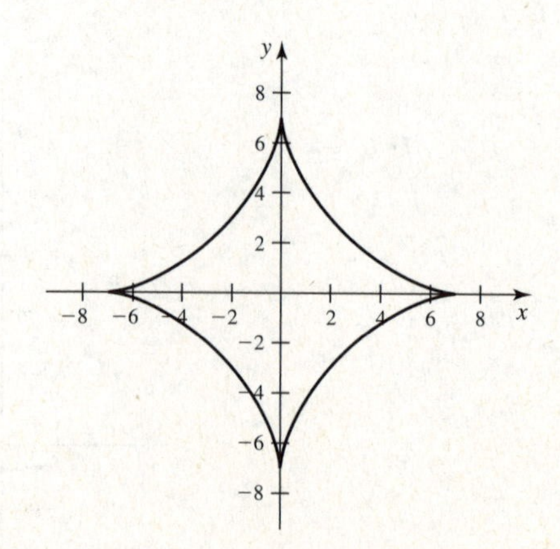

10.1.45

a. $\frac{dy}{dx} = \frac{dy/dt}{dx/dt} = \frac{-8}{4} = -2$ for all t. Since the curve is a line, the tangent line to the curve at the given point is the line itself.

b.

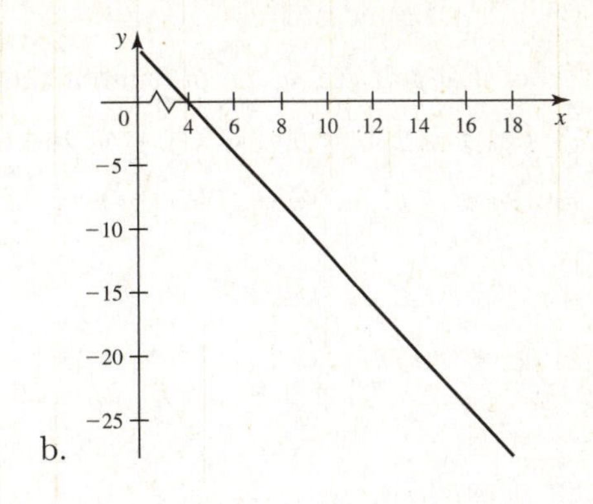

10.1.47

a. $\frac{dy}{dx} = \frac{dy/dt}{dx/dt} = \frac{8\cos t}{-\sin t} = -8\cot t$. At the given value of t, the value of $\frac{dy}{dx}$ is $-8\cot \pi/2 = 0$. The tangent line at the point $(0, 8)$ is thus the horizontal line $y = 8$.

b.

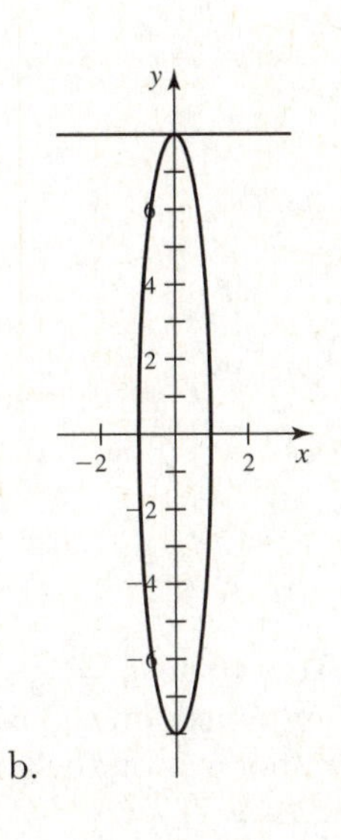

10.1.49

a. $\frac{dy}{dx} = \frac{dy/dt}{dx/dt} = \frac{1+\frac{1}{t^2}}{1-\frac{1}{t^2}} = \frac{t^2+1}{t^2-1}$. At the given value of t, the derivative doesn't exist, and the tangent line is the vertical line $x = 2$, tangent at the point $(2, 0)$.

b.

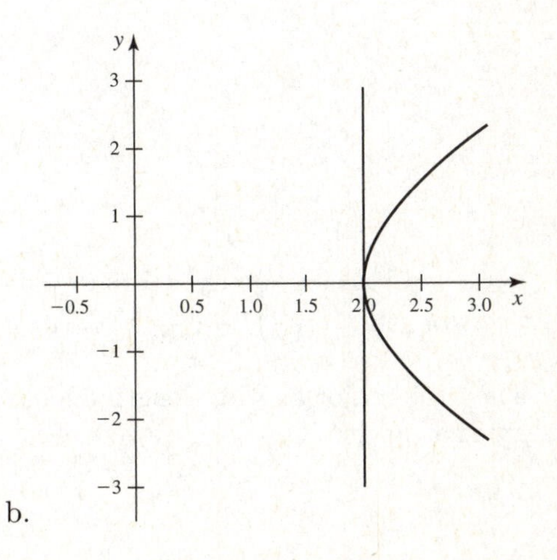

10.1.51

a. False. This generates a circle in the counterclockwise direction.

b. True. Note that when t is increased by one, the value of $2\pi t$ is increased by 2π, which is the period of both the sine and the cosine functions.

c. False. This generates only the portion of the parabola in the first quadrant, omitting the portion in the second quadrant.

d. True. They describe the portion of the unit circle in the 4th and 1st quadrants.

10.1.53 Let $x = 1 + 2t$ and $y = 1 + 4t$, for $-\infty < t < \infty$. Note that $y = 2(1 + 2t) - 1$, so $y = 2x - 1$.

10.1.55 Let $x = t^2$ and $y = t$, for $0 \le t < \infty$. Note that $x = t^2 = y^2$, and that the starting point is $(0, 0)$.

10.1.57

The entire curve is traversed for $0 \le t \le 2\pi$.

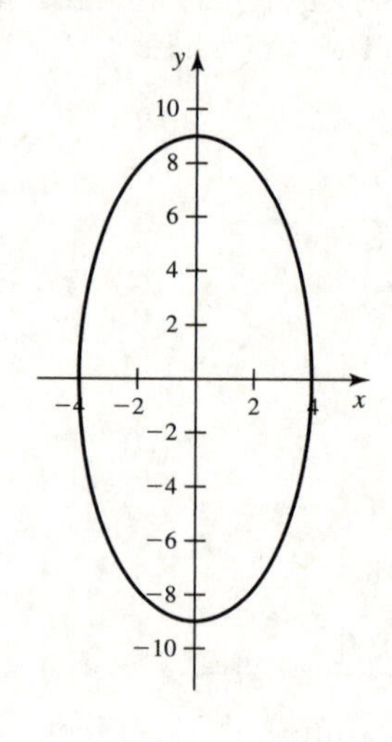

10.1.59

Let $x = 3\cos t$ and $y = \frac{3}{2}\sin t$ for $0 \le t \le 2\pi$. Then the major axis on the x-axis has length $2 \cdot 3 = 6$ and the minor axis on the y-axis has length $2 \cdot \frac{3}{2} = 3$. Note that $\left(\frac{x}{3}\right)^2 + \left(\frac{2y}{3}\right)^2 = \cos^2 t + \sin^2 t = 1$.

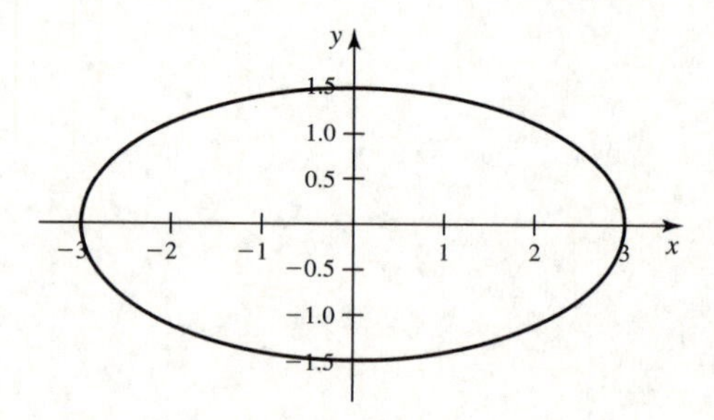

10.1.61

Let $x = 15\cos t - 2$ and $y = 10\sin t - 3$ for $0 \le t \le 2\pi$. Note that $\left(\frac{x+2}{15}\right)^2 + \left(\frac{y+3}{10}\right)^2 = \cos^2 t + \sin^2 t = 1$.
Then the major axis has length 30 and the minor axis has length 20.

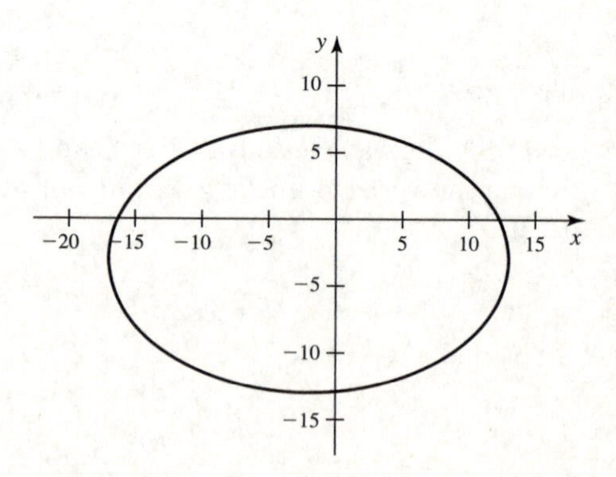

10.1.63 The lines **a** and **b.** For line **a**, note that $t = x - 3$, so $y = 4 - 2t = 4 - (2x - 6) = 10 - 2x$. For line **b**, note that $t = \frac{x-3}{4}$, so $y = 4 - 8t = 4 - 2(x - 3) = 10 - 2x$ as well.

For **c**, note that $t^3 = x - 3$, so $y = 4 - t^3 = 4 - (x - 3) = 7 - x$, so this line is not the same as the other two.

10.1.65 Note that $x^2 + y^2 = 4\sin^2 8t + 4\cos^2 8t = 4$, so the curve is $x^2 + y^2 = 4$.

10.1.67 Note that since $t = x$, we have $y = \sqrt{4 - t^2} = \sqrt{4 - x^2}$.

10.1.69 Since $\sec^2 t - 1 = \tan^2 t$, we have $y = x^2$.

10.1.71 $\frac{dy}{dx} = \frac{dy/dt}{dx/dt} = \frac{4\cos t}{-4\sin t} = -\cot t$. We seek t so that $\cot t = -1/2$, so $t = \cot^{-1}(-1/2)$. The corresponding points on the curve are $(\frac{-4\sqrt{5}}{5}, \frac{8\sqrt{5}}{5})$ and $(\frac{4\sqrt{5}}{5}, \frac{-8\sqrt{5}}{5})$.

10.1.73 $\frac{dy}{dx} = \frac{dy/dt}{dx/dt} = \frac{1+(1/t^2)}{1-(1/t^2)} = \frac{t^2+1}{t^2-1}$. We seek t so that $\frac{t^2+1}{t^2-1} = 1$, which never occurs. Thus, there are no points on this curve with slope 1.

10.1.75 Note that in equation B, the parameter is scaled by a factor of 3. Thus, the curves are the same when the corresponding interval for t is scaled by a factor of $1/3$, so for $a = 0$ and $b = \frac{2\pi}{3}$. In fact, the same curve will be generated for $a = p$, $b = p + 2\pi/3$ where p is any real number.

10.1.77

a. $\frac{dy}{dx} = \frac{dy/dt}{dx/dt} = \frac{2\cos t}{2\cos 2t}$. This is zero when $\cos t = 0$ but $\cos 2t \neq 0$, which occurs for $t = \pi/2$ and $t = 3\pi/2$. The corresponding points on the graph are $(0, 2)$ and $(0, -2)$.

b. Using the derivative obtained above, we seek points where $\cos 2t = 0$ but $\cos t \neq 0$. This occurs for $t = \pi/4$, $3\pi/4$, $5\pi/4$, and $7\pi/4$. The corresponding points on the curve are $(1, \sqrt{2})$, $(-1, \sqrt{2})$, $(-1, -\sqrt{2})$, and $(1, -\sqrt{2})$.

10.1.79

a. Let $\operatorname{sgn}(x) = \begin{cases} 1 & \text{if } x \geq 0 \\ -1 & \text{if } x < 0. \end{cases}$. Let $x = a \cdot \operatorname{sgn}(\cos t)\,|\cos(t)|^{2/n}$ and $y = b \cdot \operatorname{sgn}(\sin(t))\,|\sin(t)|^{2/n}$.

b.

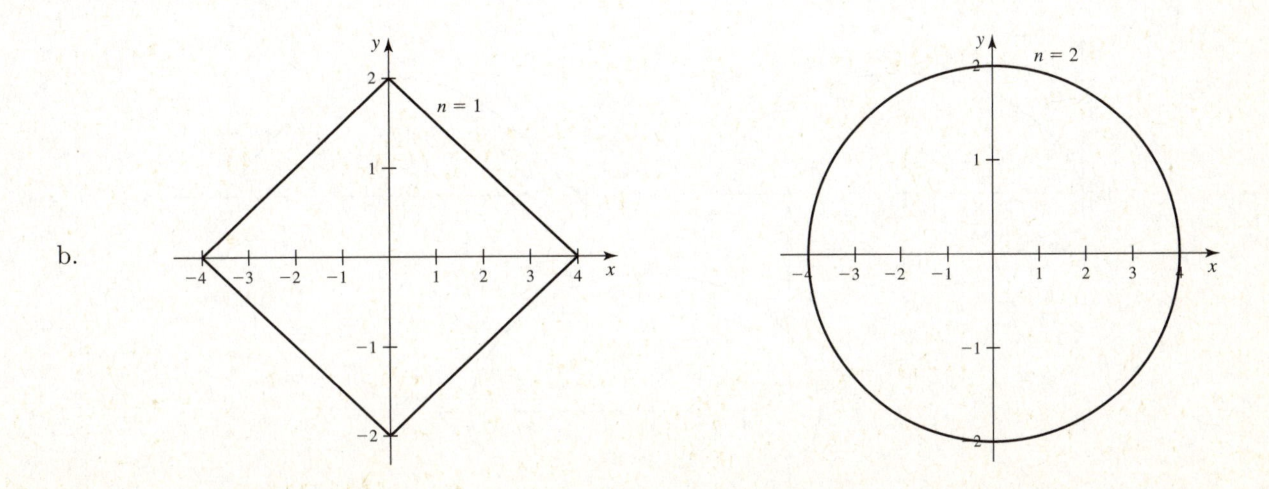

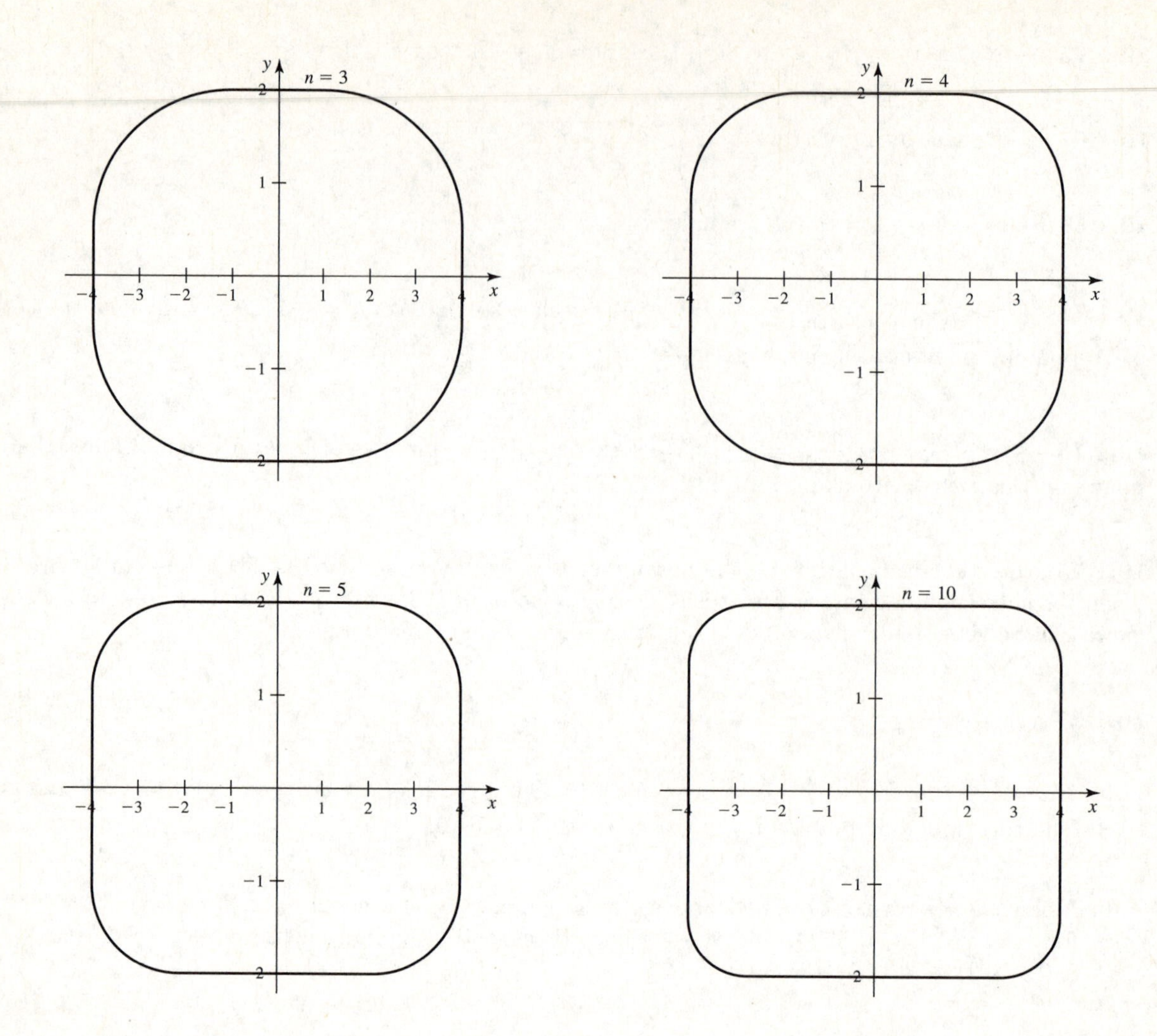

c. As n increases from near 0 to near 1, the curves change from star-shaped to a rectangular shape with corners at $(\pm a, 0)$ and $(0, \pm b)$. As n increases from 1 on, the curves become more rectangular with corners at $(\pm a, \pm b)$.

10.1.81 The first graphic shown is for $a = 1$ and $b = 1$. The second is for $a = 2$, $b = 1$, and the third is for $a = 1$, $b = 2$.

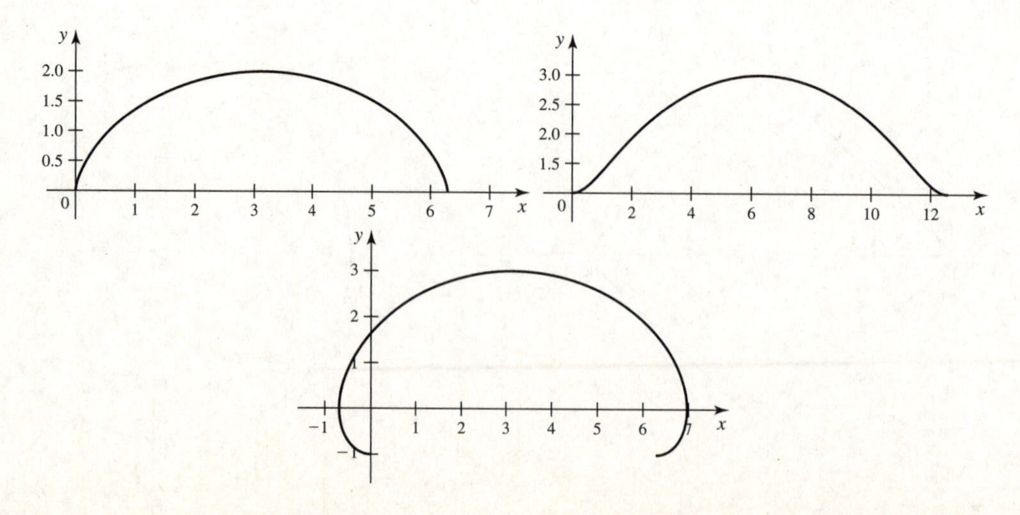

10.1.83

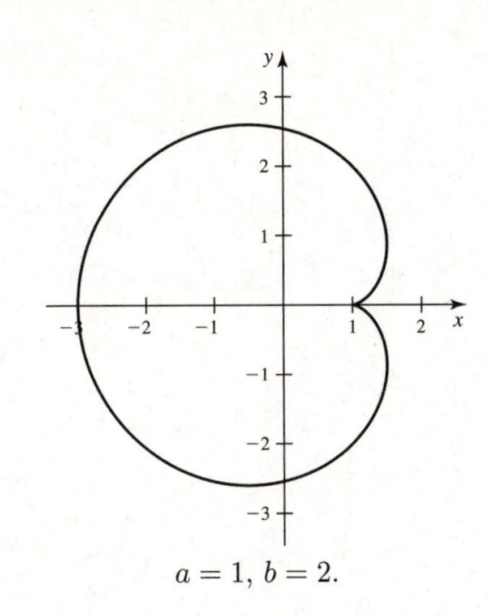

$a = 1$, $b = 2$.

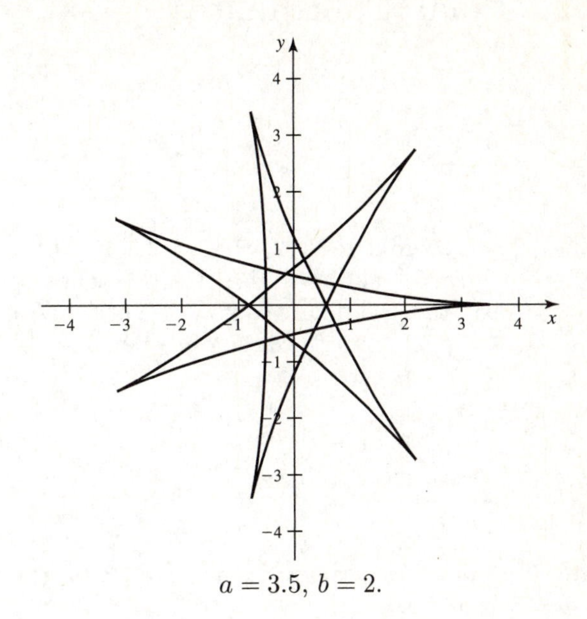

$a = 3.5$, $b = 2$.

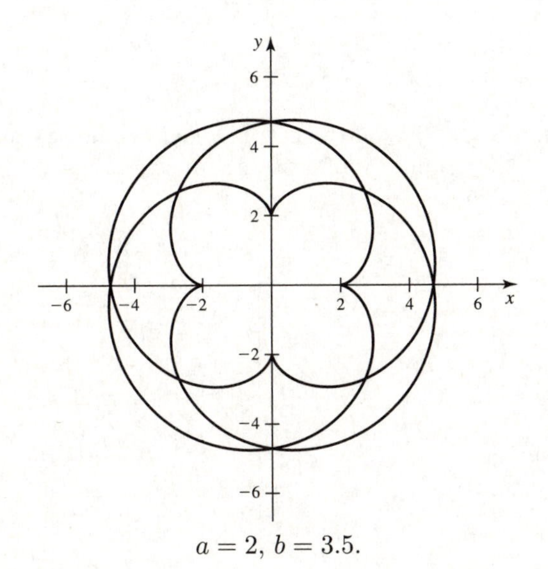

$a = 2$, $b = 3.5$.

Note that for $a < b$, we have cusps pointing inward, while for $a > b$, the cusps point outward.

10.1.85 The package lands when $y = 0$, so when $-4.9t^2 + 4000 = 0$ for $t > 0$. This occurs when $t = \sqrt{\frac{4000}{4.9}} \approx$ 28.57 seconds. At that time, $x \approx 100 \cdot 28.57 = 2857$ meters.

10.1.87 Let $x = 1 + \cos^2 t - \sin^2 t$ and $y = t$, for $-\infty < t < \infty$. Note that since $1 - \sin^2 t = \cos^2 t$, we have $x = 2\cos^2 t$, $y = t$.

10.1.89 Note that $a\cos t + b\sin t = \sqrt{a^2 + b^2}\sin(t + \alpha)$ where $\alpha = \tan^{-1}(a/b)$. This follows because $\sin(t + \tan^{-1}(a/b)) = \sin t\cos(\tan^{-1}(a/b)) + \cos t\sin(\tan^{-1}(a/b)) = \sin t \cdot \frac{b}{\sqrt{a^2+b^2}} + \cos t \cdot \frac{a}{\sqrt{a^2+b^2}}$.

Similarly, $c\cos t + d\sin t = \sqrt{c^2 + d^2} \cdot \cos(t + \beta)$, where $\beta = \tan^{-1}(-d/c)$.

Thus $x^2 + y^2 = (a\cos t + b\sin t)^2 + (c\cos t + d\sin t)^2 = (a^2 + b^2)\sin^2(t + \alpha) + (c^2 + d^2)\cos^2(t + \beta)$. This represents a circle if $a^2 + b^2 = c^2 + d^2$, and if $\alpha = \beta$, which means that $\frac{a}{b} = \frac{-d}{c}$, or $ac + bd = 0$.

10.2 Polar Coordinates

10.2.1 The coordinates $(2, \pi/6)$, $(2, -11\pi/6)$, and $(-2, 7\pi/6)$ all give rise to the same point. Also, the coordinates $(-3, -\pi/2)$, $(3, \pi/2)$ and $(-3, 3\pi/2)$ give rise to the same point.

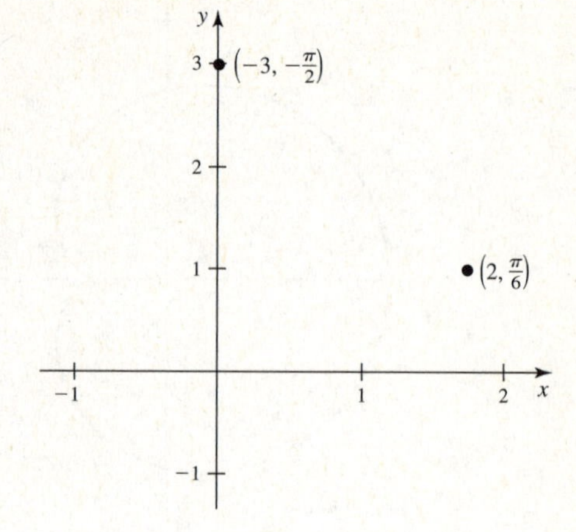

10.2.3 If a point has cartesian coordinates (x, y) then $r^2 = x^2 + y^2$ and $\tan\theta = y/x$ for $x \neq 0$. If $x = 0$, then $\theta = \pi/2$ and $r = y$.

10.2.5 Since $x = r\cos\theta$, we have that the vertical line $x = 5$ has polar equation $r = \frac{5}{\cos\theta}$.

10.2.7 x-axis symmetry occurs if (r, θ) on the graph implies $(r, -\theta)$ is on the graph. y-axis symmetry occurs if (r, θ) on the graph implies $(r, \pi - \theta) = (-r, -\theta)$ is on the graph. Symmetry about the origin occurs if (r, θ). on the graph implies $(-r, \theta) = (r, \theta + \pi)$ is on the graph.

10.2.9 The coordinates $(2, \pi/4)$, $(-2, 5\pi/4)$, and $(2, 9\pi/4)$ represent the same point.

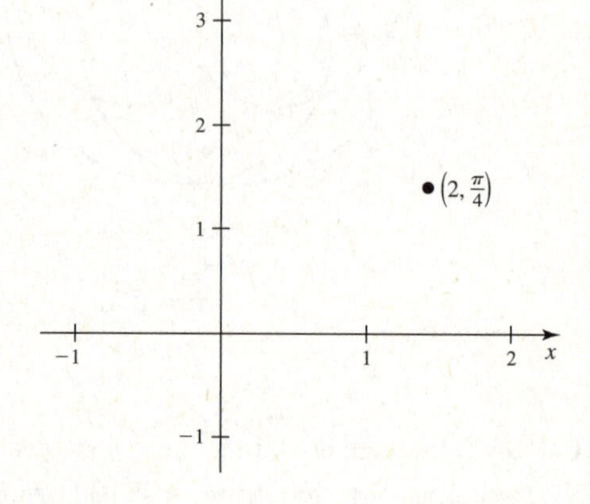

10.2.11 The coordinates $(-1, -\pi/3)$, $(1, 2\pi/3)$ and $(1, -4\pi/3)$ represent the same point.

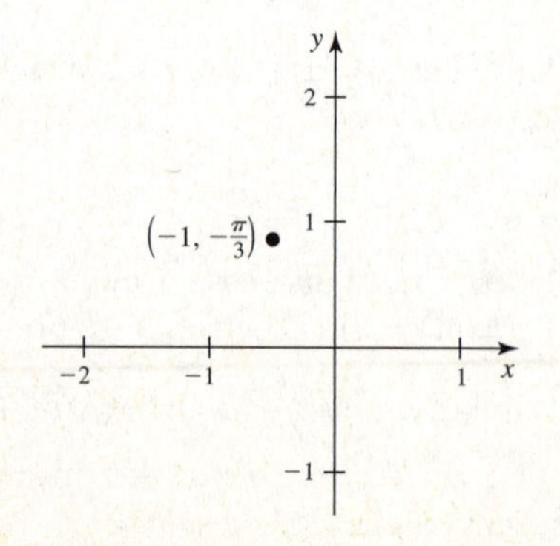

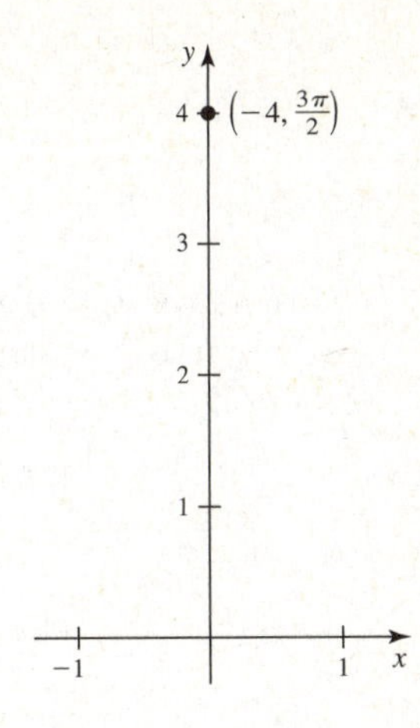

10.2.13 The coordinates $(-4, 3\pi/2)$, $(4, \pi/2)$ and $(-4, -\pi/2)$ represent the same point.

10.2.15 $x = 3\cos(\pi/4) = \frac{3\sqrt{2}}{2}$. $y = 3\sin(\pi/4) = \frac{3\sqrt{2}}{2}$.

10.2.17 $x = \cos(-\pi/3) = \frac{1}{2}$. $y = \sin(-\pi/3) = \frac{-\sqrt{3}}{2}$.

10.2.19 $x = -4\cos(3\pi/4) = 2\sqrt{2}$. $y = -4\sin(3\pi/4) = -2\sqrt{2}$.

10.2.21 $r^2 = x^2 + y^2 = 4 + 4 = 8$, so $r = \sqrt{8}$. $\tan\theta = 1$, so $\theta = \pi/4$, so $(2\sqrt{2}, \pi/4)$ is one representation of this point, and $(-2\sqrt{2}, -3\pi/4)$ is another.

10.2.23 $r^2 = x^2 + y^2 = 1 + 3 = 4$, so $r = \pm 2$. $\tan\theta = \sqrt{3}$, so $\theta = \pi/3$, $4\pi/3$. $(2, \pi/3)$ is one representation of this point, and $(-2, -2\pi/3)$ is another.

10.2.25 $r^2 = 64$, so $r = \pm 8$. $\tan\theta = -\sqrt{3}$, so $\theta = -\pi/3$, $2\pi/3$. One representation of the given point is $(8, 2\pi/3)$, and $(-8, -\pi/3)$ is another.

10.2.27

θ	0	$\pi/6$	$\pi/4$	$\pi/3$	$\pi/2$	$2\pi/3$	$3\pi/4$	$5\pi/6$	π
r	8	$4\sqrt{3}$	$4\sqrt{2}$	4	0	-4	$-4\sqrt{2}$	$-4\sqrt{3}$	-8

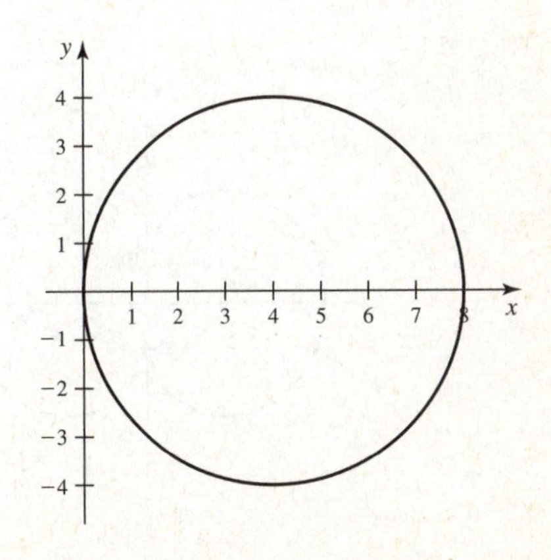

10.2.29 $r(\sin\theta - 2\cos\theta) = 0$ when $r = 0$ or when $\tan\theta = 2$, so the curve is a straight line through the origin of slope 2.

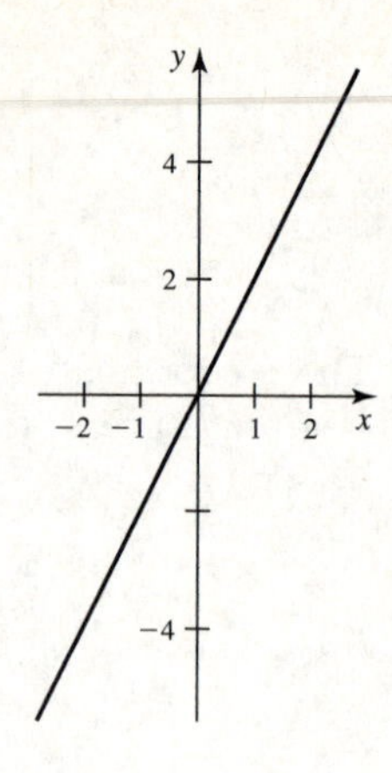

10.2.31 $x = r\cos\theta = -4$, so this is a vertical line $x = -4$ through $(-4, 0)$.

10.2.33 $r\cos\theta = \sin 2\theta = 2\sin\theta\cos\theta$. Note that if $\cos\theta = 0$, then r can be any real number, and the equation is satisfied. For $\cos\theta \neq 0$, we have $x = r\cos\theta = 2\sin\theta\cos\theta$, so $r = 2\sin\theta$, and thus $y = r\sin\theta = 2\sin^2\theta$. Thus $x^2 + y^2 - 2y = 4\sin^2\theta\cos^2\theta + 4\sin^2\theta\sin^2\theta - 4\sin^2\theta = 4\sin^2\theta(\sin^2\theta + \cos^2\theta) - 4\sin^2\theta = 4\sin^2\theta - 4\sin^2\theta = 0$. Note also that $x^2 + y^2 - 2y = 0$ is equivalent to $x^2 + (y-1)^2 = 1$, so we have a circle of radius one centered at $(0, 1)$, as well as the line $x = 0$ which is the y-axis.

10.2.35 $r = 8\sin\theta$, so $r^2 = 8r\sin\theta$, so $x^2 + y^2 = 8y$. This can be written $x^2 + (y-4)^2 = 16$, which represents a circle of radius 4 centered at $(0, 4)$.

10.2.37

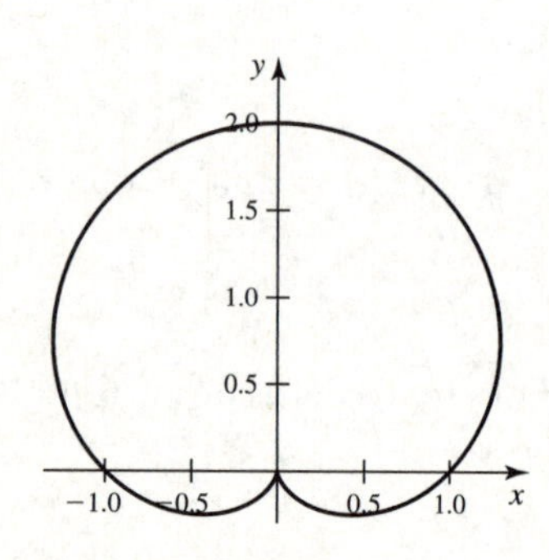

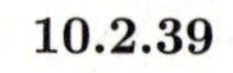
10.2.39

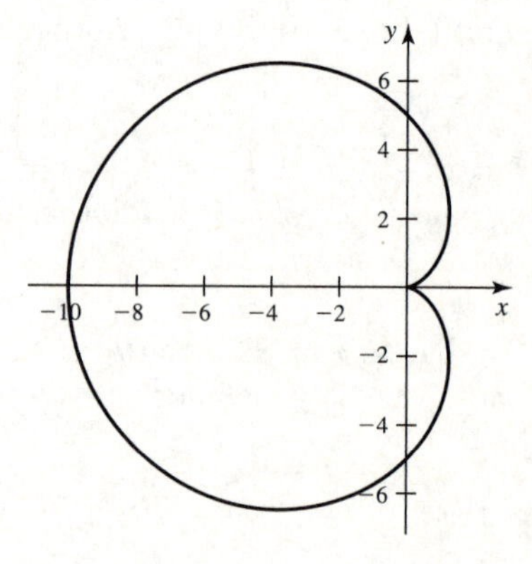

10.2.41

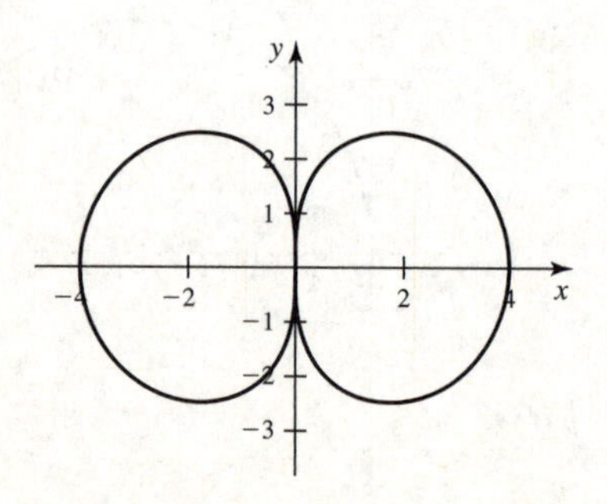

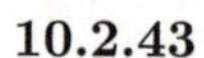
10.2.43

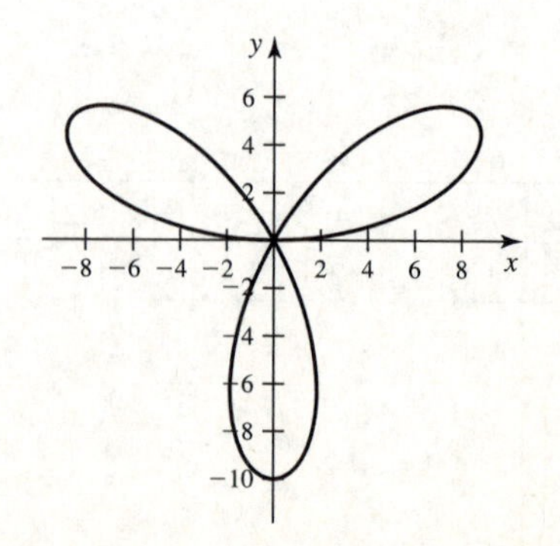

10.2.45 Points B, D, F, H, J, and L have y-coordinate 0, so the graph is at the pole for each of these points. Points E, I, and M have maximal radius, so these correspond to the points at the tips of the outer loops. The points C, G, and K correspond to the tips of the smaller loops. Point A corresponds to the polar point $(1, 0)$.

10.2.47 Points B, D, F, H, J, L, N, and P are at the origin. C, G, K, and O are on the ends of the long loops, while A, E, I, and M are at the ends of the smaller loops.

10.2.49 Since $r = \theta \sin\theta > M$ for any real number M (for suitably large θ), no finite interval can generate the entire graph.

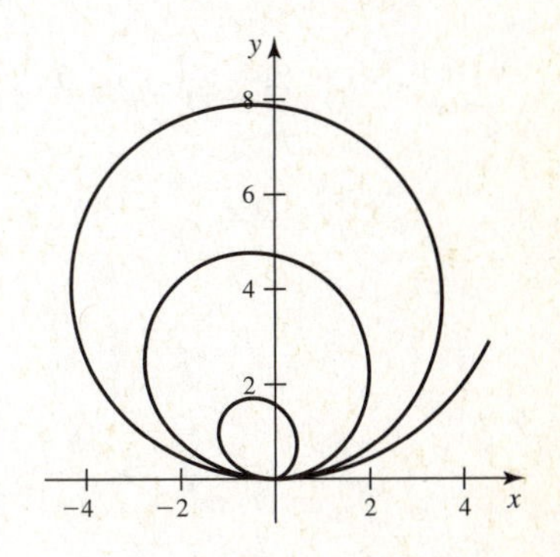

10.2.51 The interval $[0, 2\pi]$ generates the entire graph.

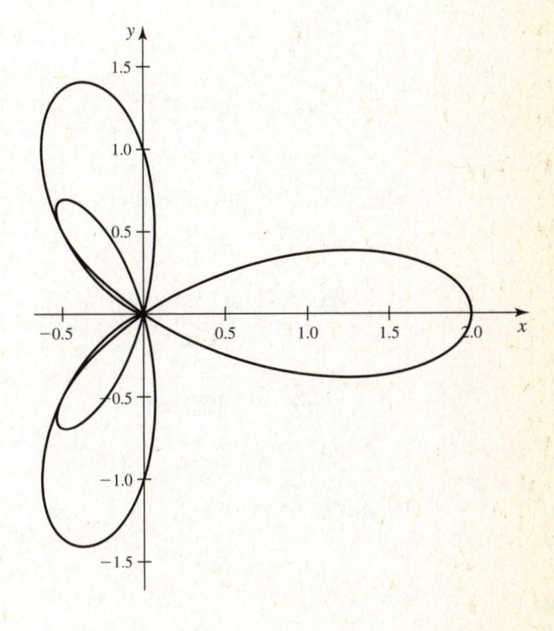

10.2.53 The interval $[0, 5\pi]$ generates the entire graph.

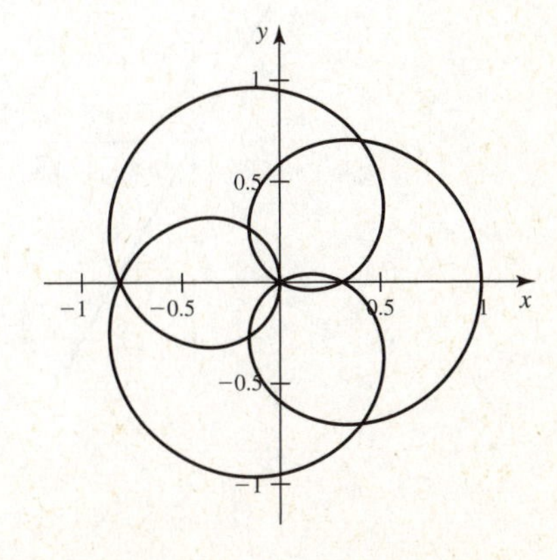

10.2.55 The interval $[0, 2\pi]$ generates the entire graph.

10.2.57

a. True. Note that $r^2 = 8$ and $\tan\theta = -1$.

b. True. Their intersection point (in Cartesian coordinates) is $(4, -2)$.

c. False. They intersect at the polar coordinates $(2, \pi/4)$ and $(2, 5\pi/4)$.

d. False. $3\cos(\pi) \neq 3$.

10.2.59

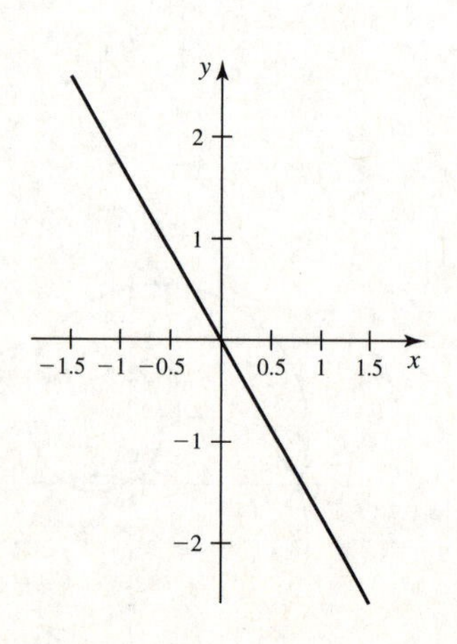

10.2.61

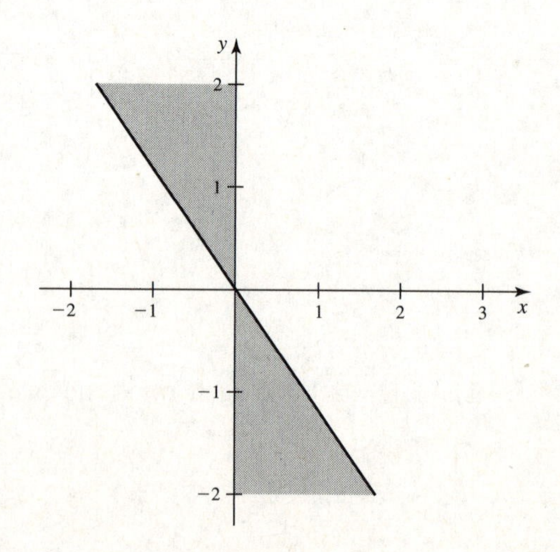

10.2.63

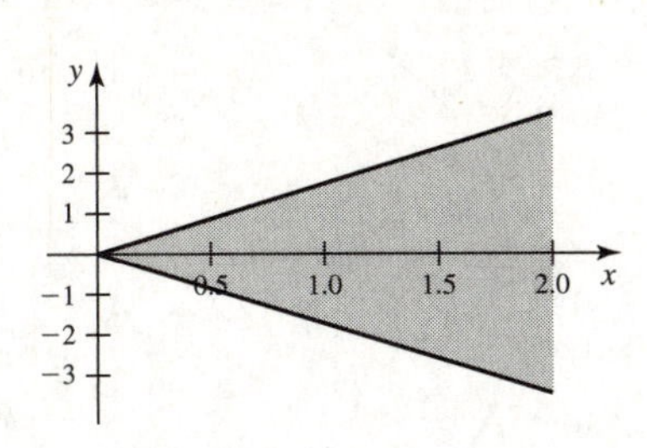

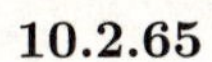
10.2.65

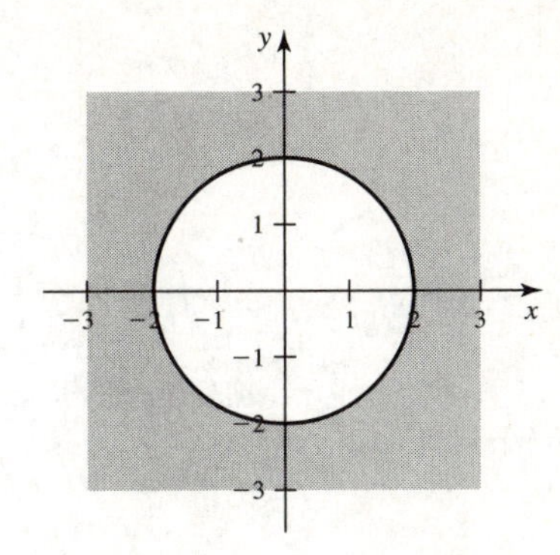

10.2.67 Consider the circle with center $C(r_0, \theta_0)$, and let A be the origin and $B(r, \theta)$ be a point on the circle not collinear with A and C. Note that the length of side BC is R, and that the angle CAB has measure $\theta - \theta_0$. Applying the law of cosines to triangle CAB yields the equation $R^2 = r^2 + r_0^2 - 2rr_0 \cos(\theta - \theta_0)$, which is equivalent to the given equation.

10.2.69 In relation to number 67, we have $r_0 = 2$ and $\theta_0 = \pi/3$, and $R^2 - 4 = 12$, so $R^2 = 16$. Thus this is a circle with polar center $(2, \pi/3)$ and radius 4.

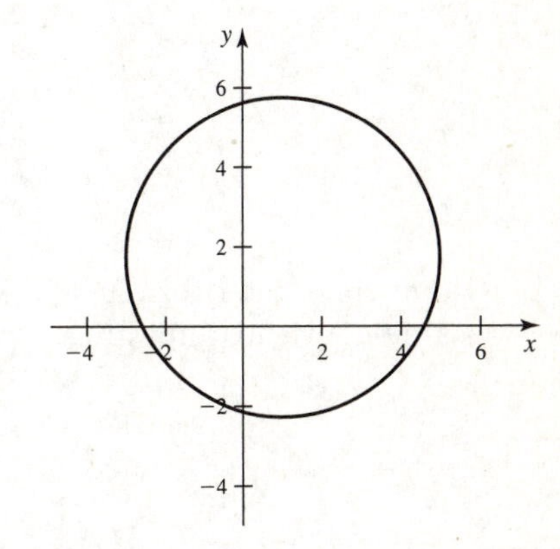

10.2.71 In relation to number 66, we have $a = 2$ and $b = 3$. So $R^2 - a^2 - b^2 = R^2 - 13 = 3$, and thus $R^2 = 16$. Thus we have a circle centered at $(2, 3)$ with radius 4.

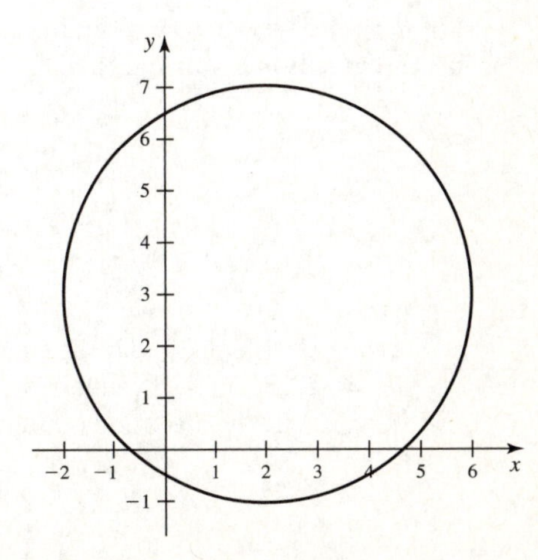

10.2.73 In relation to number 66, we have $a = -1$ and $b = 2$. So $R^2 - a^2 - b^2 = R^2 - 5 = 4$, and thus $R^2 = 9$. Thus we have a circle centered at $(-1, 2)$ with radius 3.

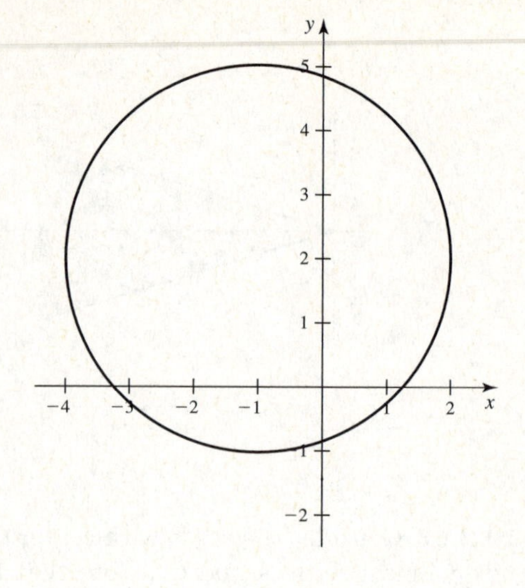

10.2.75

a. On all three intervals, the graph is the same vertical line, oriented upward.

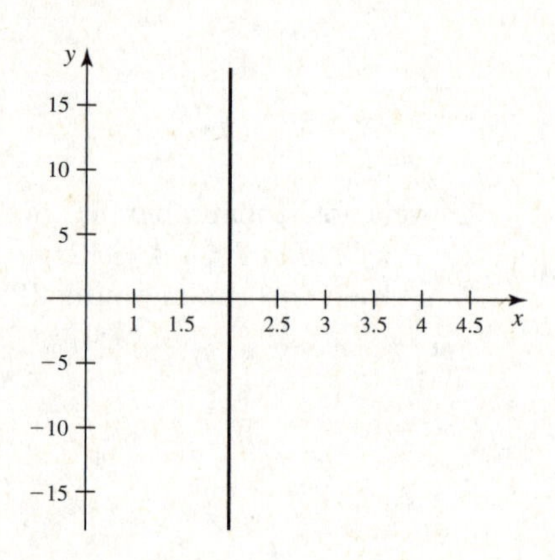

b. For $\theta \neq \frac{2m+1}{2}\pi$ where m is an integer, we have $\cos\theta \neq 0$, so the equation is equivalent to $x = r\cos\theta = 2$. So the graph is a vertical line.

10.2.77 Using problem 76b, this is the line with $r_0 = 3$ and $\theta_0 = \frac{\pi}{3}$. So it is the line through the polar point $(3, \pi/3)$ in the direction of angle $\pi/3 + \pi/2 = 5\pi/6$.

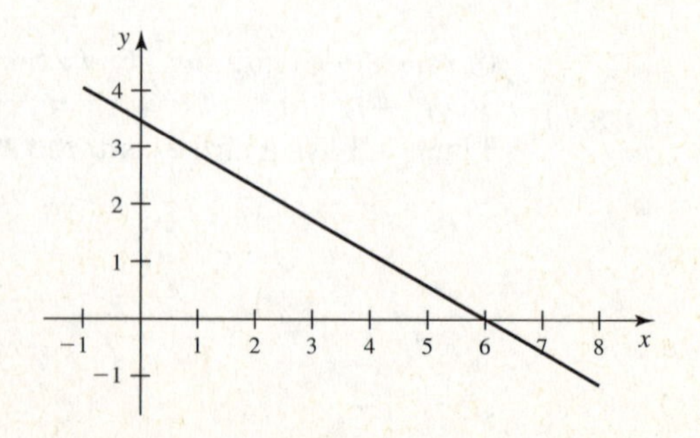

10.2.79 Using problem 76a, this is the line with $b = 3$ and $m = 4$, so $y = 4x + 3$.

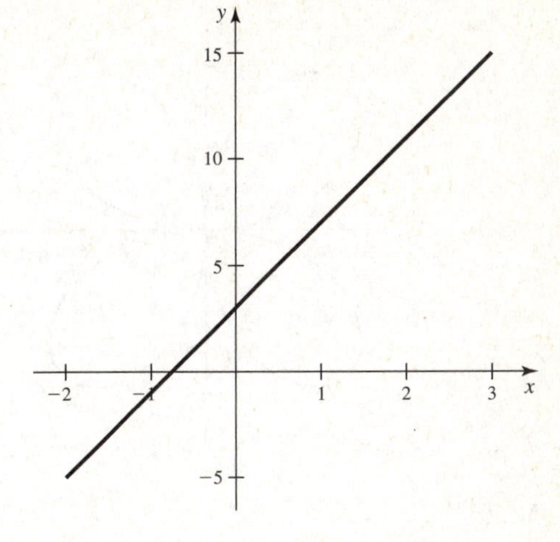

10.2.81

a. This matches (A), since we have $|a| = 1 = |b|$, and the graph is a cardiod.

b. This matches (C). This has an inner loop since $|a| = 1 < 2 = |b|$. Note that $r = 1$ when $\theta = 0$, so it can't be (D).

c. This matches (B). This has $|a| = 2 > 1 = |b|$, so it has an oval-like shape.

d. This matches (D). This has an inner loop since $|a| = 1 < 2 = |b|$. Note that $r = -1$ when $\theta = 0$, so this can't be (C).

e. This matches (E). Note that there is an inner loop since $|a| = 1 < 2 = |b|$, and that $r = 3$ when $\theta = \pi/2$.

f. This matches (F).

10.2.83

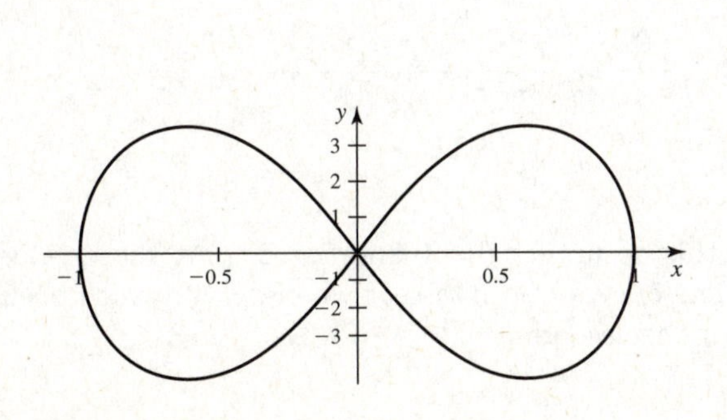

10.2.85

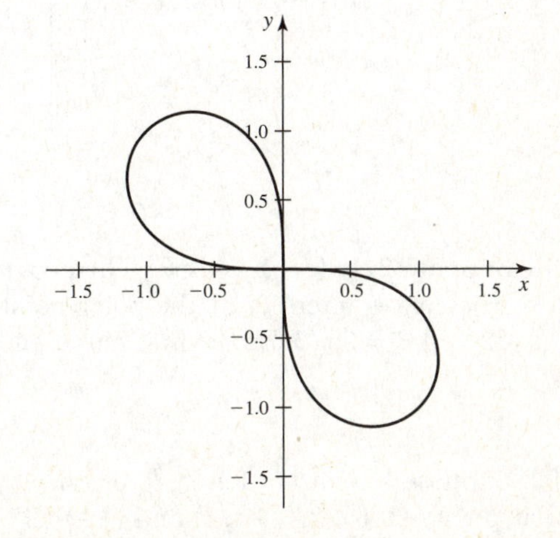

10.2.87

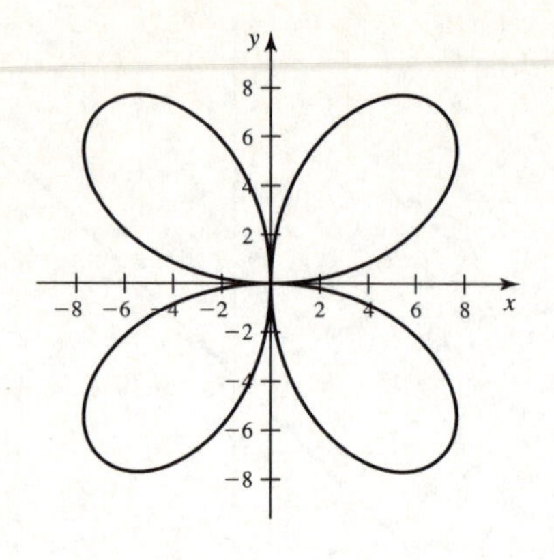

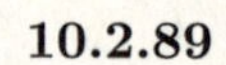

10.2.89

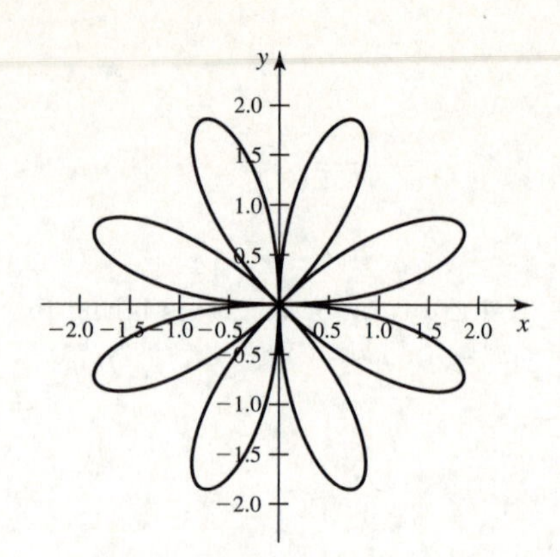

10.2.91 Note that $a \sin m\theta = 0$ for $\theta = \frac{k\pi}{m}$, $k = 1, 2, \ldots, 2m$. Thus the graph is back at the pole $r = 0$ for each of these values, and each of these gives rise to a distinct petal of the rose if m is odd. If m is even, then by symmetry, each petal for $k = 1, 2, \ldots \frac{m}{2}$ is equivalent to one for $k = \frac{m}{2} + 1, \frac{m}{2} + 2, \ldots, m$. (Note that this follows because the sine function is odd.) A similar result holds for the rose $r = a \cos \theta$.

10.2.93 For $a = 1$, the spiral winds outward counterclockwise. For $a = -1$, the spiral winds inward counterclockwise.

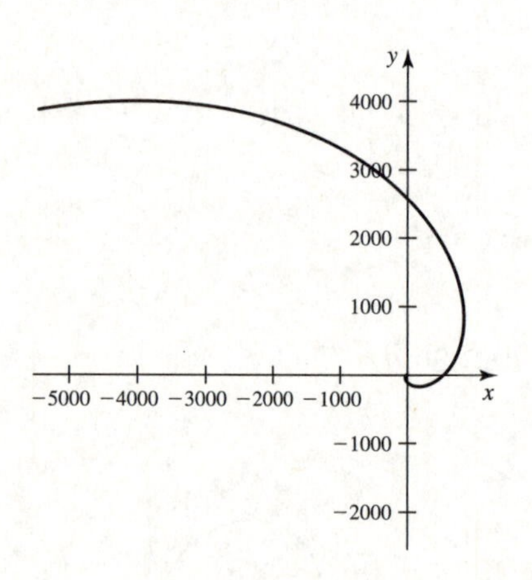

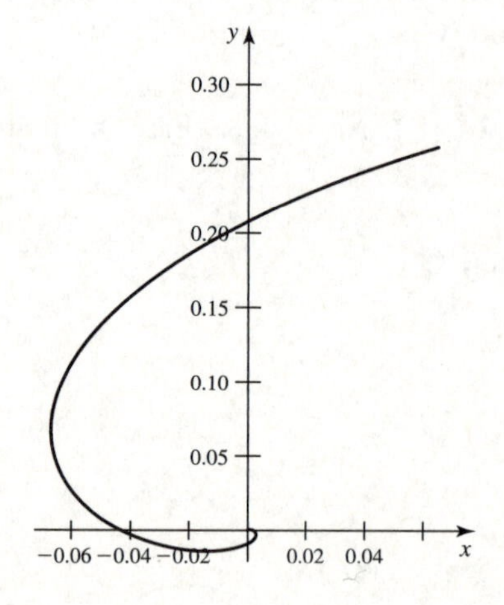

10.2.95 Suppose $2\cos\theta = 1 + \cos\theta$. Then $\cos\theta = 1$, so this occurs for $\theta = 0$ and $\theta = 2\pi$. At those values, $r = 2$, so the curves intersect at the polar point $(2, 0)$. The curves also intersect when $r = 0$, which occurs for $\theta = \pi/2$ and $\theta = 3\pi/2$ for the first curve and $\theta = \pi$ for the second.

10.2.97 Suppose $1 - \sin\theta = 1 + \cos\theta$, or $\tan\theta = -1$. Then $\theta = 3\pi/4$ or $\theta = 7\pi/4$. So the curves intersect at the polar points $(1 + \sqrt{2}/2, 7\pi/4)$ and $(1 - \sqrt{2}/2, 3\pi/4)$. They also intersect at the pole $(0, 0)$, which occurs for the first curve at $\pi/2$ and for the second curve at π.

10.2.99

a. 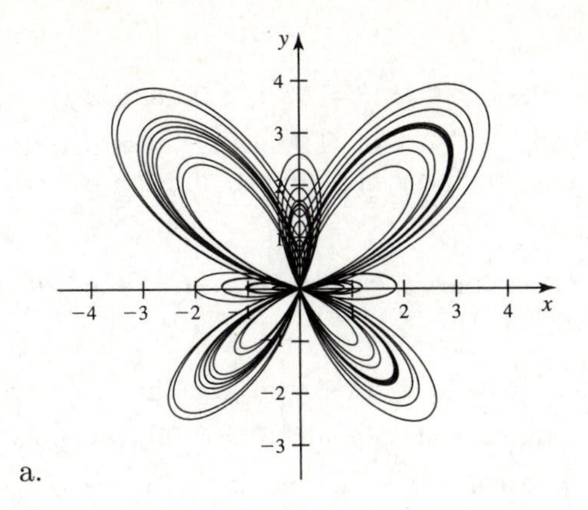

b. It adds multiple layers of the same type of curve as $\sin^5 \theta/12$ oscillates between -1 and 1 for $0 \le \theta \le 24\pi$.

10.2.101

a. 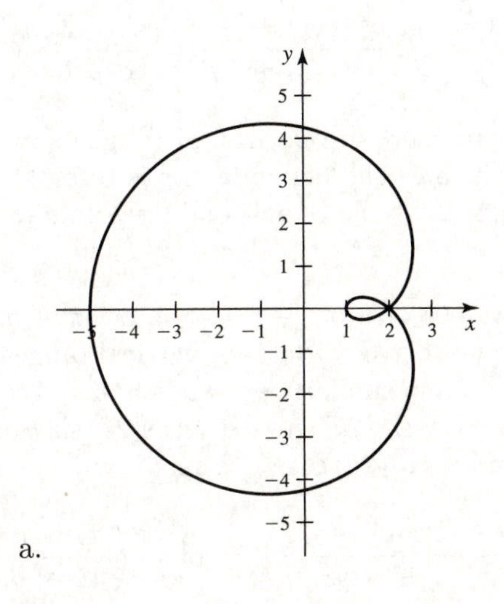

b. $r = 3 - 4\cos \pi t$ is a limaçon, and $x - 2 = r\cos \pi t$ and $y = r \sin \pi t$ is a circle, and the composition of a limaçon and a circle is a limaçon.

10.2.103 With $r = a\cos\theta + b\sin\theta$, we have $r^2 = ar\cos\theta + br\sin\theta$, or $x^2 + y^2 = ax + by$, so $\left(x - \frac{a}{2}\right)^2 + \left(y - \frac{b}{2}\right)^2 = \frac{a^2+b^2}{4}$. Thus, the center is $(a/2, b/2)$ and $r = \frac{\sqrt{a^2+b^2}}{2}$.

10.2.105 Since $\sin(\theta/2) = \sin(\pi - \theta/2) = \sin((2\pi - \theta)/2)$, we have that the graph is symmetric with respect to the x-axis.

10.3 Calculus in Polar Coordinates

10.3.1 Since $x = r\cos\theta$ and $y = r\sin\theta$, we have $x = f(\theta)\cos\theta$ and $y = f(\theta)\sin(\theta)$.

10.3.3 Since slope is given relative to the horizontal and vertical coordinates, it is given by $\frac{dy}{dx}$, not by $\frac{dr}{d\theta}$.

10.3.5 $\frac{dy}{dx} = \frac{-\cos\theta\sin\theta + (1-\sin\theta)\cos\theta}{-\cos^2\theta - (1-\sin\theta)\sin\theta}$. At $(1/2, \pi/6)$, we have $\frac{dy}{dx} = \frac{0}{-1} = 0$. The given curve intersects the origin $r = 0$ for $\theta = \pi/2$. At this point, $\frac{dy}{dx}$ does not exist, and the tangent line is vertical. (It is the line $\theta = \pi/2$.)

10.3.7 $\frac{dy}{dx} = \frac{16\cos\theta\sin\theta}{-8\sin^2\theta + 8\cos^2\theta}$. At $(4, 5\pi/6)$ we have $\frac{dy}{dx} = \frac{-4\sqrt{3}}{4} = -\sqrt{3}$. The given curve intersects the origin $r = 0$ for $\theta = 0$ and $\theta = \pi$. At these points, the derivative is 0, and the tangent line is horizontal, so $\theta = 0$ is the tangent line.

10.3.9 $\frac{dy}{dx} = \frac{-3\sin^2\theta+(6+3\cos\theta)\cos\theta}{-3\cos\theta\sin\theta-(6+3\cos\theta)\sin\theta}$. At both $(3,\pi)$ and $(9,0)$, this doesn't exist. The given curve does not intersect the origin, since $r \geq 3$ for all θ.

10.3.11 $\frac{dy}{dx} = \frac{-8\sin(2\theta)\sin\theta+4\cos(2\theta)\cos\theta}{-8\sin(2\theta)\cos\theta-4\cos(2\theta)\sin\theta}$. The tips of the leaves occur at $\theta = 0$, $\pi/2$, π and $3\pi/2$. At 0 and at π, we have that $\frac{dy}{dx}$ doesn't exist. At $\pi/2$ and $3\pi/2$ we have $\frac{dy}{dx} = 0$. The graph intersects the origin for $\theta = \pi/4$, $\theta = 3\pi/4$, $\theta = 5\pi/4$ and $\theta = 7\pi/4$, and thus the two distinct tangent lines are $\theta = \pi/4$ and $\theta = 3\pi/4$.

10.3.13 The curve hits the origin at $\pm\pi/4$, where the tangent lines are given by $\theta = \pi/4$ and $\theta = -\pi/4$. The slopes of those lines are given by $\tan(\pi/4) = 1$ and $\tan(-\pi/4) = -1$.

10.3.15 Note that the curve is at the origin at $\pi/2$, so there is vertical tangent there. Also, $\frac{dy}{dx} = \frac{-4\sin^2\theta+4\cos^2\theta}{-8\sin\theta\cos\theta} = \frac{1-2\sin^2\theta}{\sin(2\theta)}$. Thus, there are horizontal tangents at $\pi/4$ and $3\pi/4$ (at the polar points $(2\sqrt{2},\pi/4)$ and $(-2\sqrt{2},3\pi/4)$. There is a vertical tangent where $\theta = 0$, at the point $(4,0)$.

10.3.17 Using the double angle identities somewhat liberally:

$$\frac{dy}{dx} = \frac{2\cos(2\theta)\sin\theta + \sin(2\theta)\cos\theta}{2\cos(2\theta)\cos\theta - \sin(2\theta)\sin\theta} = \frac{\sin\theta(\cos 2\theta + \cos^2\theta)}{\cos\theta(\cos(2\theta) - \sin^2\theta)} = \frac{\sin\theta(3\cos^2\theta - 1)}{\cos\theta(1 - 3\sin^2\theta)} = \frac{\sin\theta(3\cos^2\theta - 1)}{\cos\theta(3\cos^2\theta - 2)}.$$

The numerator is 0 for $\theta = 0$ and for $\theta = \pm\cos^{-1}(\pm\sqrt{3}/3)$, so there are horizontal tangents at the corresponding points (all of which have either $r \approx .943$ or $r \approx -.943$). The denominator is 0 for $\theta = \pi/2$ and $3\pi/2$, and for $\theta = \pm\cos^{-1}(\pm\sqrt{6}/3)$, so there are vertical tangents at the corresponding points (all of which have either $r \approx .943$ or $r \approx -.943$.)

10.3.19 The whole curve can be generated by considering values of θ on $[-\pi/4,\pi/4] \cup [3\pi/4,5\pi/4]$. The curve intersects the origin for the endpoints of the domain, so neither horizontal or vertical tangents exist at the origin. Let $x = r\cos\theta$, so that $x^2 = r^2\cos^2\theta = 4\cos 2\theta\cos^2\theta$ and similarly $y^2 = r^2\sin^2\theta = 4\cos 2\theta\sin^2\theta$. Then $2xx' = -8\sin(2\theta)\cos^2\theta - 8\cos(2\theta)\sin\theta\cos\theta$, and $2yy' = -8\sin(2\theta)\sin^2\theta + 8\cos(2\theta)\sin\theta\cos\theta$. Thus $x' = \frac{4}{r}(-\cos(2\theta)\sin\theta - \sin(2\theta)\cos\theta)$ and $y' = \frac{4}{r}(-\sin(2\theta)\sin\theta + \cos(2\theta)\cos\theta)$. Thus

$$\frac{dy}{dx} = \frac{y'}{x'} = \frac{\sin(2\theta)\sin\theta - \cos(2\theta)\cos\theta}{\cos(2\theta)\sin\theta + \sin(2\theta)\cos\theta} = \frac{\cos\theta}{\sin\theta}\cdot\left(\frac{2\sin^2\theta - (2\cos^2\theta - 1)}{2\cos^2\theta - 1 + 2\cos^2\theta}\right) = \cot\theta\cdot\left(\frac{4\sin^2\theta - 1}{4\cos^2\theta - 1}\right),$$

where we have again made liberal use of the double angle trig identities. We have horizontal tangents on the domain at $\theta = \pm\pi/6$ and $5\pi/6$ and $7\pi/6$. There are vertical tangents on the domain only at $\theta = 0$ and $\theta = \pi$.

10.3.21 $A = \frac{1}{2}\int_0^\pi (8\sin\theta)^2\,d\theta = 32\int_0^\pi \sin^2\theta\,d\theta = 32\int_0^\pi \frac{1-\cos 2\theta}{2}\,d\theta = 32\left(\frac{1}{2}\theta - \frac{\sin\theta\cos\theta}{2}\right)\Big|_0^\pi = 16\pi.$

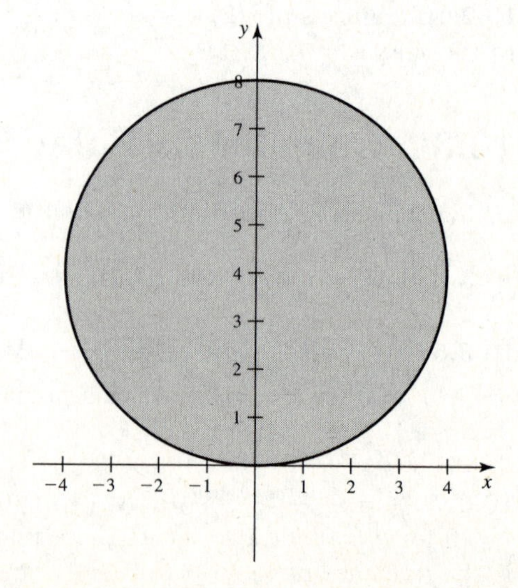

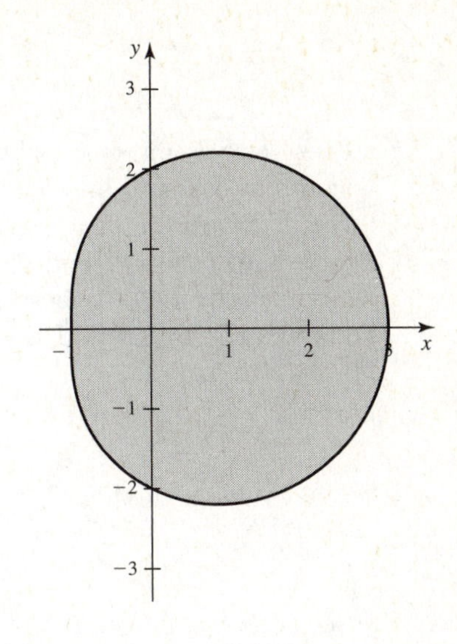

10.3.23 Using symmetry, we have $\frac{1}{2} \cdot 2 \int_0^{\pi} (2+\cos\theta)^2 \, d\theta = \int_0^{\pi} (4 + 4\cos\theta + \cos^2\theta) \, d\theta = \left(4\theta + 4\sin\theta + \frac{1}{2}\theta + \frac{\sin\theta\cos\theta}{2}\right)\Big|_0^{\pi} = \frac{9\pi}{2}$.

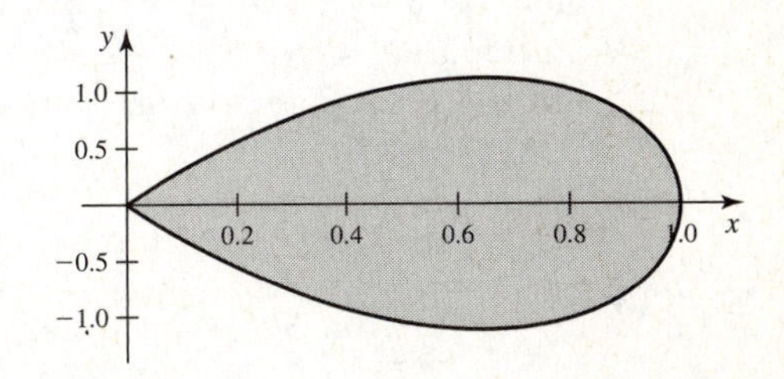

10.3.25 Using symmetry, we compute the area of 1/2 of one leaf, and then double it. We have $A = \frac{1}{2}\int_0^{\pi/10} \cos^2(5\theta) \, d\theta = \frac{1}{10}\int_0^{\pi/2} \cos^2 u \, du = \frac{1}{10}\left(\frac{1}{2}u + \frac{\cos u \sin u}{2}\right)\Big|_0^{\pi/2} = \frac{\pi}{40}$. So the area of one leaf is $2 \cdot \frac{\pi}{40} = \frac{\pi}{20}$.

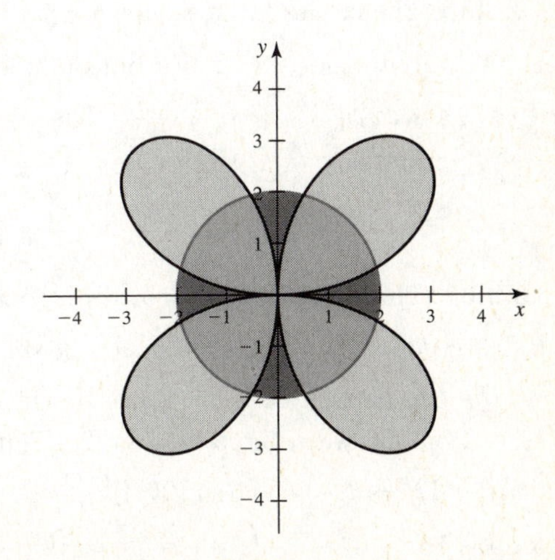

10.3.27 Note that the area inside one leaf of the rose but outside the circle is given by $\frac{1}{2}\int_{\pi/12}^{5\pi/12} (16\sin^2(2\theta) - 4) \, d\theta = (2\theta - \sin(4\theta))\big|_{\pi/12}^{5\pi/2} = \sqrt{3} + \frac{2\pi}{3}$. Also, the area inside one leaf of the rose is $\frac{1}{2}\int_0^{\pi/2} 16\sin^2(2\theta) \, d\theta = \left(4\theta - \sin(4\theta)\big|_0^{\pi/2}\right) = 2\pi$. Thus the area inside one leaf of the rose and inside the circle must be $2\pi - (\sqrt{3} + \frac{2\pi}{3}) = \frac{4\pi}{3} - \sqrt{3}$, and the total area inside the rose and inside the circle must be $4(\frac{4\pi}{3} - \sqrt{3}) = \frac{16\pi}{3} - 4\sqrt{3}$.

10.3.29 These curves intersect when $\sin\theta = \cos\theta$, which occurs at $\theta = \pi/4$ and $\theta = 5\pi/4$, and when $r = 0$ which occurs for $\theta = 0$ and $\theta = \pi$ for the first curve and $\theta = \pi/2$ and $\theta = 3\pi/2$ for the second curve. Only two of these intersection points are unique: the origin and the point $(3\sqrt{2}/2, \pi/4) = (-3\sqrt{2}/2, 5\pi/4)$.

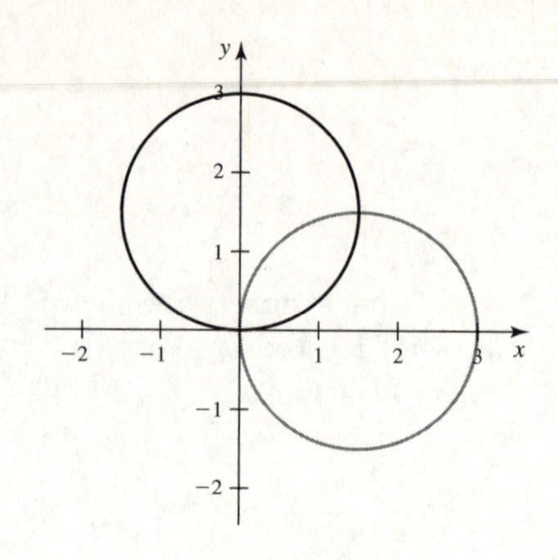

10.3.31 Suppose $4\cos\theta = 1 + 2\cos\theta + \cos^2\theta$. Then $(\cos\theta - 1)^2 = 0$, so $\theta = 0$. At that value, $r = 2$, so the curves intersect at the polar point $(2, 0)$. The curves also intersect when $r = 0$, which occurs for the first curve at $\pi/2$ and $3\pi/2$, and for the second curve at π. Also, the curves intersect when $4\cos\theta = -1 - 2\cos\theta - \cos^2\theta$, which occurs for $\cos^2\theta + 6\cos\theta + 1 = 0$, or (using the quadratic formula) $\theta = \cos^{-1}(-3+2\sqrt{2}) \approx 1.743$. This leads to the polar intersection points at approximately $(.8284, \pm 1.743)$.

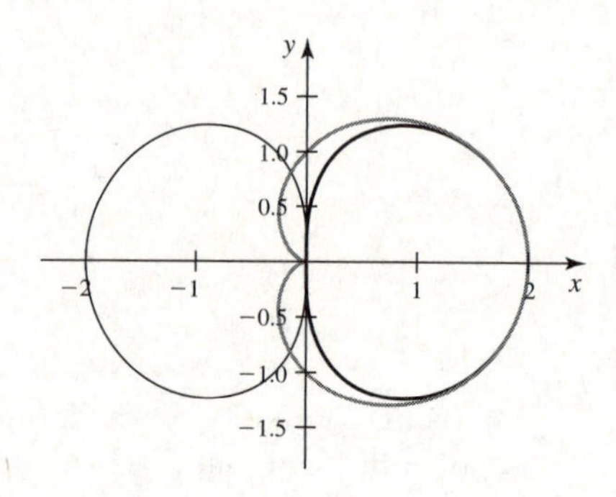

10.3.33

a. False. The area is given by $\frac{1}{2}\int_\alpha^\beta f(\theta)^2\, d\theta$.

b. False. The slope is given by $\frac{dy}{dx}$, which can be computed using the formula $\frac{dy}{dx} = \frac{dy/d\theta}{dx/d\theta} = \frac{f'(\theta)\sin\theta + f(\theta)\cos\theta}{f'(\theta)\cos\theta - f(\theta)\sin\theta}$.

10.3.35 The circles intersect for $\theta = \pi/6$ and $\theta = 5\pi/6$.

The area inside $r = 2\sin\theta$ but outside of $r = 1$ would be given by $\frac{1}{2}\int_{\pi/6}^{5\pi/6}(4\sin^2\theta - 1)\, d\theta = \frac{1}{2}\left(x - \sin(2x)\right)\Big|_{\pi/6}^{5\pi/6} = \frac{\pi}{3} + \frac{\sqrt{3}}{2}$. The total area of $r = 2\sin\theta$ is π. Thus, the area inside both circles is $\pi - \left(\frac{\pi}{3} + \frac{\sqrt{3}}{2}\right) = \frac{2\pi}{3} - \frac{\sqrt{3}}{2}$.

10.3.37 The inner loop is traced out between $\theta = \pi/6$ and $\theta = 5\pi/6$, so its area is given by $\frac{1}{2}\int_{\pi/6}^{5\pi/6}(3 - 6\sin\theta)^2\, d\theta = \frac{1}{2}\int_{\pi/6}^{5\pi/6}(9 - 36\sin\theta + 36\sin^2\theta)\, d\theta = \frac{3}{2}\left(3\theta + 12\cos\theta + 6\theta - 3\sin(2\theta)\right)\Big|_{\pi/6}^{5\pi/6} = 9\pi - \frac{27\sqrt{3}}{2}$.

We can determine the area inside the outer loop by using symmetry and doubling the area of the region traced out between $5\pi/6$ and $3\pi/2$. Thus the area inside the outer region is $2 \cdot \frac{1}{2}\int_{5\pi/6}^{3\pi/2}(3 - 6\sin\theta)^2\, d\theta = 3\left(3\theta + 12\cos\theta + 6\theta - 3\sin(2\theta)\right)\Big|_{5\pi/6}^{3\pi/2} = 18\pi + \frac{27\sqrt{3}}{2}$. So the area outside the inner loop and inside the outer loop is $18\pi + \frac{27\sqrt{3}}{2} - \left(9\pi - \frac{27\sqrt{3}}{2}\right) = 9\pi + 27\sqrt{3}$.

10.3.39 The first horizontal tangent line is at the origin. The next is at approximately $(4.0576, 2.0288)$, and the third at approximately $(9.8262, 4.9131)$. The first vertical tangent line is at approximately $(1.7206, 0.8603)$, the next is at about $(6.8512, 3.4256)$, and the next at approximately $(12.8746, 6.4373)$.

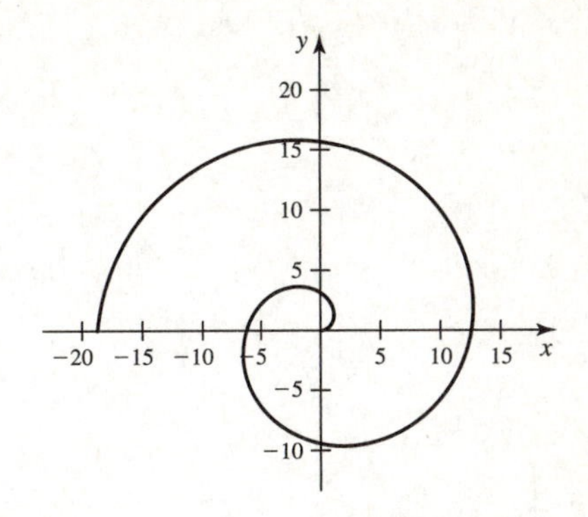

10.3.41

a. $A_n = \frac{1}{2}\int_{(2n-2)\pi}^{(2n-1)\pi} e^{-2\theta}\,d\theta - \frac{1}{2}\int_{2n\pi}^{(2n+1)\pi} e^{-2\theta}\,d\theta = \frac{-1}{4}e^{-(4n-2)\pi} + \frac{1}{4}e^{-(4n-4)\pi} + \frac{1}{4}e^{-(4n+2)\pi} - \frac{1}{4}e^{-4n\pi}$.

b. Each term tends to 0 as $n \to \infty$ so $\lim_{n\to\infty} A_n = 0$.

c. $\dfrac{A_{n+1}}{A_n} = \dfrac{e^{-(4n+2)\pi} + e^{-(4n)\pi} + e^{-(4n+6)\pi} - e^{-(4n+4)\pi}}{e^{-(4n-2)\pi} + e^{-(4n-4)\pi} + e^{-(4n+2)\pi} - e^{-4n\pi}} = e^{-4\pi}$, so $\lim_{n\to\infty} \frac{A_{n+1}}{A_n} = e^{-4\pi}$.

10.3.43 One half of the area is given by $\frac{1}{2}\int_0^{\pi/2} 6\sin 2\theta\,d\theta = \frac{-3}{2}\cos 2\theta\Big|_0^{\pi/2} = 3$, so the total area is 6.

10.3.45 The area is given by

$$\frac{1}{2}\int_0^{2\pi}(4-2\cos\theta)^2\,d\theta = \int_0^{2\pi}(8-8\cos\theta+2\cos^2\theta)\,d\theta = \left(8\theta - 8\sin\theta + \theta + \frac{1}{2}\sin(2\theta)\right)\Bigg|_0^{2\pi} = 18\pi.$$

10.3.47 Suppose that the goat is tethered at the origin, and that the center of the corral is $(1,\pi)$. The circle that the goat can graze is $r = a$, and the corral is given by $r = -2\cos\theta$. The intersection occurs for $\theta = \cos^{-1}(-a/2)$.

The area grazed by the goat is twice the area of the sector of the circle $r = a$ between $\cos^{-1}(-a/2)$ and π, plus twice the area of the circle $r = -2\cos\theta$ between $\pi/2$ and $\cos^{-1}(-a/2)$. Thus we need to compute $A = \int_{\cos^{-1}(-a/2)}^{\pi} a^2\,d\theta + \int_{\pi/2}^{\cos^{-1}(-a/2)} 4\cos^2\theta\,d\theta = a^2\pi - a^2\cos^{-1}(-a/2) + (2\cos\theta\sin\theta + 2\theta)\big|_{\pi/2}^{\cos^{-1}(-a/2)} = a^2(\pi - \cos^{-1}(-a/2)) - \pi - \frac{1}{2}a\sqrt{4-a^2} + 2\cos^{-1}(-a/2)$. Note that $\pi - \cos^{-1}(-a/2) = \cos^{-1}(a/2)$, so this can be written as $(a^2-2)\cos^{-1}(a/2) + \pi - \frac{1}{2}a\sqrt{4-a^2}$. Note that for $a = 0$ this is 0, and for $a = 2$, this is π, as desired.

10.3.49 Again, suppose that the goat is tethered at the origin, and that the center of the corral is $(1,\pi)$. The equation of the corral fence is given by $r = -2\cos\theta$. Note that to the right of the vertical line $\theta = \pi/2$, the goat can graze a half-circle of area $\pi a^2/2$. Also, there is a region in the 2nd quadrant and one in the 3rd quadrant of equal size that can also be grazed. Let this region have area A, so that the total area grazed will then be $\frac{\pi a^2}{2} + 2A$.

Imagine that the goat is walking "west" from the polar point $(a,\pi/2)$, and is keeping the rope taut until his whole rope is along the fence in the third quadrant. Let ϕ be the central angle angle from the origin to the polar point $(1,\pi)$ to the point on the fence that the goat's rope is touching as he makes this walk. When the goat is at $(a,\pi/2)$, we have $\phi = 0$. When the goat is all the way to the fence, we have $\phi = a$. Then length of the rope not along the fence is $a - \phi$. Thus, the value of A is $\frac{1}{2}\int_0^a (a-\phi)^2 d\phi = \frac{1}{2}\left(a^2\phi - a\phi^2 + \frac{\phi^3}{3}\right)\Big|_0^a = \frac{a^3}{6}$.

Thus, the goat can graze a region of area $\frac{\pi a^2}{2} + \frac{a^3}{3}$.

10.3.51

a. If $\cot\phi = \frac{f'(\theta)}{f(\theta)}$ is constant for all θ, then $\phi = \cot^{-1}\left(\frac{f'(\theta)}{f(\theta)}\right)$ is constant. Then $\frac{d}{d\theta}\ln(f(\theta)) = \frac{1}{f(\theta)}\cdot f'(\theta) = \cot\phi$ is constant.

b. If $f(\theta) = Ce^{k\theta}$, then $\cot\phi = \frac{f'(\theta)}{f(\theta)} = \frac{kCe^{k\theta}}{Ce^{k\theta}} = k$.

c. 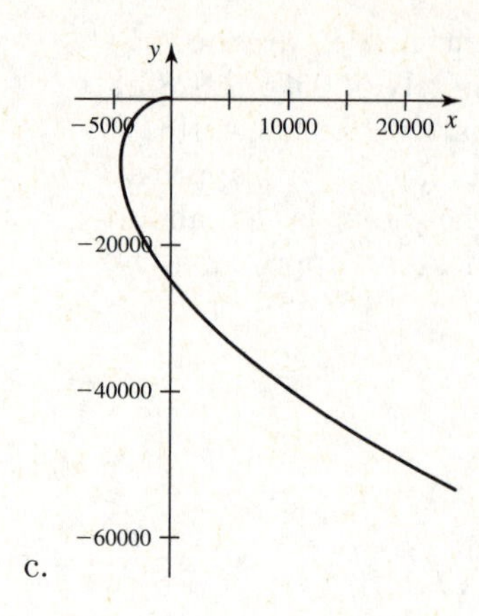

10.4 Conic Sections

10.4.1 A parabola is the set of points in the plane which are equidistant from a given fixed point and a given fixed line.

10.4.3 A hyperbola is the set of points in the plane with the property that the difference of the distances from the point to two given fixed points is a given constant.

10.4.5

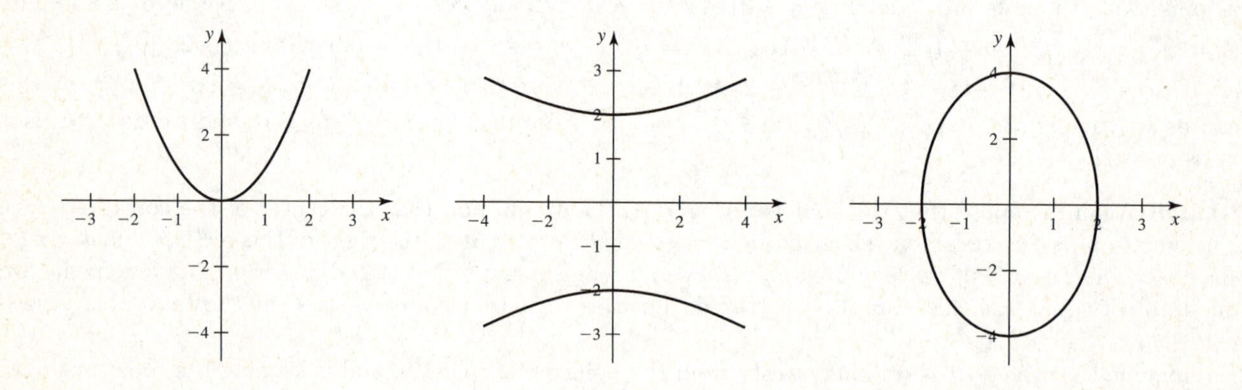

10.4.7 $\left(\frac{x}{a}\right)^2 + \frac{y^2}{a^2 - c^2} = 1$.

10.4.9 The foci for both are $(\pm ae, 0)$.

10.4.11 The asymptotes are $y = \frac{-b}{a} \cdot x$ and $y = \frac{b}{a} \cdot x$.

10.4.13 Directrix: $y = -3$. Focus: $(0, 3)$.

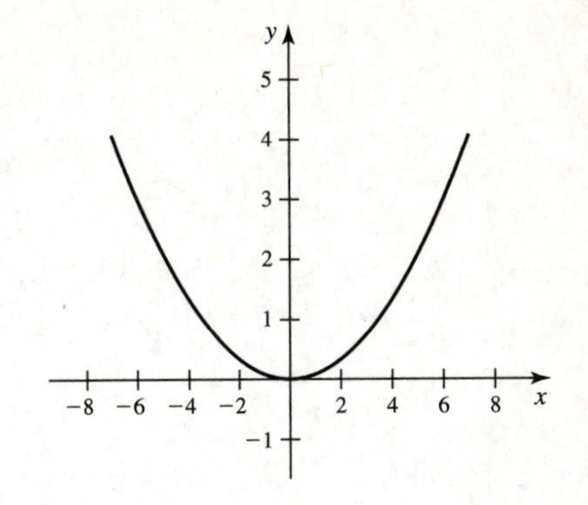

10.4.15 Directrix: $x = 4$. Focus: $(-4, 0)$.

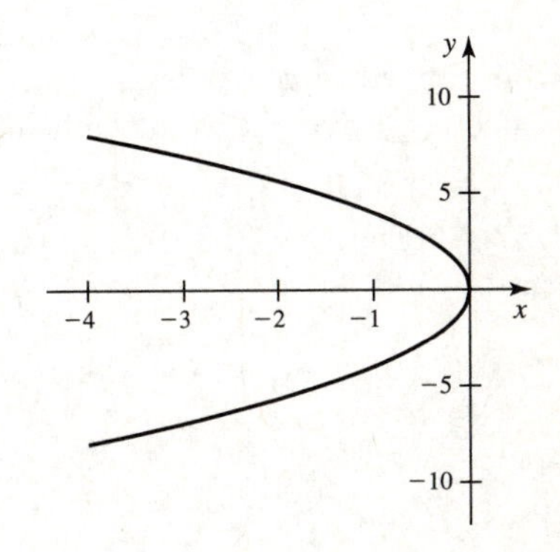

10.4.17 Directrix: $y = \frac{2}{3}$. Focus: $(0, \frac{-2}{3})$.

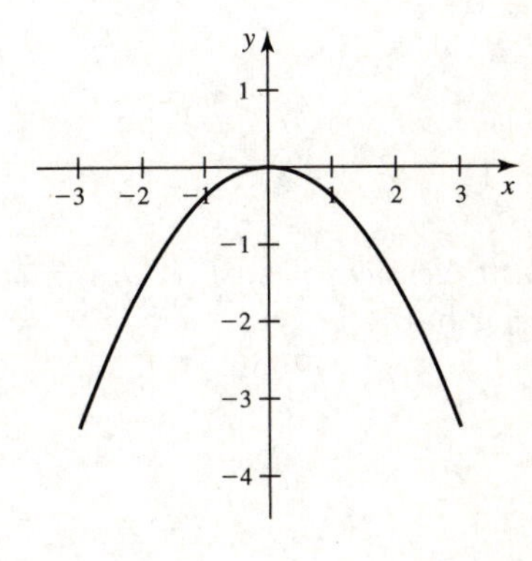

10.4.19

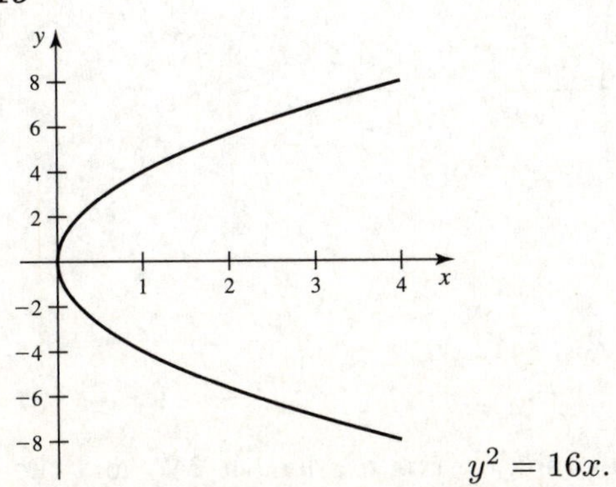

$y^2 = 16x$.

10.4.21

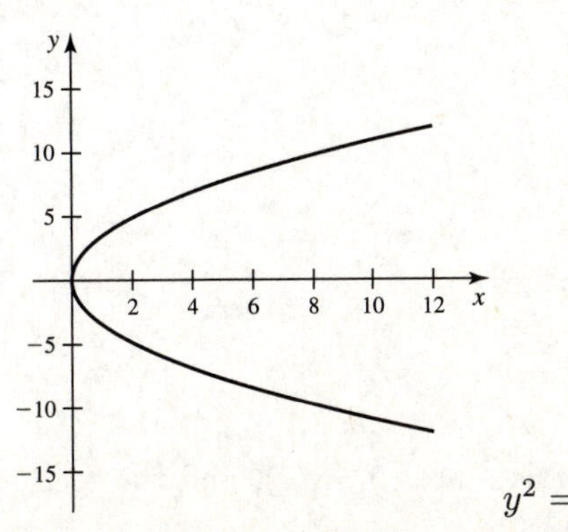

$y^2 = 12x$.

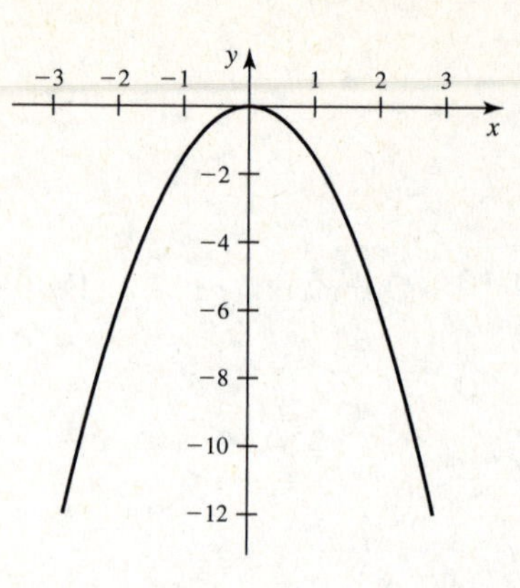

10.4.23 $x^2 = 4py$ and $4 = 4p(-6)$, so $p = \frac{-1}{6}$ and $x^2 = \frac{-2}{3} \cdot y$.

10.4.25 Since the vertex is $(-1, 0)$ and the parabola is symmetric about the x-axis, we have $y^2 = 4p(x+1)$ and since the directrix is one unit left of the vertex, we obtain $p = 1$ and $y^2 = 4(x+1)$.

10.4.27 Vertices are $(\pm 2, 0)$, and the foci are $(\pm\sqrt{3}, 0)$. The major axis has length 4 and the minor axis has length 2.

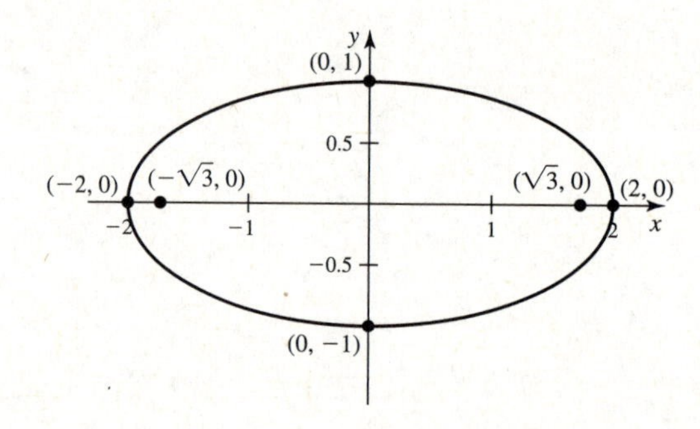

10.4.29 Vertices are $(0, \pm 4)$, and the foci are $(0, \pm 2\sqrt{3})$. The major axis has length 8 and the minor axis has length 4.

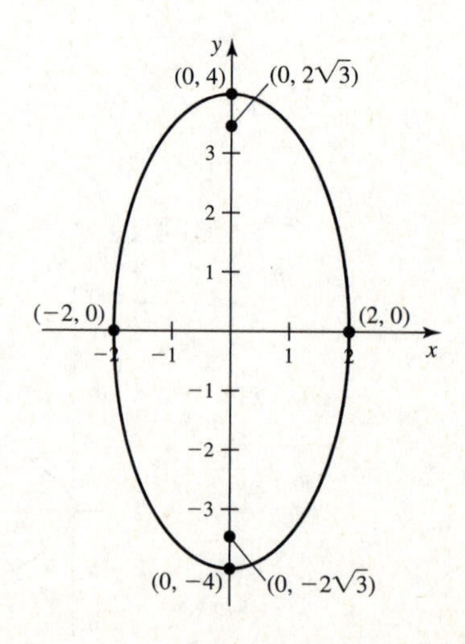

10.4.31 Vertices are $(0, \pm\sqrt{7})$, and the foci are $(0, \pm\sqrt{2})$. The major axis has length $2\sqrt{7}$ and the minor axis has length $2\sqrt{5}$.

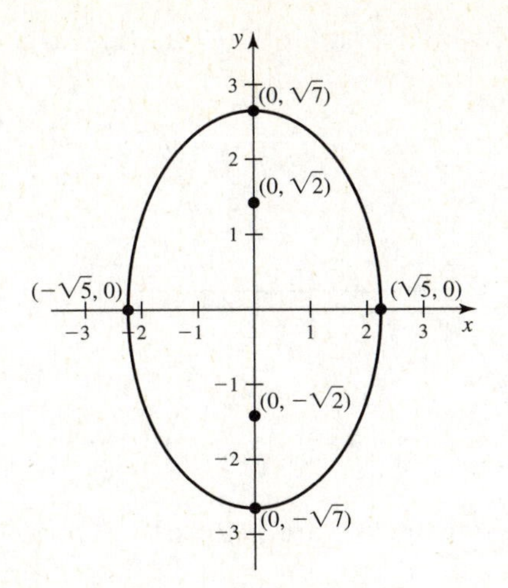

10.4.33 $a = 4$, and $b = 3$, so the equation is $\frac{x^2}{16} + \frac{y^2}{9} = 1$.

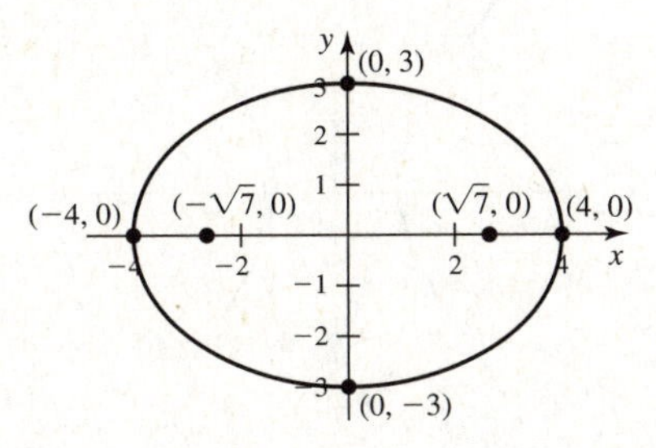

10.4.35 $a = 5$, and the equation is of the form $\frac{x^2}{25} + \frac{y^2}{b^2} = 1$. Since $(4, \frac{3}{5})$ is on the curve, we have $\frac{16}{25} + \frac{9}{25b^2} = 1$, so $b = 1$. The equation is $\frac{x^2}{25} + y^2 = 1$.

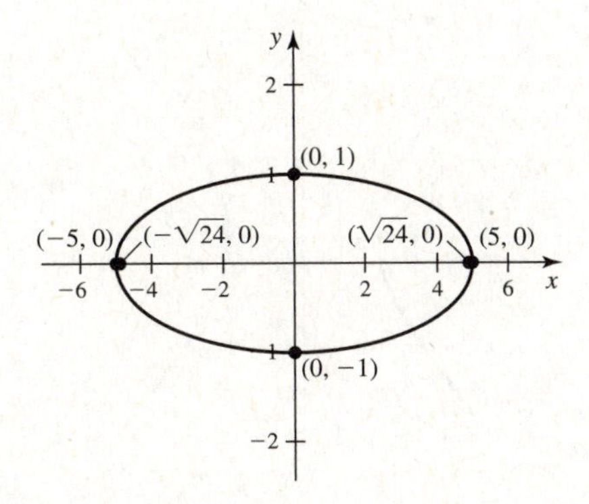

10.4.37 $a = 3$ and $b = 2$, so the equation is $\frac{x^2}{4} + \frac{y^2}{9} = 1$.

10.4.39 The vertices are $(\pm 2, 0)$, and the foci are $(\pm\sqrt{5}, 0)$. The asymptotes are $y = \frac{\pm 1}{2} \cdot x$.

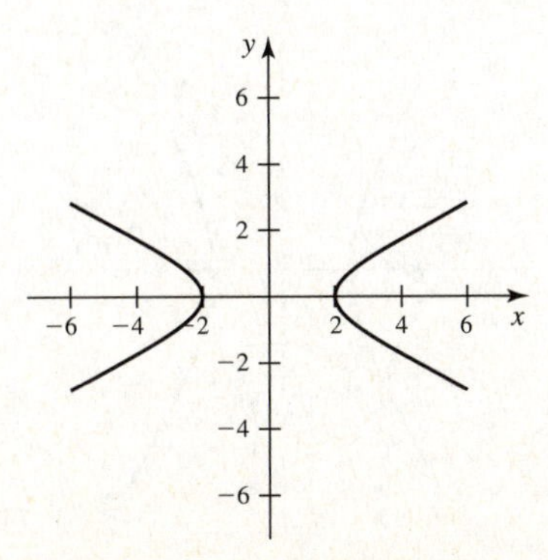

10.4.41 The vertices are $(\pm 2, 0)$, and the foci are $(\pm 2\sqrt{5}, 0)$. The asymptotes are $y = \pm 2x$.

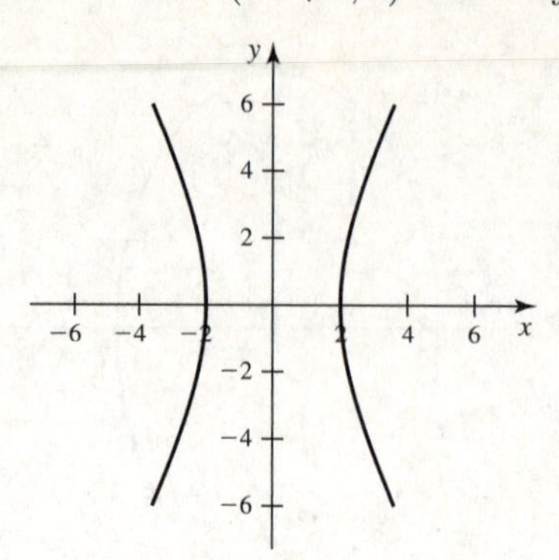

10.4.43 The vertices are $(\pm\sqrt{3}, 0)$, and the foci are $(\pm 2\sqrt{2}, 0)$. The asymptotes are $y = \pm\sqrt{\frac{5}{3}} \cdot x$.

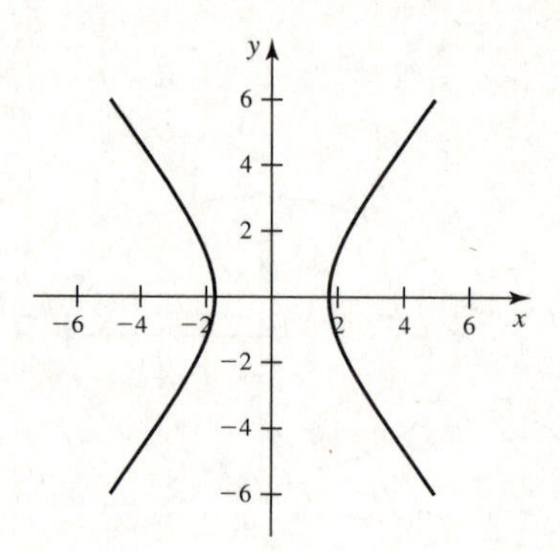

10.4.45 We have $a = 4$ and $c = 6$, so $b^2 = c^2 - a^2 = 20$, so the equation is $\frac{x^2}{16} - \frac{y^2}{20} = 1$.

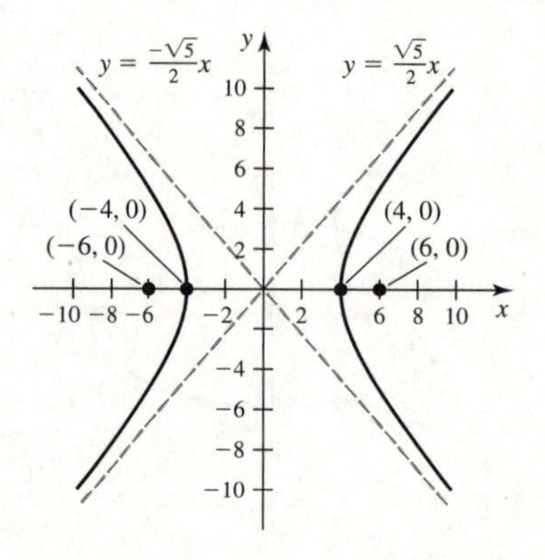

10.4.47 We have $a = 2$, and since the asymptoes are $y = \frac{\pm bx}{a}$, we have that $b = 3$, so the equation is $\frac{x^2}{4} - \frac{y^2}{9} = 1$.

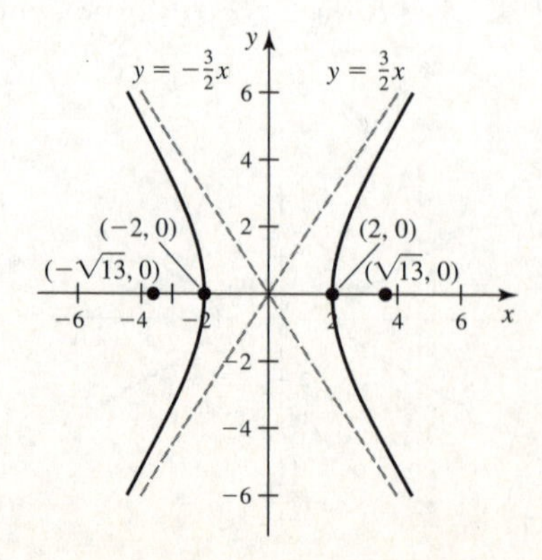

10.4.49 We have $a = 4$ and $c = 5$, so $b^2 = 25 - 16 = 9$, so $b = 3$ and the equation is $\frac{x^2}{16} - \frac{y^2}{9} = 1$.

10.4.51 We have $a = 9$ and $e = \frac{1}{3}$, so $c = ae = 3$, and $b^2 = a^2 - c^2 = 72$, so the equation is $\frac{x^2}{81} + \frac{y^2}{72} = 1$.

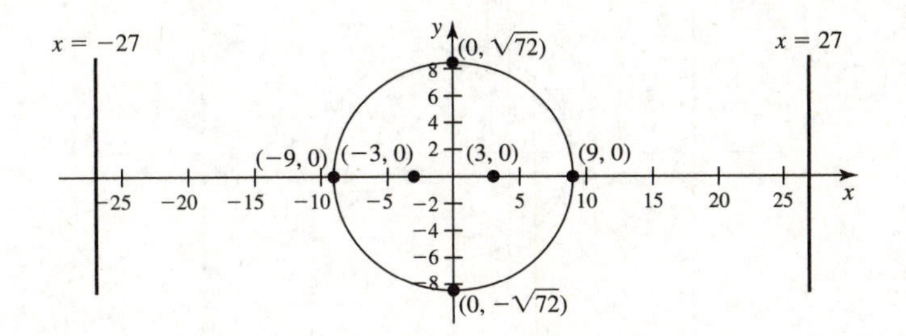

10.4.53 We have $a = 1$ and $e = 3$, so $c = ae = 3$ and $b^2 = c^2 - a^2 = 9 - 1 = 8$. Thus, the equation is $x^2 - \frac{y^2}{8} = 1$.

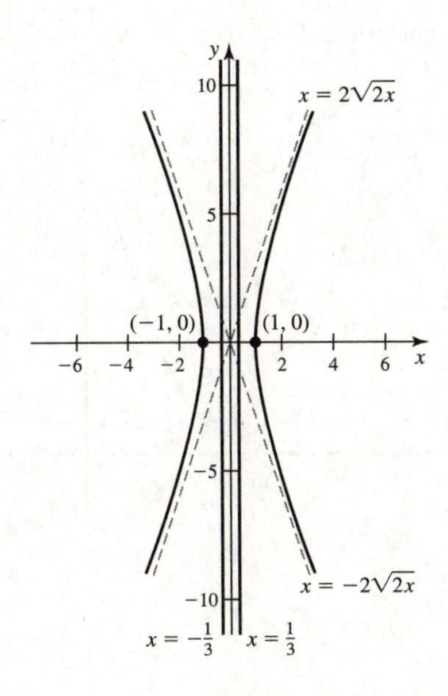

10.4.55 The vertex is $(2, 0)$. The focus is $(0, 0)$, and the directrix is the line $x = 4$.

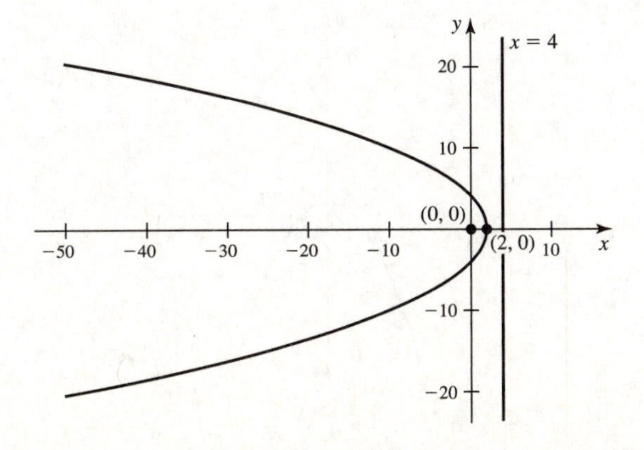

10.4.57 The vertices are $(1, 0)$ and $(-1/3, 0)$. The center is $(1/3, 0)$. The directrices are $x = -1$ and $x = 5/3$.

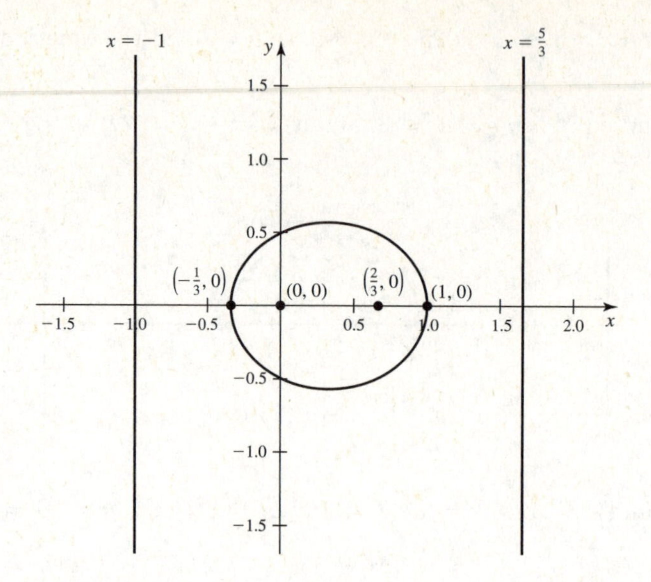

10.4.59 The vertex is $(0, -1/4)$, and the focus is $(0, 0)$. The directrix is the line $y = \frac{-1}{2}$.

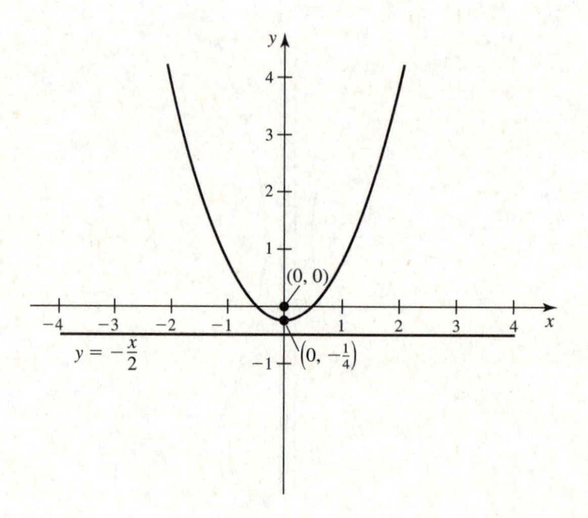

10.4.61 The parabola starts at $(1, 0)$ and goes through quadrants I, II, and III for $\theta \in [0, 3\pi/2]$. It then approaches $(1, 0)$ by traveling through quadrant IV for $\theta \in (3\pi/2, 2\pi)$.

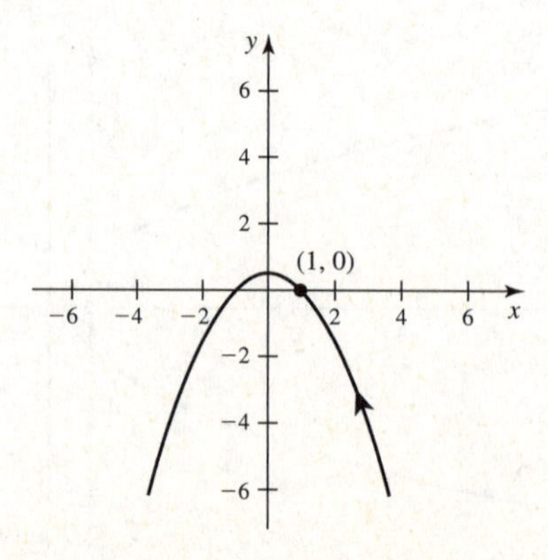

10.4.63 The parabola begins in the first quadrant and passes through the points $(0, 3)$ and then $(-3/2, 0)$ and $(0 - 3)$ as θ ranges from 0 to 2π.

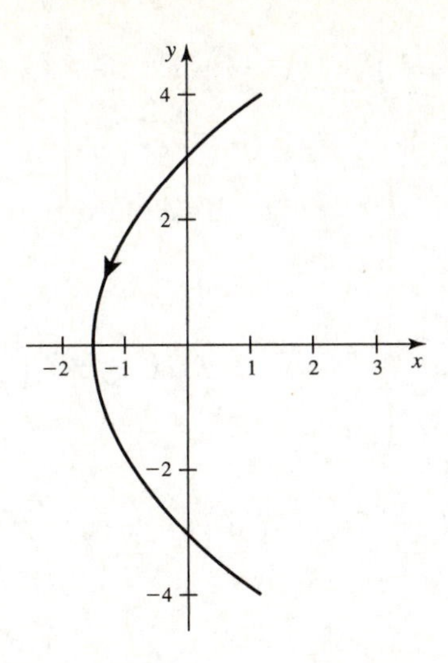

10.4.65 For negative p, the parabola opens to the left and for positive p it opens to the right. As p increases to 0, the parabola opens wider and as p decreases (for $p > 0$), it gets narrower.

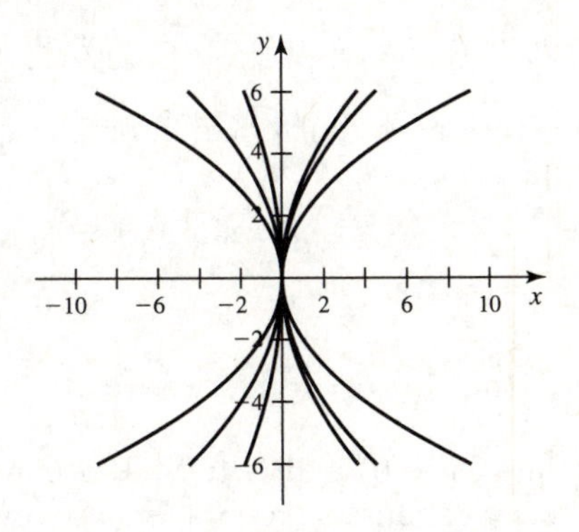

10.4.67

a. True. Note that if $x = 0$, the equation becomes $-y^2 = 9$, which has no solution.

b. True. The slopes of the tangent lines range continuously from $-\infty$ to 0 to ∞ and then back through 0 to $-\infty$ again.

c. True. Given c and d, one can compute a, b, and e. See the summary after Theorem 10.3.

d. True. The vertex is exactly halfway between the focus and the directrix.

10.4.69 Differentiating gives $2x = -6y'$, so at $(-6, -6)$ we obtain $-12 = -6y'$, so $y' = 2$. Thus $y - (-6) = 2(x - (-6))$, or $y = 2x + 6$ is the equation of the tangent line.

10.4.71 Differentiating implicitly, we have $2yy' - \frac{x}{32} = 0$, so at $(6, -5/4)$ we have $\frac{-5}{2}y' - \frac{3}{16} = 0$, so $y' = \frac{-3}{40}$. The equation of the tangent line is $y + \frac{5}{4} = \frac{-3}{40}(x - 6)$, or $y = \frac{-3x}{40} - \frac{4}{5}$.

10.4.73 We have a hyperbola with focal point at the origin and directrix $y = -2$. Furthermore $P = (0, -4/3)$ is a vertex. Thus, $e = \frac{|PF|}{|PL|} = \frac{4/3}{2/3} = 2$, and $r(\theta) = \frac{2(2)}{1-2\sin\theta} = \frac{4}{1-2\sin\theta}$.

10.4.75 The points on the intersection of the two circles are a distance of $2a + r$ from F_1 and a distance of r from F_2. So for P an intersection point, we have $|PF_1| - |PF_2| = 2a$ for all r, and the set of all such points form a hyperbola with foci F_1 and F_2.

10.4.77 Using implicit differentiation, we have $\frac{2x}{a^2} + \frac{2yy'}{b^2} = 0$, or $y' = \frac{-b^2x}{a^2y}$. If (x_0, y_0) is the point of tangency, then $\frac{-b^2x_0}{a^2y_0} = \frac{y-y_0}{x-x_0}$, so $\frac{x_0(x-x_0)}{a^2} = \frac{-y_0(y-y_0)}{b^2}$, so $\frac{x_0x}{a^2} + \frac{y_0y}{b^2} = \frac{x_0^2}{a^2} + \frac{y_0^2}{b^2} = 1$.

10.4.79 $V_x = \pi \int_{-a}^{a} \left(b^2 - \frac{b^2x^2}{a^2}\right) dx = \pi b^2 \int_{-a}^{a} \left(1 - \frac{x^2}{a^2}\right) dx = \pi b^2 \left(x - \frac{x^3}{3a^2}\right)\Big|_{-a}^{a} = \frac{4\pi b^2 a}{3}$.

$V_y = \pi \int_{-b}^{b} \left(a^2 - \frac{a^2y^2}{b^2}\right) dy = \pi a^2 \int_{-b}^{b} \left(1 - \frac{y^2}{b^2}\right) dy = \pi a^2 \left(y - \frac{y^3}{3b^2}\right)\Big|_{-b}^{b} = \frac{4\pi a^2 b}{3}$.

These are different if $a \neq b$. In the case $a = b$, both volumes give $\frac{4\pi a^3}{3}$, the volume of a sphere.

10.4.81

a. $V_x = \pi \int_a^c \left(\sqrt{\frac{b^2x^2}{a^2} - b^2}\right)^2 dx = \pi \int_a^c \left(\frac{b^2x^2}{a^2} - b^2\right) dx = \pi b^2 \left(\frac{x^3}{3a^2} - x\right)\Big|_a^c = \pi b^2 \left(\frac{c^3}{3a^2} - c - \frac{a}{3} + a\right) = \frac{\pi b^2}{3a^2}(c^3 - 3ca^2 + 2a^3) = \frac{\pi b^2}{3a^2}(a-c)^2(2a+c)$.

b. $V_y = 2 \cdot 2\pi \int_a^c a^2 b\sqrt{\frac{x^2}{a^2} - 1}\, dx = 2\pi \int_0^{b^2/a^2} a^2 b\sqrt{u}\, du = 2\pi a^2 b \left(\frac{2}{3}u^{3/2}\Big|_0^{b^2/a^2}\right) = 2\pi a^2 b \frac{2b^3}{3a^3} = \frac{4\pi b^4}{3a}$.

10.4.83

a. The slope of a line making an angle θ with the horizontal is $\tan\theta$. The slope of the tangent line at (x_0, y_0) is $y' = \frac{x}{2p}$, so $y' = \frac{x_0}{2p}$, so $\tan\theta = \frac{x_0}{2p}$.

b. The distance from $(0, y_0)$ to $(0, p)$ is $p - y_0$, and $\tan\phi = \frac{\text{opposite}}{\text{adjacent}} = \frac{p - y_0}{x_0}$.

c. Since l is perpendicular to $y = y_0$, we have $\alpha + \theta = \pi/2$, or $\alpha = \frac{\pi}{2} - \theta$, so $\tan\alpha = \cot\theta = \frac{2p}{x_0}$.

d. $\tan\beta = \tan(\theta + \phi) = \frac{\frac{x_0}{wp} + \frac{p - y_0}{x_0}}{1 - \frac{p - y_0}{2p}} = \frac{x_0^2 + 2p^2 - 2py_0}{x_0(p + y_0)}$. Now since $x_0^2 = 4py_0$, we obtain $\tan\beta = \frac{4py_0 + 2p^2 - 2py_0}{x_0(p + y_0)} = \frac{2p(p + y_0)}{x_0(p + y_0)} = \frac{2p}{x_0}$.

e. Since α and β are acute, we have that $\tan\alpha = \tan\beta$, so $\alpha = \beta$.

10.4.85 Assume the two fixed points are at $(c, 0)$ and $(-c, 0)$. Let P be the point $(0, b)$, and note that P is equidistant from the two given points, so we must have $b^2 + c^2 = a^2$ by the Pythagorean theorem. Now let $Q = (u, 0)$ be on the ellipse for $u > c$. Then $u - c + (c + u) = 2a$, so $u = a$. Now let $R = (x, y)$ be an arbitrary point on the ellipse (assume $x > 0$ and $y > 0$ – the other cases are similar.) Using the triangles formed between the foci, R, and the projection of R onto the x-axis, we have $\sqrt{(x + c)^2 + y^2} = 2a - \sqrt{(c - x)^2 + y^2}$. Squaring both sides gives $(x + c)^2 + y^2 = 4a^2 - 4a\sqrt{(c - x)^2 + y^2} + (c - x)^2 + y^2$. Isolating the root gives $\sqrt{(c - x)^2 + y^2} = \frac{1}{4a}\left((c - x)^2 + y^2 - (c + x)^2 - y^2 + 4a^2\right)$, so $\sqrt{(c - x)^2 + y^2} = a - \frac{c}{a}x$. Squaring again yields $(c - x)^2 + y^2 = a^2 - 2xc + \frac{c^2}{a^2}x^2$, so $c^2 - 2cx + x^2 + y^2 = a^2 - 2cx + \frac{c^2}{a^2}x^2$, or $x^2\left(1 - \frac{c^2}{a^2}\right) + y^2 = a^2 - c^2$. Thus $\frac{x^2}{a^2} + \frac{y^2}{a^2 - c^2} = 1$, which can be written $\frac{x^2}{a^2} + \frac{y^2}{b^2} = 1$, since $b^2 = a^2 - c^2$.

10.4.87 Let the parabola be symmetric about the y-axis with vertex at the origin. Let the circle have radius r and be centered at $(r + a, 0)$, and let the line be $y = -a$. The distance form the point $P(x, y)$ to the line is $u = y + a$. The distance from the point P to the circle is $v = \sqrt{x^2 + (r + a - y)^2} - r$. Setting $u = v$ yields $y + a = \sqrt{x^2 + (r + a - y)^2} - r$, so $y + r + a = \sqrt{x^2 + (r + a - y)^2}$, and squaring gives $y^2 + 2(r + a)y + (r + a)^2 = x^2 + (r + a - y)^2$, so $y^2 + 2(r + a)y + (r + a)^2 = x^2 + (r + a)^2 - 2(r + a)y + y^2$, and thus $4(r + a)y = x^2$, so $y = \frac{1}{4(r + a)}x^2$, the equation of a parabola.

10.4.89 Since the hyperbolas have the same asymptotes, we have that "a" and "b" are interchangeable in the formula. With $c^2 = a^2 + b^2$ and $ae = c$, $bE = c$, we have $e = \frac{\sqrt{a^2 + b^2}}{a}$ and $E = \frac{\sqrt{a^2 + b^2}}{b}$, so $\frac{1}{e^2} + \frac{1}{E^2} = \frac{a^2}{a^2 + b^2} + \frac{b^2}{a^2 + b^2} = 1$.

10.4.91 The latus rectum L intersects the parabola at $x = p$, $y = \pm 2p$. The distance between any point $P(x, y)$ on the parabola to the left of L and L is $p - x$. The distance from F to P is $\sqrt{(x - p)^2 + y^2} = \sqrt{x^2 - 2px + p^2 + 4px} = \sqrt{x^2 + 2px + p^2} = x + p$ (since both x and p are positive.) Thus $D + |FP| = p - x + x + p = 2p$.

10.4.93 Let P be a point on the intersection of the latus rectum and the ellipse. The length of the latus rectum is twice the distance from P to the focus. Let l be the length from P to the focus, and let L be the distance from P to the other focal point. Then $l+L=2a$, so $L^2=4c^2+l^2$, and thus $(2a-l)^2=4c^2+l^2$, and solving for l yields $l=a-\frac{c^2}{a}$. Since $c^2=a^2-b^2$, this can be written as $l=a-\frac{a^2-b^2}{a}=a-(a-\frac{b^2}{a})=\frac{b^2}{a}$. The length of the latus rectum is therefore $\frac{2b^2}{a}$. Now since $e=\frac{c}{a}$, we have $\sqrt{1-e^2}=\sqrt{1-\frac{a^2-b^2}{a^2}}=\sqrt{\frac{b^2}{a^2}}=\frac{b}{a}$. The length of the latus rectum can thus also be written as $2b\cdot\frac{b}{a}=2b\sqrt{1-e^2}$.

10.4.95 Let the equation of the ellipse be $\frac{x^2}{a^2}+\frac{y^2}{a^2-c^2}=1$ and let the equation of the hyperbola be $\frac{x^2}{r^2}-\frac{y^2}{c^2-r^2}=1$. Let (x_0,y_0) be a point of intersection. By evaluating both equations at the point of intersection and subtracting, we obtain the result

$$\frac{x_0^2}{a^2}-\frac{x_0^2}{r^2}+\frac{y_0^2}{a^2-c^2}+\frac{y_0^2}{c^2-r^2}=0,$$

which can be written

$$\frac{r_0^2x_0^2-a^2x_0^2}{a^2r^2}+\frac{(c^2-r^2)y_0^2+(a^2-c^2)y_0^2}{(a^2-c^2)(c^2-r^2)}=0.$$

This equation can be rewritten in the form $\frac{x_0^2}{y_0^2}=\frac{a^2r^2}{(a^2-c^2)(c^2-r^2)}$, which we will use later.

Now implicitly differentiating the equation for the ellipse yields $\frac{2x}{a^2}+\frac{2yy'}{a^2-c^2}=0$, and thus the slope of the tangent line to the ellipse at (x_0,y_0) is $y_e'=\frac{-x_0}{y_o}\cdot\frac{a^2-c^2}{a^2}$. Differentiating the equation of the hyperbola gives $\frac{2x}{r^2}-\frac{2yy'}{c^2-r^2}=0$, so the slope of the tangent line to the hyperbola at the point of intersection is $y_h'=\frac{x_0}{y_0}\cdot\frac{c^2-r^2}{r^2}$. Now consider the product

$$-1\cdot y_e'\cdot y_h'=\frac{x_0^2}{y_0^2}\cdot\frac{(a^2-c^2)(c^2-r^2)}{a^2r^2}.$$

By the result of the first paragraph, this is equal to 1, and thus the two curves are perpendicular at the point of intersection.

10.4.97

a. The curve and the line intersect when $x^2-m^2(x^2-4x+4)-1=0$, which occurs for $\frac{2m^2\pm\sqrt{1+3m^2}}{m^2-1}$, assuming $m\neq\pm1$. So there are two solutions in this case – but if $-1<m<1$, one of the solutions is negative (the intersection lies on the other branch of the hyperbola.) If $m^2=1$, then the equation becomes $4x-5=0$, and there is only the solution $x=\frac{5}{4}$. So there are two intersection points on the right branch exactly for $|m|>1$. We have $v(m)=\frac{2m^2+\sqrt{1+3m^2}}{m^2-1}$ and $u(m)=\frac{2m^2-\sqrt{1+3m^2}}{m^2-1}$.

b. $$\lim_{m\to1^+}u(m)=\lim_{m\to1^+}u(m)\cdot\frac{2m^2+\sqrt{1+3m^2}}{2m^2+\sqrt{1+3m^2}}=\lim_{m\to1^+}\frac{4m^4-3m^2-1}{(m^2-1)(2m^2+\sqrt{1+3m^2})}=$$
$$\lim_{m\to1^+}\frac{(m^2-1)(4m^2+1)}{(m^2-1)(2m^2+\sqrt{1+3m^2})}=\frac{5}{4}.$$

$$\lim_{m\to1^+}v(m)=\lim_{m\to1^+}v(m)\cdot\frac{2m^2-\sqrt{1+3m^2}}{2m^2-\sqrt{1+3m^2}}=\lim_{m\to1^+}\frac{4m^4-3m^2-1}{(m^2-1)(2m^2-\sqrt{1+3m^2})}=$$
$$\lim_{m\to1^+}\frac{(m^2-1)(4m^2+1)}{(m^2-1)(2m^2-\sqrt{1+3m^2})}=\lim_{m\to1^+}\frac{(4m^2+1)}{(2m^2-\sqrt{1+3m^2})}=\infty.$$

c. $$\lim_{m\to\infty}u(m)=\lim_{m\to\infty}\frac{2-\sqrt{\frac{1}{m^4}+\frac{3}{m^2}}}{1-\frac{1}{m^2}}=2.$$

$$\lim_{m\to\infty}v(m)=\lim_{m\to\infty}\frac{2+\sqrt{\frac{1}{m^4}+\frac{3}{m^2}}}{1-\frac{1}{m^2}}=2.$$

d. The expression $\lim_{m\to\infty} A(m)$ represents the area of the region bounded by the hyperbola and the line $x=2$. It is given by $2\int_1^2 \sqrt{x^2-1}\,dx = 2\left(\frac{x}{2}\sqrt{x^2-1} - \frac{1}{2}\ln(x+\sqrt{x^2-1})\right)\Big|_1^2 = 2\sqrt{3} - \ln(2+\sqrt{3})$.

10.4.99

a. With $x^2 = a^2\cos^2 t + 2ab\sin t\cos t + b^2\sin^2 t$, $y^2 = c^2\cos^2 t + 2cd\sin t\cos t + d^2\sin^2 t$, and $xy = ac\cos^2 t + (ad+bc)\sin t\cos t + bd\sin^2 t$, we have $Ax^2+Bxy+Cy^2 = (Aa^2+Bac+Cc^2)\cos^2 t + (2Aab + B(ad+bc)+2Ccd)\sin t\cos t + (Ab^2+Bbd+Cd^2)\sin^2 t = K$. Thus we have an equation of the desired form as long as there exist A, B, C, and K so that $A(a^2-b^2)+B(ac-bd)+C(c^2-d^2)=0$ and $2Aab+B(ad+bc)+2Ccd=0$. This turns out to be the case when $ad-bc\neq 0$. Note that the value of K is $Aa^2+Bac+Cc^2$.

b. Suppose that $ad-bc\neq 0$, but $ac+bd=0$. Then $\frac{b}{a} = \frac{-c}{d}$, and $\tan^{-1}(b/a) = \tan^{-1}(-c/d)$. Note that $x = \sqrt{a^2+b^2}\cos(t+\tan^{-1}(-b/a))$, $y = \sqrt{c^2+d^2}\sin(t+\tan^{-1}(c/d))$. This can be seen by applying the trigonometric identities for the sum of two angles. Then $\frac{x^2}{a^2+b^2} + \frac{y^2}{c^2+d^2} = \cos^2(t+\tan^1(b/a)) + \sin^2(\tan^{-1}(-c/d)) = 1$.

c. Using the work in part b), we see that the equation is $\frac{x^2}{a^2+b^2} + \frac{y^2}{c^2+d^2} = 1$, or $x^2+y^2=r^2$, where $r^2 = a^2+b^2 = c^2+d^2$.

10.5 Chapter Ten Review

10.5.1

a. False. For example, $x = r\cos t$, $y = r\sin t$ for $0\le t\le 2\pi$ and $x = r\sin t$, $y = r\cos t$ for $0\le t\le 2\pi$ generate the same circle.

b. False. Since $e^t > 0$ for all t, this only describes the portion of that line where $x > 0$.

c. True. They both describe the point whose cartesian coordinates are $(3\cos(-3\pi/4), 3\sin(-3\pi/4)) = (-3\cos(\pi/4), -3\sin(\pi/4)) = (-3/\sqrt{2}, -3/\sqrt{2})$.

d. False. The given integral counts the inner loop twice.

e. True. This follows because the equation $0 - x^2/4 = 1$ has no real solutions.

f. True. Note that the given equation can be written as $(x-1)^2+4y^2=4$, or $\frac{(x-1)^2}{4}+y^2=1$.

10.5.3

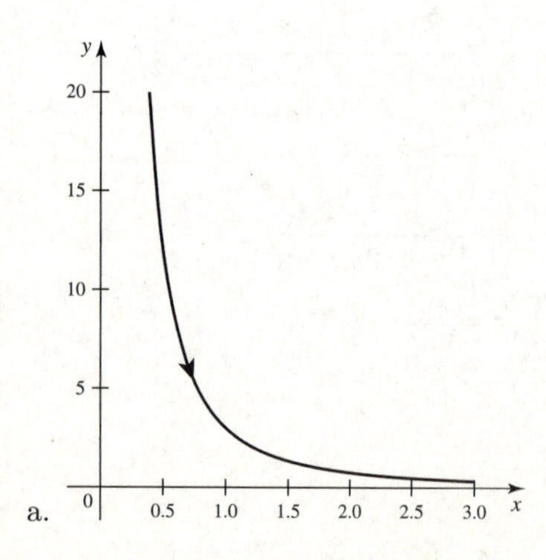

a.

b. $y = 3(e^t)^{-2} = \frac{3}{x^2}$.

c. The curve represents the portion of $\frac{3}{x^2}$ for $x > 0$.

d. $\frac{dy}{dx} = \frac{-6}{x^3}$, so at $(1,3)$ we have $\frac{dy}{dx} = -6$.

10.5.5

a. 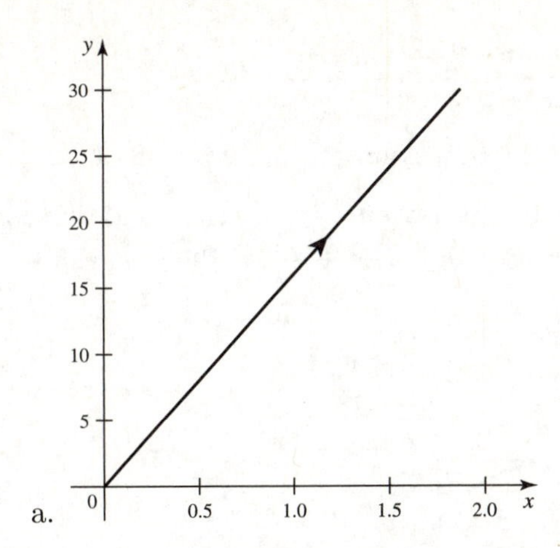

b. Since $\ln t^2 = 2\ln t$ for $t > 0$, we have $y = 16x$ for $0 \le x \le 2$.

c. The curve represents a line segment from $(0,0)$ to $(2,32)$.

d. $\frac{dy}{dx} = 16$ for all value of x.

10.5.7 $\frac{dy}{dx} = \frac{dy/dt}{dx/dt} = \frac{\sin t}{1-\cos t}$. At $t = \pi/6$, the slope of the tangent line is $\frac{1}{2-\sqrt{3}} = 2+\sqrt{3}$. So the equation of the tangent line is $y - (1 - \sqrt{3}/2) = (2+\sqrt{3})(x - (\pi/6 - 1/2))$, or $y = (2+\sqrt{3})x + (2 - \frac{\pi}{3} - \frac{\pi\sqrt{3}}{6})$.

At $t = 2\pi/3$, the slope of the tangent line is $\frac{\sqrt{3}}{3}$, so the equation of the tangent line is $y - \frac{3}{2} = \frac{\sqrt{3}}{3}(x - (\frac{2\pi}{3} - \frac{\sqrt{3}}{2}))$, or $y = \frac{x}{\sqrt{3}} + 2 - \frac{2\pi}{3\sqrt{3}}$.

10.5.9

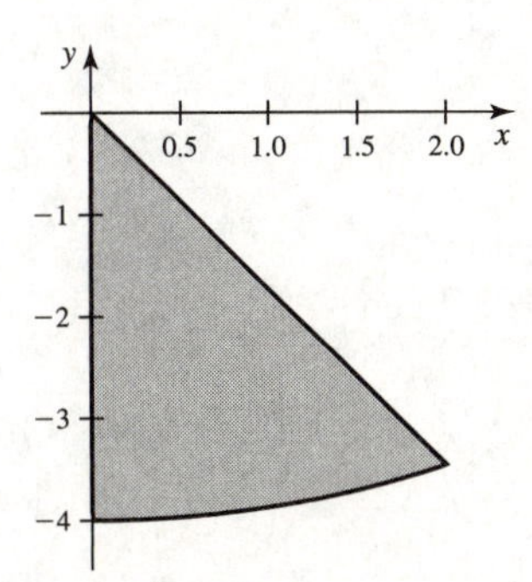

10.5.11 Letting $x = r\cos\theta$, $y = r\sin\theta$, and $r^2 = x^2 + y^2$, we have $x^2 + y^2 + 2y - 6x = 0$, which can be written as $x^2 - 6x + 9 + y^2 + 2y + 1 = 10$, or $(x-3)^2 + (y+1)^2 = 10$, so this is a circle of radius $\sqrt{10}$ centered at $(3,-1)$.

10.5.13

a. 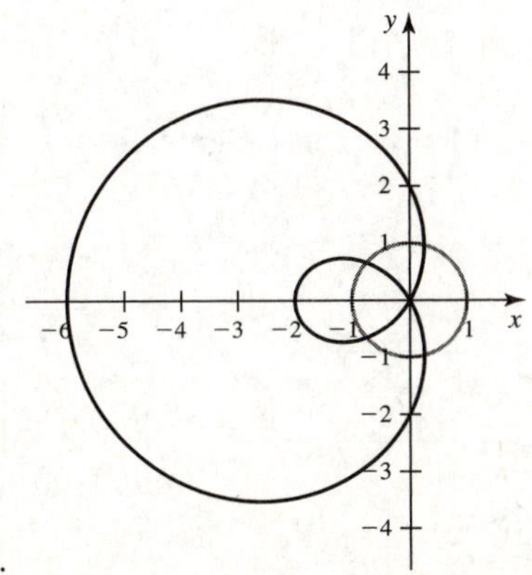

There are 4 intersection points.

b. Note that $2 - 4\cos\theta = 1$ for $\theta = \cos^{-1}(1/4) \approx 1.32$, and $2 - 4\cos\theta = -1$ for $\theta = \cos^{-1}(3/4) \approx .73$. The points of intersection (in polar form) are approximately $(1, 1.32)$, $(1, 2\pi - 1.32) \approx (1, 4.76)$, $(-1, .73)$, and $(-1, 2\pi - .73) \approx (-1, 5.56)$.

10.5.15

a. $\frac{dy}{dx} = \frac{dy/d\theta}{dx/d\theta} = \frac{2\cos\theta\sin\theta + (4+2\sin\theta)\cos\theta}{2\cos\theta\cos\theta - (4+2\sin\theta)\sin\theta} = \frac{4\cos\theta + 4\sin\theta\cos\theta}{2\cos^2\theta - 2\sin^2\theta - 4\sin\theta}$.

This is 0 when $\cos\theta = 0$, and when $4\sin\theta = -4$, so the only solutions are $\theta = \pi/2$, $3\pi/2$.

The denominator is 0 when $2 - 4\sin^2\theta - 4\sin\theta = 0$ which occurs (using the quadratic formula) for $\sin\theta = \frac{-1}{2} + \frac{\sqrt{3}}{2}$, so there are vertical tangent lines at $\theta = \sin^{-1}(\frac{-1}{2} + \frac{\sqrt{3}}{2})$ and $\theta = \pi - \sin^{-1}(\frac{-1}{2} + \frac{\sqrt{3}}{2})$.

b. The curve is never at the origin.

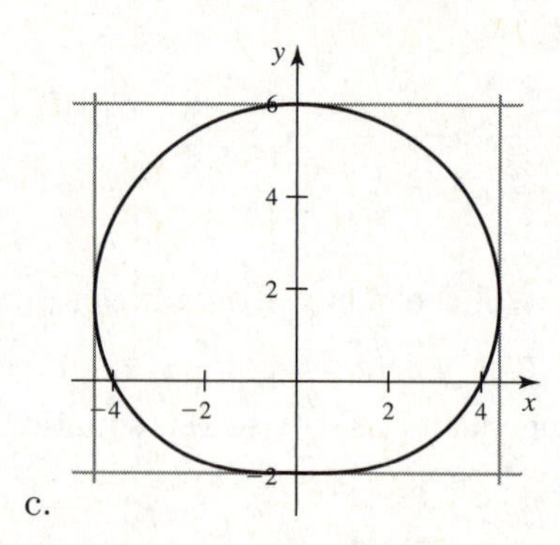

c.

.

10.5.17

a. Note that the whole curve is generated for $-\pi/4 \le \theta \le \pi/4$, so we restrict ourselves to that domain. Write the equations as $r = \sqrt{2\cos 2\theta}$. Then

$$\frac{dy}{d\theta} = \sqrt{2\cos 2\theta}\cos\theta - \sin\theta\frac{2\sin 2\theta}{\sqrt{2\cos 2\theta}} = \frac{\cos\theta}{\sqrt{2\cos 2\theta}}\left(2\cos 2\theta - 4\sin^2\theta\right) = \frac{\cos\theta}{\sqrt{2\cos 2\theta}}\left(2 - 8\sin^2\theta\right).$$

Also, $\frac{dx}{d\theta} = -\sqrt{2\cos 2\theta}\sin\theta + \cos\theta\frac{2\sin 2\theta}{\sqrt{2\cos 2\theta}} = \frac{\sin\theta}{\sqrt{2\cos 2\theta}}\left(-4\cos^2\theta - 2(\cos 2\theta)\right) = \frac{(2-8\cos^2\theta)\sin\theta}{\sqrt{2\cos(2\theta)}}$. Thus $\frac{dy}{dx} = \frac{dy/d\theta}{dx/d\theta} = \cot\theta\left(\frac{1-4\sin^2\theta}{1-4\cos^2\theta}\right)$.

This expression is 0 on the given domain only for $\sin^2\theta = \frac{1}{4}$, so there are horizontal tangent lines at $\theta = \pm\frac{\pi}{6}$. There are vertical tangent lines on the given domain only for $\theta = 0$ In cartesian coordinates, the lines are $x = \pm\sqrt{2}$.

b. The curve is at the origin for $\theta = \pm\pi/4$, and since $\tan\pi/4 = 1$, the tangent lines have the equations $y = \pm x$.

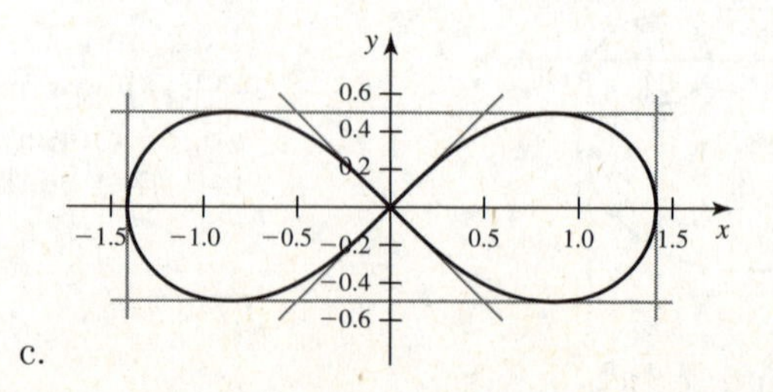

c.

.

10.5.19 The area is given by $A = \frac{1}{2}\int_0^{2\pi}(3 - \cos\theta)^2\,d\theta = \frac{1}{2}\int_0^{2\pi}(9 - 6\cos\theta + \cos^2\theta)\,d\theta = \frac{1}{2}\left(9\theta - 6\sin\theta + \frac{1}{2}(\cos\theta\sin\theta + \theta)\right)\Big|_0^{2\pi} = \frac{1}{2}(18\pi + \pi) = \frac{19\pi}{2}$.

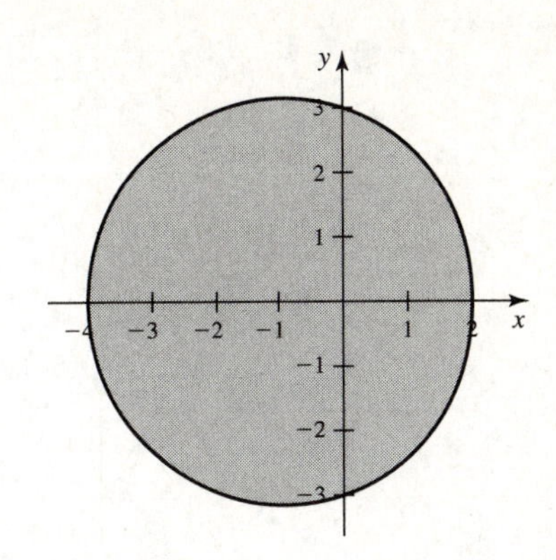

10.5.21 The curves intersect at $\theta = \pm\frac{1}{2}\cos^{-1}(1/16)$. By symmetry the total desired area is $A = 4 \cdot \frac{1}{2}\int_0^{\cos^{-1}(1/16)/2}(4\cos 2\theta - \frac{1}{4})\,d\theta = 2\left(2\sin 2\theta - \frac{\theta}{4}\right)\Big|_0^{\cos^{-1}(1/16)/2} = \frac{1}{4}\sqrt{255} - \frac{\cos^{-1}(1/16)}{4}$.

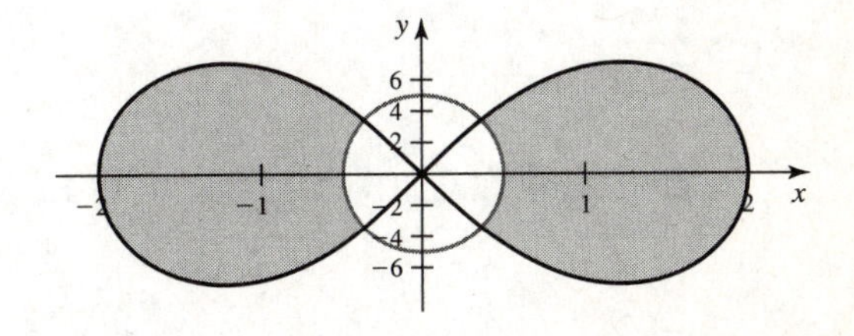

10.5.23

a. This represents a hyperbola with $a = 1$ and $b = \sqrt{2}$.

b. The vertices are $(\pm 1, 0)$, the foci are $(\pm c, 0)$ where $c^2 = a^2 + b^2 = 3$, so they are $(\pm\sqrt{3}, 0)$. The directrices are $x = \frac{\pm a^2}{c} = \frac{\pm 1}{\sqrt{3}}$.

c. The eccentricity is $e = \frac{c}{a} = \sqrt{3}$.

d.

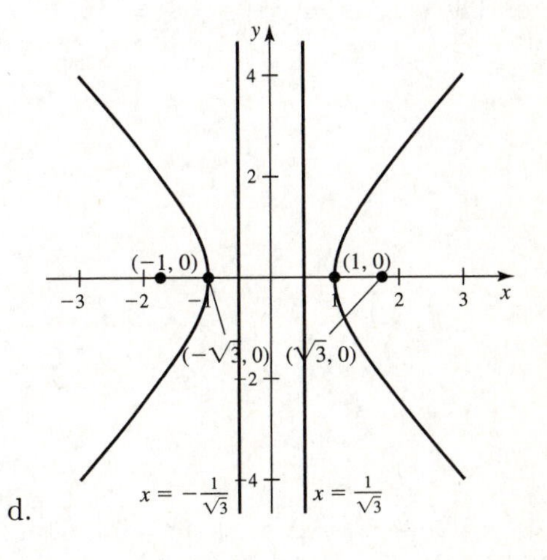

10.5.25

a. This can be written as $\frac{y^2}{16} - \frac{x^2}{4} = 1$. It is a hyperbola with $a = 4$ and $b = 2$.

b. The vertices are $(0, \pm 4)$. The foci are $(0, \pm c)$ where $c^2 = a^2 + b^2 = 16 + 4 = 20$, so they are $(0, \pm\sqrt{20})$. The directrices are $y = \frac{\pm a^2}{c} = \frac{\pm 16}{\sqrt{20}} = \frac{\pm 8}{\sqrt{5}}$.

c. The eccentricity is $e = \frac{c}{a} = \frac{\sqrt{20}}{4} = \frac{\sqrt{5}}{2}$.

d. 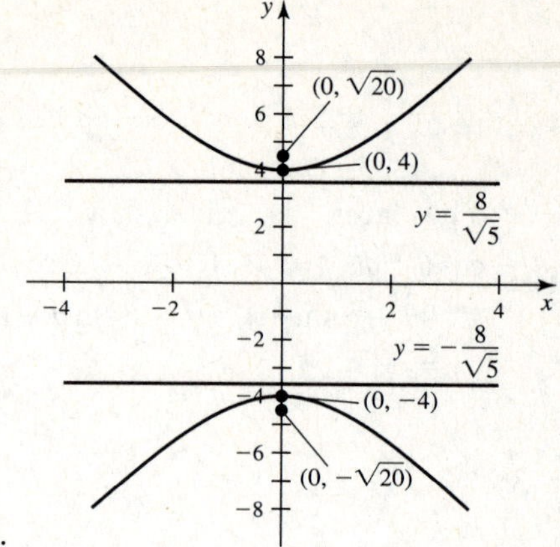

10.5.27

a. This can be written as $\frac{x^2}{4} + \frac{y^2}{2} = 1$, so it is an ellipse with $a = 2$ and $b = \sqrt{2}$.

b. The vertices are $(\pm 2, 0)$. The foci are $(\pm c, 0)$ where $c^2 = a^2 - b^2 = 4 - 2 = 2$, so they are $(\pm\sqrt{2}, 0)$. The directrices are $x = \frac{\pm a^2}{c} = \frac{\pm 4}{\sqrt{2}} = \pm 2\sqrt{2}$.

c. The eccentricity is $e = \frac{c}{a} = \frac{\sqrt{2}}{2}$.

d. 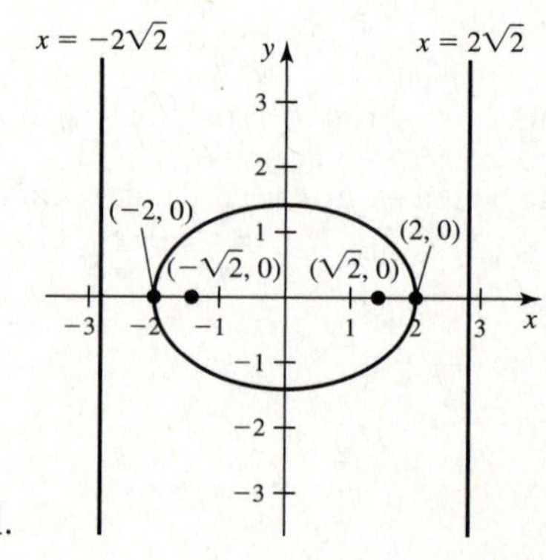

10.5.29 $2y\frac{dy}{dx} = -12$, so at the point in question, $\frac{dy}{dx} = 3/2$. So the equation of the tangent line is $y + 4 = \frac{3}{2}\left(x + \frac{4}{3}\right)$, or $y = \frac{3}{2}x - 2$.

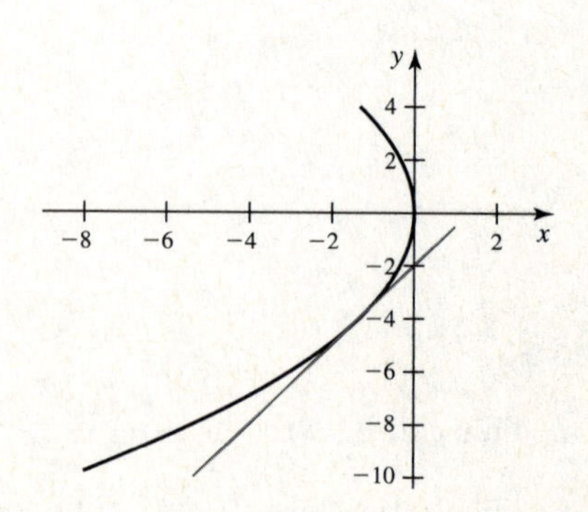

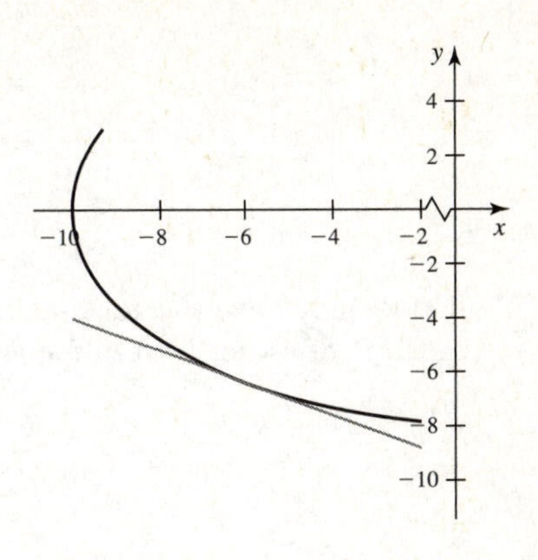

10.5.31 $\frac{x}{50} + \frac{y}{32} \cdot \frac{dy}{dx} = 0$, so at the given point, $\frac{dy}{dx} = \frac{-6}{10} = \frac{-3}{5}$. So the equation of the tangent line is $y + \frac{32}{5} = \frac{-3}{5}(x+6)$, or $y = \frac{-3}{5}x - 10$.

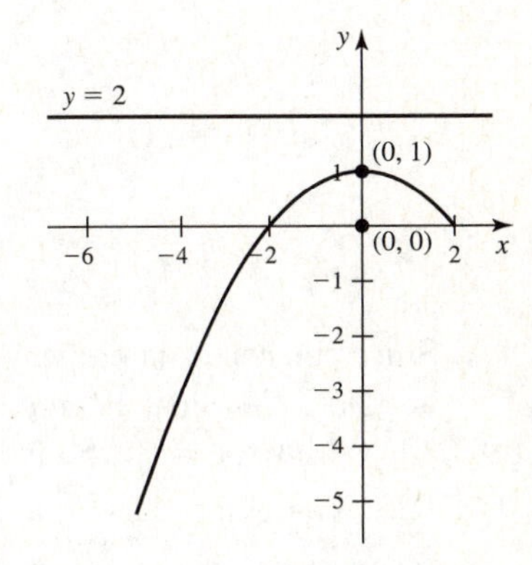

10.5.33 The eccentricity is 1, and the directrix is $y = 2$. The vertex is $(0, 1)$ and the focus is $(0, 0)$.

10.5.35 The eccentricity is $\frac{1}{2}$, and the directrices are $x = 4$ and $x = \frac{-20}{3}$. The vertices are $(\frac{4}{3}, 0)$ and $(-4, 0)$ and the foci are $(0, 0)$ and $(\frac{-8}{3}, 0)$.

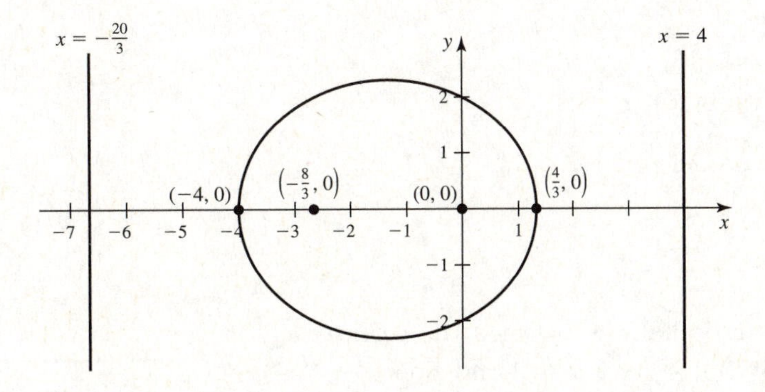

10.5.37

a. Recall that $\cos 2\theta = \cos^2\theta - \sin^2\theta$, so $r^2\cos(2\theta) = 1$ becomes $r^2(\cos^2\theta - \sin^2\theta) = x^2 - y^2 = 1$. The curve is a hyperbola.

b. With $a = b = 1$, we have $c^2 = 2$, so the vertices are $(\pm 1, 0)$ and the foci are $(\pm\sqrt{2}, 0)$. The directrices are $x = \pm\frac{a^2}{c} = \pm\frac{1}{\sqrt{2}}$. The eccentricity is $e = \frac{c}{a} = \sqrt{2}$.

c. It does not have the form as in Theorem 10.4 because it does not have a focus at the origin.

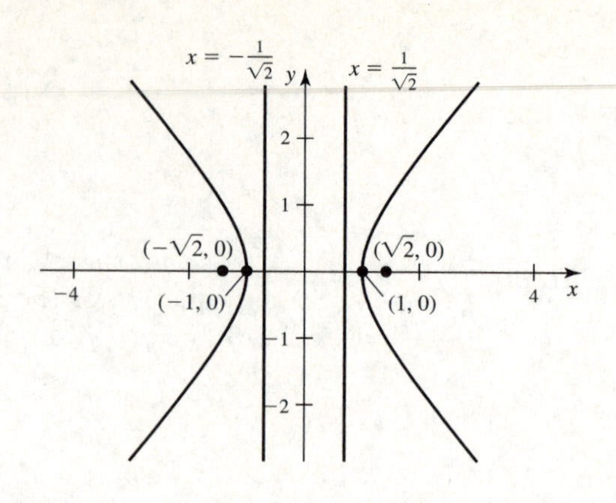

10.5.39 Since the center is halfway between the vertices, it is $(0,0)$. We must have $a = 4$ and since $\frac{a^2}{c} = d = 10$, we have $c = \frac{8}{5}$. So $b^2 = a^2 - c^2 = 16 - \frac{64}{25} = \frac{336}{25}$. The ellipse has equation $\frac{25x^2}{336} + \frac{y^2}{16} = 1$. The eccentricity is $\frac{c}{a} = \frac{8/5}{4} = \frac{2}{5}$.

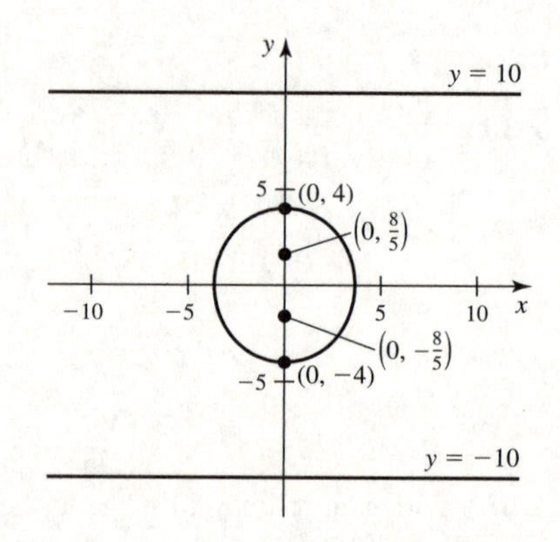

10.5.41 Since the center is halfway between the vertices, it is $(0,0)$. We must have $a = 2$ and since $\frac{a^2}{c} = d = 1$, we have $c = 4$. So $b^2 = 16 - 4 = 12$. The hyperbola has equation $\frac{y^2}{4} - \frac{x^2}{12} = 1$. The eccentricity is $\frac{c}{a} = \frac{4}{2} = 2$.

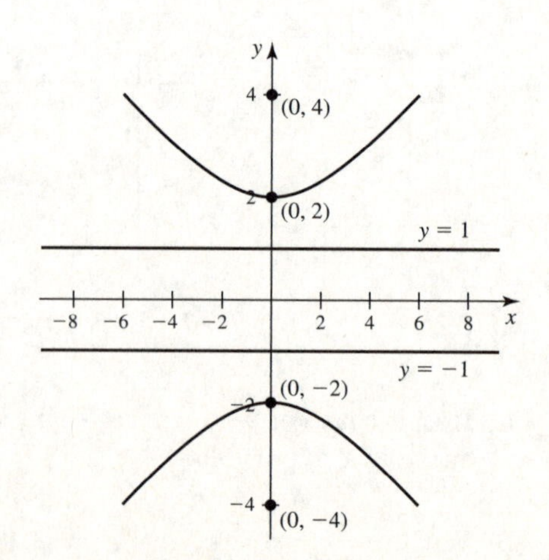

10.5.43 We have $a = 6$, $c = 4$ and $e = \frac{c}{a} = \frac{4}{6} = \frac{2}{3}$. Also, $b^2 = a^2 - c^2 = 36 - 16 = 20$, and the equation is $\frac{y^2}{36} + \frac{x^2}{20} = 1$. The vertices are $(\pm 2\sqrt{5}, 0)$. The directrices are $y = \pm\frac{a^2}{c} = \pm 9$.

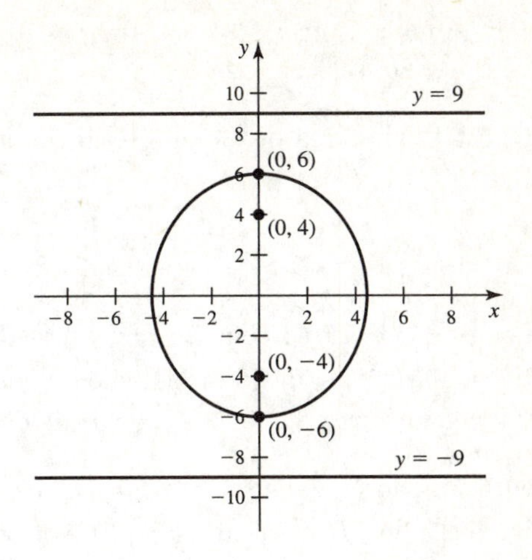

10.5.45 $\sin 2\theta = \theta^2$ when $\theta = 0$. Graphing the functions reveals a root near $\theta = 1$. A CAS reveals the intersection point to be $\theta \approx .9669$. In polar coordinates, the intersection points are $(0, 0)$ and $(.9669, .9669)$.

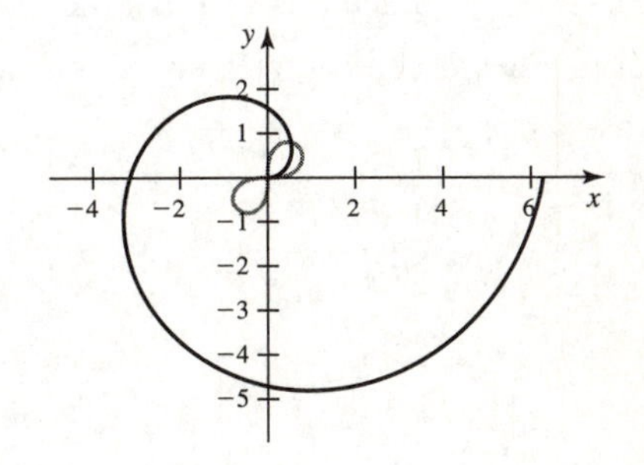

10.5.47 The curves intersect for $\theta = 0$. Note also that when $\theta = k\pi$ for k an odd integer, the curve $r = -\theta$ is at the polar point $(-k\pi, k\pi) = (k\pi, 0)$. And for $\theta = 2k\pi$, the curve $r = \frac{\theta}{2}$ is at the point $(k\pi, 0)$. So the curves intersect at these points.

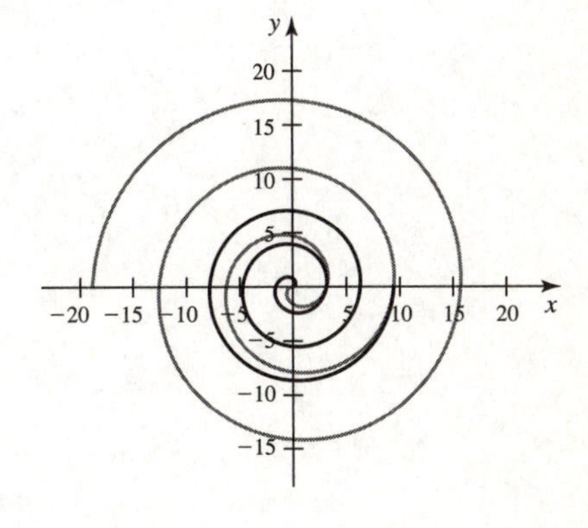

10.5.49 By symmetry, we can focus on the region in the first quadrant. That area is given by $A = xy$ where $y = \sqrt{b^2 - \frac{b^2}{a^2}x^2}$. So

$$A(x) = x\sqrt{b^2 - \frac{b^2}{a^2}x^2},$$

so

$$A'(x) = \sqrt{b^2 - \frac{b^2}{a^2}x^2} - \frac{b^2x^2}{a^2\sqrt{b^2 - \frac{b^2}{a^2}x^2}}.$$

Setting the derivative equal to 0 and clearing denominators yields $\left(b^2 - \frac{b^2}{a^2}x^2\right)a^2 - b^2x^2 = 0$, and solving for x gives the critical point $x = \frac{\sqrt{2}}{2}a$. Since this is the only critical point and it clearly does not give a minimum (since $A(0) = A(a) = 0$), it must yield a maximum. The whole rectangle has dimensions $\sqrt{2}a \times \sqrt{2}b$, and area $2ab$.

10.5.51 The area of the ellipse in the first quadrant is $\frac{\pi ab}{4}$, so we are seeking θ_0 so that

$$\frac{\pi ab}{8} = \frac{1}{2}\int_0^{\theta_0} \frac{a^2b^2}{a^2\sin^2\theta + b^2\cos^2\theta}\,d\theta = \frac{a^2}{2}\int_0^{\theta_0} \frac{\sec^2\theta}{\frac{a^2}{b^2}\tan^2\theta + 1}\,d\theta.$$

Let $u = \frac{a}{b}\tan\theta$ so that $du = \frac{a}{b}\sec^2\theta\,d\theta$. Then we have $\frac{\pi ab}{8} = \frac{ab}{2}\int_0^{\frac{a}{b}\tan\theta_0} \frac{1}{1+u^2}\,du = \frac{ab}{2}\tan^{-1}(\frac{a}{b}\tan(\theta_0))$. Note that this equation is satisfied when $\tan(\theta_0) = \frac{b}{a}$, because then the expression on the right-hand side of that equation is $\frac{ab}{2} \cdot \frac{\pi}{4} = \frac{\pi ab}{8}$. So the desired value of m is $\tan(\theta_0) = \frac{b}{a}$.

10.5.53 Note that $Q = (a\cos\theta, a\sin\theta)$ and $R = (b\cos\theta, b\sin\theta)$, where θ is the angle formed by l and the x-axis. Then $P = (a\sin\theta, b\cos\theta)$ is a point on the ellipse $\frac{x^2}{a^2} + \frac{y^2}{b^2} = 1$, since it satisfies that equation.

10.5.55 The focal point is at the origin, the directrix is $y = -d$, so we have an equation of the form $r = \frac{ed}{1-e\sin\theta}$. Since c is the distance from the center to the focal point, we have $c = 3/8$, and since a is the distance from the center to a vertex, we have $a = 9/8$. Then we have $e = \frac{c}{a} = \frac{3/8}{9/8} = \frac{1}{3}$, and $d = \frac{a^2}{c} - \frac{3}{8} = 3$. Thus $r = \frac{1}{1-\frac{1}{3}\sin\theta} = \frac{3}{3-\sin\theta}$.

Chapter 11

11.1 Vectors in the Plane

11.1.1 The coordinates of a point determine its location, but a given point has no width or breadth, so it has no size or direction. A nonzero vector has size (magnitude) and direction, but it has no location in the sense that it can be translated to a different initial point and be considered the same vector.

11.1.3

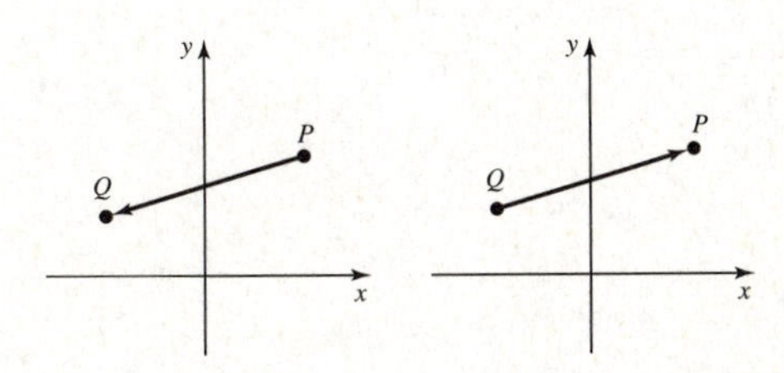

11.1.5 Two vectors are equal if they have the same magnitude and direction. Given a position vector, any translation of that vector to a different initial point yields an equivalent vector. Since there are infinitely many such translations which don't change the given vector's direction or magnitude, there are infinitely many vectors equivalent to the given one.

11.1.7 If $c > 0$ is given, the scalar multiple $c\mathbf{v}$ of the vector $\mathbf{v}$ is obtained by scaling the magnitude of $\mathbf{v}$ by a factor of c, and keeping the direction the same. If $c < 0$, then the head and tail of $\mathbf{v}$ are interchanged, and then the vector's magnitude is scaled by a factor of $|c|$.

11.1.9 $\mathbf{u}+\mathbf{v} = \langle u_1, u_2\rangle + \langle v_1, v_2\rangle = \langle u_1+v_1, u_2+v_2\rangle$.

11.1.11 $|\mathbf{v}| = |\langle v_1, v_2\rangle| = \sqrt{v_1^2+v_2^2}$.

11.1.13 If $P(p_1,p_2)$ and $Q(q_1,q_2)$ are given, then $\left|\overrightarrow{PQ}\right| = |\langle q_1-p_1, q_2-p_2\rangle| = \sqrt{(q_1-p_1)^2+(q_2-p_2)^2}$.

11.1.15 The vector $10\cdot\frac{\mathbf{v}}{|\mathbf{v}|} = 10\cdot\frac{1}{\sqrt{9+4}}\cdot\langle 3,-2\rangle = \langle \frac{30}{\sqrt{13}}, \frac{-20}{\sqrt{13}}\rangle$ has the desired properties.

11.1.17

a. $3\mathbf{v}$ b. $2\mathbf{u}$ c. $-3\mathbf{u}$ d. $-2\mathbf{u}$ e. $\mathbf{v}$

11.1.19

a. $3\mathbf{u}+3\mathbf{v}$ b. $\mathbf{u}+2\mathbf{v}$ c. $2\mathbf{u}+5\mathbf{v}$ d. $-2\mathbf{u}+3\mathbf{v}$ e. $3\mathbf{u}+2\mathbf{v}$

f. $-3\mathbf{u}-2\mathbf{v}$ g. $-2\mathbf{u}-4\mathbf{v}$ h. $\mathbf{u}-4\mathbf{v}$ i. $-\mathbf{u}-6\mathbf{v}$

11.1.21

a. $\overrightarrow{OP}$

i.

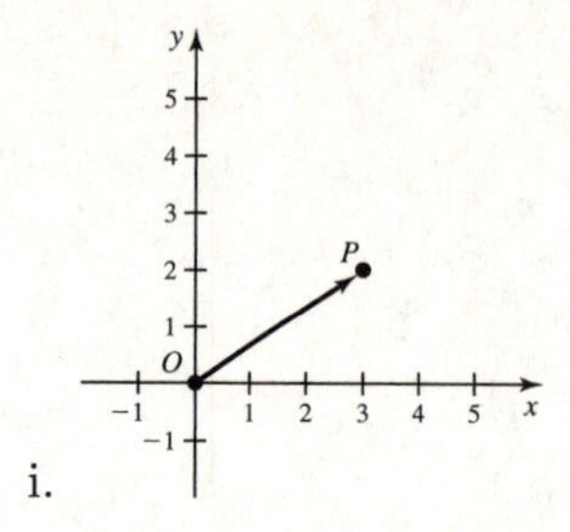

ii. $|3\mathbf{i}+2\mathbf{j}|=\sqrt{13}$.

b. $\overrightarrow{QP}$

i.

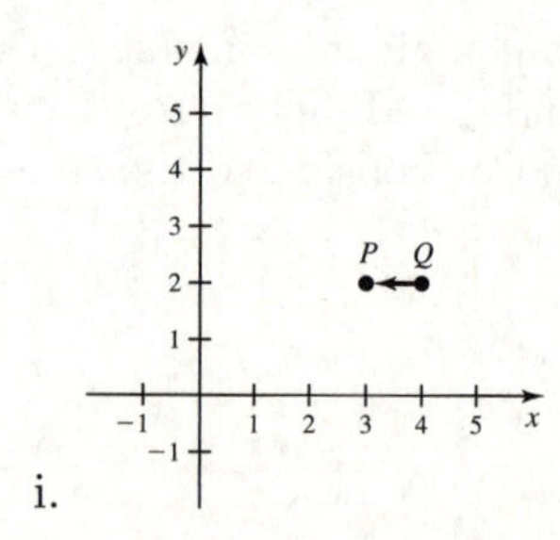

ii. $|-\mathbf{i}+0\cdot\mathbf{j}|=1$.

c. $\overrightarrow{RQ}$

i.

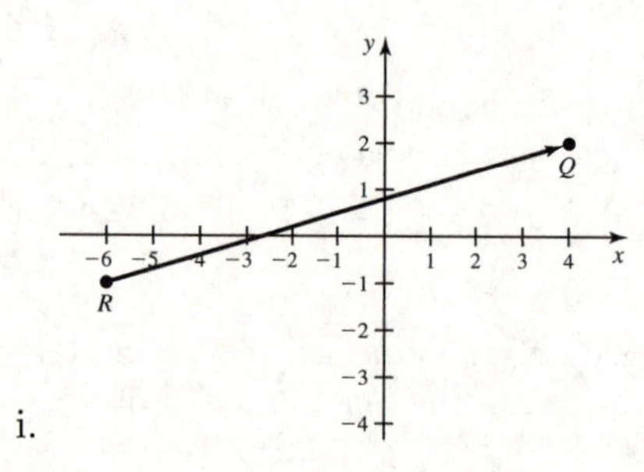

ii. $|10\mathbf{i}+3\mathbf{j}|=\sqrt{109}$.

11.1.23 $\overrightarrow{QU}=\langle 7,2\rangle$, $\overrightarrow{PT}=\langle 7,3\rangle$, $\overrightarrow{RS}=\langle 2,3\rangle$.

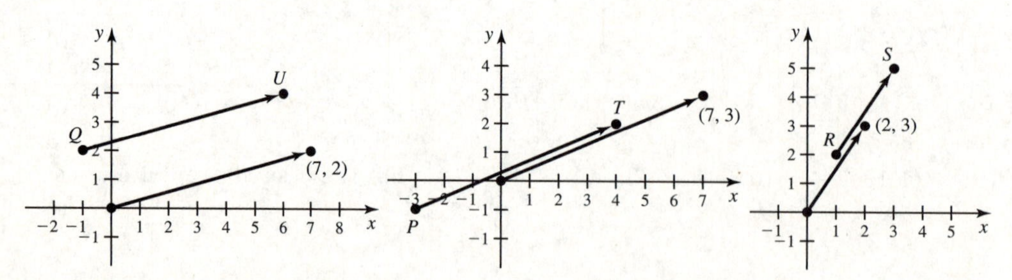

11.1.25 $\overrightarrow{QT}=\langle 5,0\rangle$, while $\overrightarrow{SU}=\langle 3,-1\rangle$.

11.1.27 $\mathbf{w}-\mathbf{u}=\langle 0,8\rangle-\langle 4,-2\rangle=\langle -4,10\rangle$.

11.1.29 $\mathbf{w}-3\mathbf{v}=\langle 0,8\rangle-3\langle -4,6\rangle=\langle 12,-10\rangle$.

11.1.31 $8\mathbf{w}+\mathbf{v}-6\mathbf{u}=8\langle 0,8\rangle+\langle -4,6\rangle-6\langle 4,-2\rangle=\langle -28,82\rangle$.

11.1.33 $|2\mathbf{u}+3\mathbf{v}-4\mathbf{w}| = |2\langle 8,-4\rangle + 3\langle 2,6\rangle - 4\langle 5,0\rangle| = |\langle 2,10\rangle| = \sqrt{104} = 2\sqrt{26}$.

11.1.35 $\mathbf{v_1} = 3\mathbf{v} = 3\langle 2,6\rangle = \langle 6,18\rangle$. $\mathbf{v_2} = -3\mathbf{v_1} = -3\langle 2,6\rangle = \langle -6,-18\rangle$.

11.1.37 $|\mathbf{u}-\mathbf{v}| = |\langle 8,-4\rangle - \langle 2,6\rangle| = |\langle 6,-10\rangle| = \sqrt{36+100} = 2\sqrt{34}$. $|\mathbf{w}-\mathbf{u}| = |\langle 5,0\rangle - \langle 8,-4\rangle| = |\langle -3,4\rangle| = 5$. Thus, $\mathbf{u}-\mathbf{v}$ has the greater magnitude.

11.1.39 $\overrightarrow{QR} = \langle 2,6\rangle - \langle 3,-4\rangle = \langle -1,10\rangle = -\mathbf{i}+10\mathbf{j}$.

11.1.41 $\mathbf{u} = \frac{\overrightarrow{PR}}{\left|\overrightarrow{PR}\right|} = \frac{\langle 2,6\rangle - \langle -4,1\rangle}{\left|\overrightarrow{PR}\right|} = \frac{\langle 6,5\rangle}{\sqrt{36+25}} = \langle 6/\sqrt{61}, 5/\sqrt{61}\rangle$.

11.1.43 $\overrightarrow{QP} = \langle -4,1\rangle - \langle 3,-4\rangle = \langle -7,5\rangle$. A unit vector parallel to $\overrightarrow{QP}$ is $\frac{1}{\sqrt{74}}\langle -7,5\rangle$. So the desired vectors are $\frac{4}{\sqrt{74}}\langle -7,5\rangle$ and $\frac{-4}{\sqrt{74}}\langle -7,5\rangle$.

11.1.45 The plane's vector is given by $\mathbf{u} = -320\mathbf{i} + -20\sqrt{2}(\mathbf{i}+\mathbf{j}) = (-320-20\sqrt{2})\mathbf{i} - 20\sqrt{2}\mathbf{j}$. The magnitude of $\mathbf{u}$ is $\sqrt{(-320-20\sqrt{2})^2 + (-20\sqrt{2})^2} \approx 349.43$ miles per hour. $\theta = \tan^{-1}\left(\frac{20\sqrt{2}}{320+20\sqrt{2}}\right) \approx .0810$ radians, or about 4.64 degrees south of west.

11.1.47 Let $\mathbf{u} = 4\mathbf{i}$ represent the wind and $\mathbf{v} = 4\sqrt{3}\cos(\pi/6)\mathbf{i} + 4\sqrt{3}\sin(\pi/6)\mathbf{j} = 6\mathbf{i} + 2\sqrt{3}\mathbf{j}$ represent the boat relative to land. If $\mathbf{w}$ represents the wind, then $\mathbf{u}+\mathbf{w} = \mathbf{v}$, so $\mathbf{w} = \mathbf{v}-\mathbf{u} = 2\mathbf{i}+2\sqrt{3}\mathbf{j}$. Then $\theta = \tan^{-1}(\sqrt{3}) = \pi/3$, or 60 degrees. The speed of the wind is 4 meters per second in the direction 60 degrees north of east (or 30 degrees east of north.)

11.1.49

a. $\mathbf{F} = 40\cos(\pi/3)\mathbf{i} + 40\sin(\pi/3)\mathbf{j} = 20\mathbf{i} + 20\sqrt{3}\mathbf{j}$, so the horizontal component is 20 and the vertical is $20\sqrt{3}$.

b. Yes. If it is 45 degrees, the horizontal component would be $40\cos(\pi/4) = 20\sqrt{2} > 20$.

c. No. If it is 45 degrees, the vertical component would be $40\sin(\pi/4) = 20\sqrt{2} < 20\sqrt{3}$.

11.1.51 Let the magnitude of the force on the two chains be f. Let $\mathbf{F_1} = (-\frac{\sqrt{2}}{2}\mathbf{i} + \frac{\sqrt{2}}{2}\mathbf{j})f$ and let $\mathbf{F_2} = (\frac{\sqrt{2}}{2}\mathbf{i} + \frac{\sqrt{2}}{2}\mathbf{j})f$. Then $\mathbf{F_1} + \mathbf{F_2} - 500\mathbf{j} = 0$, and solving for f yields $f = 250\sqrt{2}$ pounds.

11.1.53

a. True. This follows because $(\mathbf{u}+\mathbf{v})+\mathbf{w} = (\mathbf{w}+\mathbf{u})+\mathbf{v}$ (vector addition is commutative and associative.)

b. True. This is because $\mathbf{u} + (-\mathbf{u}) = \mathbf{0}$.

c. False. For example, if $\mathbf{u} = \langle 3,4\rangle$ and $\mathbf{v} = \langle -3,-1\rangle$, then $|\mathbf{u}+\mathbf{v}| = |\langle 0,3\rangle| = 3$, while $|\mathbf{u}| = 5$.

d. False. For example, if $\mathbf{u} = \langle 3,4\rangle$ and $\mathbf{v} = \langle -1,-4\rangle$, then $|\mathbf{u}+\mathbf{v}| = |\langle 2,0\rangle| = 2$, while $|\mathbf{u}|+|\mathbf{v}| = 5+\sqrt{17}$.

e. False. For example, if $\mathbf{u} = \langle 3,0\rangle$ and $\mathbf{v} = \langle 6,0\rangle$, then $\mathbf{u}$ and $\mathbf{v}$ are parallel, but have different lengths.

f. False. For example, given $A(0,0)$, $B(3,4)$, $C(1,1)$ and $D(4,5)$, we have $\overrightarrow{AB} = \langle 3,4\rangle$ and $\overrightarrow{CD} = \langle 3,4\rangle$, but $A \neq C$ and $B \neq D$.

g. False. For example, $\mathbf{u} = \langle 0,1\rangle$ and $\mathbf{v} = \langle -1,0\rangle$ are perpendicular, but $|\mathbf{u}+\mathbf{v}| = \sqrt{2}$, while $|\mathbf{u}|+|\mathbf{v}| = 2$.

h. True. Suppose $\mathbf{v} = k\mathbf{u}$ with $k > 0$. Then

$$|\mathbf{u}+\mathbf{v}| = |\mathbf{u}+k\mathbf{u}| = |(1+k)\mathbf{u}| = (1+k)\,|\mathbf{u}| = |\mathbf{u}| + k\,|\mathbf{u}| = |\mathbf{u}| + |k\mathbf{u}| = |\mathbf{u}| + |\mathbf{v}|.$$

11.1.55

a. Since the magnitude of $\mathbf{v}$ is $\sqrt{36+64} = 10$, the two desired vectors are $\langle 6/10, -8/10\rangle = \langle 3/5, -4/5\rangle$ and $\langle -3/5, 4/5\rangle$.

b. If the magnitude of $\mathbf{v}$ is 1, then $\sqrt{\frac{1}{9}+b^2} = 1$, so $b^2 = \frac{8}{9}$, so $b = \pm\frac{2\sqrt{2}}{3}$.

c. If the magnitude of $\mathbf{w}$ is 1, then $\sqrt{a^2+\frac{a^2}{9}} = 1$, so $\frac{10a^2}{9} = 1$, so $a = \pm\frac{3}{\sqrt{10}}$.

11.1.57 $10\langle a,b\rangle = \langle 2,-3\rangle$, so $10a = 2$, and $a = 1/5$. Also, $10b = -3$, so $b = -3/10$. Thus $\mathbf{x} = \langle 1/5, -3/10\rangle$.

11.1.59 $3\langle a,b\rangle - 4\langle 2,-3\rangle = \langle -4,1\rangle$, so $\langle a,b\rangle = \frac{1}{3}\langle 4,-11\rangle = \mathbf{x}$.

11.1.61 $\langle 4,-8\rangle = 4\mathbf{i} + -8\mathbf{j}$.

11.1.63 Let $\langle a,b\rangle = c_1\mathbf{u}+c_2\mathbf{v}$. Then $c_1 - c_2 = a$ and $c_1 + c_2 = b$. Adding these two equations to each other yields $2c_1 = a+b$, so $c_1 = \frac{a+b}{2}$. And thus $c_2 = \frac{b-a}{2}$. We have $\langle a,b\rangle = \frac{a+b}{2}\mathbf{u} + \frac{b-a}{2}\mathbf{v}$.

11.1.65 Since $2\mathbf{u}+3\mathbf{v} = \mathbf{i}$ and $-2(\mathbf{u}-\mathbf{v}) = -2\mathbf{j}$, we can conclude that $3\mathbf{v}+2\mathbf{v} = \mathbf{i}-2\mathbf{j}$ (by adding), so $\mathbf{v} = \frac{1}{5}\mathbf{i} - \frac{2}{5}\mathbf{j}$. It then follows that $\mathbf{u} = \mathbf{v}+\mathbf{j} = \frac{1}{5}\mathbf{i} - \frac{2}{5}\mathbf{j} + \mathbf{j} = \frac{1}{5}\mathbf{i} + \frac{3}{5}\mathbf{j}$.

11.1.67 $\mathbf{u} = 3\frac{\langle 5,-12\rangle}{\sqrt{25+144}} = \frac{3}{13}\langle 5,-12\rangle$.

11.1.69 $\mathbf{u} = \mathbf{u_1}+\mathbf{u_2} = \langle 4,-6\rangle + \langle 5,9\rangle = \langle 9,3\rangle$.

11.1.71

a. The sum is $\mathbf{0}$ since each vector has exactly one additive inverse in the set among the 12 vectors.

b. The 6:00 vector, since the others cancel in pairs, but this vector remains.

c. If we remove the 1:00 through 6:00 vectors, the sum is as large as possible, since all the vectors are pointing toward the left side of the clock. Removing any 6 consecutive vectors gives a sum whose magnitude is as large as possible.

d. Let $\mathbf{w}$ be the vector that points from 12:00 toward 6:00 but which has length r equal to the radius of the clock. The sum we are seeking is $12\mathbf{w}$. The sum of the vectors pointing to 1:00 and 11:00 add up to $(2-\sqrt{3})\mathbf{w}$, the sum of the vectors pointing to 2:00 and 10:00 is $\mathbf{w}$, the vectors pointing to 3:00 and 9:00 add up to $2\mathbf{w}$, the vectors pointing to 4:00 and 8:00 add up to $3\mathbf{w}$, and the vectors pointing to 5:00 and 7:00 add up to $(\sqrt{3}+2)\mathbf{w}$. Finally, the single vector pointing to 6:00 is $2\mathbf{w}$. The sum of all of these is $12\mathbf{w}$.

11.1.73 The magnitude of the net force is $|\mathbf{F}| = \sqrt{40^2+30^2} = 50$ pounds. $\alpha = \tan^{-1}(\frac{3}{4}) \approx .6435$ radians or 36.87 degrees. The net force has magnitude 50 pounds in the direction 36.87 degrees north of east.

11.1.75 $\mathbf{u}+\mathbf{v} = \langle u_1,u_2\rangle + \langle v_1,v_2\rangle = \langle u_1+v_1, u_2+v_2\rangle = \langle v_1+u_1, v_2+u_2\rangle = \langle v_1,v_2\rangle + \langle u_1,u_2\rangle = \mathbf{v}+\mathbf{u}$.

11.1.77 $a(c\mathbf{v}) = a(c\langle v_1,v_2\rangle) = a\langle cv_1,cv_2\rangle = \langle a(cv_1), a(cv_2)\rangle = \langle (ac)v_1, (ac)v_2\rangle = (ac)\langle v_1,v_2\rangle = (ac)\mathbf{v}$.

11.1.79 $(a+c)\mathbf{v} = (a+c)\langle v_1,v_2\rangle = \langle (a+c)v_1, (a+c)v_2\rangle = \langle av_1+cv_1, av_2+cv_2\rangle = \langle av_1,av_2\rangle + \langle cv_1,cv_2\rangle = a\mathbf{v}+c\mathbf{v}$.

11.1.81 $|c\mathbf{v}| = |c\langle v_1,v_2\rangle| = |\langle cv_1,cv_2\rangle| = \sqrt{(cv_1)^2+(cv_2)^2} = \sqrt{c^2}\sqrt{v_1^2+v_2^2} = |c|\,|\mathbf{v}|$.

11.1.83

a. Note that $-6\mathbf{u} = \mathbf{v}$, so $\mathbf{u}$ and $\mathbf{v}$ are linearly dependent. But there is no scalar c so that $c\mathbf{u} = \mathbf{w}$, nor any scalar d so that $d\mathbf{v} = \mathbf{w}$ so $\mathbf{w}$ is independent of both $\mathbf{u}$ and $\mathbf{v}$.

b. Two nonzero vectors are linearly independent when they are not parallel, and are linearly dependent when they are parallel.

c. Suppose **u** and **v** are linearly independent. Consider the equation $c_1\mathbf{u}+c_2\mathbf{v}=\mathbf{w}$ for a given vector **w**. We are seeking a solution for the system of linear equatons $c_1u_1+c_2v_1=w_1$ and $c_1u_2+c_2v_2=w_2$. The solution for this system is given by $c_1=\frac{1}{u_1v_2-u_2v_1}(v_2w_1-v_1w_2)$ and $c_2=\frac{1}{u_1v_2-u_2v_1}(-u_2w_1+u_1w_2)$, provided $u_1v_2-u_2v_1\neq 0$. This condition is equivalent to saying that **v** is not a multiple of **u**. Thus a solution to the system of linear equations exists exactly when the vectors **u** and **v** are linearly independent.

11.1.85

a. If **u** and **v** are parallel, we must have $\frac{a}{2}=\frac{5}{6}$, so $a=\frac{5}{3}$.

b. If **u** and **v** are perpendicular, we must have $2a+30=0$, so $a=-15$.

11.2 Vectors in Three Dimensions

11.2.1 Starting at the origin $(0,0,0)$, move 3 units in the positive x-direction, 2 units in the negative y-direction, and 1 unit in the positive z-direction, to arrive at the point $(3,-2,1)$.

11.2.3 The plane $x=4$ is parallel to the yz-plane, but contains all of the points with x-coordinate 4. It is perpendicular to the x-axis.

11.2.5 $\mathbf{u}+\mathbf{v}=\langle 3+6,5+(-5),-7+1\rangle=\langle 9,0,-6\rangle$. $3\mathbf{u}-\mathbf{v}=\langle 9,15,-21\rangle-\langle 6,-5,1\rangle=\langle 3,20,-22\rangle$.

11.2.7 Since $\sqrt{3^2+(-1)^2+2^2}=\sqrt{14}<\sqrt{0^2+0^2+(-4)^2}=4$, the point $(0,0,-4)$ is further from the origin.

11.2.9 $A(3,0,5)$, $B(3,4,0)$, $C(0,4,5)$.

11.2.11 $A(3,-4,5)$, $B(0,-4,0)$, $C(0,-4,5)$.

11.2.13

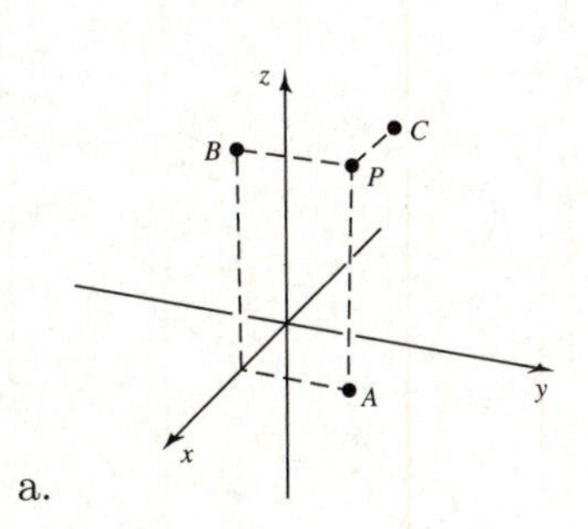

a.

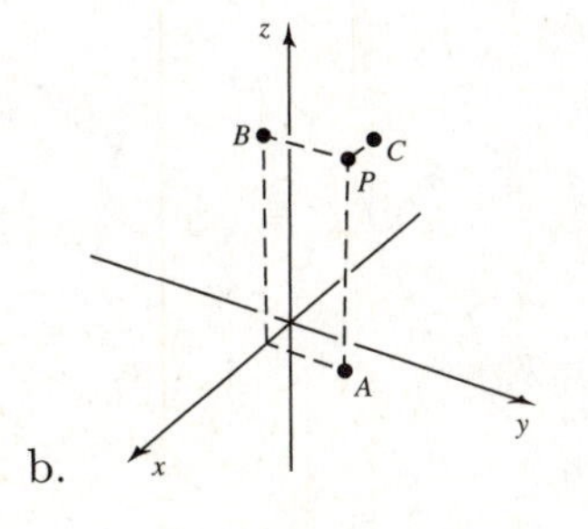

b.

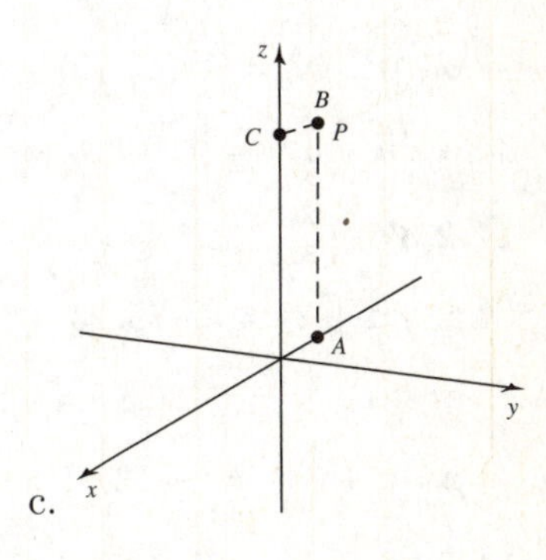

c.

11.2.15

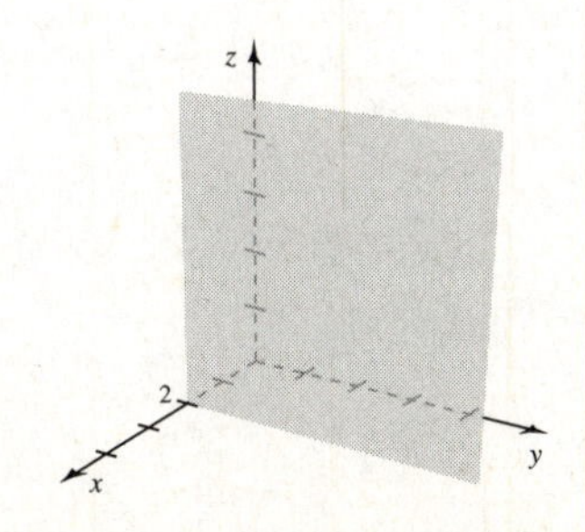

11.2.17

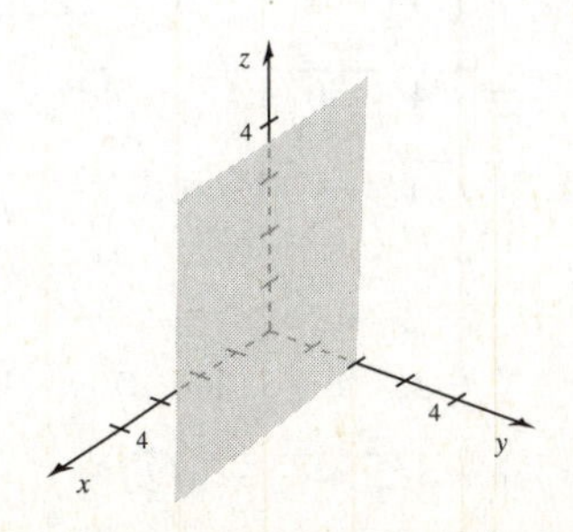

11.2.19

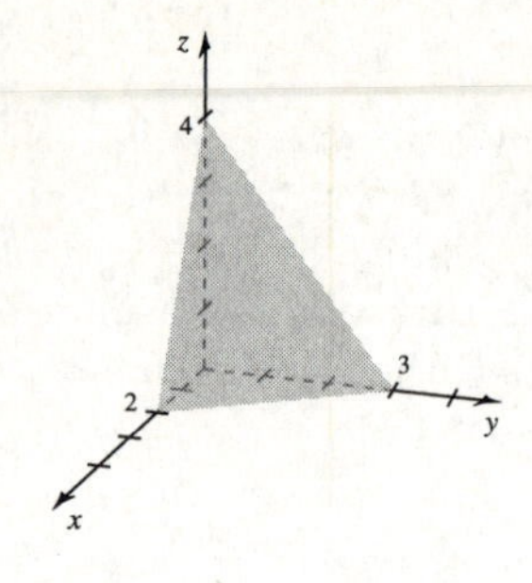

11.2.21

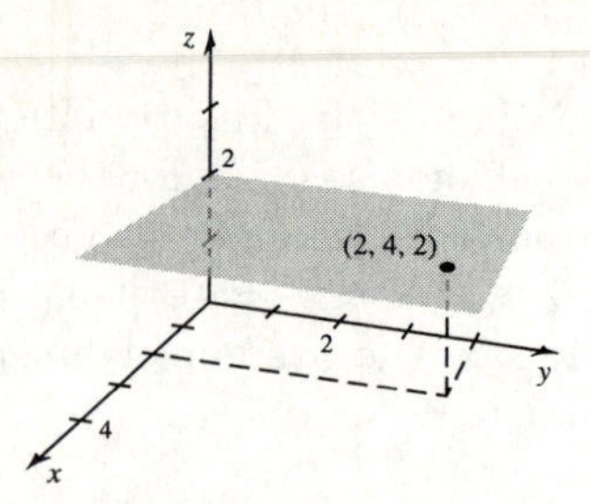

The plane $z = 2$.

11.2.23 $(x-1)^2 + (y-2)^2 + (z-3)^2 = 16$.

11.2.25 $(x+2)^2 + y^2 + (z-4)^2 \le 1$.

11.2.27 The midpoint of the line segment $\overline{PQ}$ is $(\frac{1+2}{2}, \frac{0+3}{2}, \frac{5+9}{2}) = (3/2, 3/2, 7)$. The radius of the sphere is $r = \frac{1}{2}\sqrt{(2-1)^2 + (3-0)^2 + (9-5)^2} = \frac{\sqrt{26}}{2}$. The equation of the sphere is therefore $(x - 3/2)^2 + (y - 3/2)^2 + (z-7)^2 = \frac{13}{2}$.

11.2.29 Completing the squares, we have $x^2 + (y^2 - 2y + 1) + (z^2 - 4z + 4) = 4 + 5$, so we have $x^2 + (y-1)^2 + (z-2)^2 = 3^2$, which describes a sphere of radius 3 centered at $(0, 1, 2)$.

11.2.31 Completing the square, we have $x^2 + (y^2 - 14y + 49) + z^2 \ge -13 + 49 = 36$, which can be written as $x^2 + (y-7)^2 + z^2 \ge 6^2$. This is the outside of a ball centered at $(0, 7, 0)$ with radius 6. (Including the sphere itself.)

11.2.33 Completing the squares, we have $(x^2 - 8x + 16) + (y^2 - 14y + 49) + (z^2 - 18z + 81) \le 65 + 16 + 49 + 81 = 211$, which can be written as $(x-4)^2 + (y-7)^2 + (z-9)^2 \le 211$. This is the inside of a ball centered at $(4, 7, 9)$ with radius $\sqrt{211}$. (Including the sphere itself.)

11.2.35

a. $3\mathbf{u} + 2\mathbf{v} = \langle 9, 9, 4 \rangle$.

b. $4\mathbf{u} - \mathbf{v} = \langle 1, 12, -2 \rangle$.

c. $|\mathbf{u} + 3\mathbf{v}| = |\langle 10, 3, 6 \rangle| = \sqrt{145}$.

11.2.37

a. $3\mathbf{u} + 2\mathbf{v} = \langle -25, 23, 3 \rangle$.

b. $4\mathbf{u} - \mathbf{v} = \langle -26, 16, 4 \rangle$.

c. $|\mathbf{u} + 3\mathbf{v}| = |\langle -13, 17, 1 \rangle| = 3\sqrt{51}$.

11.2.39

a. $\overrightarrow{PQ} = \langle 3-1, 11-5, 2-0 \rangle = \langle 2, 6, 2 \rangle = 2\mathbf{i} + 6\mathbf{j} + 2\mathbf{k}$.

b. $|\langle 2, 6, 2 \rangle| = \sqrt{4 + 36 + 4} = \sqrt{44} = 2\sqrt{11}$.

c. $\langle 1/\sqrt{11}, 3/\sqrt{11}, 1/\sqrt{11} \rangle$ and $\langle -1/\sqrt{11}, -3/\sqrt{11}, -1/\sqrt{11} \rangle$.

11.2.41

a. $\overrightarrow{PQ} = \langle -3+3, -4-1, 1-0 \rangle = \langle 0, -5, 1 \rangle = -5\mathbf{j} + 1\mathbf{k}$.

b. $|\langle 0, -5, 1 \rangle| = \sqrt{25 + 1} = \sqrt{26}$.

c. $\langle 0, -5/\sqrt{26}, 1/\sqrt{26}\rangle$ and $\langle 0, 5/\sqrt{26}, -1/\sqrt{26}\rangle$.

11.2.43

a. $\overrightarrow{PQ} = \langle -2-0, 4-0, 0-2\rangle = \langle -2, 4, -2\rangle = -2\mathbf{i} + 4\mathbf{j} - 2\mathbf{k}$.

b. $|\langle -2, 4, -2\rangle| = \sqrt{4+16+4} = 2\sqrt{6}$.

c. $\langle -1/\sqrt{6}, 2/\sqrt{6}, -1/\sqrt{6}\rangle$ and $\langle 1/\sqrt{6}, -2/\sqrt{6}, 1/\sqrt{6}\rangle$.

11.2.45 The airplane's velocity is $\mathbf{v_1} = 250\mathbf{i}$. The crosswind is blowing $\mathbf{v_2} = -25\sqrt{2}\mathbf{i} - 25\sqrt{2}\mathbf{j}$. The updraft is $\mathbf{v_3} = 30\mathbf{k}$. We have $|\mathbf{v_1} + \mathbf{v_2} + \mathbf{v_3}| = |\langle 250 - 25\sqrt{2}, -25\sqrt{2}, 30\rangle| \approx 219.596$ miles per hour. The direction is sketched in the diagram—it is slightly south of east and upward.

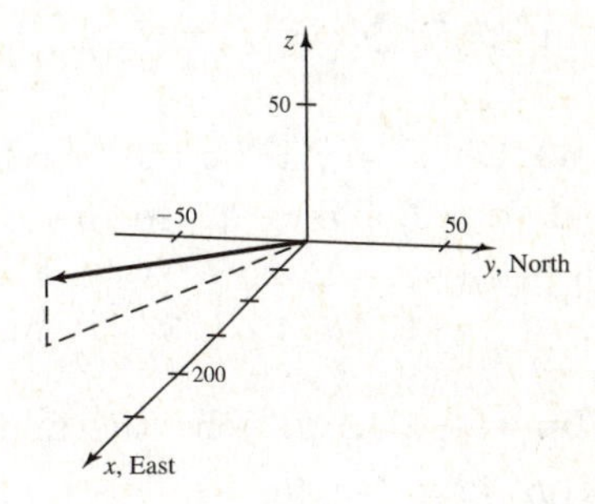

11.2.47 The component in the east direction is $(20\cos 30°)(\cos 45°) = 5\sqrt{6}$ knots. In the north direction, it is $(20\cos 30°)(\sin 45°) = 5\sqrt{6}$ knots. In the vertical direction, it is $20\sin 30° = 10$ knots.

11.2.49

a. False. For example, let $\mathbf{u} = \langle 1,0,0\rangle$, $\mathbf{v} = \langle 0,1,0\rangle$ and $\mathbf{w} = \langle 1,1,0\rangle$. Then both $\mathbf{u}$ and $\mathbf{v}$ make a 45 degree angle with $\mathbf{w}$, but $\mathbf{u}+\mathbf{v} = \mathbf{w}$ makes a zero degree angle with $\mathbf{w}$.

b. False. For example, $\mathbf{i}$ and $\mathbf{j}$ form a 90 degree angle with $\mathbf{k}$, as does $\mathbf{i}+\mathbf{j}$.

c. False. $\mathbf{i}+\mathbf{j}+\mathbf{k} = \langle 1,1,1\rangle \neq \langle 0,0,0\rangle$.

d. True. They intersect at the point $(1,1,1)$.

11.2.51

This represents all the points in 3-space, excluding the three axes.

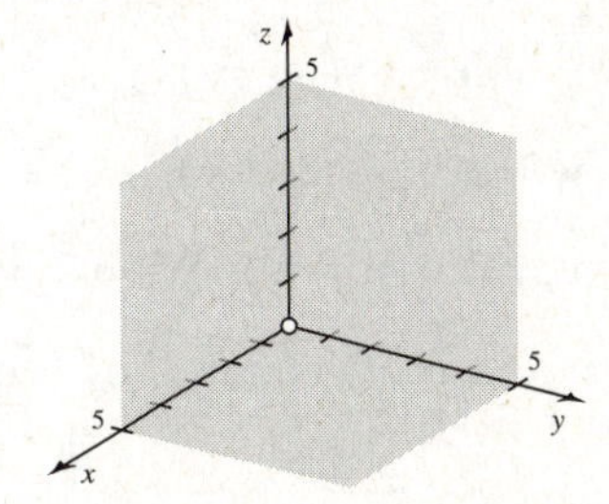

11.2.53 Since the magnitude of $\mathbf{v}$ is $\sqrt{36+64+0} = 10$, the desired vectors are $\pm 20\langle .6, -.8, 0\rangle = \pm\langle 12, -16, 0\rangle$.

11.2.55 Since the magnitude of $\mathbf{v}$ is $\sqrt{1+1+1} = \sqrt{3}$, the desired vectors are $\pm 3\langle -1/\sqrt{3}, -1/\sqrt{3}, 1/\sqrt{3}\rangle = \pm\langle -\sqrt{3}, -\sqrt{3}, \sqrt{3}\rangle$.

11.2.57

a. Since $\overrightarrow{PQ} = \langle 1,-1,2\rangle$ and $\overrightarrow{PR} = \langle 3,-3,6\rangle = 3\langle 1,-1,2\rangle$ they are collinear. Q is between P and R.

b. Since $\overrightarrow{PQ} = \langle 4,8,-8\rangle$ and $\overrightarrow{PR} = \langle -1,-2,2\rangle = \frac{-1}{4}\langle 4,8,-8\rangle$ they are collinear. P is between Q and R.

c. Since $\overrightarrow{PQ} = \langle 1, -5, 3\rangle$ and $\overrightarrow{PR} = \langle 2, -3, 6\rangle$ are not parallel, the given points are not collinear.

d. Since $\overrightarrow{PQ} = \langle 2, 13, 3\rangle$ and $\overrightarrow{PR} = \langle -3, -2, -1\rangle$ are not parallel, the given points are not collinear.

11.2.59 The diagonal of the box has magnitude $\sqrt{2^2+3^2+4^2} = \sqrt{29}$, so the longest rod that will fit in the box has length $\sqrt{29}$ feet.

11.2.61 Let $P(1, -\sqrt{3}, 0)$, $Q(1, \sqrt{3}, 0)$, $R(-2, 0, 0)$, and $S(0, 0, -2\sqrt{3})$ be the given points. Note that $\overrightarrow{PS} = \langle -1, \sqrt{3}, -2\sqrt{3}\rangle$, $\overrightarrow{QS} = \langle -1, -\sqrt{3}, -2\sqrt{3}\rangle$, $\overrightarrow{RS} = \langle 2, 0, -2\sqrt{3}\rangle$. Let $x(\overrightarrow{PS} + \overrightarrow{QS} + \overrightarrow{RS}) = -500\mathbf{k}$, then $-6\sqrt{3}x = -500$, so $x = \frac{250}{3\sqrt{3}}$. Then $x\overrightarrow{PS} = \frac{250}{3\sqrt{3}}\langle -1, \sqrt{3}, -2\sqrt{3}\rangle = \frac{250}{3}\langle -1/\sqrt{3}, 1, -2\rangle$. $x\overrightarrow{QS} = \frac{250}{3\sqrt{3}}\langle -1, -\sqrt{3}, -2\sqrt{3}\rangle = \frac{250}{3}\langle -1/\sqrt{3}, -1, 2\rangle$. $x\overrightarrow{RS} = \frac{250}{3\sqrt{3}}\langle 2, 0, -2\sqrt{3}\rangle = \frac{250}{3}\langle 2/\sqrt{3}, 0, -2\rangle$.

11.2.63 Let $R(x, y, z)$ be the fourth vertex. Then perhaps $\overrightarrow{OQ} = \overrightarrow{RP}$, so $\langle 2, 4, 3\rangle = \langle 1-x, 4-y, 6-z\rangle$, so $x = -1$, $y = 0$, and $z = 3$, so $R(-1, 0, 3)$ is one possible desired vertex. We could also have $\overrightarrow{RP} = -\overrightarrow{OQ}$, in which case $\langle -2, -4, -3\rangle = \langle 1-x, 4-y, 6-z\rangle$, so $R(3, 8, 9)$ is the other vertex. We could also have $\overrightarrow{OP} = \overrightarrow{RQ}$, so $\langle 1, 4, 6\rangle = \langle 2-x, 4-y, 3-z\rangle$ and $R(1, 0, -3)$ is the desired point.

11.2.65 Let $M(x, y, z)$ be the midpoint. Since $\overrightarrow{OM} = \overrightarrow{OP} + \frac{1}{2}\overrightarrow{PQ}$, we have $\langle x, y, z\rangle = \langle x_1, y_1, z_1\rangle + \frac{1}{2}(\langle x_2, y_2, z_2\rangle - \langle x_1, y_1, z_1\rangle) = \langle x_1, y_1, z_1\rangle + \langle \frac{1}{2}x_2, \frac{1}{2}y_2, \frac{1}{2}z_2\rangle + \langle \frac{-1}{2}x_1, \frac{-1}{2}y_1, \frac{-1}{2}z_1\rangle = \langle \frac{x_1+x_2}{2}, \frac{y_1+y_2}{2}, \frac{z_1+z_2}{2}\rangle$.

11.2.67

a. $\mathbf{u} + \mathbf{v} = -\mathbf{w}$ (by the geometric definition of vector addition), so $\mathbf{u} + \mathbf{v} + \mathbf{w} = \mathbf{0}$.

b. Let $\mathbf{M_1} = \overrightarrow{EB}$, $\mathbf{M_2} = \overrightarrow{FO}$, and $\mathbf{M_3} = \overrightarrow{GA}$. Consider triangle EAB. We have $\overrightarrow{EA} + \overrightarrow{AB} + \overrightarrow{BE} = \mathbf{0}$, so $\frac{1}{2}\mathbf{u} + \mathbf{v} = -\overrightarrow{BE} = \overrightarrow{EB} = \mathbf{M_1}$. Using similar arguments, we have $\mathbf{M_2} = \frac{1}{2}\mathbf{v} + \mathbf{w}$ and $\mathbf{M_3} = \frac{1}{2}\mathbf{w} + \mathbf{u}$.

c. Let $\mathbf{a}$, $\mathbf{b}$, and $\mathbf{c}$ be the vectors from O to the points 1/3 of the way along $\mathbf{M_1}$, $\mathbf{M_2}$ and $\mathbf{M_3}$ respectively. Since $-\mathbf{w} = \mathbf{u} + \mathbf{v}$, we have $\frac{\mathbf{u}-\mathbf{w}}{3} = \frac{\mathbf{u}}{3} + \frac{\mathbf{u}+\mathbf{v}}{3} = \frac{2}{3}\mathbf{u} + \frac{1}{3}\mathbf{v}$. Also, $\mathbf{a} = \frac{1}{2}\mathbf{u} + \frac{1}{3}\mathbf{M_1} = \frac{1}{2}\mathbf{u} + \frac{1}{3}\left(\frac{1}{2}\mathbf{u} + \mathbf{v}\right) = \frac{1}{2}\mathbf{u} + \frac{1}{6}\mathbf{u} + \frac{1}{3}\mathbf{v} = \frac{2}{3}\mathbf{u} + \frac{1}{3}\mathbf{v}$. Thus $\mathbf{a} = \frac{\mathbf{u}-\mathbf{w}}{3}$. Also, $\mathbf{b} = \frac{-2}{3}\mathbf{M_2} = \frac{-2}{3}(\frac{1}{2}\mathbf{v} + (-\mathbf{u} - \mathbf{v})) = \frac{2}{3}\mathbf{u} + \frac{1}{3}\mathbf{v}$. We also have $\mathbf{c} = -\frac{1}{2}\mathbf{w} + \frac{1}{3}\mathbf{M_3} = -\frac{1}{2}\mathbf{w} + \frac{1}{3}(\frac{1}{2}\mathbf{w} + \mathbf{u}) = \frac{1}{3}\mathbf{u} + \frac{-1}{3}(-\mathbf{u} - \mathbf{v}) = \frac{2}{3}\mathbf{u} + \frac{1}{3}\mathbf{v}$. Thus $\mathbf{a} = \mathbf{b} = \mathbf{c}$.

d. Since $\mathbf{a} = \mathbf{b} = \mathbf{c}$, the medians all meet at a point that divides each median in a 2:1 ratio.

11.2.69

a. $\mathbf{u} + \mathbf{v} = \overrightarrow{PR}$ and $\mathbf{w} + \mathbf{x} = \mathbf{x} + \mathbf{w} = \overrightarrow{PR}$, so $\mathbf{u} + \mathbf{v} = \mathbf{w} + \mathbf{x}$.

b. $\frac{1}{2}\mathbf{u} + \frac{1}{2}\mathbf{v} = \mathbf{m} = \frac{1}{2}(\mathbf{u} + \mathbf{v})$.

c. $\frac{1}{2}\mathbf{x} + \frac{1}{2}\mathbf{w} = \mathbf{n} = \frac{1}{2}(\mathbf{x} + \mathbf{w})$.

d. We have $\mathbf{n} = \frac{1}{2}(\mathbf{x} + \mathbf{w}) = \frac{1}{2}(\mathbf{u} + \mathbf{v}) = \mathbf{m}$.

e. Since $\mathbf{m}$ and $\mathbf{n}$ are equal, they are parallel. A similar argument will show that the other two sides are parallel as well.

11.3 Dot Products

11.3.1 $\mathbf{u} \cdot \mathbf{v} = |\mathbf{u}|\,|\mathbf{v}|\cos\theta$, where θ is the angle between the two vectors.

11.3.3 $\langle 2, 3, -6\rangle \cdot \langle 1, -8, 3\rangle = 2 \cdot 1 + 3 \cdot (-8) + (-6) \cdot 3 = -40$.

11.3.5 Given non-zero vectors $\mathbf{u}$ and $\mathbf{v}$, the angle between them is $\cos^{-1}(\frac{\mathbf{u}\cdot\mathbf{v}}{|\mathbf{u}||\mathbf{v}|})$.

11.3.7 The scalar component of $\mathbf{u}$ in the direction of $\mathbf{v}$ is the number $|\mathbf{u}|\cos\theta$ where θ is the angle between the vectors. This number represents the signed length of the "shadow" that $\mathbf{u}$ casts on $\mathbf{v}$. Thus, referring to the diagram in the previous problem, the scalar projection is the length of the base of the shaded triangle.

11.3.9 $\mathbf{u}\cdot\mathbf{v} = 4\cdot 6\cdot\cos(\pi/2) = 0$.

11.3.11 The angle between these vectors is $\pi/4$. Thus, their dot product is $10\cdot 10\sqrt{2}\cdot\frac{\sqrt{2}}{2} = 100$.

11.3.13 $\mathbf{u}\cdot\mathbf{v} = 4\cdot 4 + 3\cdot(-6) = -2$. The angle between the vectors is thus $\cos^{-1}(\frac{-2}{|\mathbf{u}||\mathbf{v}|}) = \cos^{-1}(\frac{-2}{5\cdot 2\sqrt{13}}) \approx$ 1.627 radians.

11.3.15 $\mathbf{u}\cdot\mathbf{v} = -10+0+12 = 2$. The angle between the vectors is thus $\cos^{-1}(\frac{2}{|\mathbf{u}||\mathbf{v}|}) = \cos^{-1}(\frac{2}{\sqrt{116}\cdot\sqrt{14}}) \approx$ 1.521 radians.

11.3.17 $\mathbf{u}\cdot\mathbf{v} = 2+0-6 = -4$. The angle between the vectors is thus $\cos^{-1}(\frac{-4}{|\mathbf{u}||\mathbf{v}|}) = \cos^{-1}(\frac{-4}{\sqrt{13}\cdot\sqrt{21}}) \approx 1.815$ radians.

11.3.19 $\text{proj}_{\mathbf{v}}\mathbf{u} = 3\mathbf{i}$. $\text{scal}_{\mathbf{v}}\mathbf{u} = 3$.

11.3.21 $\text{proj}_{\mathbf{v}}\mathbf{u} = 3\mathbf{j}$. $\text{scal}_{\mathbf{v}}\mathbf{u} = 3$.

11.3.23 $\text{proj}_{\mathbf{v}}\mathbf{u} = \frac{\mathbf{u}\cdot\mathbf{v}}{\mathbf{v}\cdot\mathbf{v}}\mathbf{v} = \frac{12}{20}\langle -4, 2\rangle = \langle -12/5, 6/5\rangle$. $\text{scal}_{\mathbf{v}}\mathbf{u} = \frac{\mathbf{u}\cdot\mathbf{v}}{|\mathbf{v}|} = \frac{12}{\sqrt{20}} = \frac{6}{\sqrt{5}}$.

11.3.25 $\text{proj}_{\mathbf{v}}\mathbf{u} = \frac{\mathbf{u}\cdot\mathbf{v}}{\mathbf{v}\cdot\mathbf{v}}\mathbf{v} = \frac{-14}{19}\langle 1, 3, -3\rangle = \langle -14/19, -42/19, 42/19\rangle$. $\text{scal}_{\mathbf{v}}\mathbf{u} = \frac{\mathbf{u}\cdot\mathbf{v}}{|\mathbf{v}|} = \frac{-14}{\sqrt{19}}$.

11.3.27 $\text{proj}_{\mathbf{v}}\mathbf{u} = \frac{\mathbf{u}\cdot\mathbf{v}}{\mathbf{v}\cdot\mathbf{v}}\mathbf{v} = \frac{10}{85}\langle 9, 0, 2\rangle = \frac{2}{17}\langle 9, 0, 2\rangle$. $\text{scal}_{\mathbf{v}}\mathbf{u} = \frac{\mathbf{u}\cdot\mathbf{v}}{|\mathbf{v}|} = \frac{10}{\sqrt{85}}$.

11.3.29 $w = 30\cdot 50\cos\pi/6 = 750\sqrt{3}$ foot-pounds.

11.3.31 $w = (40\mathbf{i} + 30\mathbf{j})\cdot 10\mathbf{i} = 400$ J.

11.3.33 Parallel to: use $\mathbf{v} = \langle\sqrt{2}/2, \sqrt{2}/2\rangle$. $\text{proj}_{\mathbf{v}}\mathbf{F} = \frac{\mathbf{F}\cdot\mathbf{v}}{\mathbf{v}\cdot\mathbf{v}}\mathbf{v} = \frac{-5\sqrt{2}}{1}\langle\sqrt{2}/2, \sqrt{2}/2\rangle = \langle -5, -5\rangle$.
Normal to: use $\mathbf{v} = \langle -\sqrt{2}/2, \sqrt{2}/2\rangle$. $\text{proj}_{\mathbf{v}}\mathbf{F} = \frac{\mathbf{F}\cdot\mathbf{v}}{\mathbf{v}\cdot\mathbf{v}}\mathbf{v} = \frac{-5\sqrt{2}}{1}\langle -\sqrt{2}/2, \sqrt{2}/2\rangle = \langle 5, -5\rangle$.
Note that $\langle -5, -5\rangle + \langle 5, -5\rangle = \langle 0, -10\rangle$.

11.3.35 Parallel to: use $\mathbf{v} = \langle 1/2, \sqrt{3}/2\rangle$. $\text{proj}_{\mathbf{v}}\mathbf{F} = \frac{\mathbf{F}\cdot\mathbf{v}}{\mathbf{v}\cdot\mathbf{v}}\mathbf{v} = \frac{-5\sqrt{3}}{1}\langle 1/2, \sqrt{3}/2\rangle = \langle -5\sqrt{3}/2, -15/2\rangle$.
Normal to: use $\mathbf{v} = \langle -\sqrt{3}/2, 1/2\rangle$. $\text{proj}_{\mathbf{v}}\mathbf{F} = \frac{\mathbf{F}\cdot\mathbf{v}}{\mathbf{v}\cdot\mathbf{v}}\mathbf{v} = \frac{-5}{1}\langle -\sqrt{3}/2, 1/2\rangle = \langle 5\sqrt{3}/2, -5/2\rangle$.
Note that $\langle -5\sqrt{3}/2, -15/2\rangle + \langle 5\sqrt{3}/2, -5/2\rangle = \langle 0, -10\rangle$.

11.3.37

a. False. One is a vector in the same direction as $\mathbf{u}$ and the other is a vector in the direction of $\mathbf{v}$, so if these vectors aren't in the same direction, they can't be equal.

b. True. This follows because $\mathbf{u}\cdot(\mathbf{u}+\mathbf{v}) = |\mathbf{u}|^2 + \mathbf{u}\cdot\mathbf{v}$ and $\mathbf{v}\cdot(\mathbf{u}+\mathbf{v}) = \mathbf{v}\cdot\mathbf{u} + |\mathbf{v}|^2$, and these are equal if $\mathbf{u}$ and $\mathbf{v}$ have the same magnitude.

c. True. Let $\mathbf{u} = \langle a, b, c\rangle$. Then $(\mathbf{u}\cdot\mathbf{i})^2 + (\mathbf{u}\cdot\mathbf{j})^2 + (\mathbf{u}\cdot\mathbf{k})^2 = a^2 + b^2 + c^2 = |\mathbf{u}|^2$.

d. False. For example, consider $\mathbf{u} = \langle 1, 0\rangle$, $\mathbf{v} = \langle 0, 1\rangle$, and $\mathbf{w} = \langle 2, 0\rangle$.

e. False. Consider $\langle 1, -1, 0\rangle$, $\langle 2, -1, -1\rangle$ and $\langle 3, -2, -1\rangle$. These are all orthogonal to $\langle 1, 1, 1\rangle$, but don't all lie in the same line.

f. True. If $\mathbf{u}$ and $\mathbf{v}$ are nonzero vectors, then $\text{proj}_{\mathbf{v}}\mathbf{u} = \frac{\mathbf{u}\cdot\mathbf{v}}{\mathbf{v}\cdot\mathbf{v}}\mathbf{v}$, and this can't be zero unless $\mathbf{u}\cdot\mathbf{v} = 0$.

11.3.39 We must have $4 - 8a + 2b = 0$, so $b = 4a - 2$. These vectors have the form $\langle 1, a, 4a-2\rangle$ where a can be any real number.

11.3.41 Let $\mathbf{u} = \pm\langle\sqrt{2}/2, \sqrt{2}/2, 0\rangle$, $\mathbf{v} = \pm\langle -\sqrt{2}/2, \sqrt{2}/2, 0\rangle$ and $\mathbf{w} = \pm\langle 0, 0, 1\rangle$.

11.3.43

a. $\text{proj}_{\mathbf{k}}\mathbf{u} = \frac{\mathbf{u}\cdot\mathbf{k}}{\mathbf{k}\cdot\mathbf{k}}\mathbf{k} = \frac{|\mathbf{u}||\mathbf{k}|\cos\theta}{\mathbf{k}\cdot\mathbf{k}}\mathbf{k} = \frac{1}{2}\mathbf{k}$, which is independent of $\mathbf{u}$.

b. Yes, since the scalar projection is the length of the vector projection. In fact, using the above result, it is equal to $1/2$.

11.3.45 Using the idea from the last problem, any vector of the form $\langle x, y\rangle$ with $x+y=3$ will work, so any vector of the form $\langle x, 3-x\rangle$.

11.3.47 Note that $\text{proj}_{\mathbf{v}}\mathbf{u} = \frac{\langle 1,2,3\rangle\cdot\langle 0,0,1\rangle}{1}\langle 0,0,1\rangle = \langle 0,0,3\rangle$. We are seeking $\langle x,y,z\rangle$ so that $\frac{z}{1}\langle 0,0,1\rangle = \langle 0,0,3\rangle$, so we require $z=3$. Any vector of the form $\langle x,y,3\rangle$ will suffice.

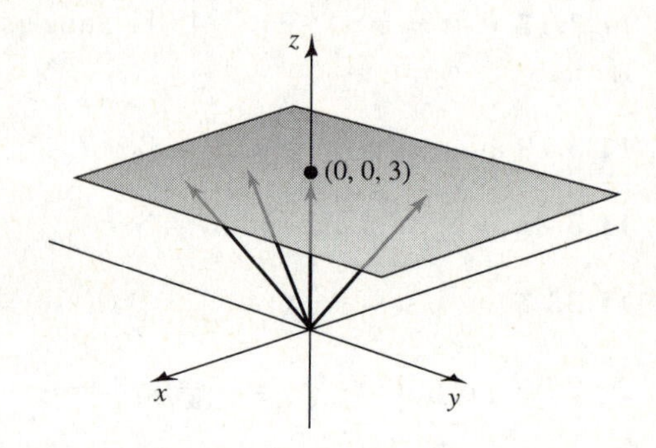

11.3.49 Let $\mathbf{p} = \text{proj}_{\mathbf{v}}\mathbf{u} = \frac{-2}{5}\langle 2,1\rangle$. Then let $\mathbf{n} = \mathbf{u} - \mathbf{p} = \langle -2,2\rangle - \frac{-2}{5}\langle 2,1\rangle = \langle -6/5, 12/5\rangle$.

11.3.51 Let $\mathbf{p} = \text{proj}_{\mathbf{v}}\mathbf{u} = \frac{3}{6}\langle 2,1,1\rangle$. Then let $\mathbf{n} = \mathbf{u} - \mathbf{p} = \langle -1,2,3\rangle - \frac{1}{2}\langle 2,1,1\rangle = \langle -2, 3/2, 5/2\rangle$.

11.3.53

a. $\mathbf{v} = \langle 1,2\rangle$.

b. $\mathbf{u} = \langle -12,4\rangle$.

c. $\text{proj}_{\mathbf{v}}\mathbf{u} = \frac{-4}{5}\langle 1,2\rangle$.

d. $\mathbf{w} = \mathbf{u} - \text{proj}_{\mathbf{v}}\mathbf{u} = \langle -12,4\rangle - \langle -4/5, -8/5\rangle = \langle -56/5, 28/5\rangle$. Note that $\mathbf{w}\cdot\mathbf{v} = 0$, and has length equal to the distance between P and l.

e. $|\mathbf{w}| = \frac{1}{5}\sqrt{(56)^2 + (28)^2} = \frac{28\sqrt{5}}{5}$. $|\mathbf{w}|$ is the component of $\mathbf{u}$ orthogonal to $\mathbf{v}$, so it is the distance from P to l.

11.3.55

a. $\mathbf{v} = \langle -6,8,3\rangle$.

b. $\mathbf{u} = \langle 1,1,-1\rangle$.

c. $\text{proj}_{\mathbf{v}}\mathbf{u} = \frac{-1}{109}\langle -6,8,3\rangle$.

d. $\mathbf{w} = \mathbf{u} - \text{proj}_{\mathbf{v}}\mathbf{u} = \langle 1,1,-1\rangle - \frac{-1}{109}\langle -6,8,3\rangle = \langle 103/109, 117/109, -106/109\rangle$. Note that $\mathbf{w}\cdot\mathbf{v} = 0$, and has length equal to the distance between P and l.

e. $|\mathbf{w}| = \frac{1}{109}\sqrt{(103)^2 + (117)^2 + (-106)^2} = \sqrt{\frac{326}{109}}$. $|\mathbf{w}|$ is the component of $\mathbf{u}$ orthogonal to $\mathbf{v}$, so it is the distance from P to l.

11.3.57 $\mathbf{I} = \langle 1/\sqrt{2}, 1/\sqrt{2}\rangle = \frac{1}{\sqrt{2}}\mathbf{i} + \frac{1}{\sqrt{2}}\mathbf{j}$. $\mathbf{J} = \langle -1/\sqrt{2}, 1/\sqrt{2}\rangle = \frac{-1}{\sqrt{2}}\mathbf{i} + \frac{1}{\sqrt{2}}\mathbf{j}$. $\mathbf{i} = \langle 1,0\rangle = \frac{\sqrt{2}}{2}(\mathbf{I} - \mathbf{J})$. $\mathbf{j} = \frac{\sqrt{2}}{2}(\mathbf{I} + \mathbf{J})$.

11.3.59

a. $|\mathbf{I}| = \sqrt{1/4+1/4+1/2} = 1$. $|\mathbf{J}| = \sqrt{1/2+1/2+0} = 1$. $|\mathbf{K}| = \sqrt{1/4+1/4+1/2} = 1$.

b. $\mathbf{I}\cdot\mathbf{J} = -1/2\sqrt{2}+1/2\sqrt{2} = 0$. $\mathbf{I}\cdot\mathbf{K} = 1/4+1/4-1/2 = 0$, and $\mathbf{J}\cdot\mathbf{K} = -1/2\sqrt{2}+1/2\sqrt{2} = 0$.

c. Let $\langle 1,0,0\rangle = a\mathbf{I}+b\mathbf{J}+c\mathbf{K}$. Then $\frac{1}{2}a - \frac{1}{\sqrt{2}}b + \frac{1}{2}c = 1$, $\frac{1}{2}a + \frac{1}{\sqrt{2}}b + \frac{1}{2}c = 0$, and $\frac{1}{\sqrt{2}}a - \frac{1}{\sqrt{2}}c = 0$. Solving this system of linear equations yields $a = \frac{1}{2}$, $b = \frac{-1}{\sqrt{2}}$, and $c = \frac{1}{2}$. Thus, $\langle 1,0,0\rangle = \frac{1}{2}\mathbf{I} + \frac{-1}{\sqrt{2}}\mathbf{J} + \frac{1}{2}\mathbf{K}$.

11.3.61 Note that $\overrightarrow{PQ} = \langle 2,3,-2\rangle$, $\overrightarrow{QR} = \langle -4,0,3\rangle$, and $\overrightarrow{RP} = \langle 2,-3,-1\rangle$, and these have lengths $\sqrt{17}$, 5 and $\sqrt{14}$ respectively.

The angle at P measures $\cos^{-1}(\frac{3}{\sqrt{17}\sqrt{14}}) \approx 78.8$ degrees. The angle at Q measures $\cos^{-1}(\frac{14}{5\sqrt{17}}) \approx 47.2$ degrees, and the angle at R measures $\cos^{-1}(\frac{11}{5\sqrt{14}}) \approx 54$ degrees.

11.3.63

a. The faces on $y = 0$ and $z = 0$.

b. The faces on $y = 1$ and $z = 1$.

c. The faces on $x = 0$ and $x = 1$.

d. Since $\mathbf{Q}$ is tangential on this face, the scalar component of $\mathbf{Q}$ normal to the face is 0.

e. The scalar component of $\mathbf{Q}$ normal to $z = 1$ is 1. Note that a vector normal to $z = 1$ is $\langle 0,0,1\rangle$.

f. The scalar component of $\mathbf{Q}$ normal to $y = 0$ is 2. Note that a vector normal to $y = 0$ is $\langle 0,1,0\rangle$.

11.3.65

a. Let the coordinates of R be (x,y,z). By symmetry, we have $y = 0$. We must have $x^2+y^2+z^2 = (x-\sqrt{3})^2+(y+1)^2+z^2 = (x-\sqrt{3})^2+(y-1)^2+z^2 = 4$. The first equality gives $x^2+z^2 = x^2-2\sqrt{3}x+3+1+z^2$, so $2\sqrt{3}x = 4$, and $x = \frac{2}{\sqrt{3}}$. It then follows that $z = \frac{2\sqrt{2}}{\sqrt{3}}$.

b. We have $\mathbf{r}_{OP} = \langle\sqrt{3},-1,0\rangle$, $\mathbf{r}_{OQ} = \langle\sqrt{3},1,0\rangle$, $\mathbf{r}_{PQ} = \langle 0,2,0\rangle$, $\mathbf{r}_{OR} = \langle 2/\sqrt{3},0,2\sqrt{2}/\sqrt{3}\rangle$, and $\mathbf{r}_{PR} = \langle -\sqrt{3}/3,1,2\sqrt{2}/3\rangle$.

11.3.67 $\mathbf{u}\cdot\mathbf{v} = u_1v_1+u_2v_2+u_3v_3 = v_1u_1+v_2u_2+v_3u_3 = \mathbf{v}\cdot\mathbf{u}$.

11.3.69 $\mathbf{u}(\mathbf{v}+\mathbf{w}) = \langle u_1,u_2,u_3\rangle\cdot\langle v_1+w_1, v_2+w_2, v_3+w_3\rangle = u_1(v_1+w_1)+u_2(v_2+w_2)+u_3(v_3+w_3) = u_1v_1+u_1w_1+u_2v_2+u_2w_2+u_3v_3+u_3w_3 = (u_1v_1+u_2v_2+u_3v_3)+(u_1w_1+u_2w_2+u_3w_3) = (\mathbf{u}\cdot\mathbf{v})+(\mathbf{u}\cdot\mathbf{w})$.

11.3.71 The statement is true. We have $\text{proj}_{\langle ka,kb\rangle}\langle c,d\rangle = \frac{\langle c,d\rangle\cdot\langle ka,kb\rangle}{(ka)^2+(kb)^2}\langle ka,kb\rangle = \frac{k(\langle a,b\rangle\cdot\langle c,d\rangle)}{k^2(a^2+b^2)}\cdot k\langle a,b\rangle = \frac{\langle a,b\rangle\cdot\langle c,d\rangle}{(a^2+b^2)}\langle a,b\rangle = \text{proj}_{\langle a,b\rangle}\langle c,d\rangle$.

11.3.73

a. $\cos\alpha = \frac{a}{\sqrt{a^2+b^2+c^2}}$, $\cos\beta = \frac{b}{\sqrt{a^2+b^2+c^2}}$, and $\cos\gamma = \frac{c}{\sqrt{a^2+b^2+c^2}}$. Thus, $\cos^2\alpha+\cos^2\beta+\cos^2\gamma = \frac{a^2}{a^2+b^2+c^2}+\frac{b^2}{a^2+b^2+c^2}+\frac{c^2}{a^2+b^2+c^2} = 1$.

b. We require $\cos^2\alpha+\cos^2\beta+\cos^2\gamma = \frac{1}{2}+\frac{1}{2}+\cos^2\gamma = 1$, so $\gamma = 90°$. The vector could be $\langle 1,1,0\rangle$; it makes a 90 degree angle with $\mathbf{k}$.

c. We require $\cos^2\alpha+\cos^2\beta+\cos^2\gamma = \frac{1}{4}+\frac{1}{4}+\cos^2\gamma = 1$, so $\gamma = 45°$. The vector could be $\langle 1,1,\sqrt{2}\rangle$; it makes a 45 degree angle with $\mathbf{k}$.

d. No. If so, we would have $\cos^2\alpha+\cos^2\beta+\cos^2\gamma = \frac{3}{4}+\frac{3}{4}+\cos^2\gamma = 1$, which would imply that $\cos^2\gamma = \frac{-1}{2}$, which can't occur.

e. If $\alpha = \beta = \gamma$, then $3\cos^2\alpha = 1$, and $\alpha = \cos^{-1}(\sqrt{3}/3) \approx 54.7356$ degrees. The vector could be $\langle 1, 1, 1\rangle$.

11.3.75 $\mathbf{u}\cdot\mathbf{v} = -24 - 15 + 6 = -33$. $|\mathbf{u}| = \sqrt{3^2 + (-5)^2 + 6^2} = \sqrt{70}$. $|\mathbf{v}| = \sqrt{(-8)^2 + 3^2 + 1^2} = \sqrt{74}$. Note that

$$33 < \sqrt{70}\sqrt{74},$$

so $|\mathbf{u}\cdot\mathbf{v}| < |\mathbf{u}|\,|\mathbf{v}|$.

11.3.77

a. We have $|\mathbf{u}+\mathbf{v}|^2 = (\mathbf{u}+\mathbf{v})\cdot(\mathbf{u}+\mathbf{v}) = (\mathbf{u}+\mathbf{v})\cdot\mathbf{u} + (\mathbf{u}+\mathbf{v})\cdot\mathbf{v} = \mathbf{u}\cdot(\mathbf{u}+\mathbf{v}) + \mathbf{v}\cdot(\mathbf{u}+\mathbf{v}) = \mathbf{u}\cdot\mathbf{u} + \mathbf{u}\cdot\mathbf{v} + \mathbf{v}\cdot\mathbf{u} + \mathbf{v}\cdot\mathbf{v} = |\mathbf{u}|^2 + 2(\mathbf{u}\cdot\mathbf{v}) + |\mathbf{v}|^2$

b. Note that $2(\mathbf{u}\cdot\mathbf{v}) \le 2\,|\mathbf{u}\cdot\mathbf{v}| \le 2\,|\mathbf{u}|\,|\mathbf{v}|$. Thus (using the previous part) we have,

$$|\mathbf{u}+\mathbf{v}|^2 = |\mathbf{u}|^2 + 2(\mathbf{u}\cdot\mathbf{v}) + |\mathbf{v}|^2 \le |\mathbf{u}|^2 + 2\,|\mathbf{u}|\,|\mathbf{v}| + |\mathbf{v}|^2 \le (|\mathbf{u}| + |\mathbf{v}|)^2.$$

c. Taking square roots of the previous result, and using the fact that the square root function is strictly increasing, we have $|\mathbf{u}+\mathbf{v}| \le |\mathbf{u}| + |\mathbf{v}|$.

d. Since the vectors $\mathbf{u}$, $\mathbf{v}$ and $\mathbf{u}+\mathbf{v}$ form a triangle, we can interpret this as meaning that the sum of the lengths of any two sides of a triangle is greater than or equal to the length of the other side.

11.3.79

a. One diagonal consists of the sum of one side ($\mathbf{u}$) and the side opposite the side adjacent to $\mathbf{u}$, but since it is a parallelogram, the side opposite $\mathbf{v}$ is also $\mathbf{v}$. So the diagonal is $\mathbf{u}+\mathbf{v}$. The other diagonal is the difference of two adjacent sides, so it is $\mathbf{u}-\mathbf{v}$.

b. The two diagonals are equal when $|\mathbf{u}+\mathbf{v}| = |\mathbf{u}-\mathbf{v}|$. Squaring both sides, we see that this is equivalent to requiring $|\mathbf{u}|^2 + 2(\mathbf{u}\cdot\mathbf{v}) + |\mathbf{v}|^2 = |\mathbf{u}|^2 - 2(\mathbf{u}\cdot\mathbf{v}) + |\mathbf{v}|^2$, which would imply that $2(\mathbf{u}\cdot\mathbf{v}) = -2(\mathbf{u}\cdot\mathbf{v})$, or $4(\mathbf{u}\cdot\mathbf{v}) = 0$. So if the diagonals are equal, the vectors are orthogonal. These steps are reversible, so the converse is also true.

c. $|\mathbf{u}+\mathbf{v}|^2 + |\mathbf{u}-\mathbf{v}|^2 = |\mathbf{u}|^2 + 2(\mathbf{u}\cdot\mathbf{v}) + |\mathbf{v}|^2 + |\mathbf{u}|^2 - 2(\mathbf{u}\cdot\mathbf{v}) + |\mathbf{v}|^2 = 2(|\mathbf{u}|^2 + |\mathbf{v}|^2)$.

11.4 Cross Products

11.4.1 $|\mathbf{u}\times\mathbf{v}| = |\mathbf{u}|\,|\mathbf{v}|\sin\theta$, where θ is the angle between $\mathbf{u}$ and $\mathbf{v}$.

11.4.3 Two parallel vectors have $\sin\theta = 0$ where θ is the angle between them. Thus, $|\mathbf{u}\times\mathbf{v}| = |\mathbf{u}|\,|\mathbf{v}|\sin\theta = 0$.

11.4.5 If $\mathbf{u} = \langle u_1, u_2, u_3\rangle$ and $\mathbf{v} = \langle v_1, v_2, v_3\rangle$, then $\mathbf{u}\times\mathbf{v}$ can be thought of as the determinant of the matrix

$$\begin{bmatrix} \mathbf{i} & \mathbf{j} & \mathbf{k} \\ u_1 & u_2 & u_3 \\ v_1 & v_2 & v_3 \end{bmatrix}.$$

11.4.7 $|\mathbf{u}\times\mathbf{v}| = |\mathbf{u}|\,|\mathbf{v}|\sin\pi/2 = 15$.

11.4.9 $\mathbf{u}\times\mathbf{v} = \langle 0, 0, 0\rangle$.

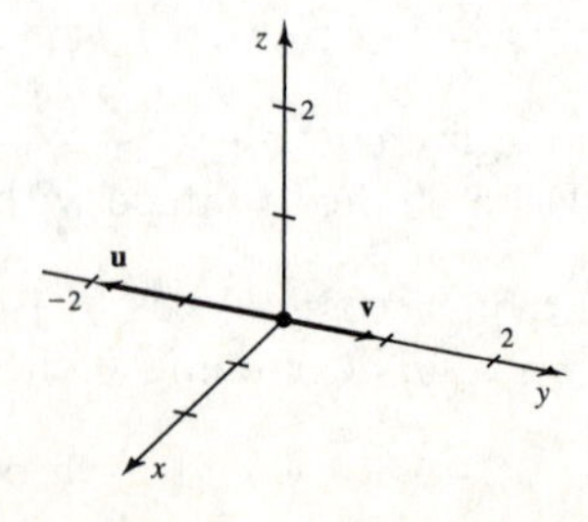

11.4.11 $\mathbf{u} \times \mathbf{v} = \langle 9\sqrt{2}, -9\sqrt{2}, 0 \rangle$.

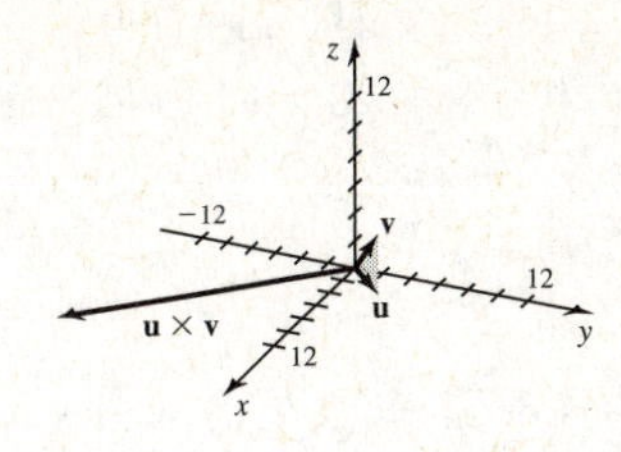

11.4.13 $\mathbf{j} \times \mathbf{k} = \mathbf{i}$.

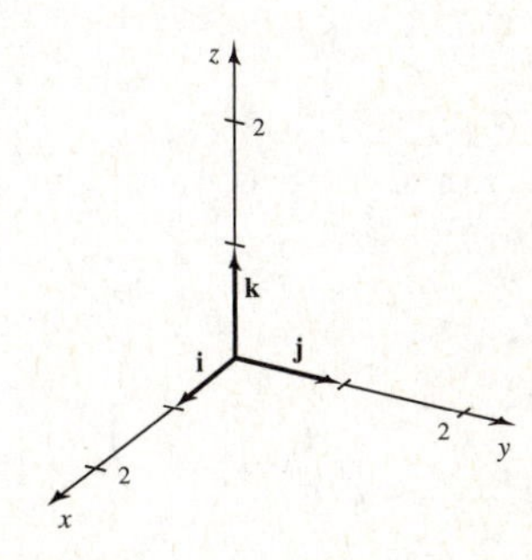

11.4.15 $-\mathbf{j} \times \mathbf{k} = -\mathbf{i}$.

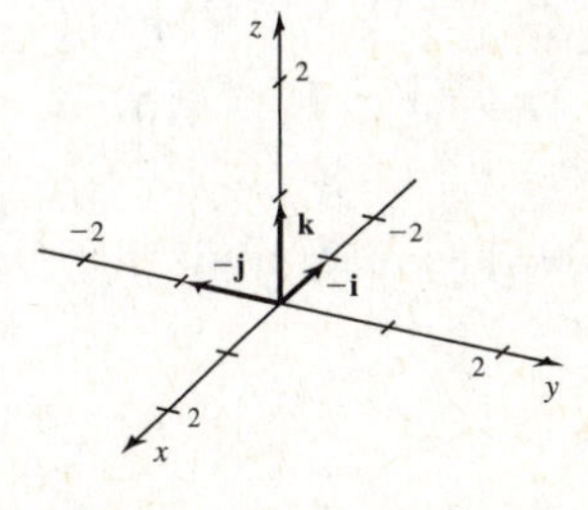

11.4.17 $-2\mathbf{i} \times 3\mathbf{k} = 6\mathbf{j}$.

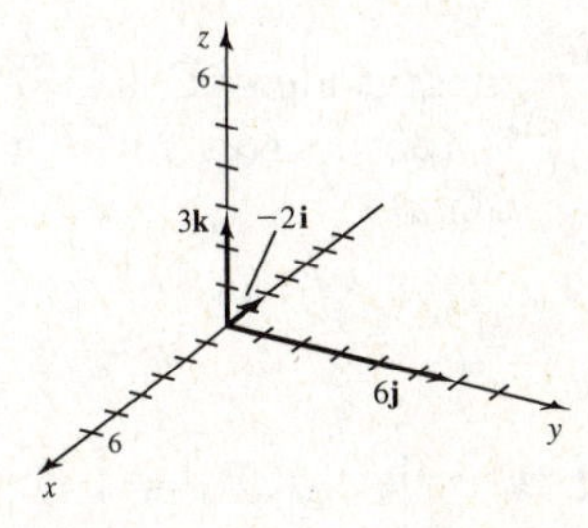

11.4.19 $|\mathbf{u} \times \mathbf{v}| = |\langle -2, -6, 9 \rangle| = \sqrt{4 + 36 + 81} = 11$.

11.4.21 $|\mathbf{u} \times \mathbf{v}| = |\langle 5, -4, 7 \rangle| = \sqrt{25 + 16 + 49} = 3\sqrt{10}$.

11.4.23 $\mathbf{u} \times \mathbf{v} = \begin{vmatrix} \mathbf{i} & \mathbf{j} & \mathbf{k} \\ 3 & 5 & 0 \\ 0 & 3 & -6 \end{vmatrix} = \langle -30, 18, 9 \rangle$. $\mathbf{v} \times \mathbf{u} = \langle 30, -18, -9 \rangle$.

11.4.25 $\mathbf{u} \times \mathbf{v} = \begin{vmatrix} \mathbf{i} & \mathbf{j} & \mathbf{k} \\ 2 & 3 & -9 \\ -1 & 1 & -1 \end{vmatrix} = \langle 6, 11, 5 \rangle$. $\mathbf{v} \times \mathbf{u} = \langle -6, -11, -5 \rangle$.

11.4.27 $\mathbf{u}\times\mathbf{v} = \begin{vmatrix} \mathbf{i} & \mathbf{j} & \mathbf{k} \\ 3 & -1 & -2 \\ 1 & 3 & -2 \end{vmatrix} = \langle 8,4,10\rangle$. $\mathbf{v}\times\mathbf{u} = \langle -8,-4,-10\rangle$.

11.4.29 Let $\mathbf{u} = \langle 0,1,2\rangle$ and $\mathbf{v} = \langle -2,0,3\rangle$. $\mathbf{u}\times\mathbf{v} = \begin{vmatrix} \mathbf{i} & \mathbf{j} & \mathbf{k} \\ 0 & 1 & 2 \\ -2 & 0 & 3 \end{vmatrix} = \langle 3,-4,2\rangle$ is perpendicular to both $\mathbf{u}$ and $\mathbf{v}$.

11.4.31 Let $\mathbf{u} = \langle 8,0,4\rangle$ and $\mathbf{v} = \langle -8,2,1\rangle$. $\mathbf{u}\times\mathbf{v} = \begin{vmatrix} \mathbf{i} & \mathbf{j} & \mathbf{k} \\ 8 & 0 & 4 \\ -8 & 2 & 1 \end{vmatrix} = \langle -8,-40,16\rangle$ is perpendicular to both $\mathbf{u}$ and $\mathbf{v}$.

11.4.33 $\boldsymbol{\tau} = \mathbf{r}\times\mathbf{F} = \begin{vmatrix} \mathbf{i} & \mathbf{j} & \mathbf{k} \\ 1 & 1 & 1 \\ 20 & 0 & 0 \end{vmatrix} = \langle 0,20,-20\rangle$.

11.4.35 $\boldsymbol{\tau} = \mathbf{r}\times\mathbf{F} = \begin{vmatrix} \mathbf{i} & \mathbf{j} & \mathbf{k} \\ 10 & 0 & 0 \\ 5 & 0 & -5 \end{vmatrix} = \langle 0,50,0\rangle$ has magntitude 50, while $\boldsymbol{\tau} = \mathbf{r}\times\mathbf{F} = \begin{vmatrix} \mathbf{i} & \mathbf{j} & \mathbf{k} \\ 10 & 0 & 0 \\ 4 & -3 & 0 \end{vmatrix} =$ $\langle 0,0,-30\rangle$ has magnitude 30, so the first force has greater magnitude.

11.4.37 $\mathbf{F} = 1\cdot(\mathbf{v}\times\mathbf{B}) = \begin{vmatrix} \mathbf{i} & \mathbf{j} & \mathbf{k} \\ 0 & 0 & 20 \\ 1 & 1 & 0 \end{vmatrix} = \langle -20,20,0\rangle$. The magnitude of $\mathbf{F}$ is $20\sqrt{2}$ and the angle of the force is 135 degrees with the positive x axis in the xy-plane.

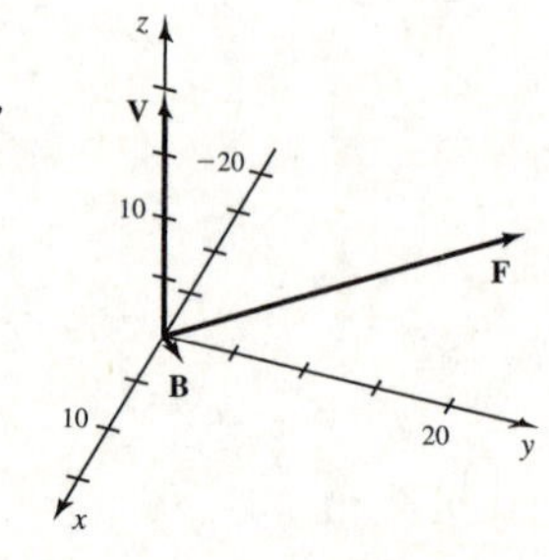

11.4.39 $|\mathbf{F}| = |q(\mathbf{v}\times\mathbf{B})| = \left|-1.6\cdot 10^{-19}\right|\ \mathrm{C}\cdot 2\cdot 10^5\cdot 2\cdot\sin 45^\circ = 4.53\cdot 10^{-14}\,\mathrm{kg}\cdot\mathrm{m/s}^2$.

11.4.41

a. False. For example $\mathbf{i}\times\mathbf{i} = \langle 0,0,0\rangle$, even though $\mathbf{i} \neq \langle 0,0,0\rangle$.

b. False. For example, $2\mathbf{i}\times 4\mathbf{j} = 8\mathbf{k}$ has magnitude 8, while $2\mathbf{i}$ has magnitude 2 and $4\mathbf{j}$ has magnitude 4.

c. False. If the compass directions are thought to lie in a plane, $\mathbf{u}\times\mathbf{v}$ doesn't lie in that plane, so it can't be a compass direction.

d. True. If both were nonzero, the first statement implies that the vectors are parallel, and the second that they are perpendicular, which can't both occur. So at least one of the vectors must be the zero vector.

e. False. $\mathbf{i}\times 2\mathbf{i} = \langle 0,0,0\rangle = \mathbf{i}\times 3\mathbf{i}$, but $2\mathbf{i} \neq 3\mathbf{i}$.

11.4.43 The area is $|\mathbf{u}\times\mathbf{v}| = \left\| \begin{vmatrix} \mathbf{i} & \mathbf{j} & \mathbf{k} \\ -1 & 1 & 1 \\ 0 & -1 & 1 \end{vmatrix} \right\| = |\langle 2,1,1\rangle| = \sqrt{6}$.

11.4.45 Two of the sides are $\mathbf{u} = \langle 2,4,8\rangle$ and $\mathbf{v} = \langle 1,4,10\rangle$.

The area is $|\mathbf{u}\times\mathbf{v}| = \left\| \begin{vmatrix} \mathbf{i} & \mathbf{j} & \mathbf{k} \\ 2 & 4 & 8 \\ 1 & 4 & 10 \end{vmatrix} \right\| = |\langle 8,-12,4\rangle| = 4\sqrt{14}$.

11.4.47 The area is $\frac{1}{2}|\mathbf{u}\times\mathbf{v}| = \frac{1}{2}\left\| \begin{vmatrix} \mathbf{i} & \mathbf{j} & \mathbf{k} \\ 3 & 3 & 3 \\ 6 & 0 & 6 \end{vmatrix} \right\| = \frac{1}{2}|\langle 18,0,-18\rangle| = 9\sqrt{2}$.

11.4.49 Two of the sides are $\mathbf{u} = \langle 1,2,3\rangle$ and $\mathbf{v} = \langle 6,5,4\rangle$.

The area is $\frac{1}{2}|\mathbf{u}\times\mathbf{v}| = \frac{1}{2}\left\| \begin{vmatrix} \mathbf{i} & \mathbf{j} & \mathbf{k} \\ 1 & 2 & 3 \\ 6 & 5 & 4 \end{vmatrix} \right\| = \frac{1}{2}|\langle -7,14,-7\rangle| = \frac{7\sqrt{6}}{2}$.

11.4.51 Let $\mathbf{u} = \langle u_1, u_2, u_3\rangle$. Then we have

$$\begin{vmatrix} \mathbf{i} & \mathbf{j} & \mathbf{k} \\ 1 & 1 & 1 \\ u_1 & u_2 & u_3 \end{vmatrix} = \langle -1,-1,2\rangle,$$

so $u_3 - u_2 = -1$, $u_1 - u_3 = -1$, and $u_2 - u_1 = 2$. The solutions to this system of linear equation can be characterized by letting u_1 be arbitrary, and by letting $u_2 = u_1 + 2$ and $u_3 = u_1 + 1$. Thus, $\mathbf{u} = \langle u_1, u_1 + 2, u_1 + 1\rangle$ for any real number u_1.

11.4.53 Two of the sides of the triangle are $\mathbf{u} = \langle -a, b, 0\rangle$ and $\mathbf{v} = \langle -a, 0, c\rangle$.

The area is $\frac{1}{2}|\mathbf{u}\times\mathbf{v}| = \frac{1}{2}\left\| \begin{vmatrix} \mathbf{i} & \mathbf{j} & \mathbf{k} \\ -a & b & 0 \\ -a & 0 & c \end{vmatrix} \right\| = \frac{1}{2}|\langle bc, ac, ab\rangle| = \frac{1}{2}\sqrt{b^2c^2 + a^2c^2 + a^2b^2}$.

11.4.55 $|\mathbf{u}\cdot(\mathbf{v}\times\mathbf{w})| = |\mathbf{u}|\,|\mathbf{v}\times\mathbf{w}|\,|\cos\theta|$. Since $|\mathbf{v}\times\mathbf{w}|$ represents the area of the base, we just need to see that the height of the parallelepiped is $|\mathbf{u}|\,|\cos\theta|$. Note that the height is given by the scalar projection of $\mathbf{u}$ on $\mathbf{v}\times\mathbf{w}$, which has value $|\cos\theta|\,|\mathbf{u}|$. Thus the given expression represents the volume of the parallelepiped.

11.4.57 Note that $\mathbf{r} = .66\mathbf{k}$, and $\mathbf{F} = 40\mathbf{j}$. $\boldsymbol{\tau} = \mathbf{r}\times\mathbf{F} = \begin{vmatrix} \mathbf{i} & \mathbf{j} & \mathbf{k} \\ 0 & 0 & 0.66 \\ 0 & 40 & 0 \end{vmatrix} = \langle -26.4, 0, 0\rangle$. The magnitude of the torque is 26.4 Newton-meters and the direction is on the negative x axis.

11.4.59 Since $\mathbf{F} = q(\mathbf{v}\times\mathbf{B})$, we have $|\mathbf{F}| = |q|\,|\mathbf{v}|\,|\mathbf{B}|\sin\theta$. Thus, $\frac{m|\mathbf{v}|^2}{R} = |q|\,|\mathbf{v}|\,|\mathbf{B}|\sin\pi/2$. Therefore,

$$|\mathbf{v}| = \frac{R\,|q|\,|\mathbf{B}|}{m} = \frac{0.002\cdot 1.6\cdot 10^{-19}\cdot .05}{9\cdot 10^{-31}} \approx 1.758\cdot 10^7\ \text{m/s}.$$

11.4.61 The result is trivial if either $a = 0$ or $b = 0$, so assume $ab \neq 0$. Note that the sine of the angle between $a\mathbf{u}$ and $b\mathbf{v}$ is the same as the sine of the angle between $\mathbf{u}$ and $\mathbf{v}$, as is demonstrated in the following diagrams.

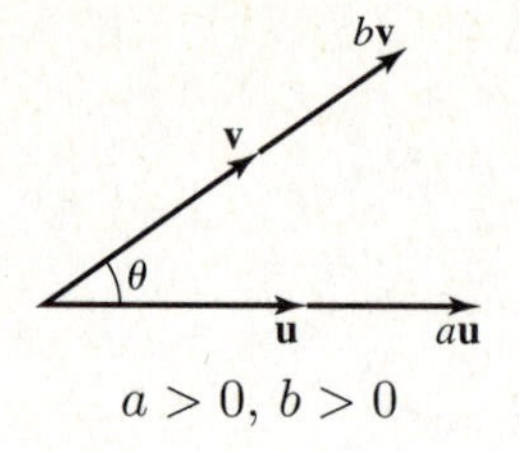

$a > 0,\ b > 0$

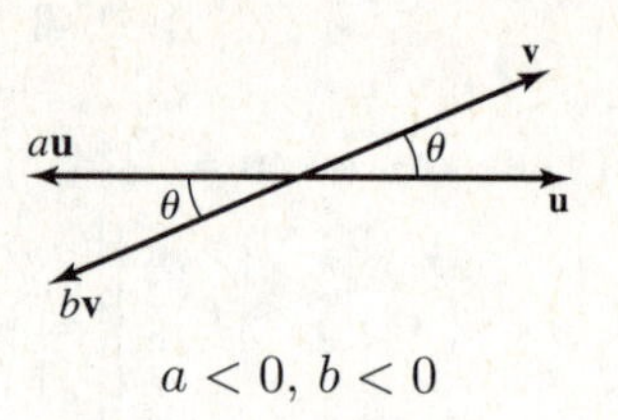

$a < 0,\ b < 0$

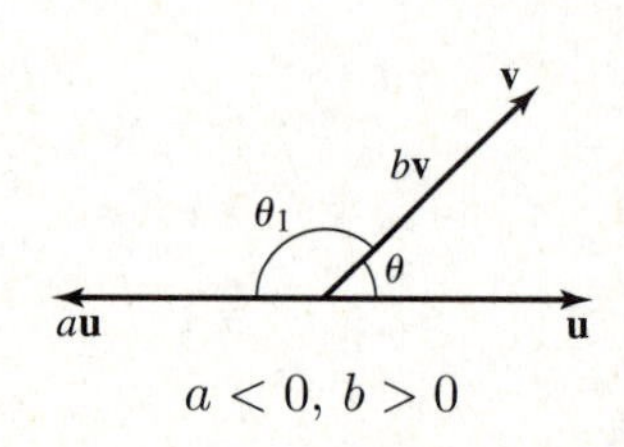

$a < 0,\ b > 0$

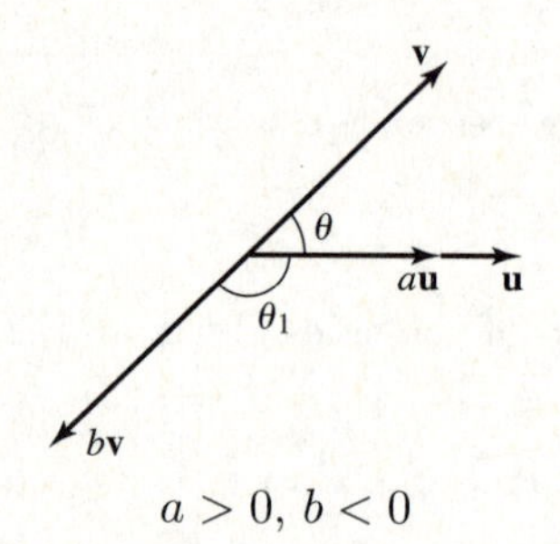

$a > 0,\ b < 0$

By the definition: $|(a\mathbf{u}) \times (b\mathbf{v})| = |a\mathbf{u}|\,|b\mathbf{v}| \sin\theta$, where θ is the angle between $a\mathbf{u}$ and $b\mathbf{v}$. But this is equal to $|a|\,|\mathbf{u}|\,|b|\,|\mathbf{v}| \sin\theta = |ab|\,(|\mathbf{u}|\,|\mathbf{v}| \sin\theta) = |ab|\,(|\mathbf{u} \times \mathbf{v}|)$. When a and b have the same sign, the directions are also the same, since they are determined by the right-hand rule (see diagrams above.) When a and b have opposite signs, the directions are opposite, but then $ab < 0$.

Using the determinant formula:

$$(a\mathbf{u}) \times (b\mathbf{v}) = \begin{vmatrix} \mathbf{i} & \mathbf{j} & \mathbf{k} \\ au_1 & au_2 & au_3 \\ bv_1 & bv_2 & bv_3 \end{vmatrix} = ab \begin{vmatrix} \mathbf{i} & \mathbf{j} & \mathbf{k} \\ u_1 & u_2 & u_3 \\ v_1 & v_2 & v_3 \end{vmatrix} = ab(\mathbf{u} \times \mathbf{v}).$$

11.4.63 True. $(\mathbf{u} - \mathbf{v}) \times (\mathbf{u} + \mathbf{v}) = \mathbf{u} \times \mathbf{u} + \mathbf{u} \times \mathbf{v} - (\mathbf{v} \times \mathbf{u}) - (\mathbf{v} \times \mathbf{v}) = 2(\mathbf{u} \times \mathbf{v}) = (2\mathbf{u} \times \mathbf{v})$.

11.4.65

$$\begin{aligned}
\mathbf{u} \times (\mathbf{v} \times \mathbf{w}) &= \begin{vmatrix} \mathbf{i} & \mathbf{j} & \mathbf{k} \\ u_1 & u_2 & u_3 \\ v_2w_3 - v_3w_2 & v_3w_1 - v_1w_3 & v_1w_2 - v_2w_1 \end{vmatrix} \\
&= \langle u_2(v_1w_2 - v_2w_1) - u_3(v_3w_1 - v_1w_3), u_3(v_2w_3 - v_3w_2) \\
&\quad - u_1(v_1w_2 - v_2w_1), u_1(v_3w_1 - v_1w_3) - u_2(v_2w_3 - v_3w_2)\rangle \\
&= \langle v_1(u_2w_2 + u_3w_3) - w_1(u_2v_2 + u_3v_3), v_2(u_1w_1 + u_3w_3) \\
&\quad - w_2(u_1v_1 + u_3v_3), v_3(u_1w_1 + u_2w_2) - w_3(u_1v_1 + u_2v_2)\rangle \\
&= \langle v_1(\mathbf{u} \cdot \mathbf{w}) - w_1(\mathbf{u} \cdot \mathbf{v}), v_2(\mathbf{u} \cdot \mathbf{w}) - w_2(\mathbf{u} \cdot \mathbf{v}), v_3(\mathbf{u} \cdot \mathbf{w}) - w_3(\mathbf{u} \cdot \mathbf{w})\rangle \\
&= (\mathbf{u} \cdot \mathbf{w})\mathbf{v} - (\mathbf{u} \cdot \mathbf{w})\mathbf{w}.
\end{aligned}$$

11.4.67

a. Suppose $\mathbf{u} \times \mathbf{z} = \mathbf{v}$. Then $\mathbf{v} \times (\mathbf{u} \times \mathbf{z}) = \mathbf{v} \times \mathbf{v} = \langle 0, 0, 0\rangle$. Now $\mathbf{v} \times (\mathbf{u} \times \mathbf{z}) = \mathbf{u}(\mathbf{v} \cdot \mathbf{z}) - \mathbf{z}(\mathbf{v} \cdot \mathbf{u})$ by exercise 65. If $\mathbf{u} \cdot \mathbf{v} = 0$, then we have $\mathbf{u}(\mathbf{v} \cdot \mathbf{z}) = \langle 0, 0, 0\rangle$. Any vector $\mathbf{z}$ which is perpendicular to $\mathbf{v}$ is a solution to this equation.

   Now suppose that the equation $\mathbf{u} \times \mathbf{z} = \mathbf{v}$ has a nonzero solution. Since the cross product of any two vectors is perpendicular to both of the vectors, we must have that $\mathbf{u} \times \mathbf{z} \cdot \mathbf{u} = 0$. But this means that $\mathbf{v} \cdot \mathbf{u} = 0$, as desired.

b. If there exists a vector **z** so that $\mathbf{u} \times \mathbf{z} = \mathbf{v}$, then **u** and **v** must be perpendicular. If **u** and **v** are perpendicular nonzero vectors, then there must be a plane which contains **u** and a nonzero vector **z** so that $\mathbf{u} \times \mathbf{z} = \mathbf{v}$.

11.5 Lines and Curves in Space

11.5.1 It has one, namely t.

11.5.3 For every real number t that is put into the function, the output is a vector $\mathbf{r}(t)$.

11.5.5 First find the direction vector **d** as in the previous problem, and then let $\mathbf{r}(t) = \langle x_0, y_0, z_0 \rangle + t\mathbf{d}$.

11.5.7 Compute $\lim_{t\to a} f(t) = L_1$, $\lim_{t\to a} g(t) = L_2$, and $\lim_{t\to a} h(t) = L_3$. Then $\lim_{t\to a} \mathbf{r}(t) = \langle L_1, L_2, L_3 \rangle$.

11.5.9 The direction is $\langle 0, 1, 0 \rangle$, so the line l_1 is $\mathbf{r}(t) = \langle 0, 0, 1 \rangle + t\langle 0, 1, 0 \rangle$.

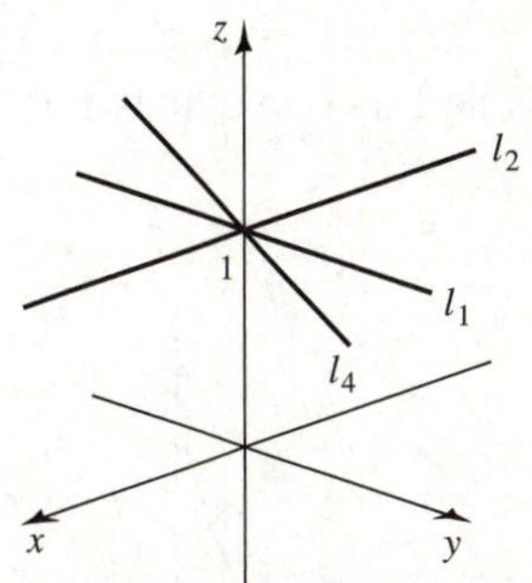

This diagram shows the lines for problems 9, 10, and 12.

11.5.11 The direction is $\langle 2, -2, 2 \rangle$, so the line is $\mathbf{r}(t) = \langle 0, 1, 1 \rangle + t\langle 2, -2, 2 \rangle$.

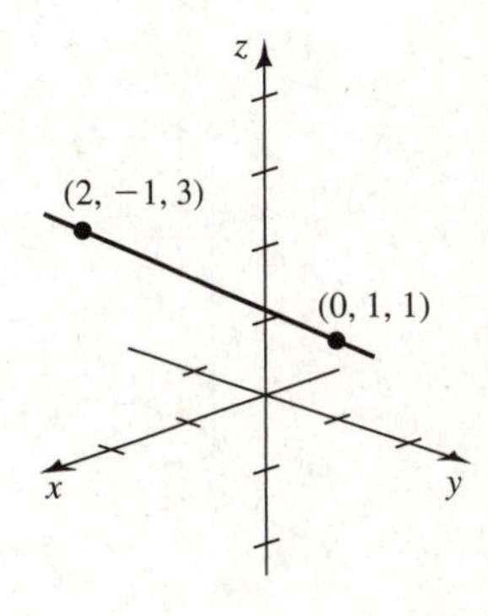

11.5.13 The direction is $\langle 1, 2, 3 \rangle$, so the line is $\mathbf{r}(t) = t\langle 1, 2, 3 \rangle$.

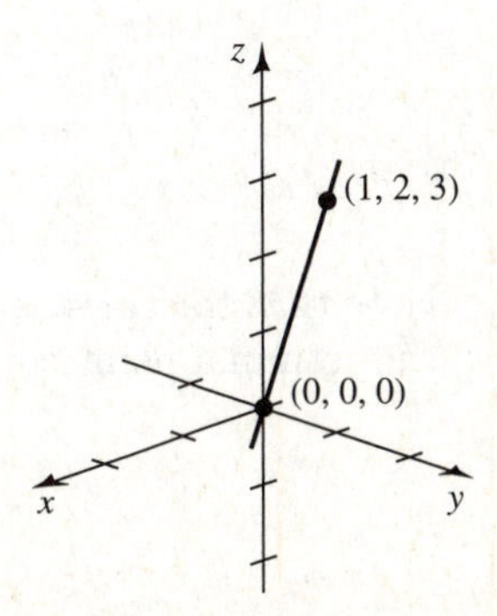

11.5.15 The direction is $\langle 8, -5, -6\rangle$, so the line is $\mathbf{r}(t) = \langle -3, 4, 6\rangle + t\langle 8, -5, -6\rangle$.

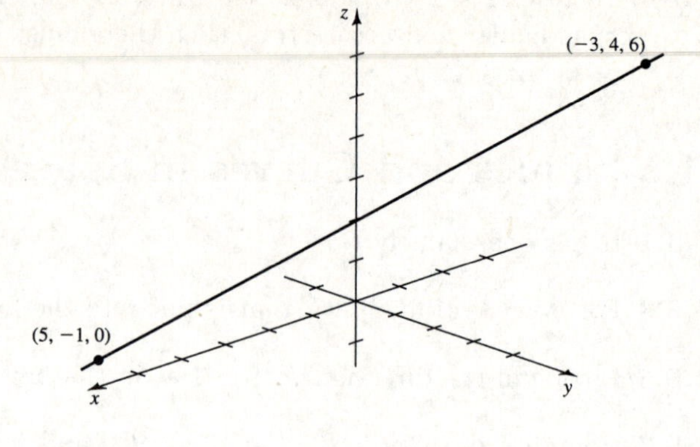

11.5.17 The line segment is $\mathbf{r}(t) = t\langle 1, 2, 3\rangle$, where $0 \le t \le 1$.

11.5.19 The line segment is $\mathbf{r}(t) = \langle 2, 4, 8\rangle + t\langle 5, 1, -5\rangle$, where $0 \le t \le 1$.

11.5.21

11.5.23

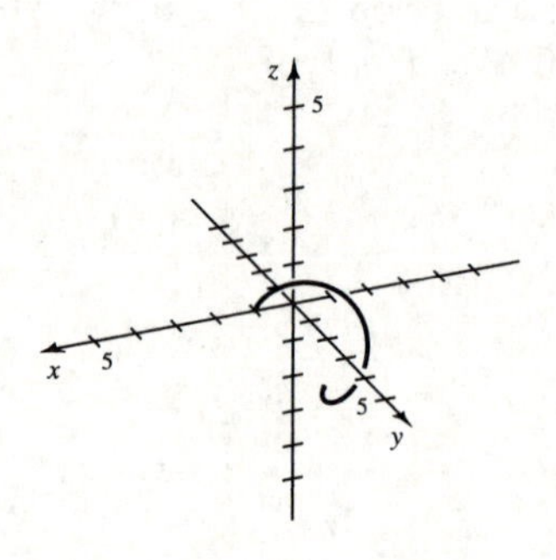

11.5.25

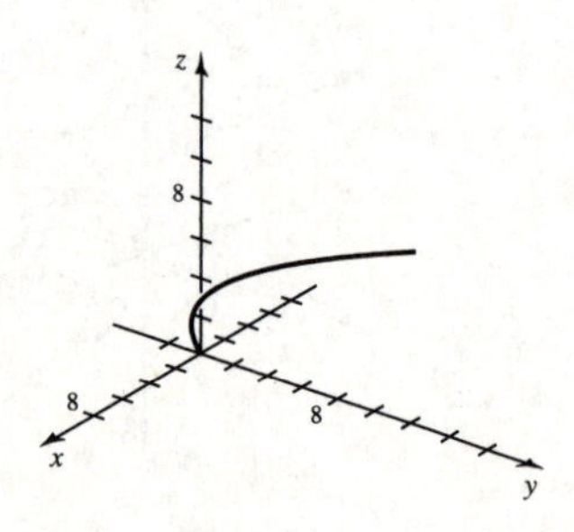

11.5.27

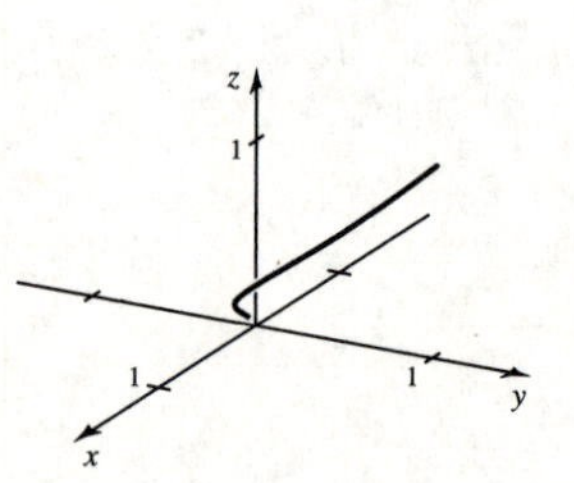

11.5.29 Note that the curve is closed (the initial point and the terminal point coincide), and is very "wavy."

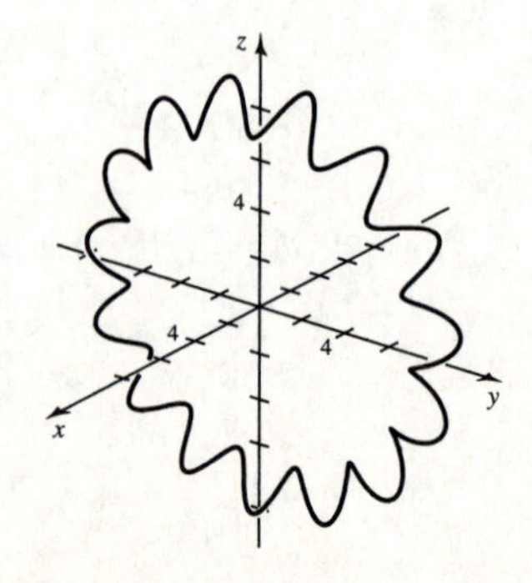

11.5.31 When viewed from the top, the curve is circular looking.

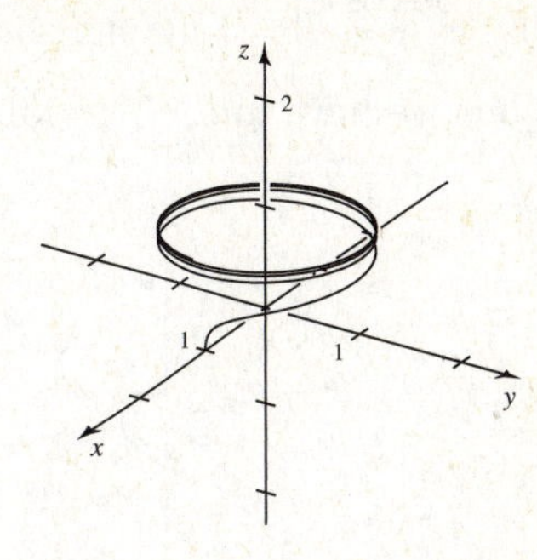

11.5.33 $\lim_{t\to\pi/2}\langle\cos 2t, -4\sin t, \frac{2t}{\pi}\rangle = \langle\cos\pi, -4\sin\pi/2, \frac{2\cdot\pi/2}{\pi}\rangle = \langle -1, -4, 1\rangle$.

11.5.35 $\lim_{t\to\infty}\langle e^{-t}, \frac{-2t}{t+1}, \tan^{-1}(t)\rangle = \langle 0, -2, \pi/2\rangle$.

11.5.37

a. True. This curve passes through the origin at $t = -1/2$.

b. False. For example, the x axis is not parallel to the line $\langle 0,0,1\rangle + t\langle 0,1,0\rangle$, but neither do they intersect.

c. True. The first component function approaches 0 as $t\to\infty$, while the others are periodic. The parametric equations $y = \sin t$ and $z = -\cos t$ form a circle in the yz-plane.

d. True. Both have limit $\langle 0,0,0\rangle$.

11.5.39 The first component function has domain $[-2,\infty)$ and the second has domain $(-\infty, 2]$, so the domain of $\mathbf{r}(t)$ is $[-2,2]$.

11.5.41 The first component function has domain $[-2,2]$, the second has domain $[0,\infty)$, and the third has domain $(-1,\infty)$, so the domain of $\mathbf{r}(t)$ is $[0,2]$.

11.5.43 The intersection occurs for $z = 4 = t - 6$, so for $t = 10$. The point of intesection is $(21,-6,4)$.

11.5.45 The intersection occurs for $z = -8 = -2t + 4$, so for $t = 6$. The point of intersection is $(16, 0, -8)$.

11.5.47 The intersection occurs for $z = 16 = 4 + 3t$, so for $t = 4$. The point of intersection is $(4, 8, 16)$.

11.5.49

a. This matches graph E. (It is a straight line.)

b. This matches graph D. (It is parabolic-like.)

c. This matches graph F. (It is a circle.)

d. This matches graph C. (It is a circular helix, elongated along the x-axis.)

e. This matches graph A. (It is a closed curve which isn't a circle.)

f. This matches graph B. (It is a circular helix, elongated along the y-axis.)

11.5.51

a. $\mathbf{r}(0) = \langle 50, 0, 0\rangle$.

b. $\lim_{t\to\infty} \frac{50\cos t}{e^t} = 0$ by the squeeze theorem, and likewise $\lim_{t\to\infty} \frac{50\sin t}{e^t} = 0$. Also, $\lim_{t\to\infty}\left(5 - \frac{5}{e^t}\right) = 5$. Thus we have $\lim_{t\to\infty} \mathbf{r}(t) = \langle 0, 0, 5\rangle$.

c. 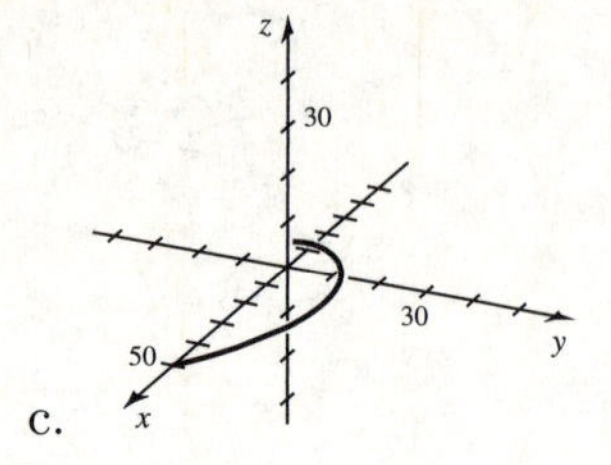

d. Let $x = 50e^{-t}\cos t$ and $y = 50e^{-t}\sin t$ and $z = 5 - 5e^{-t}$. Note that $x^2 + y^2 = 2500e^{-2t}$, so $r = 50e^{-t}$. We have $z = 5 - 5e^{-t} = 5 - \frac{r}{10}$.

11.5.53 This has the form mentioned in exercise 52, with $a = 1/\sqrt{2}$, $b = 1/\sqrt{3}$, $c = -1/\sqrt{2}$, $d = 1/\sqrt{3}$, $e = 0$, and $f = 1/\sqrt{3}$. Note that $a^2 + c^2 + e^2 = \frac{1}{2} + \frac{1}{2} + 0 = 1 = \frac{1}{3} + \frac{1}{3} + \frac{1}{3} = b^2 + d^2 + f^2$, and $ab + cd + ef = 0$. So this is a circle of radius 1 centered at the origin.

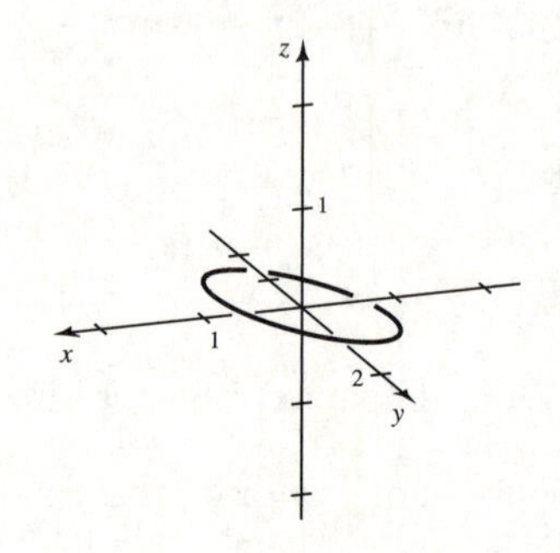

11.5.55 Note that $\mathbf{r}(0) = \langle a, c, e\rangle$ and $\mathbf{r}(\pi/2) = \langle b, d, f\rangle$, and $\mathbf{r}(\pi) = \langle -a, -c, -e\rangle$ have their terminal points on the curve. So $\langle a, c, e\rangle - \langle -a, -c, -e\rangle = \langle 2a, 2c, 2e\rangle = 2\mathbf{r}(0)$ lies in the plane containing the curve, which implies that $\mathbf{r}(0)$ lies in the plane containing the curve, and that implies that the point $(0, 0, 0)$ is in the plane containing the curve. So a normal to the curve is $\mathbf{r}(0) \times \mathbf{r}(\pi/2) = \langle a, c, e\rangle \times \langle b, d, f\rangle = \langle cf - de, be - af, ad - bc\rangle$.

11.5.57 First note that $x = \frac{1}{2}\sin 2t = \sin t\cos t$ and $y = \frac{1}{2}(1 - \cos 2t) = \sin^2 t$. Then $x^2 + y^2 + z^2 = \sin^2 t\cos^2 t + \sin^4 t + \cos^2 t = \sin^2 t(\cos^2 t + \sin^2 t) + \cos^2 t = \sin^2 t \cdot 1 + \cos^2 t = 1$. So all points on the curve are equidistant from the origin, so they lie on the sphere of radius one centered at the origin.

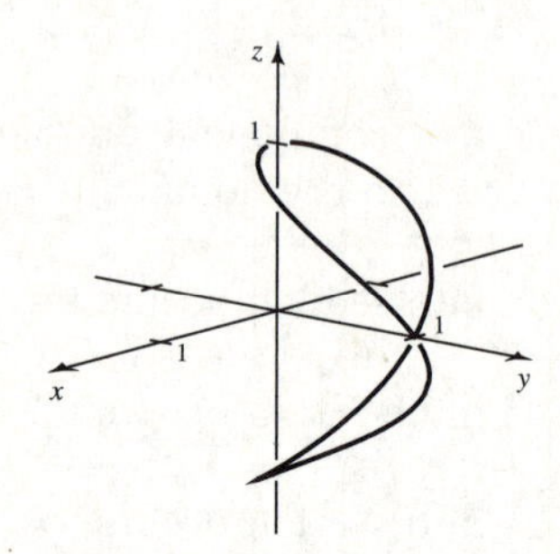

11.5.59 In order for $\sin(mt + mT)\cos(nt + nT) = \sin mt\cos nt$ and $\sin(mt + mT)\sin(nt + nT) = \sin mt\sin nt$ and $\cos(mt + mT) = \cos mt$ we would need $T = \frac{2\pi}{m}$ or a multiple of it, and then it would be necessary for $\sin(nt + nT) = \sin nt$, which would require $T = \frac{2\pi}{n}$, or a multiple of it. Thus, the smallest such T would be $\frac{2\pi}{(m,n)}$, where (m, n) represents the greatest common factor of m and n.

11.6 Calculus of Vector-Valued Functions

11.6.1 Compute the derivative of each of the component functions, and then $\mathbf{r}'(t) = \langle f'(t), g'(t), h'(t)\rangle$.

11.6.3 Divide the vector by its length, so if the vector is $\mathbf{r}'(t)$, form $\frac{\mathbf{r}'(t)}{|\mathbf{r}'(t)|}$.

11.6.5 Compute the indefinite integral of each of the component functions, and then

$$\int \mathbf{r}(t)\, dt = \langle \int f(t)\, dt, \int g(t)\, dt, \int h(t)\, dt\rangle.$$

11.6.7 $\mathbf{r}'(t) = \langle 6t, 3/\sqrt{t}, -3/t^2\rangle$.

11.6.9 $\mathbf{r}'(t) = \langle e^t, -2e^{-t}, -8e^{2t}\rangle$.

11.6.11 $\mathbf{r}'(t) = \langle e^{-t}(1-t), 1 + \ln t, \cos t - t\sin t\rangle$.

11.6.13 $\mathbf{r}'(t) = \langle 1, -2\sin 2t, 2\cos t\rangle$. At $t = \pi/2$ we have $\langle 1, 0, 0\rangle$.

11.6.15 $\mathbf{r}'(t) = \langle 8t^3, 9\sqrt{t}, -10/t^2\rangle$. At $t = 1$ we have $\langle 8, 9, -10\rangle$.

11.6.17 $\mathbf{r}'(t) = \langle 0, -2\sin 2t, 4\cos 2t\rangle$, so

$$\frac{\mathbf{r}'(t)}{|\mathbf{r}'(t)|} = \frac{1}{\sqrt{4\sin^2 2t + 16\cos^2 2t}}\langle 0, -2\sin 2t, 4\cos 2t\rangle = \frac{1}{\sqrt{1 + 3\cos^2 2t}}\langle 0, -\sin 2t, 2\cos 2t\rangle.$$

11.6.19 $\mathbf{r}'(t) = \langle 1, 0, -2/t^2\rangle$, so

$$\frac{\mathbf{r}'(t)}{|\mathbf{r}'(t)|} = \frac{1}{\sqrt{1 + (4/t^4)}}\langle 1, 0, -2/t^2\rangle = \frac{1}{\sqrt{t^4 + 4}}\langle t^2, 0, -2\rangle.$$

11.6.21 $\mathbf{r}'(t) = \langle -2\sin 2t, 0, 6\cos 2t\rangle$, so at $t = \pi/2$, we have $\mathbf{r}'(\pi/2) = \langle 0, 0, -6\rangle$. Thus, the unit tangent at $\pi/2$ is $\langle 0, 0, -1\rangle$.

11.6.23 $\mathbf{r}'(t) = \langle 6, 0, -3/t^2\rangle$, so at $t = 1$, we have $\mathbf{r}'(1) = \langle 6, 0, -3\rangle$. Thus, the unit normal at 1 is $\langle 2/\sqrt{5}, 0, -1/\sqrt{5}\rangle$.

11.6.25 $(t^{12} + 3t)\mathbf{u}'(t) + \mathbf{u}(t)(12t^{11} + 3) = (t^{12} + 3t)\langle 6t^2, 2t, 0\rangle + \langle 2t^3, (t^2 - 1), -8\rangle(12t^{11} + 3) = \langle 30t^{14} + 24t^3, 14t^{13} - 12t^{11} + 9t^2 - 3, -96t^{11} - 24\rangle$.

11.6.27 $\mathbf{u}'(t^4 - 2t)\cdot(4t^3 - 2) = \langle 6(t^4 - 2t)^2, 2(t^4 - 2t), 0\rangle(4t^3 - 2) = \langle 6(t^4 - 2t)^2(4t^3 - 2), 2(t^4 - 2t)(4t^3 - 2), 0\rangle$.

11.6.29 $\mathbf{u}(t)\cdot\mathbf{v}'(t) + \mathbf{v}(t)\cdot\mathbf{u}'(t) = \langle 2t^3, (t^2 - 1), -8\rangle\cdot\langle e^t, -2e^{-t}, -2e^{2t}\rangle + \langle e^t, 2e^{-t}, -e^{2t}\rangle\cdot\langle 6t^2, 2t, 0\rangle = 2t^3e^t - 2(t^2 - 1)e^{-t} + 16e^{2t} + 6t^2e^t + 4te^{-t} + 0 = e^t(2t^3 + 6t^2) - 2e^{-t}(t^2 - 2t - 1) + 16e^{2t}$.

11.6.31 $\langle t^2, 2t^2, -2t^3\rangle\cdot\langle e^t, 2e^t, 3e^{-t}\rangle + \langle 2t, 4t, -6t^2\rangle\cdot\langle e^t, 2e^t, -3e^{-t}\rangle = t^2e^t + 4t^2e^t - 6t^3e^{-t} + 2te^t + 8te^t + 18t^2e^{-t} = 5t^2e^t + 10te^t - 6t^3e^{-t} + 18t^2e^{-t}$.

11.6.33 $\langle 3t^2, \sqrt{t}, -2/t\rangle\cdot\langle -\sin t, 2\cos 2t, -3\rangle + \langle 6t, 1/(2\sqrt{t}), 2/t^2\rangle\cdot\langle\cos t, \sin 2t, -3t\rangle = -3t^2\sin t + 2\sqrt{t}\cos 2t + 6t\cos t + \sin(2t)/(2\sqrt{t})$.

11.6.35 $\mathbf{r}'(t) = \langle 2t, 1, 0\rangle$. $\mathbf{r}''(t) = \langle 2, 0, 0\rangle$. $\mathbf{r}'''(t) = \langle 0, 0, 0\rangle$.

11.6.37 $\mathbf{r}'(t) = \langle -3\sin 3t, 4\cos 4t, -6\sin 6t\rangle$. $\mathbf{r}''(t) = \langle -9\cos 3t, -16\sin 4t, -36\cos 6t\rangle$. $\mathbf{r}'''(t) = \langle 27\sin 3t, -64\cos 4t, 216\sin 6t\rangle$.

11.6.39 $\mathbf{r}'(t) = \langle \frac{1}{2\sqrt{t+4}}, \frac{1}{(t+1)^2}, 2e^{-t^2}t\rangle$. $\mathbf{r}''(t) = \langle -\frac{1}{4(t+4)^{3/2}}, -\frac{2}{(t+1)^3}, e^{-t^2}(2 - 4t^2)\rangle$. $\mathbf{r}'''(t) = \langle \frac{3}{8(t+4)^{5/2}}, \frac{6}{(t+1)^4}, 4e^{-t^2}t(2t^2 - 3)\rangle$.

11.6.41 $\int\langle t^4 - 3t, 2t - 1, 10\rangle\,dt = \langle t^5/5 - 3t^2/2, t^2 - t, 10t\rangle + \mathbf{C}$.

11.6.43 $\int\langle 2\cos t, 2\sin 3t, 4\cos 8t\rangle\,dt = \langle 2\sin t, (-2/3)\cos 3t, (1/2)\sin 8t\rangle + \mathbf{C}$.

11.6.45 $\int\langle 1, 2t, 3t^2\rangle\,dt = \langle t, t^2, t^3\rangle + \mathbf{C}$. Since $\mathbf{r}(1) = \langle 4, 3, -5\rangle$, we have $\langle 1, 1, 1\rangle + \mathbf{C} = \langle 4, 3, -5\rangle$, so $\mathbf{C} = \langle 3, 2, -6\rangle$, and $\mathbf{r}(t) = \langle t + 3, t^2 + 2, t^3 - 6\rangle$.

11.6.47 $\int \langle e^{2t}, 1-2e^{-t}, 1-2e^t\rangle\,dt = \langle e^{2t}/2, t+2e^{-t}, t-2e^t\rangle + \mathbf{C}$. Since $\mathbf{r}(0) = \langle 1,1,1\rangle$, we have $\langle 1/2, 2, -2\rangle + \mathbf{C} = \langle 1,1,1\rangle$, so $\mathbf{C} = \langle 1/2, -1, 3\rangle$, and $\mathbf{r}(t) = \langle e^{2t}/2 + 1/2, t + 2e^{-t} - 1, t - 2e^t + 3\rangle$.

11.6.49 $\int_{-1}^{1} \langle 1, t, 3t^2\rangle\,dt = \langle t, t^2/2, t^3\rangle\big|_{-1}^{1} = \langle 2, 0, 2\rangle$.

11.6.51 $\int_{-\pi}^{\pi} \langle \sin t, \cos t, 2t\rangle\,dt = \langle -\cos t, \sin t, t^2\rangle\big|_{-\pi}^{\pi} = \langle 0,0,0\rangle$.

11.6.53 $\int_0^2 \langle te^t, 2te^t, -te^t\rangle\,dt = \langle (t-1)e^t, 2(t-1)e^t, -(t-1)e^t\rangle\big|_0^2 = \langle e^2+1, 2e^2+2, -e^2-1\rangle = (e^2+1)\langle 1, 2, -1\rangle$.

11.6.55

a. False. For example, if $\mathbf{r}(t) = \langle \cos t, \sin t\rangle$, then $\mathbf{r}'(t) = \langle -\sin t, \cos t\rangle$ is not parallel to $\mathbf{r}(t)$, and is in fact perpendicular to it.

b. False. $\mathbf{r}'(t) = \langle 0, 2t-2, -\pi\sin \pi t\rangle = \langle 0,0,0\rangle$ for $t = 1$, so the curve is not smooth by definition.

c. True. This follows because $\int_{-a}^{a} o(x)\,dx = 0$ for any odd function $o(x)$.

11.6.57 $\mathbf{v}'(e^t)\cdot e^t = e^t\langle 2e^t, -2, 0\rangle = \langle 2e^{2t}, -2e^t, 0\rangle$.

11.6.59 $\mathbf{v}'(g(t))g'(t) = \langle 4\sqrt{t}, -2, 0\rangle \left(\frac{1}{\sqrt{t}}\right) = \langle 4, -2/\sqrt{t}, 0\rangle$.

11.6.61 $\mathbf{u}(t)\times\mathbf{v}'(t) + \mathbf{u}'(t)\times\mathbf{v}(t) = \langle 1, t, t^2\rangle \times \langle 2t, -2, 0\rangle + \langle 0, 1, 2t\rangle \times \langle t^2, -2t, 1\rangle = \langle 2t^2, 2t^3, -2t^2-2\rangle + \langle 4t^2+1, 2t^3, -t^2\rangle = \langle 6t^2+1, 4t^3, -3t^2-2\rangle$

11.6.63 $\mathbf{r}'(t) = \langle 2at, 1\rangle$. We have $\mathbf{r}(t)\cdot\mathbf{r}'(t) = 0$ when $(2at)(at^2+1) + t(1) = 0$, which occurs only for $t = 0$ since $2a^2t^2 + 2a + 1 > 0$ for all t. The corresponding point on the parabola is $(1, 0)$.

11.6.65 $\mathbf{r}'(t) = \langle -\sin t, \cos t, 1\rangle$. We have $\mathbf{r}(t)\cdot\mathbf{r}'(t) = 0$ when $-\sin t\cos t + \sin t \cos t + t = 0$, which occurs only for $t = 0$. So the only point on the helix where these vectors are orthogonal is at $t = 0$. This corresponds to the point $(1, 0, 0)$.

11.6.67 Note that $\mathbf{r}(t) = \langle a_1 t, a_2 t, a_3 t\rangle = t\langle a_1, a_2, a_3\rangle$ where the a_i's are real numbers has this property since $\mathbf{r}'(t) = \langle a_1, a_2, a_3\rangle$, and $\mathbf{r}(t)$ is a multiple of $\mathbf{r}'(t)$.

Also, $\mathbf{r}(t) = \langle a_1 e^{kt}, a_2 e^{kt}, a_3 e^{kt}\rangle$ where k is a real number has this property, as its derivative is k times itself.

11.6.69

a. $\mathbf{r}(t)\cdot\mathbf{r}'(t) = (a^2+b^2+c^2)t = |\mathbf{r}(t)|\,|\mathbf{r}'(t)|\cos\theta$, so $\cos\theta = \frac{(a^2+b^2+c^2)t}{\sqrt{a^2+b^2+c^2}t\cdot\sqrt{a^2+b^2+c^2}} = 1$, so $\theta = 0$.

b. $\mathbf{r}(t)\cdot\mathbf{r}'(t) = ax_0 + by_0 + cz_0 + (a^2+b^2+c^2)t = |\mathbf{r}(t)|\,|\mathbf{r}'(t)|\cos\theta$, so

$$\cos\theta = \frac{ax_0 + by_0 + cz_0 + (a^2+b^2+c^2)t}{\sqrt{(x_0+at)^2 + (y_0+bt)^2 + (z_0+ct)^2}\cdot\sqrt{a^2+b^2+c^2}}.$$

Since x_0, y_0, and z_0 are not all 0, $\cos\theta$ depends on t.

c. In part a, the curve is a straight line through the origin, so the position vector and the tangent vector are parallel for all t. In part b, the line is not through the origin, so the tangent vector (which is the direction vector for the line) is not parallel to the position vector.

11.6.71

$$\begin{aligned}\frac{d}{dt}(f(t)\mathbf{u}(t)) &= \frac{d}{dt}\langle f(t)u_1(t), f(t)u_2(t), f(t)u_3(t)\rangle \\ &= \langle f'(t)u_1(t) + f(t)u_1'(t)+, f'(t)u_2(t) + f(t)u_2'(t), f'(t)u_3(t) + f(t)u_3'(t)\rangle \\ &= f'(t)\mathbf{u}(t) + f(t)\mathbf{u}'(t).\end{aligned}$$

11.6.73

a. $\mathbf{r}'(t) = \langle 3t^2, 3t^2\rangle$, so $\mathbf{r}'(0) = \langle 0,0\rangle$. There is no cusp because $\lim_{t\to 0}\frac{dy}{dx} = \frac{dy/dt}{dx/dt} = \frac{3t^2}{3t^2} = 1$ exists.

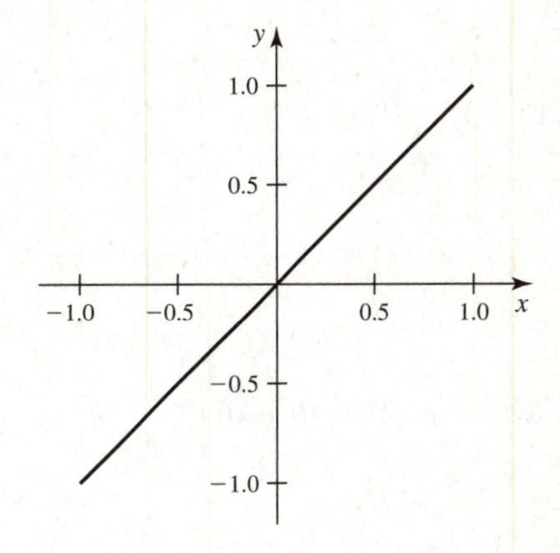

b. $\mathbf{r}'(t) = \langle 3t^2, 2t\rangle$, so $\mathbf{r}'(0) = \langle 0,0\rangle$. There is a cusp because $\lim_{t\to 0}\frac{dy}{dx} = \lim_{t\to 0}\frac{dy/dt}{dx/dt} = \lim_{t\to 0}\frac{2t}{3t^2} = \lim_{t\to 0}\frac{2}{3t}$ does not exist.

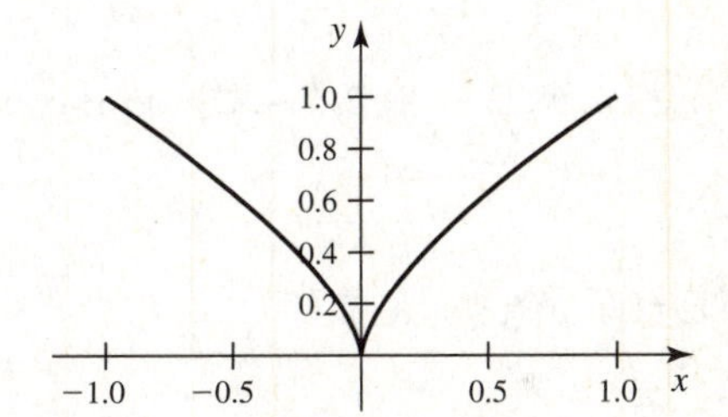

c. The curve $\mathbf{r}(t)$ for $-\infty < t < \infty$ traces out the whole curve $y = x^2$, while the curve $\mathbf{p}(t)$ only traces out the part in the first quadrant, because $x = t^2 > 0$ for all t.

d. No. For example, $\mathbf{r}(t) = \langle t^2, t^4, t^6\rangle$ does not have a cusp at the origin, since the branches for $t < 0$ and $t > 0$ are identical.

11.7 Motion in Space

11.7.1 The velocity is the derivative of position, the speed is the magnitude of velocity, and the acceleration is the derivative of velocity.

11.7.3 $m\mathbf{a}(t) = \mathbf{F}(t)$.

11.7.5 Integrate the acceleration to find an expression for the velocity plus a constant, and then use the initial velocity condition to find the constant.

11.7.7

a. $\mathbf{v}(t) = \mathbf{r}'(t) = \langle 2, -4\rangle$, so the speed is $|\mathbf{r}'(t)| = \sqrt{20} = 2\sqrt{5}$.

b. $\mathbf{a}(t) = \mathbf{r}''(t) = \langle 0,0\rangle$.

11.7.9

a. $\mathbf{v}(t) = \mathbf{r}'(t) = \langle 8\cos t, -8\sin t\rangle$, so the speed is $|\mathbf{r}'(t)| = 8$.

b. $\mathbf{a}(t) = \mathbf{r}''(t) = \langle -8\sin t, -8\cos t\rangle$.

11.7.11

a. $\mathbf{v}(t) = \mathbf{r}'(t) = \langle 1, -4, 6\rangle$, so the speed is $|\mathbf{r}'(t)| = \sqrt{1+16+36} = \sqrt{53}$.

b. $\mathbf{a}(t) = \mathbf{r}''(t) = \langle 0, 0, 0\rangle$.

11.7.13

a. $\mathbf{v}(t) = \mathbf{r}'(t) = \langle 0, 2t, -e^{-t}\rangle$, so the speed is $|\mathbf{r}'(t)| = \sqrt{4t^2 + e^{-2t}}$.

b. $\mathbf{a}(t) = \mathbf{r}''(t) = \langle 0, 2, e^{-t}\rangle$.

11.7.15

a. The interval must be shrunk by a factor of 1/3, so $[c, d] = [0, 2\pi/3]$.

b. $\mathbf{r}'(t) = \langle -\sin t, 4\cos t\rangle$, and $\mathbf{R}'(t) = \langle -3\sin 3t, 12\cos 3t\rangle$.

c. $|\mathbf{r}'(t)| = \sqrt{\sin^2 t + 16\cos^2 t}$ and $|\mathbf{R}'(t)| = 3\sqrt{\sin^2 3t + 16\cos^2 3t}$.

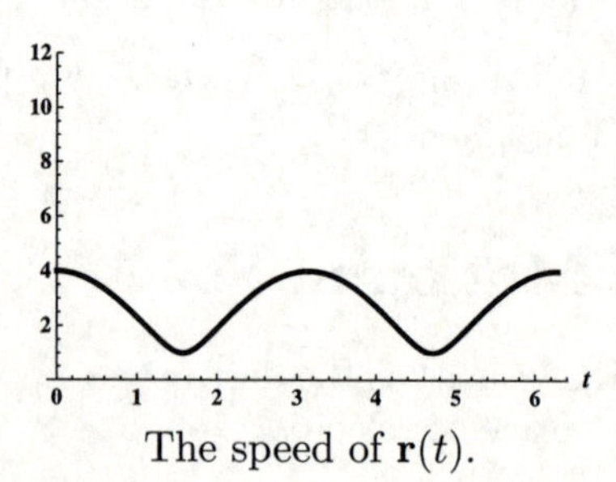

The speed of $\mathbf{r}(t)$.

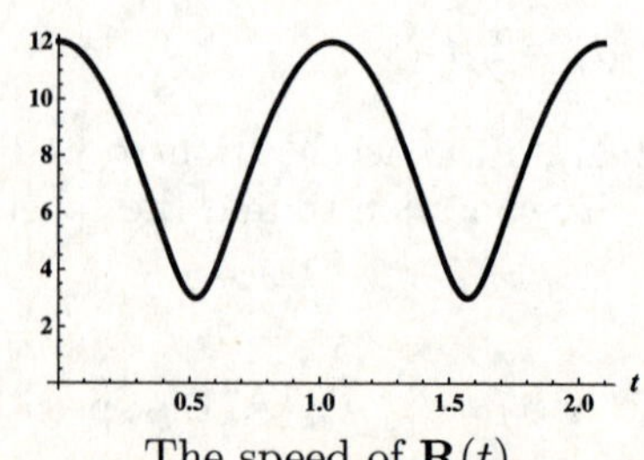

The speed of $\mathbf{R}(t)$.

11.7.17

a. Since $e^{0^2} = 1$ and $e^{6^2} = e^{36}$, we have $[c, d] = [1, e^{36}]$.

b. $\mathbf{r}'(t) = \langle 2t, -8t^3, 18t^5 \rangle$, and $\mathbf{R}'(t) = \langle 1/t, (-4\ln t)/t, (9\ln^2 t)/t \rangle$.

c. $|\mathbf{r}'(t)| = 2t\sqrt{1 + 16t^4 + 81t^8}$ and $|\mathbf{R}'(t)| = \frac{1}{t}\sqrt{1 + 16\ln^2 t + 81\ln^4 t}$.

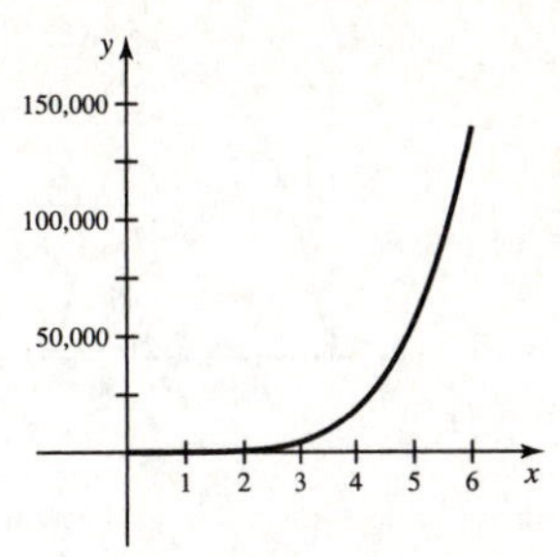

The speed of $\mathbf{r}(t)$.

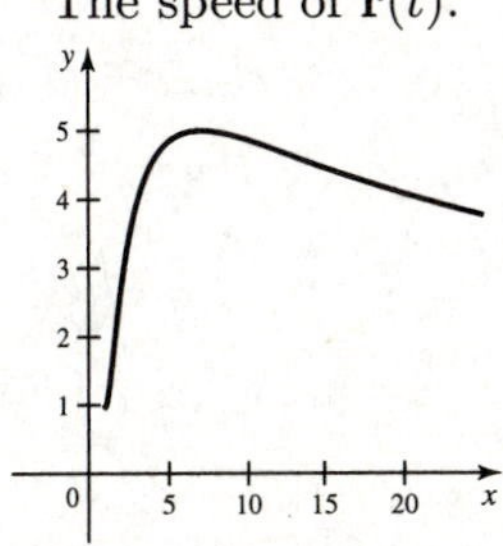

The speed of $\mathbf{R}(t)$.

11.7.19 Note that $x^2 + y^2 = 64$, so the trajectory lies on a circle centered at the origin of radius 8. $\mathbf{r}(t) \cdot \mathbf{r}'(t) = \langle 8\cos 2t, 8\sin 2t \rangle \cdot \langle -16\sin 2t, 16\cos 2t \rangle = -128\sin 2t\cos 2t + 128\sin 2t\cos 2t = 0$.

11.7.21 Note that $x^2 + y^2 = (\sin t + \sqrt{3}\cos t)^2 + (\sqrt{3}\sin t - \cos t)^2 = (\sin^2 t + 2\sin t\sqrt{3}\cos t + 3\cos^2 t) + (3\sin^2 t - 2\cos t\sqrt{3}\sin t + \cos^2 t) = 4$, so the trajectory lies on a circle centered at the origin of radius 2. $\mathbf{r}(t) \cdot \mathbf{r}'(t) = \langle \sin t + \sqrt{3}\cos t, \sqrt{3}\sin t - \cos t \rangle \cdot \langle \cos t - \sqrt{3}\sin t, \sqrt{3}\cos t + \sin t \rangle = (\sin t\cos t - \sqrt{3}\sin^2 t + \sqrt{3}\cos^2 t - 3\sin t\cos t) + (3\sin t\cos t + \sqrt{3}\sin^2 t - \sqrt{3}\cos^2 t - \sin t\cos t) = 0$.

11.7.23 $x^2 + y^2 + z^2 = \sin^2 t + \cos^2 t + \cos^2 t = 1 + \cos^2 t$, which is not a constant, so the trajectory does not lie on a sphere centered at the origin.

11.7.25 $\mathbf{v}(t) = \int \mathbf{a}(t)\,dt = \int \langle 0, 10 \rangle\,dt = \langle 0, 10t \rangle + \mathbf{C}$. Since $\mathbf{v}(0) = \langle 0, 5 \rangle$, we have $\mathbf{v}(t) = \langle 0, 10t + 5 \rangle$.

Also, $\mathbf{r}(t) = \int \mathbf{v}(t)\,dt = \int \langle 0, 10t + 5 \rangle\,dt = \langle 0, 5t^2 + 5t \rangle + \mathbf{D}$, and since $\mathbf{r}(0) = \langle 1, -1 \rangle$, we have $\mathbf{r}(t) = \langle 1, 5t^2 + 5t - 1 \rangle$.

11.7.27 $\mathbf{v}(t) = \int \mathbf{a}(t)\,dt = \int \langle \cos t, 2\sin t \rangle\,dt = \langle \sin t, -2\cos t \rangle + \mathbf{C}$. Since $\mathbf{v}(0) = \langle 0, 1 \rangle$, we have $\mathbf{v}(t) = \langle \sin t, 3 - 2\cos t \rangle$.

Also, $\mathbf{r}(t) = \int \mathbf{v}(t)\,dt = \int \langle \sin t, 3 - 2\cos t \rangle\,dt = \langle -\cos t, 3t - 2\sin t \rangle + \mathbf{D}$, and since $\mathbf{r}(0) = \langle 1, 0 \rangle$, we have $\mathbf{r}(t) = \langle 2 - \cos t, 3t - 2\sin t \rangle$.

11.7.29

a. $\mathbf{v}(t) = \int \langle 0, -9.8 \rangle\,dt = \langle 0, -9.8t \rangle + \mathbf{C}$, and since $\mathbf{v}(0) = \langle 30, 6 \rangle$, we have $\mathbf{v}(t) = \langle 30, 6 - 9.8t \rangle$.

Also, $\mathbf{r}(t) = \int \mathbf{v}(t)\,dt = \int \langle 30, 6 - 9.8t \rangle\,dt = \langle 30t, 6t - 4.9t^2 \rangle + \mathbf{D}$, and since $\mathbf{r}(0) = \langle 0, 0 \rangle$, we have $\mathbf{r}(t) = \langle 30t, 6t - 4.9t^2 \rangle$.

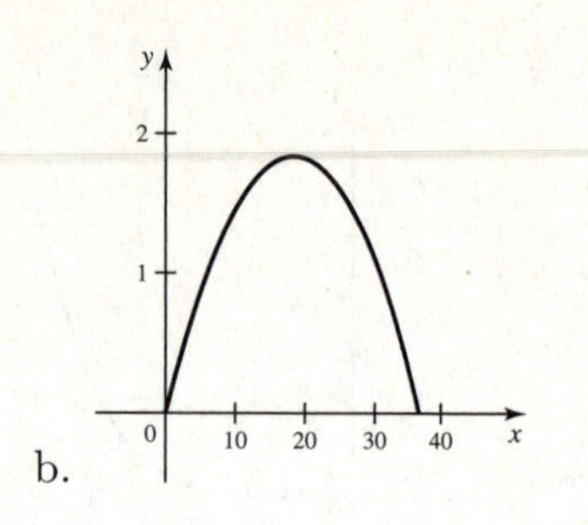

b.

c. The ball hits the ground when $6t - 4.9t^2 = 0$, which occurs for $t = 6/4.9 \approx 1.22$ seconds. The range of the ball is approximately $30 \cdot 1.22 \approx 36.7$ meters.

d. The maximum height occurs at time $T \approx 1.22/2 = .61$ seconds, and is $6T - 4.9T^2 \approx 1.84$ meters.

11.7.31

a. $\mathbf{v}(t) = \int \langle 0, -32 \rangle \, dt = \langle 0, -32t \rangle + \mathbf{C}$, and since $\mathbf{v}(0) = 250\langle 1/2, \sqrt{3}/2 \rangle = \langle 125, 125\sqrt{3} \rangle$, we have $\mathbf{v}(t) = \langle 125, 125\sqrt{3} - 32t \rangle$.

Also, $\mathbf{r}(t) = \int \mathbf{v}(t) \, dt = \int \langle 125, 125\sqrt{3} - 32t \rangle \, dt = \langle 125t, 125\sqrt{3}t - 16t^2 \rangle + \mathbf{D}$, and since $\mathbf{r}(0) = \langle 0, 20 \rangle$, we have $\mathbf{r}(t) = \langle 125t, 20 + 125\sqrt{3}t - 16t^2 \rangle$.

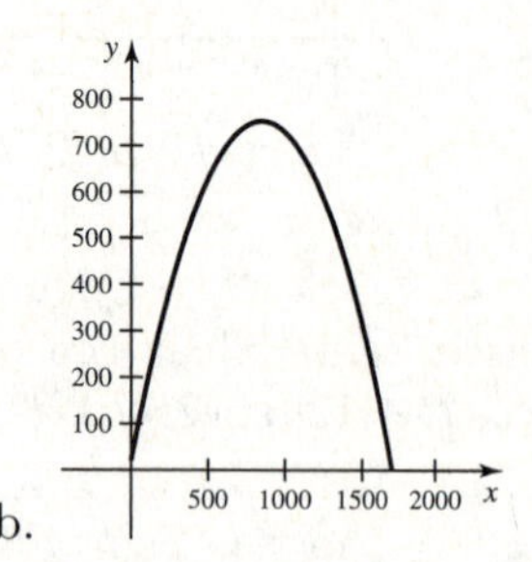

b.

c. The ball hits the ground when $20 + 125\sqrt{3}t - 16t^2 = 0$, which occurs for $t \approx 13.62$ seconds. The range of the ball is approximately $125 \cdot 13.62 \approx 1702$ feet.

d. The maximum height occurs when $125\sqrt{3} - 32t = 0$, which is when $t \approx 6.77$ and it is about $20 + 125\sqrt{3}(6.77) - 16(6.77)^2 \approx 752.4$ feet.

11.7.33 $\mathbf{v}(t) = \int \mathbf{a}(t) \, dt = \int \langle 0, 0, 10 \rangle \, dt = \langle 0, 0, 10t \rangle + \mathbf{C}$. Since $\mathbf{v}(0) = \langle 1, 5, 0 \rangle$, we have $\mathbf{v}(t) = \langle 1, 5, 10t \rangle$.

Also, $\mathbf{r}(t) = \int \mathbf{v}(t) \, dt = \int \langle 1, 5, 10t \rangle \, dt = \langle t, 5t, 5t^2 \rangle + \mathbf{D}$, and since $\mathbf{r}(0) = \langle 0, 5, 0 \rangle$, we have $\mathbf{r}(t) = \langle t, 5t + 5, 5t^2 \rangle$.

11.7.35 $\mathbf{v}(t) = \int \mathbf{a}(t) \, dt = \int \langle \sin t, \cos t, 1 \rangle \, dt = \langle -\cos t, \sin t, t \rangle + \mathbf{C}$. Since $\mathbf{v}(0) = \langle 0, 2, 0 \rangle$, we have $\mathbf{v}(t) = \langle 1 - \cos t, \sin t + 2, t \rangle$.

Also, $\mathbf{r}(t) = \int \mathbf{v}(t) \, dt = \int \langle 1 - \cos t, \sin t + 2, t \rangle \, dt = \langle t - \sin t, -\cos t + 2t, t^2/2 \rangle + \mathbf{D}$, and since $\mathbf{r}(0) = \langle 0, 0, 0 \rangle$, we have $\mathbf{r}(t) = \langle t - \sin t, 1 - \cos t + 2t, t^2/2 \rangle$.

11.7.37

a. $\mathbf{v}(t) = \int \mathbf{a}(t) \, dt = \int \langle 0, 0, -9.8 \rangle \, dt = \langle 0, 0, -9.8t \rangle + \mathbf{C}$. Since $\mathbf{v}(0) = \langle 200, 200, 0 \rangle$, we have $\mathbf{v}(t) = \langle 200, 200, -9.8t \rangle$.

Also, $\mathbf{r}(t) = \int \mathbf{v}(t) \, dt = \int \langle 200, 200, -9.8t \rangle \, dt = \langle 200t, 200t, -4.9t^2 \rangle + \mathbf{D}$, and since $\mathbf{r}(0) = \langle 0, 0, 1 \rangle$, we have $\mathbf{r}(t) = \langle 200t, 200t, -4.9t^2 + 1 \rangle$.

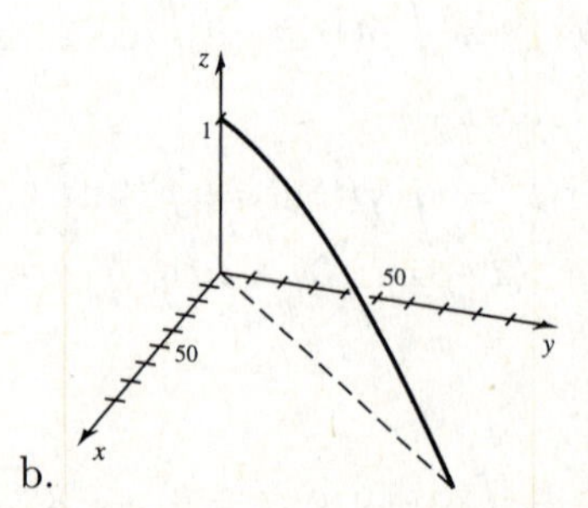

b.

c. The bullet hits the ground when $-4.9t^2 + 1 = 0$, which occurs for $t \approx .452$ seconds. At this time, the bullet is approximately at the point $(200 \cdot 0.452, 200 \cdot 0.452, 0) \approx (90.35, 90.35, 0)$. So its range is approximately $\sqrt{90.35^2 + 90.35^2} \approx 127.8$ meters.

d. The maximum height of the bullet is its initial height of 1 meter.

11.7.39

a. $\mathbf{v}(t) = \int \mathbf{a}(t)\,dt = \int \langle 0, 2.5, -9.8\rangle\,dt = \langle 0, 2.5t, -9.8t\rangle + \mathbf{C}$. Since $\mathbf{v}(0) = \langle 300, 400, 500\rangle$, we have $\mathbf{v}(t) = \langle 300, 2.5t + 400, 500 - 9.8t\rangle$.
Also, $\mathbf{r}(t) = \int \mathbf{v}(t)\,dt = \int \langle 300, 2.5t + 400, 500 - 9.8t\rangle\,dt = \langle 300t, 1.25t^2 + 400t, 500t - 4.9t^2\rangle + \mathbf{D}$, and since $\mathbf{r}(0) = \langle 0, 0, 10\rangle$, we have $\mathbf{r}(t) = \langle 300t, 1.25t^2 + 400t, 10 + 500t - 4.9t^2\rangle$.

b. 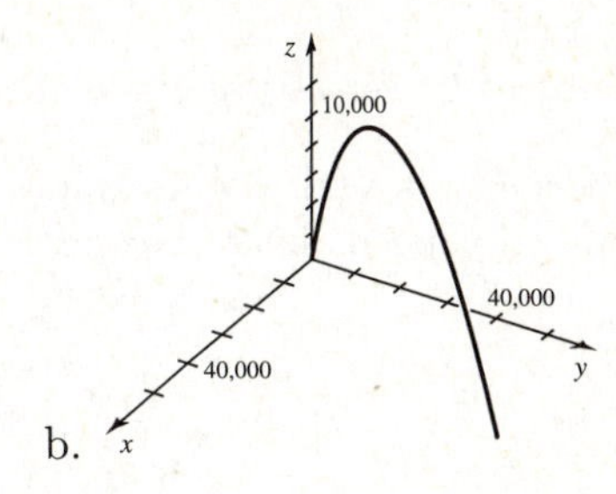

c. The rocket hits the ground when $-4.9t^2 + 500t + 10 = 0$, which occurs for $t \approx 102.1$ seconds. At this time, the rocket is at the point $(30630, 53870.5, 0)$. So its range is approximately $\sqrt{30630^2 + 53870.5^2} \approx 61969.6$ meters.

d. The maximum height of the ball occurs when $-9.8t + 500 = 0$, or when $t = 500/9.8 \approx 51.02$ seconds. At this time the rocket's height is about $10 + 500(51.02) - 4.9(51.02)^2 \approx 12,765$ meters.

11.7.41

a. False. For example, if $\mathbf{v}(t) = \langle \cos t, \sin t\rangle$, then its speed is constantly 1 even though its components aren't constant.

b. True. They both generate $\{(x, y)\,|\,x^2 + y^2 = 1\}$.

c. False. For example, $\langle t, t, t\rangle$ has variable magnitude but constant direction.

d. True. If $\mathbf{a}(t) = \langle 0, 0, 0\rangle$, then $\mathbf{v}(t) = \langle 0, 0, 0\rangle + \mathbf{C}$ for a constant vector $\mathbf{C}$.

e. False. Recall that for two-dimensional motion the range is given by $\frac{|\mathbf{v}_0|^2 \sin 2\alpha}{g}$, so doubling the speed should quadruple the range.

f. True. The time of flight is given by $T = \frac{2|\mathbf{v}_0|\sin\alpha}{g}$, so doubling the speed doubles the time of flight.

g. True. For example, if $\mathbf{v}(t) = \langle e^t, e^t, e^t\rangle$, then $\mathbf{a}(t) = \langle e^t, e^t, e^t\rangle$ as well.

11.7.43 Note that $\langle u_0, v_0\rangle = 75\langle\sqrt{3}, 1\rangle$
The time of flight is $T = \frac{2|\mathbf{v}_0|\sin\alpha}{g} = \frac{2\cdot 75}{9.8} \approx 15.3$ seconds.
The range of the flight is $\frac{|\mathbf{v}_0|^2 2\sin\alpha\cos\alpha}{g} = \frac{11250\sqrt{3}}{9.8} \approx 1988.3$ meters.
The maximum height is given by $\frac{(|\mathbf{v}_0|\sin\alpha)^2}{2g} = \frac{75^2}{19.6} \approx 287$ meters.

11.7.45 Note that $\langle u_0, v_0\rangle = 200\langle 1, \sqrt{3}\rangle$ The time of flight is $T = \frac{2|\mathbf{v}_0|\sin\alpha}{g} = \frac{2\cdot 200\sqrt{3}}{32} = 21.65$ seconds.
The range of the flight is $\frac{|\mathbf{v}_0|^2 2\sin\alpha\cos\alpha}{g} = \frac{80{,}000\sqrt{3}}{32} = 4330.13$ feet.
The maximum height is given by $\frac{(|\mathbf{v}_0|\sin\alpha)^2}{2g} = \frac{(200\sqrt{3})^2}{64} = 1875$ feet.

11.7.47 We desire $\frac{|\mathbf{v}_0|^2 2\sin\alpha\cos\alpha}{g} = 300$ meters, so we require $\sin 2\alpha = 300 \cdot \frac{9.8}{60^2} \approx .81666$. So $2\alpha = \sin^{-1}(.81666)$, and $\alpha \approx 27.4$ degrees or $\alpha \approx 62.62$ degrees.

11.7.49

a. If $t_1 > t_0$ are two values of t, we have $\mathbf{r}(t_1) - \mathbf{r}(t_0) = (f(t_1) - f(t_0))\langle a, b, c\rangle$, which is always a vector in the same direction, regardless of the values of t_1 and t_0.

b. $\mathbf{r}'(t) = f'(t)\langle a, b, c\rangle$ is a multiple of $\langle a, b, c\rangle$, so the tangent vector is always a multiple of the vector $\langle a, b, c\rangle$, so the motion of the object doesn't vary in direction, although it might vary in speed.

11.7.51

a. The object traverses the circle once over the interval $[0, 2\pi/\omega]$.

b. The velocity is $\mathbf{v}(t) = \langle -A\omega\sin(\omega t), A\omega\cos(\omega t)\rangle$. The velocity is not constant in direction, but it is constant in speed, since the speed is $|A\omega|$.

c. The acceleration is $\mathbf{a}(t) = -A\omega^2\langle \cos\omega t, \sin\omega t\rangle$.

d. The position and velocity are orthogonal. The position and acceleration are in opposite directions.

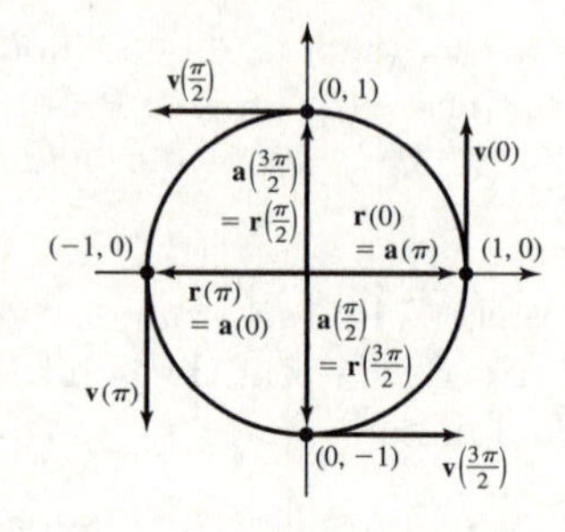

11.7.53

a. Consider $\mathbf{r}(t) = \langle 5\sin(\pi t/6), 5\cos(\pi t/6)\rangle$ for $0 \le t \le 12$. Note that the speed is the constant $5\pi/6$, and that $\mathbf{r}(0) = (0, 5) = \mathbf{r}(12)$.

b. Consider $\mathbf{r}(t) = \langle 5\sin((1-e^{-t})/5), 5\cos((1-e^{-t})/5)\rangle$ for $-\ln(10\pi+1) \le t \le 0$. Note that $\mathbf{r}(-\ln(10\pi+1)) = (0, 5) = \mathbf{r}(0)$. Also note that the speed is $|\mathbf{r}'(t)| = |e^{-t}\langle\cos((1-e^{-t})/5), -\sin((1-e^{-t})/5)\rangle| = e^{-t}$.

11.7.55

a. The velocity is $\mathbf{v}(t) = \langle -a\sin t, b\cos t\rangle$ and the speed is $\sqrt{a^2\sin^2 t + b^2\cos^2 t}$.

b. 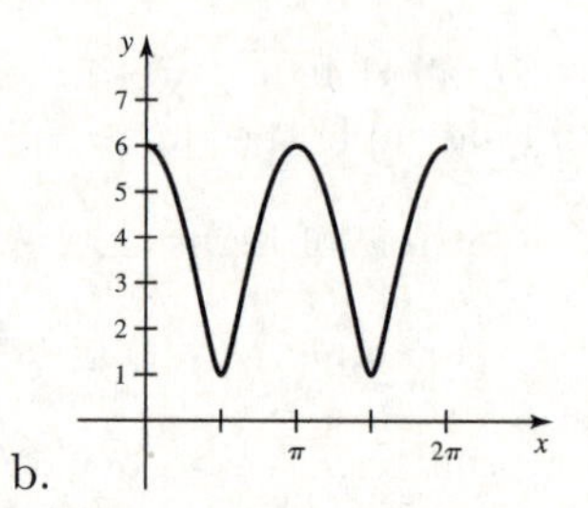

c. 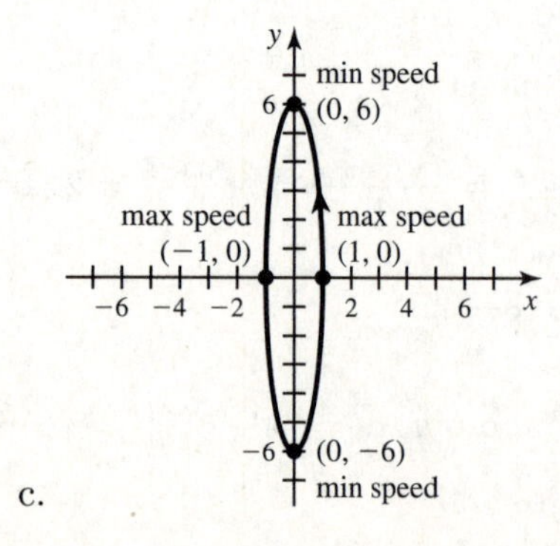

Yes, as the diagram indicates.

d. Assume $a > b > 0$. Then the maximum speed occurs at $\pi/2$ and is equal to a, while the minimum speed occurs at π and is equal to b. So the ratio is $\frac{a}{b}$. In the case $b > a > 0$, the ratio is $\frac{b}{a}$.

11.7.57

a. The initial point is $\mathbf{r}(0) = \langle 50, 0, 0\rangle$, and the "terminal" point is $\langle 0, 0, 5\rangle$ since $\lim_{t\to\infty} e^{-t} = 0$, while $\sin t$ and $\cos t$ are bounded between -1 and 1 as $t \to \infty$.

b. The speed is $5e^{-t}\,|\langle -10(\cos t + \sin t), 10(\cos t - \sin t), 1\rangle| = 5e^{-t}\sqrt{201}$.

c. 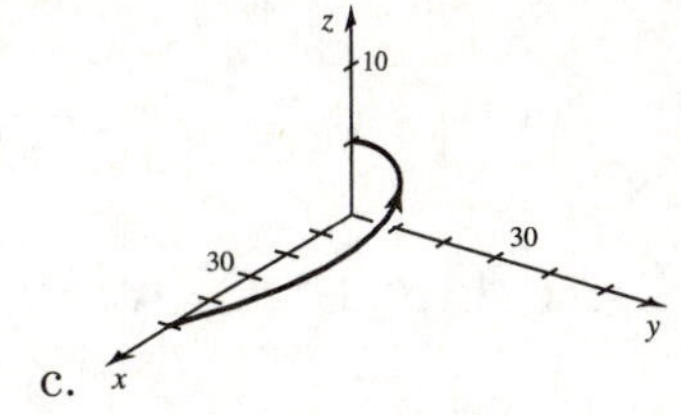

11.7.59 Let the angle be α. Then $\mathbf{v}(t) = \langle 120\cos\alpha, 120\sin\alpha - 32t\rangle$, and $\mathbf{r}(t) = \langle 120t\cos\alpha, 120t\sin\alpha - 16t^2\rangle$. Since we require the ball to land in the hole, we need the point $(420, -50)$ to be on the curve. So $120t\cos\alpha = 420$ and $120t\sin\alpha - 16t^2 = -50$. Thus $t = \frac{420}{120\cos\alpha}$, and therefore $120\cdot\frac{420\cdot\sin\alpha}{120\cos\alpha} - 16\left(\frac{420}{120\cos\alpha}\right)^2 = -50$. This can be written $420\tan\alpha - 196\sec^2\alpha + 50 = -196\tan^2\alpha + 420\tan\alpha - 146 = 0$. By the quadratic formula, we have

$$\tan\alpha = \frac{1}{14}\left(15 - \sqrt{79}\right) \text{ and } \tan\alpha = \frac{1}{14}\left(15 + \sqrt{79}\right).$$

Applying the inverse tangent and then writing the answer in degrees, we obtain $\alpha = 59.63$ degrees and $\alpha = 23.58$ degrees.

11.7.61

a. $\mathbf{v} = \langle 130, 0, -3 - 32t\rangle$ and $\mathbf{r}(t) = \langle 130t, 0, 6 - 3t - 16t^2\rangle$. When $x = 60$, $t = 6/13$, so $z = 6 - 3(6/13) - 16(6/13)^2 \approx 1.207$ feet. The flight lasts $t = 6/13$ seconds.

b. Suppose that the initial velocity is $\langle 130, 0, b\rangle$, so that $\mathbf{v}(t) = \langle 130t, 0, 6 + bt - 16t^2\rangle$. So $z = 3 = 6 + b(6/13) - 16(6/13)^2$, which when solved for b gives $b = .8846$.

c. In this scenario, we have $\mathbf{v}(t) = \langle 130, 8t, -3 - 32t\rangle$ and $\mathbf{r}(t)\langle 130t, 4t^2, 6 - 3t - 16t^2\rangle$. As before, $x = 60$ when $t = 6/13$, and at that time $y = 4(6/13)^2 \approx .8521$ feet.

d. It moves more in the second half, because of the factor of t^2. This makes life more difficult for the batter!

e. In this case $\mathbf{v}(t) = \langle 130, ct, -3 - 32t\rangle$ and $\mathbf{r}(t) = \langle 130t, -3 + ct^2/2, 6 - 3t - 16t^2\rangle$. Again, we have $t = 6/13$, and so we require $-3 + c\left(\frac{6}{13}\right)^2\cdot\frac{1}{2} = 0$, so $c \approx 28.17$.

11.7.63 We have $\mathbf{v}(t) = \langle v_0\cos\alpha, v_0\sin\alpha - gt\rangle$, and $\mathbf{r}(t) = \langle v_0t\cos\alpha, y_0 + v_0t\sin\alpha - \frac{1}{2}gt^2\rangle$. Suppose that object hits the ground at $(a, 0)$. Then $a = v_0T\cos\alpha$ and $0 = y_0 + v_0T\sin\alpha - \frac{1}{2}gT^2$ where T is the time of the flight. So by the quadratic formula, $T = \frac{v_0\sin\alpha + \sqrt{v_0^2\sin^2\alpha + 2gy_0}}{g}$. Thus $a = v_0T\cos\alpha = v_0(\cos\alpha)\left(\frac{v_0\sin\alpha + \sqrt{v_0^2\sin^2\alpha + 2gy_0}}{g}\right)$ is the range. Since the maximum height when $y_0 = 0$ is $\frac{v_0^2\sin^2\alpha}{2g}$, the maximum height in this scenario is $y_0 + \frac{v_0^2\sin^2\alpha}{2g}$.

11.7.65 Note that $x^2 + y^2 = \cos^2 t + \sin^2 t = 1$, and $z = cy$, so this curve is the intersection of the cylinder $x^2 + y^2 = 1$ with the plane $z = cy$, which results in an ellipse in that plane.

11.7.67 $\mathbf{r}(t)$ can be written $\langle R\cos\phi\cos t, R\sin\phi\cos t, R\sin t\rangle$ where R is the radius of the sphere and ϕ is a real constant. Note that

$$\mathbf{v}(t) = \langle -R\cos\phi\sin t, -R\sin\phi\sin t, R\cos t\rangle$$

and $\mathbf{a}(t) = \langle -R\cos\phi\cos t, -R\sin\phi\cos t, -R\sin t\rangle$, and that $\mathbf{r}(t)\cdot\mathbf{a}(t) = -R^2\cos^2\phi\cos^2 t - R^2\sin^2\phi\cos^2 t - R^2\sin^2 t = -R^2(\cos^2\phi + \sin^2\phi)\cos^2 t - R^2\sin^2 t = -R^2(\cos^2 t + \sin^2 t) = -R^2 = -|\mathbf{v}(t)|^2$.

11.7.69

a. $|\mathbf{r}(t)|^2 = (a\cos t + b\sin t)^2 + (c\cos t + d\sin t)^2 + (e\cos t + f\sin t)^2 = (a^2+c^2+e^2)\cos^2 t + (2ab+2cd+2ef)\sin t\cos t + (b^2+d^2+f^2)\sin^2 t$ In order for the path to be a circle, it would be sufficient that $a^2+c^2+e^2 = b^2+d^2+f^2$ and that $ab+cd+ef=0$.

b. In order for the path to be an ellipse, it would be sufficient that $ab+cd+ef=0$.

11.8 Lengths of Curves

11.8.1 $L = \int_a^b \sqrt{x'(t)^2 + y'(t)^2}\,dt = \int_a^b \sqrt{1+4}\,dt = \sqrt{5}(b-a)$.

11.8.3 The arc length is $L = \int_a^b |\mathbf{r}'(t)|\,dt = \int_a^b |\mathbf{v}(t)|\,dt$.

11.8.5 $L = \int_0^\pi |\langle -20\sin(2t), 20\cos(2t)\rangle|\,dt = 20\pi$.

11.8.7 $L = \int_0^\pi \sqrt{(-3\sin t)^2 + (3\cos t)^2}\,dt = \int_0^\pi 3\,dt = 3\pi$.

11.8.9 Note that $(\cos t + t\sin t)' = -\sin t + \sin t + t\cos t = t\cos t$, and $(\sin t - t\cos t)' = \cos t - (\cos t - t\sin t) = t\sin t$. $L = \int_0^{\pi/2} \sqrt{(t\cos t)^2 + (t\sin t)^2}\,dt = \int_0^{\pi/2} t\,dt = \left.\frac{t^2}{2}\right|_0^{\pi/2} = \frac{\pi^2}{8}$.

11.8.11 $L = \int_1^6 \sqrt{3^2 + (-4)^2 + 3^2}\,dt = \sqrt{34}(6-1) = 5\sqrt{34}$.

11.8.13 $L = \int_0^{4\pi} \sqrt{1 + 64\cos^2 t + 64\sin^2 t}\,dt = \sqrt{65}(4\pi)$.

11.8.15

$$\begin{aligned} L &= \int_0^2 \sqrt{t^2+16t+16}\,dt = \int_0^2 \sqrt{(t+8)^2 - 48}\,dt \\ &= \left(\frac{t}{2}+4\right)\sqrt{t^2+16t+16} - 24\ln\left(2\left(\sqrt{t^2+16t+16} + t + 8\right)\right)\Big|_0^2 \\ &= (10\sqrt{13} - 16) - 24\ln\left(\frac{5+\sqrt{13}}{6}\right). \end{aligned}$$

The integral was computed with the help of the table of integrals from the back of the book.

11.8.17 $L = \int_0^{\pi/2} \sqrt{9\cos^4 t\sin^2 t + 9\sin^4 t\cos^2 t}\,dt = \int_0^{\pi/2} 3\sin t\cos t\,dt = \frac{3}{2}\cdot \sin^2 t\big|_0^{\pi/2} = \frac{3}{2}$.

11.8.19 The speed is $\sqrt{36t^4 + 9t^4 + 225t^4} = \sqrt{270}\,t^2$. The length is thus $L = \int_0^4 \sqrt{270}\,t^2\,dt = \sqrt{270}\left.\frac{t^3}{3}\right|_0^4 = \sqrt{30}(64-0) = 64\sqrt{30}$.

11.8.21 The speed is $2\sqrt{(13\cos 2t)^2 + (-12\sin 2t)^2 + (-5\sin 2t)^2} = 2\cdot\sqrt{169} = 26$. The length is thus $L = \int_0^\pi 26\,dt = 26\pi$.

11.8.23 $L = \int_0^{2\pi} \sqrt{4\sin^2 t + 16\cos^2 t}\,dt = \int_0^{2\pi} \sqrt{4 + 12\cos^2 t}\,dt = 2\int_0^{2\pi} \sqrt{1 + 3\cos^2 t}\,dt \approx 19.38$.

11.8.25 $L = \int_{-2}^2 \sqrt{1+64t^2}\,dt \approx 32.5$.

11.8.27 Note that the diameter of the circle is a, and that the complete circle is traversed for $0 \le t \le \pi$. $L = \int_0^\pi \sqrt{(a\sin\theta)^2 + (a\cos\theta)^2}\,d\theta = \int_0^\pi a\,d\theta = \pi a$.

11.8.29 Using symmetry, $L = 8\int_0^{\pi} \sqrt{(1+\cos\theta)^2 + (-\sin\theta)^2}\,d\theta = 8\int_0^{\pi} \sqrt{2+2\cos\theta}\,d\theta = 8\sqrt{2}\int_0^{\pi} \sqrt{1+\cos\theta}\cdot \frac{\sqrt{1-\cos\theta}}{\sqrt{1-\cos\theta}}\,d\theta = 8\sqrt{2}\int_0^{\pi} \frac{\sin\theta}{\sqrt{1-\cos\theta}}\,d\theta$. Let $u = 1-\cos\theta$ so that $du = \sin\theta\,d\theta$. Then $L = 8\sqrt{2}\int_0^2 \frac{1}{\sqrt{u}}\,du = 16\sqrt{2}\left(\sqrt{u}\big|_0^2\right) = 32$.

11.8.31 $L = \int_0^{\ln 8} \sqrt{4e^{4\theta} + 16e^{4\theta}}\,d\theta = 2\sqrt{5}\int_0^{\ln 8} e^{2\theta}\,d\theta = \sqrt{5}\cdot e^{2\theta}\big|_0^{\ln 8} = \sqrt{5}(64-1) = 63\sqrt{5}$.

11.8.33 $L = \int_0^{\pi/2} \sqrt{\sin^6(\theta/3) + \sin^4(\theta/3)\cos^2(\theta/3)}\,d\theta = \int_0^{\pi/2} \sin^2(\theta/3)\,d\theta = \frac{1}{2}\int_0^{\pi/2} 1 - \cos(2\theta/3)\,d\theta = \frac{1}{2}\left(\theta - \frac{3}{2}\sin(2\theta/3)\big|_0^{\pi/2}\right) = \frac{1}{2}\left(\frac{\pi}{2} - \frac{3\sqrt{3}}{4}\right) = \frac{2\pi - 3\sqrt{3}}{8}$.

11.8.35

a. True. $L = \int_a^b S\,dt = S(b-a)$.

b. True. Both have length $L = \int_a^b \sqrt{f'(t)^2 + g'(t)^2}\,dt = \int_a^b \sqrt{g'(t)^2 + f'(t)^2}\,dt$.

c. True. Both have length $L = \int_a^b \sqrt{f'(t)^2 + g'(t)^2}\,dt = \int_{\sqrt{a}}^{\sqrt{b}} \sqrt{(f'(u^2)(2u))^2 + (g'(u^2)(2u))^2}\,du$. The equality can be seen via the substitution $u^2 = t$.

11.8.37

a. Let $x = a\cos t$, $y = b\sin t$ and $z = c\sin t$. Then $x^2 + y^2 + z^2 = a^2\cos^2 t + b^2\sin^2 t + c^2\sin^2 t = a^2\cos^2 t + (b^2 + c^2)\sin^2 t = a^2$, assuming $a^2 = b^2 + c^2$. So the curve lies on a sphere, but also note that $cy = bz$, so the curve also lies in the plane $cy - bz = 0$. So the curve is a circle centered at the origin.

b. The circle has arc length $L = \int_0^{2\pi} \sqrt{a^2\sin^2 t + b^2\cos^2 t + c^2\cos^2 t}\,dt = \sqrt{a^2}\int_0^{2\pi} \sqrt{\sin^2 t + \cos^2 t}\,dt = 2a\pi$.

c. As in exercise 69a from section 11.7, the curve describes a circle when $a^2 + c^2 + e^2 = b^2 + d^2 + f^2 = R^2$ and $ab + cd + ef = 0$. Note that $|\mathbf{r}(t)|^2 = (a\cos t + b\sin t)^2 + (c\cos t + d\sin t)^2 + (e\cos t + f\sin t)^2 = (a^2 + c^2 + e^2)\cos^2 t + (2ab + 2cd + 2ef)\sin t\cos t + (b^2 + d^2 + f^2)\sin^2 t$, so if the conditions are met, then the curve describes a circle of radius R and $L = \int_0^{2\pi} \sqrt{R^2}\,dt = 2\pi R$.

11.8.39

a. $\mathbf{r}'(t) = h'(t)\langle A, B\rangle$, so $|\mathbf{r}'(t)| = |h'(t)|\sqrt{A^2 + B^2}$. Thus, $L = \sqrt{A^2 + B^2}\int_a^b |h'(t)|\,dt$

b. $L = \sqrt{2^2 + 5^2}\int_0^4 3t^2\,dt = \sqrt{29}\,t^3\big|_0^4 = 64\sqrt{29}$.

c. $L = \sqrt{4^2 + 10^2}\int_1^8 \left|-1/t^2\right|\,dt = \sqrt{116}\,\frac{1}{t}\big|_8^1 = \frac{7\sqrt{29}}{4}$.

11.8.41 $r'(\theta) = -ae^{-a\theta}$. Thus,

$$\begin{aligned} L = \int_0^{\infty} \sqrt{e^{-2a\theta} + a^2 e^{-2a\theta}}\,d\theta &= \sqrt{1+a^2}\int_0^{\infty} e^{-a\theta}\,d\theta \\ &= \lim_{b\to\infty} \sqrt{1+a^2}\int_0^b e^{-a\theta}\,d\theta = \frac{\sqrt{1+a^2}}{-a}\lim_{b\to\infty} e^{-a\theta}\big|_0^b \\ &= \frac{\sqrt{1+a^2}}{-a}(0-1) = \frac{\sqrt{1+a^2}}{a}. \end{aligned}$$

11.8.43 Using symmetry, we are seeking 4 times the curve traversed for $0 \le \theta \le \pi/4$. Thus, $L = 4\displaystyle\int_0^{\pi/4} \sqrt{6\sin 2\theta + 6\cot 2\theta\cos 2\theta}\,d\theta = 12.85$.

11.8.45 $L = \displaystyle\int_0^{2\pi} \sqrt{(4 - 2\cos\theta)^2 + 4\sin^2\theta}\,d\theta = \int_0^{2\pi} \sqrt{20 - 16\cos\theta}\,d\theta = 26.73$.

11.8.47

a. $y = -4.9t^2 + 25t$ is 0 for $t > 0$ when $-4.9t + 25 = 0$, or $t = 25/4.9 \approx 5.102$ seconds.

b. $L \approx \int_0^{5.102} \sqrt{400 + (25 - 9.8t)^2}\, dt$.

c. Let $u = -9.8t + 25$ so that $du = -9.8dt$. Then $L \approx \frac{1}{-9.8} \int_{25}^{-24.9996} \sqrt{400 + u^2}\, du \approx 124.43$ meters.

d. $x = u_0(5.102) = 20(5.102) = 102.04$ meters.

11.8.49 $\int_a^b |\mathbf{r}'(t)|\, dt = \int_a^b \sqrt{(cf'(t))^2 + (cg'(t))^2}\, dt = |c| \int_a^b \sqrt{f'(t)^2 + g'(t)^2}\, dt = |c|\, L$.

11.8.51 The curve can be parametrized by $x = t$ and $y = f(t)$. Then $\mathbf{r}'(t) = \langle 1, f'(t)\rangle$, so $|\mathbf{v}(t)| = \sqrt{1 + f'(t)^2}$. So $L = \int_a^b \sqrt{1 + f'(t)^2}\, dt = \int_a^b \sqrt{1 + f'(x)^2}\, dx$.

11.9 Curvature and Normal Vectors

11.9.1 If the parameter t used to describe a trajectory also measures the arc length s of the curve that is generated, then we say that the curve is parametrized by its arc length. This occurs when $|\mathbf{v}(t)| = 1$.

11.9.3 Note that $|\mathbf{v}(t)| = \sqrt{1+1+1} = \sqrt{3} \neq 1$, so it isn't parametrized by arc length.

11.9.5 $\kappa = \frac{1}{|\mathbf{v}|}\left|\frac{d\mathbf{T}}{dt}\right|$ or $\kappa = \frac{|\mathbf{a} \times \mathbf{v}|}{|\mathbf{v}|^3}$.

11.9.7 $\mathbf{N} = \frac{d\mathbf{T}/dt}{|d\mathbf{T}/dt|}$.

11.9.9 Note that $|\mathbf{v}| = \sqrt{1+4} \neq 1$, so it doesn't use arc length as a parameter.
Consider $\mathbf{r}(s) = \langle s/\sqrt{5}, 2s/\sqrt{5}\rangle$ for $0 \le s \le 3\sqrt{5}$. This has $|\mathbf{r}'(s)| = \frac{1}{\sqrt{5}}\sqrt{1+4} = 1$, so it does use arc length as its parameter.

11.9.11 Note that $|\mathbf{v}| = \sqrt{4\sin^2 t + 4\cos^2 t} \neq 1$, so it doesn't use arc length as a parameter.

Consider $\mathbf{r}(s) = \langle 2\cos(s/2), 2\sin(s/2)\rangle$ for $0 \le s \le 4\pi$. This has $|\mathbf{r}'(s)| = \sqrt{\sin^2(s/2) + \cos^2(s/2)} = 1$, so it does use arc length as its parameter.

11.9.13 Note that $|\mathbf{v}| = \sqrt{4t^2\sin^2(t^2) + 4t^2\cos^2(t^2)} = 2t \neq 1$, so it doesn't use arc length as a parameter.

Consider $\mathbf{r}(s) = \langle \cos s, \sin s\rangle$ for $0 \le s \le \pi$. This has $|\mathbf{r}'(s)| = \sqrt{\sin^2 s + \cos^2 s} = 1$, so it does use arc length as its parameter.

11.9.15 $\mathbf{r}'(t) = \langle 2, 4, 6\rangle$, so $\mathbf{T} = \frac{1}{2\sqrt{14}}\langle 2, 4, 6\rangle = \frac{1}{\sqrt{14}}\langle 1, 2, 3\rangle$.
So $\frac{d\mathbf{T}}{dt} = \langle 0, 0, 0\rangle$, and $\kappa = 0$.

11.9.17 $\mathbf{r}'(t) = \langle 2, 4\cos t, -4\sin t\rangle$, so $\mathbf{T}(t) = \frac{1}{\sqrt{5}}\langle 1, 2\cos t, -2\sin t\rangle$. So

$$\kappa = \frac{1}{|\mathbf{v}|}\left|\frac{d\mathbf{T}}{dt}\right| = \frac{1}{2\sqrt{5}}\left|\frac{1}{\sqrt{5}}\langle 0, -2\sin t, -2\cos t\rangle\right| = \frac{1}{5}.$$

11.9.19 $\mathbf{r}'(t) = \langle \sqrt{3}\cos t, \cos t, -2\sin t\rangle$, so $|\mathbf{v}(t)| = \sqrt{3\cos^2 t + \cos^2 t + 4\sin^2 t} = 2$.
Thus, $\mathbf{T}(t) = \frac{1}{2}\langle \sqrt{3}\cos t, \cos t, -2\sin t\rangle$. So

$$\kappa = \frac{1}{|\mathbf{v}|}\left|\frac{d\mathbf{T}}{dt}\right| = \frac{1}{2}\left|\frac{1}{2}\langle -\sqrt{3}\sin t, -\sin t, -2\cos t\rangle\right| = \frac{1}{2} \cdot \frac{1}{2} \cdot 2 = \frac{1}{2}.$$

11.9.21 $\mathbf{r}'(t) = \langle 1, 4t \rangle$, so $|\mathbf{v}(t)| = \sqrt{16t^2+1}$. Thus, $\mathbf{T}(t) = \frac{1}{\sqrt{16t^2+1}}\langle 1, 4t \rangle$.

$$\kappa = \frac{|\mathbf{a}\times\mathbf{v}|}{|\mathbf{v}|^3} = \frac{1}{(16t^2+1)^{3/2}}\,|\langle 0,4,0\rangle \times \langle 1,4t,0\rangle| = \frac{1}{(16t^2+1)^{3/2}}\,|\langle 0,0,-4\rangle| = \frac{4}{(16t^2+1)^{3/2}}.$$

11.9.23 $\mathbf{r}'(t) = \langle 3\sin t, 3\cos t\rangle$ and $\mathbf{r}''(t) = \langle 3\cos t, -3\sin t\rangle$. So

$$\kappa = \frac{|\mathbf{a}\times\mathbf{v}|}{|\mathbf{v}|^3} = \frac{1}{27}\,|\langle 3\cos t, -3\sin t, 0\rangle \times \langle 3\sin t, 3\cos t, 0\rangle| = \frac{1}{27}\,|\langle 0,0,9\rangle| = \frac{1}{3}.$$

11.9.25 $\mathbf{r}'(t) = \langle 2t, 1\rangle$ and $\mathbf{r}''(t) = \langle 2, 0\rangle$. So

$$\kappa = \frac{|\mathbf{a}\times\mathbf{v}|}{|\mathbf{v}|^3} = \frac{1}{(4t^2+1)^{3/2}}\,|\langle 2,0,0\rangle \times \langle 2t,1,0\rangle| = \frac{1}{(4t^2+1)^{3/2}}\,|\langle 0,0,2\rangle| = \frac{2}{(4t^2+1)^{3/2}}.$$

11.9.27 $\mathbf{r}'(t) = \langle -7\sin t, \sqrt{3}\cos t, -2\sin t\rangle$ and $\mathbf{r}''(t) = \langle -7\cos t, -\sqrt{3}\sin t, -2\cos t\rangle$. So

$$\kappa = \frac{|\mathbf{a}\times\mathbf{v}|}{|\mathbf{v}|^3} = \frac{1}{(53\sin^2 t+3)^{3/2}}\left|\langle -7\cos t, -\sqrt{3}\sin t, -2\cos t\rangle \times \langle -7\sin t, \sqrt{3}\cos t, -2\sin t\rangle\right|$$

$$= \frac{1}{(53\sin^2 t+3)^{3/2}}\left|\langle 2\sqrt{3}, 0, -7\sqrt{3}\rangle\right| = \frac{\sqrt{159}}{(53\sin^2 t+3)^{3/2}}.$$

11.9.29 $\mathbf{r}'(t) = \langle 2\cos t, -2\sin t\rangle$, so $\mathbf{T} = \langle \cos t, -\sin t\rangle$ and $|\mathbf{T}| = 1$. $\mathbf{N}(t) = \langle -\sin t, -\cos t\rangle$. Note that $|\mathbf{N}| = 1$ and $\mathbf{T}\cdot\mathbf{N} = -\sin t\cos t + \sin t\cos t = 0$.

11.9.31 $\mathbf{r}'(t) = \langle t, -3, 0\rangle$, and $|\mathbf{r}'(t)| = \sqrt{t^2+9}$, so $\mathbf{T} = \frac{1}{\sqrt{t^2+9}}\langle t, -3, 0\rangle$ and $|\mathbf{T}| = 1$. We have $\mathbf{T}'(t) = \langle \frac{9}{(\sqrt{t^2+9})^3}, \frac{3t}{(\sqrt{t^2+9})^3}, 0\rangle$, so $\mathbf{N}(t) = \frac{1}{\sqrt{t^2+9}}\langle 3, t, 0\rangle$. Note that $|\mathbf{N}| = 1$ and $\mathbf{T}\cdot\mathbf{N} = 0$.

11.9.33 $\mathbf{r}'(t) = \langle -2t\sin t^2, 2t\cos t^2\rangle$, and $|\mathbf{r}'(t)| = 2t$, so $\mathbf{T} = \langle -\sin t^2, \cos t^2\rangle$ and $|\mathbf{T}| = 1$. $\mathbf{T}'(t) = \langle -2t\cos t^2, -2t\sin t^2\rangle$, so $\mathbf{N}(t) = \langle -\cos t^2, -\sin t^2\rangle$. Note that $|\mathbf{N}| = 1$ and $\mathbf{T}\cdot\mathbf{N} = 0$.

11.9.35 $\mathbf{r}'(t) = \langle 2t, 1\rangle$, and $|\mathbf{r}'(t)| = \sqrt{4t^2+1}$, so $\mathbf{T} = \frac{1}{\sqrt{4t^2+1}}\langle 2t, 1\rangle$ and $|\mathbf{T}| = 1$.
$\mathbf{T}'(t) = \langle \frac{2}{(\sqrt{4t^2+1})^3}, \frac{-4t}{(\sqrt{4t^2+1})^3}\rangle$, so $\mathbf{N}(t) = \frac{1}{\sqrt{4t^2+1}}\langle 1, -2t\rangle$. Note that $|\mathbf{N}| = 1$ and $\mathbf{T}\cdot\mathbf{N} = 0$.

11.9.37 $\mathbf{r}'(t) = \langle 1, 4, -6\rangle$ and $\mathbf{r}''(t) = \langle 0,0,0\rangle$. So $\kappa = 0$ and thus $a_N = 0$. Also, $a_T = 0$. We have $\mathbf{a} = 0\cdot\mathbf{T} + 0\cdot\mathbf{N} = \langle 0,0,0\rangle$.

11.9.39 $\mathbf{r}'(t) = \langle -\sin t, 6\cos t, -\sqrt{5}\sin t\rangle$ and $\mathbf{r}''(t) = \langle -\cos t, -6\sin t, -\sqrt{5}\cos t\rangle$. Note that $\mathbf{a}\cdot\mathbf{v} = -30\sin t\cos t$, so $a_T = \frac{-30\sin t\cos t}{\sqrt{6}\sqrt{1+5\cos^2 t}}$. $\mathbf{a}\times\mathbf{v} = \langle 6\sqrt{5}, 0, -6\rangle$, so $a_N = \frac{6\sqrt{6}}{\sqrt{6}\sqrt{1+5\cos^2 t}} = \frac{6}{\sqrt{1+5\cos^2 t}}$. We have $\mathbf{a} = \langle -\cos t, -6\sin t, -\sqrt{5}\cos t\rangle = \frac{-30\sin t\cos t}{\sqrt{6}\sqrt{1+5\cos^2 t}}\cdot\mathbf{T} + \frac{6}{\sqrt{1+5\cos^2 t}}\cdot\mathbf{N}$.

11.9.41 $\mathbf{r}'(t) = \langle 3t^2, 2t\rangle$ and $\mathbf{r}''(t) = \langle 6t, 2\rangle$. Note that $\mathbf{a}\cdot\mathbf{v} = 18t^3 + 4t$, so $a_T = \frac{18t^3+4t}{\sqrt{9t^4+4t^2}} = \frac{18t^2+4}{\sqrt{9t^2+4}}$. $\mathbf{a}\times\mathbf{v} = \langle 0,0,6t^2\rangle$, so $a_N = \frac{6t^2}{t\sqrt{9t^2+4}} = \frac{6t}{\sqrt{9t^2+4}}$. We have $\mathbf{a} = \langle 6t, 2\rangle = \frac{18t^2+4}{\sqrt{9t^2+4}}\cdot\mathbf{T} + \frac{6t}{\sqrt{9t^2+4}}\cdot\mathbf{N}$.

11.9.43

a. False. For example, consider $\mathbf{r}(t) = \langle \cos t, \sin t, 1\rangle$. Then $\mathbf{T} = \langle -\sin t, \cos t, 0\rangle$ and also $\mathbf{N} = \langle -\cos t, -\sin t, 0\rangle$. Note that $\mathbf{T}$ and $\mathbf{N}$ lie in the xy-plane, but $\mathbf{r}$ doesn't.

b. False. $\mathbf{T}$ does depend on the orientation, but $\mathbf{N}$ doesn't. Reversing the orientation changes $\mathbf{T}$ to $-\mathbf{T}$, but leaves $\mathbf{N}$ alone.

c. False. Note that $|\mathbf{T}|$ is independent of orientation, so $\left|\frac{d\mathbf{T}}{ds}\right|$ is too.

d. True. As we have already seen for the circle, $\mathbf{v}\cdot\mathbf{a}=0$. Thus $\mathbf{a}=a_T\mathbf{T}+a_N\mathbf{N}=0\cdot\mathbf{T}+\kappa\,|\mathbf{v}|^2\,\mathbf{N}=\frac{1}{R}\mathbf{N}$.

e. False. For example, if the car's motion is given by $\mathbf{r}(t)=\langle 60\cos t, 60\sin t\rangle$, then the speed is a constant 60, but $\mathbf{a}=\langle -60\cos t, -60\sin t\rangle \neq \langle 0,0\rangle$.

11.9.45 $f'(x)=2x$ and $f''(x)=2$, so

$$\kappa=\frac{|f''(x)|}{(1+f'(x)^2)^{3/2}}=\frac{2}{(1+4x^2)^{3/2}}$$

11.9.47 $f'(x)=1/x$ and $f''(x)=-1/x^2$, so

$$\kappa=\frac{|f''(x)|}{(1+f'(x)^2)^{3/2}}=\frac{1/x^2}{(1+(1/x)^2)^{3/2}}=\frac{x}{(x^2+1)^{3/2}}.$$

11.9.49 $\mathbf{r}'(t)=\langle f'(t),g'(t)\rangle$ and $\mathbf{r}''(t)=\langle f''(t),g''(t)\rangle$. So $\mathbf{a}\times\mathbf{v}=\langle f''(t),g''(t),0\rangle\times\langle f'(t),g'(t),0\rangle=\langle 0,0,g'(t)f''(t)-f'(t)g''(t)\rangle$. Since $|\mathbf{v}|=\sqrt{(f'(t))^2+(g'(t))^2}$, we have

$$\kappa=\frac{|\mathbf{a}\times\mathbf{v}|}{|\mathbf{v}|^3}=\frac{|f'g''-g'f''|}{((f')^2+(g')^2)^{3/2}}.$$

11.9.51 $f'(t)=a\cos t$ and $f''(t)=-a\sin t$, while $g'(t)=-b\sin t$ and $g''(t)=-b\cos t$. So

$$\kappa=\frac{|\mathbf{a}\times\mathbf{v}|}{|\mathbf{v}|^3}=\frac{|f'g''-g'f''|}{((f')^2+(g')^2)^{3/2}}=\frac{|-ab\cos^2 t-ab\sin^2 t|}{(a^2\cos^2 t+b^2\sin^2 t)^{3/2}}=\frac{|ab|}{(a^2\cos^2 t+b^2\sin^2 t)^{3/2}}.$$

11.9.53 $f'(t)=1$ and $f''(t)=0$, while $g'(t)=2at$ and $g''(t)=2a$. So

$$\kappa=\frac{|f'g''-g'f''|}{((f')^2+(g')^2)^{3/2}}=\frac{|2a-0|}{(1+4a^2t^2)^{3/2}}=\frac{2\,|a|}{(1+4a^2t^2)^{3/2}}.$$

11.9.55

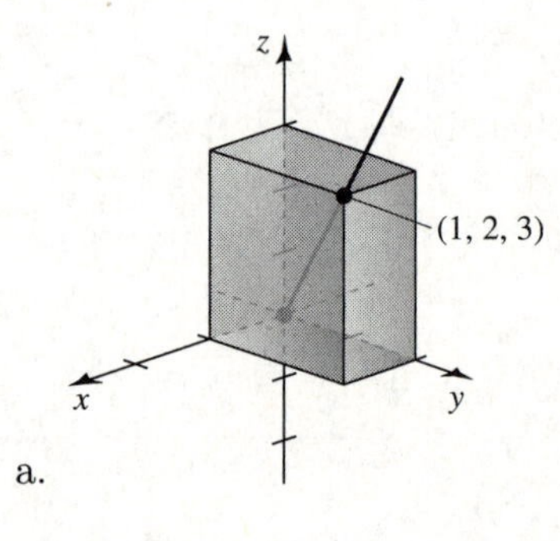

a.

b. For line A: $\mathbf{v}(t)=\langle 1,2,3\rangle$ and $\mathbf{a}(t)=\langle 0,0,0\rangle$.
For line B: $\mathbf{v}(t)=\langle 2t,4t,6t\rangle$ and $\mathbf{a}(t)=\langle 2,4,6\rangle$.
A has constant velocity and zero acceleration, while B has linearly increasing velocity and constant acceleration.

c. For A, we have $a_T=a_N=0$. For B, we have $a_T=\frac{56t}{\sqrt{56t}}=2\sqrt{14}$ and $a_N=0$ (since $\mathbf{v}$ and $\mathbf{a}$ are in the same direction).

11.9.57

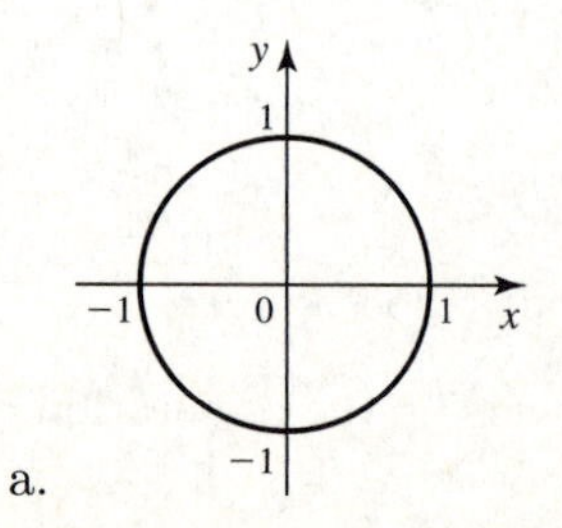

a.

b. For curve A: $\mathbf{v}(t)=\langle -\sin t,\cos t\rangle$ and $\mathbf{a}(t)=\langle -\cos t,-\sin t\rangle$.
For curve B: $\mathbf{v}(t)=2t\langle -\sin t^2,\cos t^2\rangle$ and $\mathbf{a}(t)=\langle -4t^2\cos t^2-2\sin t^2, -4t^2\sin t^2+2\cos t^2\rangle$.
A has constant velocity 1 while B does not have constant velocity.

c. For A, we have $\mathbf{a}=\mathbf{N}$, so $a_T=0$ and $a_N=1$. For B, we have $a_T=2$ and $a_N=4t^2$.

11.9.59

a. 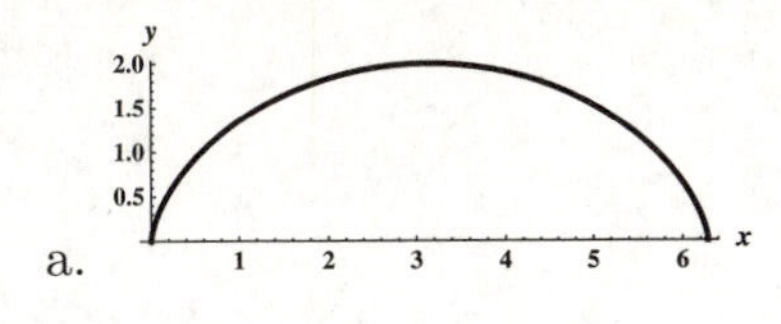

b. $\kappa = \frac{|f'g''-f''g'|}{(f'(x)^2+g'(x)^2)^{3/2}} = \frac{|(1-\cos t)\cos t - \sin t \sin t|}{((1-\cos t)^2+\sin^2 t)^{3/2}} = \frac{1-\cos t}{2\sqrt{2}(1-\cos t)^{3/2}} = \frac{1}{2\sqrt{2}\sqrt{1-\cos t}}$.

c. 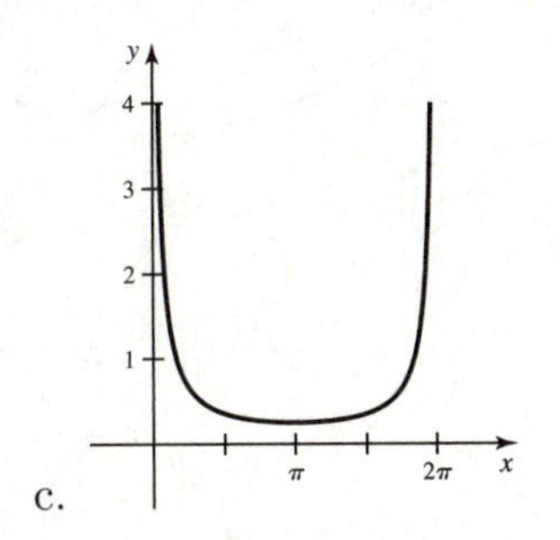

d. $\kappa'(t) = \frac{-\sin(t)}{4\sqrt{2}(1-\cos(t))^{3/2}}$. κ has a minimum at π. $\kappa''(t) = \frac{\sin^2(\frac{t}{2})(\cos(t)+3)}{4\sqrt{2}(1-\cos(t))^{5/2}}$. κ has no inflection points on the given interval.

e. Symmetry of the given curve about π (on the interval $(0, 2\pi)$ implies symmetry in κ, which does occur. The curve is flatter near π and more curved near 0 and 2π.

11.9.61

a. 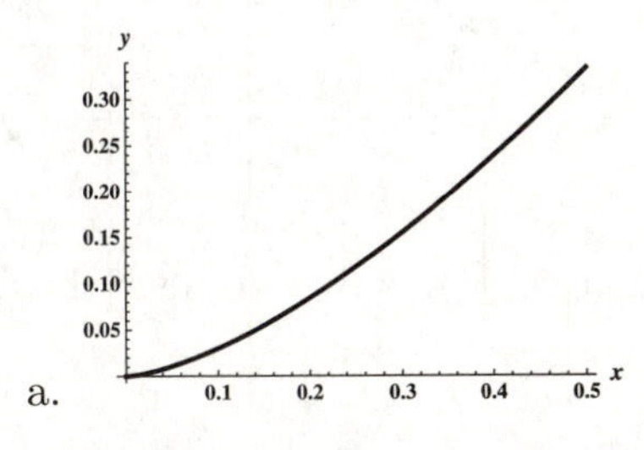

b. $\kappa = \frac{|f'g''-f''g'|}{(f'(x)^2+g'(x)^2)^{3/2}} = \frac{|t\cdot 2t-1\cdot t^2|}{(t^2+t^4)^{3/2}} = \frac{1}{t(1+t^2)^{3/2}}$.

c. 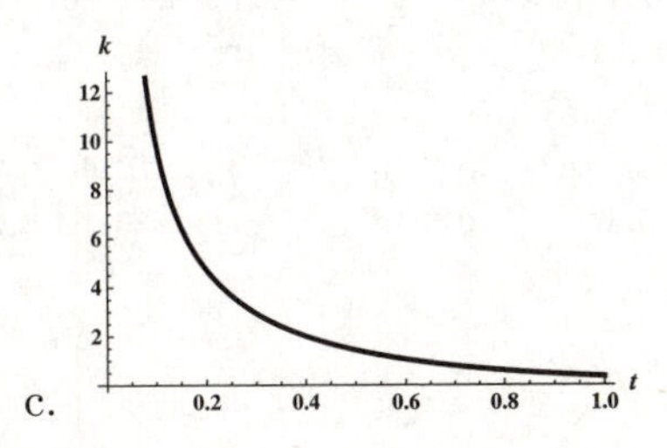

d. $\kappa'(t) = \frac{-4t^2-1}{t^2(t^2+1)^{5/2}}$. κ has no extrema. $\kappa''(t) = \frac{20t^4+7t^2+2}{t^3(t^2+1)^{7/2}}$. κ has no inflection points.

e. The curve gets flatter as $t \to \infty$.

11.9.63 $y'(x) = e^x$ and $y''(x) = e^x$. So $\kappa = \frac{|f''(x)|}{(1+f'(x)^2)^{3/2}} = \frac{e^x}{(1+(e^{2x}))^{3/2}}$.
$\kappa'(x) = \frac{e^x-2e^{3x}}{(e^{2x}+1)^{5/2}}$, which is 0 for $x = \frac{-\ln 2}{2}$. The first derivative test shows that this is where the maximum curvature exists, and the value of the maximum curvature is $\kappa((-\ln 2)/2) = 2\sqrt{3}/9$.

11.9.65 $\mathbf{r}'(t) = \langle 1, 2t\rangle$ and $\mathbf{r}''(t) = \langle 0, 2\rangle$. Thus $\kappa = \frac{2}{(1+4t^2)^{3/2}}$. At $t = 0$, we have $\kappa = 2$. So we are seeking a circle of radius 1/2. The center of the osculating circle is the point along the normal line to the curve at $(0,0)$ which is 1/2 unit from $(0,0)$ and is on the inside of the curve, so it is $(0, 1/2)$. The equation is $x^2 + (y-(1/2))^2 = \frac{1}{4}$.

11.9.67 $\mathbf{r}'(t) = \langle 1-\cos t, \sin t\rangle$ and $\mathbf{r}''(t) = \langle \sin t, -\cos t\rangle$. Thus $\kappa(\pi) = \frac{(2-0)}{(2^2+0^2)^{3/2}} = \frac{2}{8} = \frac{1}{4}$, so the radius of the osculating circle is 4. The center of the osculating circle is the point along the normal line to the curve at $(\pi, 2)$ which is 4 units from $(\pi, 2)$ and is on the inside of the curve, so it is $(\pi, -2)$. The equation of the osculating circle is $(x-\pi)^2 + (y+2)^2 = 16$.

11.9.69 $y' = n\cos nx$ and $y'' = -n^2 \sin nx$, so $\kappa(\pi/2n) = \frac{|-n^2\sin(\pi/2)|}{(1+(n^2\cos^2(\pi/2))^{3/2}} = \frac{n^2}{(1+0)^{3/2}} = n^2$. This increases as n increases.

11.9.71

a. $\mathbf{r}'(t) = \langle V_0\cos\alpha, V_0\sin\alpha - gt\rangle$, so the speed is

$$\sqrt{V_0^2\cos^2\alpha + V_0^2\sin^2\alpha - 2gtV_0\sin\alpha + g^2t^2},$$

which can be written

$$\sqrt{V_0^2 - 2gtV_0\sin\alpha + g^2t^2}.$$

The graph shown is for $V_0 = 1$, $g = 32$, and $\alpha = 45$ degrees.

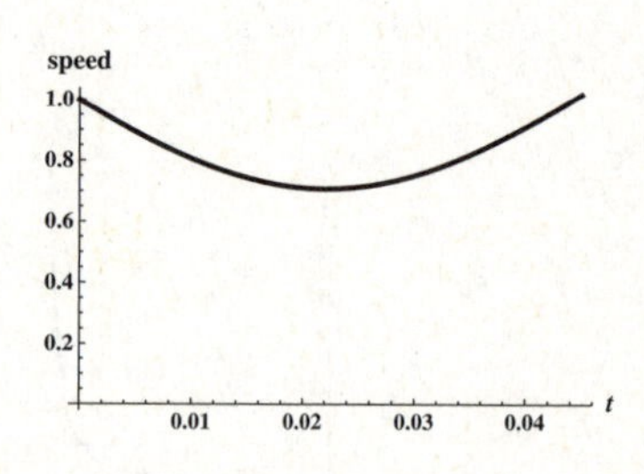

b. $\mathbf{a} = \langle 0, -g\rangle$. We have $\mathbf{a}\times\mathbf{v} = \langle 0, 0, gV_0\cos\alpha\rangle$. So $\kappa = \frac{|gV_0\cos\alpha|}{(V_0^2 - 2gtV_0\sin\alpha + g^2t^2)^{3/2}}$. The graph shown is for $V_0 = 1$, $g = 32$, and $\alpha = 45$ degrees.

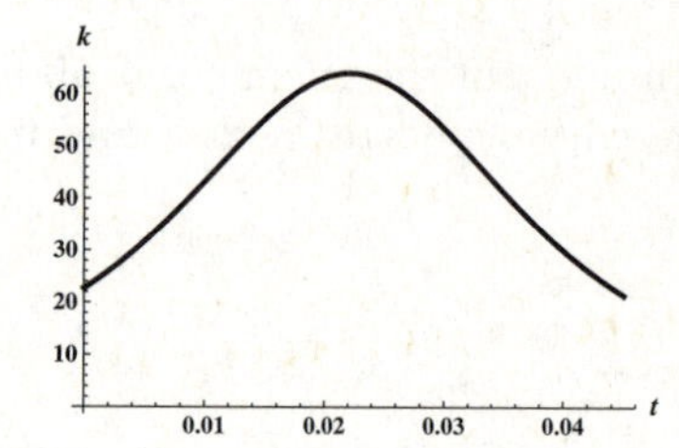

c. The speed has a minimum and the curvature has a maximum at the halfway time of the flight, namely at $\frac{V_0\sin\alpha}{g}$.

11.9.73 Recall that a curve is parametrized by arc length exactly when $|\mathbf{v}| = 1$. (Theorem 11.9) We have $\mathbf{v} = \langle a, b, c\rangle$, so $|\mathbf{v}| = \sqrt{a^2+b^2+c^2}$, which is equal to one if and only if $a^2+b^2+c^2 = 1$.

11.9.75 $\mathbf{r}'(t) = \langle pbt^{p-1}, pdt^{p-1}, pft^{p-1}\rangle = pt^{p-1}\langle b, d, f\rangle$. $\mathbf{r}''(t) = p(p-1)t^{p-2}\langle b, d, f\rangle$. Since for any t, $\mathbf{r}''(t)$ and $\mathbf{r}'(t)$ are multiples of $\langle b, d, f\rangle$, their cross product is $\langle 0, 0, 0\rangle$. Thus $\kappa = \frac{0}{|\mathbf{v}|^3} = 0$. The given curve represents a straight line, which has zero curvature.

11.9.77

a. $f_n'(x) = 2nx^{2n-1}$ and $f_n''(x) = (2n)(2n-1)x^{2n-2}$. $\kappa = \frac{|2n(2n-1)x^{2n-2}|}{(1+4n^2x^{4n-2})^{3/2}}$. So $\kappa_1(x) = \frac{2}{(1+4x^2)^{3/2}}$, $\kappa_2(x) = \frac{12x^2}{(1+16x^6)^{3/2}}$, and $\kappa_3(x) = \frac{30x^4}{(1+36x^{10})^{3/2}}$.

b.

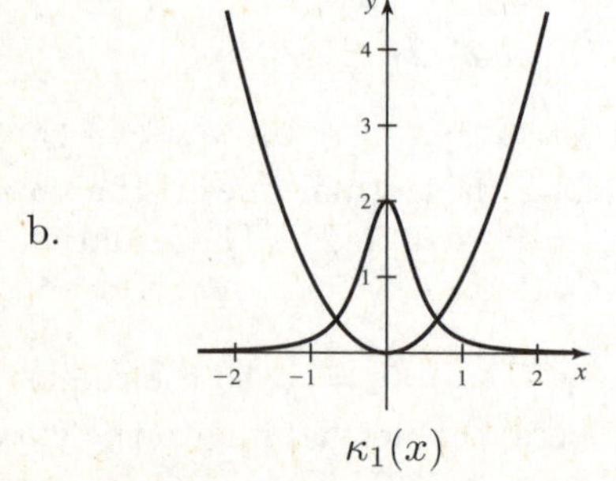

$\kappa_1(x)$

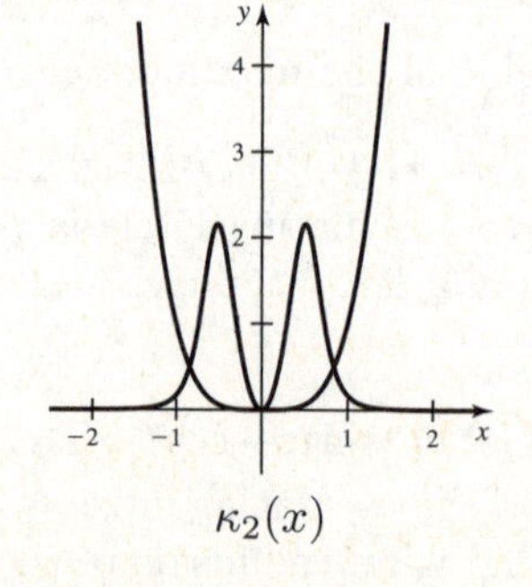

$\kappa_2(x)$

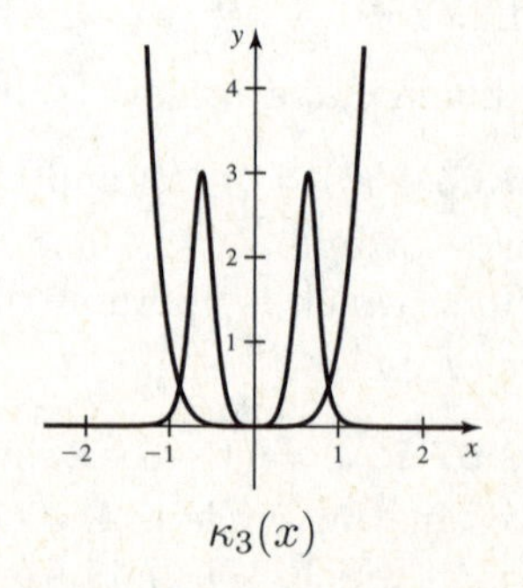

$\kappa_3(x)$

Note that the curves are symmetric about the y-axis.

c. $\kappa'(x) = -\frac{4n(2n-1)x^{2n-1}(2n^2(4n-1)x^{4n}-(n-1)x^2)}{(4n^2x^{4n}+x^2)^2\sqrt{4n^2x^{4n-2}+1}}$. By symmetry, we can concentrate on the critical points for $x > 0$. We have $\kappa'(x) = 0$ for $x > 0$ and $n > 1$, when $x = 2^{\frac{1}{2-4n}}\left(\frac{\sqrt{n^2(4n-1)}}{\sqrt{n-1}}\right)^{\frac{1}{1-2n}}$. For $n = 1$, the maximum occurs at 0. For $n = 2$, it occurs at $\frac{1}{\sqrt{2}\cdot 7^{1/6}}$. For $n = 3$, it occurs at $\frac{1}{3^{1/5}11^{1/10}}$.

d. If the maximum curvature for f_n occurs at $\pm z_n$, then $\lim_{n\to\infty} z_n = 1$.

11.10 Chapter Eleven Review

11.10.1

a. Yes – addition of vectors is commutative.

b. False. For example, the vector in the direction of $\mathbf{i}$ with the length of $\mathbf{j}$ is $\mathbf{i}$, but the vector in the direction of $\mathbf{j}$ with the length of $\mathbf{i}$ is $\mathbf{j}$, and $\mathbf{i} \neq \mathbf{j}$.

c. True, because it then follows that $\mathbf{u} = -\mathbf{v}$, so the two are parallel.

d. True. This follows because $\int \mathbf{r}'(t)\,dt = \int \langle 0,0,0\rangle\,dt = \langle 0,0,0\rangle + \mathbf{C} = \langle a,b,c\rangle$ for some real numbers a, b, and c.

e. False. Its length is $\sqrt{169\sin^2 t + 169\cos^2 t} = 13 \neq 1$.

f. False. For example, for the curve $\mathbf{r}(t) = \langle t^2, t\rangle$, we have $\mathbf{N} = \frac{1}{\sqrt{1+4t^2}}\langle 1, -2t\rangle$. So if, for example, $t = 2$, we have $\mathbf{r}(2) = \langle 4, 2\rangle$ and $\mathbf{N} = \frac{1}{\sqrt{17}}\langle 1, -4\rangle$, which aren't parallel.

11.10.3

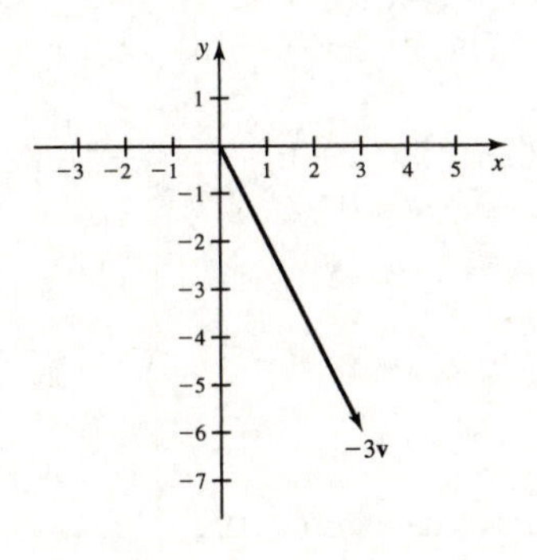

11.10.5

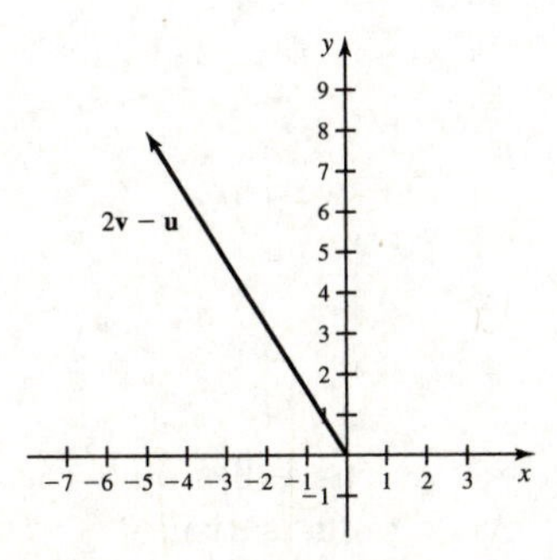

11.10.7 $|\mathbf{u}+\mathbf{v}| = |\langle -4, 14, -3\rangle| = \sqrt{16+196+9} = \sqrt{221}$.

11.10.9 $|\mathbf{v}| = \sqrt{36+100+4} = \sqrt{140}$. So the desired vector is $\frac{20}{\sqrt{140}}\langle -6, 10, 2\rangle = \frac{20}{\sqrt{35}}\langle -3, 5, 1\rangle$. The vector $\frac{-20}{\sqrt{35}}\langle -3, 5, 1\rangle$ also has the desired property.

11.10.11

a. $\mathbf{v} = 550\langle -\sqrt{2}/2, \sqrt{2}/2\rangle = \langle -275\sqrt{2}, 275\sqrt{2}\rangle$.

b. $\mathbf{v} = 550\langle -\sqrt{2}/2, \sqrt{2}/2\rangle + \langle 0, 40\rangle = \langle -275\sqrt{2}, 275\sqrt{2}+40\rangle$.

11.10.13 $\{(x,y,z) \,|\, (x-1)^2 + y^2 + (z+1)^2 = 16\}$.

11.10.15 $\{(x,y,z) \,|\, x^2 + (y-1)^2 + z^2 > 4\}$.

11.10.17 The magnitude of $\langle 0, 4, -50\rangle$ is $\sqrt{2516} \approx 50.16$ meters per second. The direction is about $\cos^{-1}(4/\sqrt{2516}) \approx 85.4$ degrees below the horizontal in the northerly horizontal direction.

11.10.19

a. $\mathbf{u}\cdot\mathbf{v} = 48+0+3 = 51$. $|\mathbf{u}| = \sqrt{37}$ and $|\mathbf{v}| = \sqrt{77}$. Thus the angle between the vectors is $\cos^{-1}(\frac{51}{\sqrt{37}\sqrt{77}}) \approx$ .3 radians.

b. $\text{proj}_{\mathbf{v}}\mathbf{u} = \frac{51}{77}\langle 8, 2, -3\rangle$, and $\text{scal}_{\mathbf{v}}\mathbf{u} = \frac{51}{\sqrt{77}}$.

c. $\text{proj}_{\mathbf{u}}\mathbf{v} = \frac{51}{37}\langle 6, 0, -1\rangle$, and $\text{scal}_{\mathbf{u}}\mathbf{v} = \frac{51}{\sqrt{37}}$.

11.10.21 $\langle 2, -6, 9\rangle \times \langle -1, 0, 6\rangle = \langle -36, -21, -6\rangle$. The length of this vector is $3\sqrt{12^2 + 7^2 + 2^2} = 3\sqrt{197}$. Thus the unit normals are $\frac{\pm 1}{\sqrt{197}}\langle 12, 7, 2\rangle$.

11.10.23 $T(\theta) = (.4)\cdot 98 \cdot \sin\theta \approx 39.2\sin\theta$ Newton-meters. This has a maximum of 39.2 when $\sin\theta = 1$ (at $\theta = \pi/2$) and a minimum of 0 at $\theta = 0$. The direction of the torque does not change as the knee is lifted.

11.10.25 The direction of the line segment is $\langle 2-0, -8-(-3), 1-9\rangle = \langle 2, -5, -8\rangle$. The line segment is described by $\langle x, y, z\rangle = \langle 0, -3, 9\rangle + t\langle 2, -5, -8\rangle$, $0 \le t \le 1$.

11.10.27 The curve is a circle of radius 4 with center $(0, 1, 0)$ sitting in the plane $y = 1$. It is the intersection of the plane $y = 1$ and the cylinder $x^2 + z^2 = 16$.

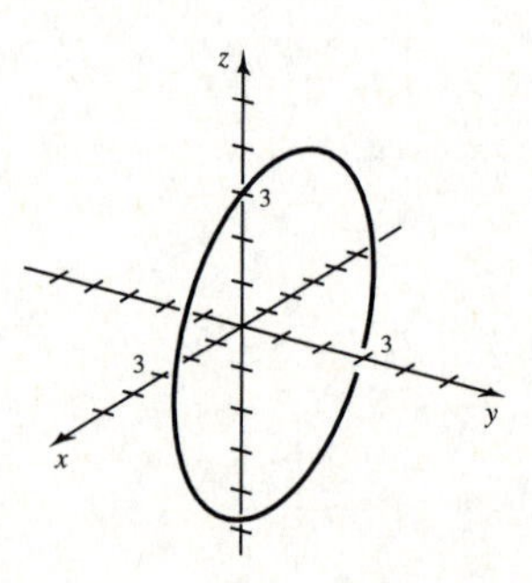

11.10.29 Note that $x^2 + y^2 + z^2 = 2$, so this curve lies on a sphere of radius 2. Also, every point satisfies $z = x$, so it is a circle centered at the origin of radius 2, sitting in the plane $z = x$.

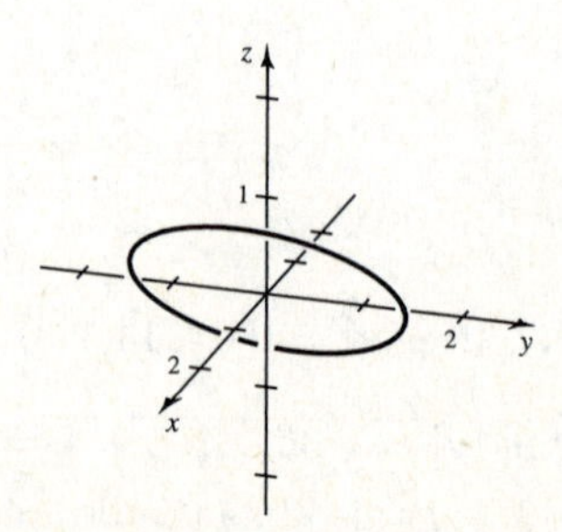

11.10.31

a. The trajectory is given by $\mathbf{r}(t) = \langle 50t, 50t - 16t^2 \rangle$. The projectile is at $y = 30$ when $-16t^2 + 50t - 30 = 0$, which occurs at $t = \frac{1}{16}\left(25 \pm \sqrt{145}\right) \approx .81$ and 2.32. At these times, $x = 50t \approx 40.5$ and 116. The first time represents when the projectile has not yet reached the cliff, while the second time represents when the projectile lands on the cliff, so the coordinates of the landing spot are approximately $(116, 30)$.

b. The maximum height occurs where $y' = 0$, which occurs for $50 - 32t = 0$, or $t = 25/16$. The maximum height is $50 \cdot \frac{25}{16} - 16\left(\frac{25}{16}\right)^2 = \frac{625}{16} \approx 39.06$ feet.

c. As mentioned above, the flight ends at $t \approx 2.32$ seconds.

d. The length of the trajectory is $\int_0^{2.32} \sqrt{x'(t)^2 + y'(t)^2}\, dt = \int_0^{2.32} \sqrt{2500 + (50 - 32t)^2}\, dt$.

e. $L \approx 129$ feet.

f. Suppose the launch angle is α. Then $\mathbf{r}(t) = \langle 50\sqrt{2}t\cos\alpha, 50\sqrt{2}t\sin\alpha - 16t^2 \rangle$. We want $y \geq 30$ when $x = 50$. We know that $x = 50$ when $t = \frac{\sec\alpha}{\sqrt{2}}$. At this time, we have $y = 50\tan\alpha - 8\sec^2\alpha$. This expression is greater than or equal to 30 for approximately $41.5 \leq \alpha \leq 79.4$.

11.10.33 $L = \int_0^{2\pi} \sqrt{(3 - 6\cos\theta)^2 + (6\sin\theta)^2}\, d\theta = \int_0^{2\pi} 3\sqrt{5 - 4\cos(\theta)}\, d\theta \approx 40.09$.

11.10.35

a. $\mathbf{r}'(t) = \langle -6\sin t, 3\cos t \rangle$, so $\mathbf{T} = \frac{1}{\sqrt{1+3\sin^2 t}}\langle -2\sin t, \cos t \rangle$.

b. $\kappa = \frac{|\mathbf{r}''(t)\times\mathbf{r}'(t)|}{(3\sqrt{1+3\sin^2 t})^3} = \frac{|\langle 0,0,-18\rangle|}{(3\sqrt{1+3\sin^2 t})^3} = \frac{2}{3(\sqrt{1+3\sin^2 t})^3}$.

c. Note that $\frac{1}{\sqrt{1+3\sin^2 t}}\langle -\cos t, -2\sin t\rangle$ has length one, and is perpendicular to $\mathbf{T}$ (see part [d]), and points to the inside of the curve, so it is $\mathbf{N}$.

d. $\left|\frac{1}{\sqrt{1+3\sin^2 t}}\langle -\cos t, -2\sin t\rangle\right| = \frac{\sqrt{1+3\sin^2 t}}{\sqrt{1+3\sin^2 t}} = 1$ and $\mathbf{T}\cdot\mathbf{N} = \frac{1}{\sqrt{1+3\sin^2 t}}(2\sin t\cos t - 2\sin t\cos t) = 0$.

e.

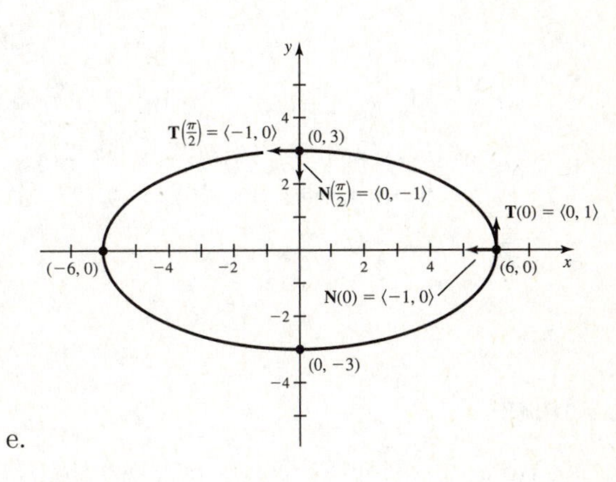

11.10.37

a. $\mathbf{r}'(t) = \langle -\sin t, -2\sin t, \sqrt{5}\cos t\rangle$, so

$$\mathbf{T} = \frac{1}{\sqrt{5}}\langle -\sin t, -2\sin t, \sqrt{5}\cos t\rangle.$$

b. $\kappa = \frac{|\mathbf{r}''(t)\times\mathbf{r}'(t)|}{(\sqrt{5})^3} = \frac{|\langle -2\sqrt{5}, \sqrt{5}, 0\rangle|}{(\sqrt{5})^3} = \frac{1}{\sqrt{5}}$.

c. $\frac{dT}{dt} = \frac{1}{\sqrt{5}}\langle -\cos t, -2\cos t, -\sqrt{5}\sin t\rangle = \mathbf{N}$.

d. $\left|\frac{1}{\sqrt{5}}\langle -\cos t, -2\cos t, -\sqrt{5}\sin t\rangle\right| = \frac{\sqrt{5}}{\sqrt{5}} = 1$ and

$$\mathbf{T}\cdot\mathbf{N} = \frac{1}{\sqrt{5}}(\sin t\cos t + 4\sin t\cos t - 5\sin t\cos t) = 0.$$

e.

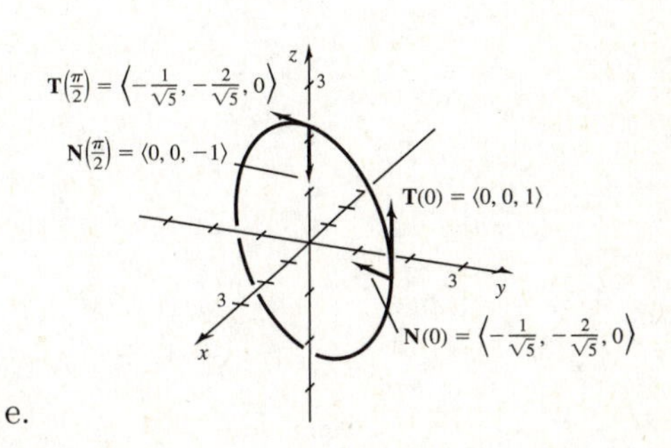

11.10.39

a. $\mathbf{v}(t) = \langle -2\sin t, 2\cos t\rangle$, and $\mathbf{a}(t) = \langle -2\cos t, -2\sin t\rangle$. Since $\mathbf{v}\cdot\mathbf{a} = 0$, we have $a_T = 0$. Note that $\mathbf{a}\times\mathbf{v} = \langle -2\cos t, -2\sin t, 0\rangle \times \langle -2\sin t, 2\cos t, 0\rangle = \langle 0,0,-4\rangle$, so $a_N = \frac{4}{2} = 2$. We have $\mathbf{a} = \langle -2\cos t, -2\sin t\rangle = 2\mathbf{N} + 0\cdot\mathbf{T}$.

b. Note that at $t = 0$ we have $\mathbf{a} = \langle -2, 0\rangle = 2\langle -1, 0\rangle = 2\mathbf{N}$, and at $t = \pi/2$ we have $\mathbf{a} = \langle 0,-2\rangle = 2\langle 0,-1\rangle = 2\mathbf{N}$.

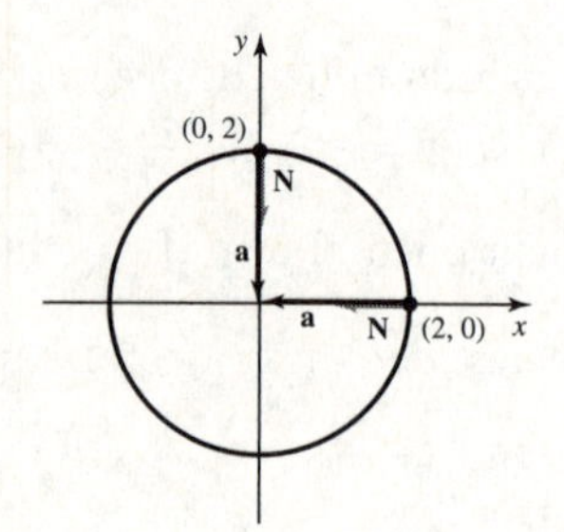

11.10.41

a. $\mathbf{v}(t) = \langle 2t, 2\rangle$, and $\mathbf{a}(t) = \langle 2, 0\rangle$. Since $\mathbf{v}\cdot\mathbf{a} = 4t$, we have $a_T = \frac{4t}{2\sqrt{t^2+1}} = \frac{2t}{\sqrt{t^2+1}}$. Note that $\mathbf{a}\times\mathbf{v} = \langle 0,0,4\rangle$, so $a_N = \frac{4}{2\sqrt{t^2+1}} = \frac{2}{\sqrt{t^2+1}}$. We have that $\mathbf{a} = \langle 2,0\rangle = \frac{2}{\sqrt{t^2+1}}\mathbf{N} + \frac{2t}{\sqrt{t^2+1}}\mathbf{T}$.

b. At $t = 1$, we have $\mathbf{a} = \langle 2,0\rangle = \frac{2}{\sqrt{2}}\mathbf{T} + \frac{2}{\sqrt{2}}\mathbf{N} = \frac{2}{\sqrt{2}}\langle\sqrt{2}/2, \sqrt{2}/2\rangle + \frac{2}{\sqrt{2}}\langle\sqrt{2}/2, -\sqrt{2}/2\rangle$.
At $t = 2$, we have $\mathbf{a} = \langle 2,0\rangle = \frac{4}{\sqrt{5}}\langle 2/\sqrt{5}, 1/\sqrt{5}\rangle + \frac{2}{\sqrt{5}}\langle 1/\sqrt{5}, -2/\sqrt{5}\rangle$.

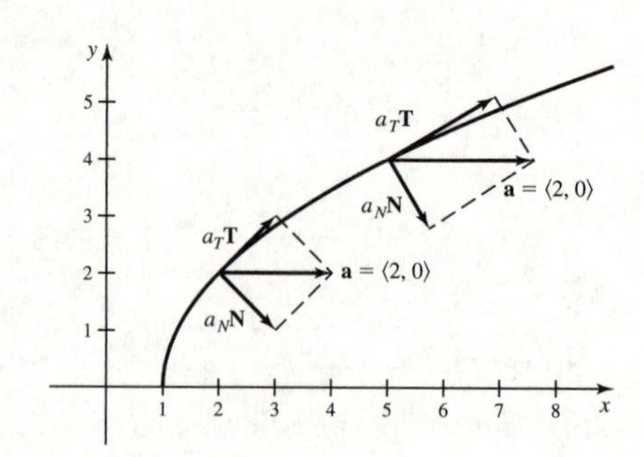

11.10.43

a. We are looking for points (x, y) so that $\langle x - x_0, y - y_0\rangle \cdot \langle a, b\rangle = 0$, so $a(x - x_0) + b(y - y_0) = 0$, or $ax + by = ax_0 + by_0$.

b. Note that $\langle a, b, 0\rangle \times \langle x - x_0, y - y_0, 0\rangle = \langle 0, 0, a(y - y_0) - b(x - x_0)\rangle$. This is equal to the zero vector when $ay - ay_0 = bx - bx_0$, or $ay - bx = ay_0 - bx_0$. So the equation of a line passing through (x_0, y_0) and parallel to $\langle a, b\rangle$ is given by $ay - bx = ay_0 - bx_0$.

Chapter 12

12.1 Planes and Surfaces

12.1.1 One point and a normal vector determine a plane.

12.1.3 The point $(x,\ y,\ z)$ where this plane intersects the x-axis has $y = z = 0$; substituting in the equation of the plane gives $x = -6$. Similarly, we see that the plane meets the y-axis at $y = -4$ and the z-axis at $z = 3$.

12.1.5 Since z is absent from the equation $x^2 + 2y^2 = 8$, this cylinder is parallel to the z-axis. Similarly, $z^2 + 2y^2 = 8$ is parallel to the x-axis and $x^2 + 2z^2 = 8$ is parallel to the y-axis.

12.1.7 The traces of a surface are the sets of points at which the surface intersects a plane that is parallel to one of the coordinate planes.

12.1.9 This is an ellipsoid.

12.1.11 Substituting in the general equation for a plane gives $1(x-0)+1(y-2)-1(z-(-2))=0$, which simplifies to $x+y-z=4$.

12.1.13 Substituting in the general equation for a plane gives $-1(x-2)+2(y-3)-3(z-0)=0$, which simplifies to $-x+2y-3z=4$.

12.1.15 Let $P = (1,\ 0,\ 3)$, $Q = (0,\ 4,\ 2)$ and $R = (1,\ 1,\ 1)$. Then the vectors $\overrightarrow{PQ} = \langle -1,\ 4,\ -1\rangle$ and $\overrightarrow{PR} = \langle 0,\ 1,\ -2\rangle$ lie in the plane, so

$$\mathbf{n} = \overrightarrow{PQ} \times \overrightarrow{PR} = \begin{vmatrix} \mathbf{i} & \mathbf{j} & \mathbf{k} \\ -1 & 4 & -1 \\ 0 & 1 & -2 \end{vmatrix} = -7\mathbf{i} - 2\mathbf{j} - \mathbf{k}$$

is normal to the plane. The plane has equation $7(x-1)+2(y-0)+1(z-3)=0$, which simplifies to $7x+2y+z=10$.

12.1.17 Let $P = (2,\ -1,\ 4)$, $Q = (1,\ 1,\ -1)$ and $R = (-4,\ 1,\ 1)$. Then the vectors $\overrightarrow{PQ} = \langle -1,\ 2,\ -5\rangle$ and $\overrightarrow{PR} = \langle -6,\ 2,\ -3\rangle$ lie in the plane, so

$$\mathbf{n} = \overrightarrow{PQ} \times \overrightarrow{PR} = \begin{vmatrix} \mathbf{i} & \mathbf{j} & \mathbf{k} \\ -1 & 2 & -5 \\ -6 & 2 & -3 \end{vmatrix} = 4\mathbf{i} + 27\mathbf{j} + 10\mathbf{k}$$

is normal to the plane. The plane has equation $4(x-2)+27(y-(-1))+10(z-4)=0$, which simplifies to $4x+27y+10z=21$.

12.1.19
The x-intercept is found by setting $y = z = 0$ and solving $3x = 6$ to get $x = 2$. Similarly, we see that the y-intercept is -3 and the z-intercept is 6. The xy-trace is found by setting $z = 0$, which gives $3x - 2y = 6$. Similarly, the xz-trace is $3x + z = 6$ and the yz-trace is $-2y + z = 6$.

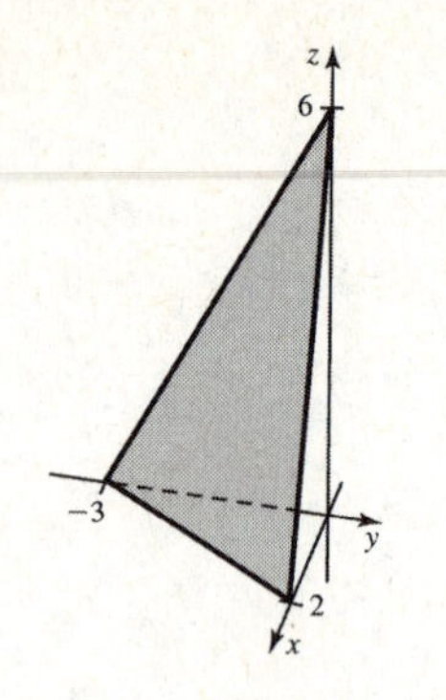

12.1.21
The x-intercept is found by setting $y = z = 0$ which gives $x = 30$. Similarly, we see that the y-intercept is 10 and the z-intercept is -6. The xy-trace is found by setting z = 0, which gives $x + 3y = 30$. Similarly, the xz-trace is $x - 5z = 30$ and the yz-trace is $3y - 5z = 30$.

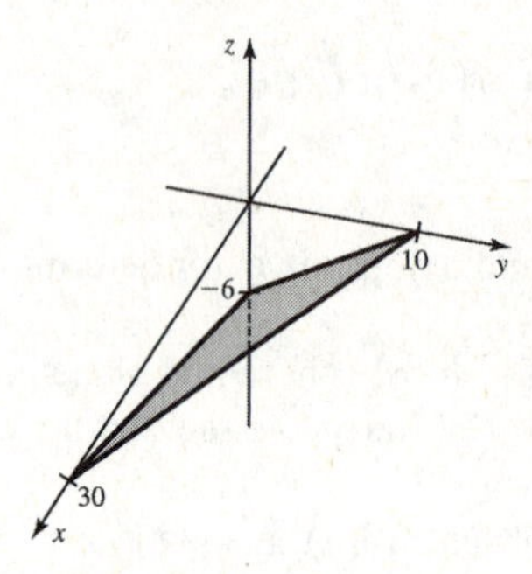

12.1.23 Rewrite R and T so we have

$$\begin{aligned} Q &: \ 3x - 2y + z = 12 \\ R &: \ 3x - 2y + z = 0 \\ T &: \ 3x - 2y + z = 12 \end{aligned}$$

This shows that Q and T are identical, and Q, R and T are parallel. Note that $\langle 3, -2, 1\rangle \cdot \langle -1, 2, 7\rangle = 0$ so S is orthogonal to Q, R and T.

12.1.25 The plane Q has normal vector $\langle -1, 2, -4\rangle$; therefore the parallel plane passing through the point $P_0(1, 0, 4)$ has equation $-1(x-1) + 2(y-0) - 4(z-4) = 0$, which simplifies to $-x + 2y - 4z = -17$.

12.1.27 The plane Q has normal vector $\langle 4, 3, -2\rangle$; therefore the parallel plane passing through the point $P_0(1, -1, 3)$ has equation $4(x-1) + 3(y-(-1)) - 2(z-3) = 0$, which simplifies to $4x + 3y - 2z = -5$.

12.1.29 First, note that the vectors normal to the planes, $\mathbf{n}_Q = \langle -1, 2, 1\rangle$ and $\mathbf{n}_R = \langle 1, 1, 1\rangle$, are not multiples of each other; therefore these planes are not parallel and they intersect in a line ℓ. We need to find a point on ℓ and a vector in the direction of ℓ. Setting $x = 0$ in the equations of the planes gives equations of the lines in which the planes intersect the yz-plane:

$$\begin{aligned} 2y + z &= 1 \\ y + z &= 0 \end{aligned}$$

Solving these equations simultaneously gives $y = 1$ and $z = -1$, so $(0, 1, -1)$ is a point on ℓ. A vector in the direction of ℓ is

$$\mathbf{n}_Q \times \mathbf{n}_R = \begin{vmatrix} \mathbf{i} & \mathbf{j} & \mathbf{k} \\ -1 & 2 & 1 \\ 1 & 1 & 1 \end{vmatrix} = \mathbf{i} + 2\mathbf{j} - 3\mathbf{k} = \langle 1, 2, -3\rangle$$

Therefore ℓ has equation $\mathbf{r}(t) = \langle 0, 1, -1\rangle + t\langle 1, 2, -3\rangle = \langle t, 1+2t, -1-3t\rangle$, or $x = t$, $y = 1 + 2t$, $z = -1 - 3t$.

12.1.31 First, note that the vectors normal to the planes, $\mathbf{n}_Q = \langle 2, -1, 3\rangle$ and $\mathbf{n}_R = \langle -1, 3, 1\rangle$, are not multiples of each other; therefore these planes are not parallel and they intersect in a line ℓ. We need to find a point on ℓ and a vector in the direction of ℓ. Setting $z = 0$ in the equations of the planes gives equations of the lines in which the planes intersect the xy-plane:

$$\begin{aligned} 2x - \ \ y &= 1 \\ -x + 3y &= 4 \end{aligned}$$

Solving these equations simultaneously gives $x = \dfrac{7}{5}$ and $y = \dfrac{9}{5}$, so $\left(\dfrac{7}{5}, \dfrac{9}{5}, 0\right)$ is a point on ℓ. A vector in the direction of ℓ is

$$\mathbf{n}_Q \times \mathbf{n}_R = \begin{vmatrix} \mathbf{i} & \mathbf{j} & \mathbf{k} \\ 2 & -1 & 3 \\ -1 & 3 & 1 \end{vmatrix} = -10\mathbf{i} - 5\mathbf{j} + 5\mathbf{k} = -5\langle 2, 1, -1\rangle$$

Therefore ℓ has equation $\mathbf{r}(t) = \left\langle \dfrac{7}{5}, \dfrac{9}{5}, 0\right\rangle + t\langle 2, 1, -1\rangle = \left\langle \dfrac{7}{5} + 2t, \dfrac{9}{5} + t, -t\right\rangle$, or $x = \dfrac{7}{5} + 2t$, $y = \dfrac{9}{5} + t$, $z = -t$.

12.1.33

a. Since z is absent from the equation, this cylinder is parallel to the z-axis.

b.

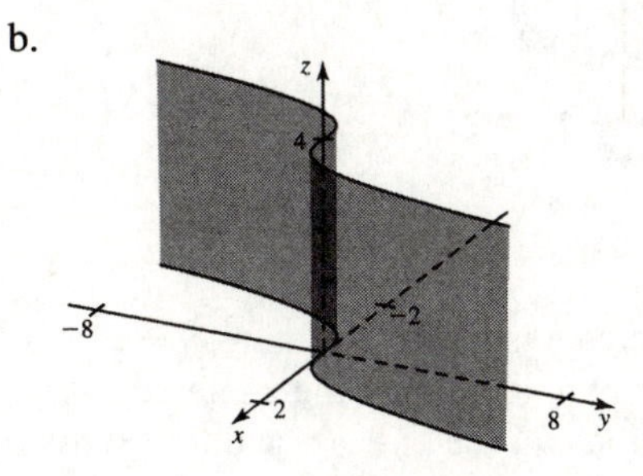

12.1.35

a. Since x is absent from the equation, this cylinder is parallel to the x-axis.

b.

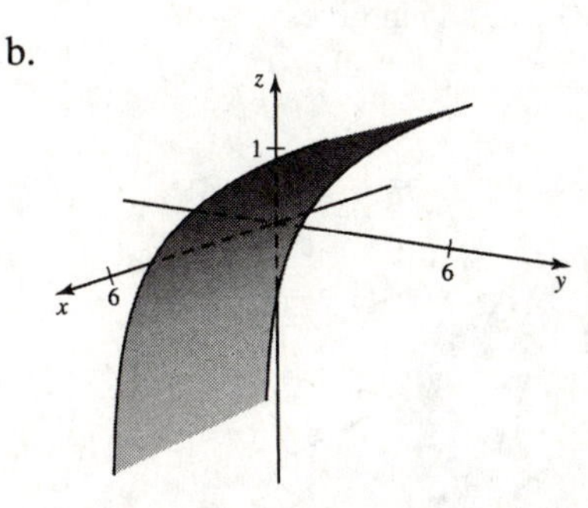

12.1.37

a. The x-intercept is found by setting $y = z = 0$ in the equation of this surface, which gives $x^2 = 1$, so the x-intercepts are $x = \pm 1$. Similarly we see that the y-intercepts are $y = \pm 2$ and the z-intercepts are $z = \pm 3$.

b. The equations for the xy-, xz- and yz-traces are found be setting $z = 0$, $y = 0$ and $x = 0$ respectively in the equation of the surface, which gives

$$x^2 + \frac{y^2}{4} = 1, \quad x^2 + \frac{z^2}{9} = 1, \quad \frac{y^2}{4} + \frac{z^2}{9} = 1.$$

c.

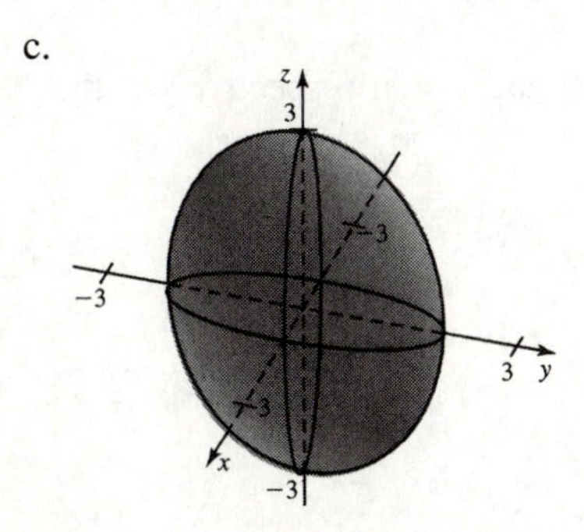

12.1.39

a. The x-intercept is found by setting $y = z = 0$ in the equation of this surface, which gives $x^2 = 9$, so the x-intercepts are $x = \pm 3$. Similarly we see that the y-intercepts are $y = \pm 1$ and the z-intercepts are $z = \pm 6$.

b. The equations for the xy-, xz- and yz-traces are found be setting $z=0$, $y=0$ and $x=0$ respectively in the equation of the surface, which gives

$$\frac{x^2}{3}+3y^2=3,\quad \frac{x^2}{3}+\frac{z^2}{12}=3,\quad 3y^2+\frac{z^2}{12}=3.$$

c.

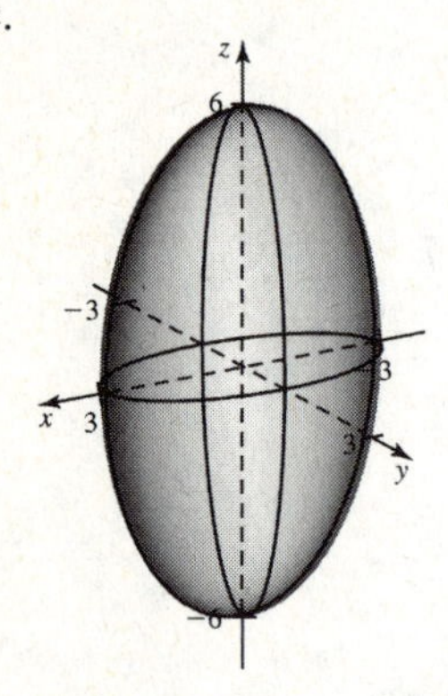

12.1.41

a The x-intercept is found by setting $y=z=0$ in the equation of this surface, which gives $x=0$. Similarly we see that the y-intercept is $y=0$ and the z-intercept is $z=0$.

b. The equations for the xy-, xz- and yz-traces are found be setting $z=0$, $y=0$ and $x=0$ respectively in the equation of the surface, which gives

$$x=y^2,\quad x=z^2,\quad y^2+z^2=0 \iff x=y=z=0.$$

c.

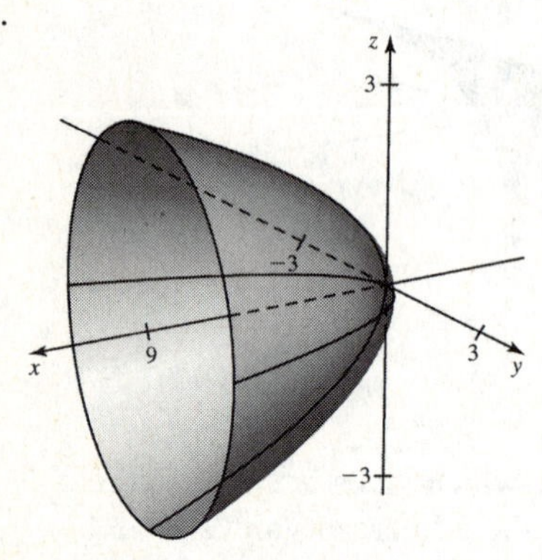

12.1.43

a. The x-intercept is found by setting $y=z=0$ in the equation of this surface, which gives $x=0$. Similarly we see that the y-intercept is $y=0$ and the z-intercept is $z=0$.

b. The equations for the xy-, xz- and yz-traces are found be setting $z=0$, $y=0$ and $x=0$ respectively in the equation of the surface, which gives

$$x-9y^2=0,\quad 9x-\frac{z^2}{4}=0,\quad 81y^2+\frac{z^2}{4}=0 \iff x=y=z=0.$$

c.

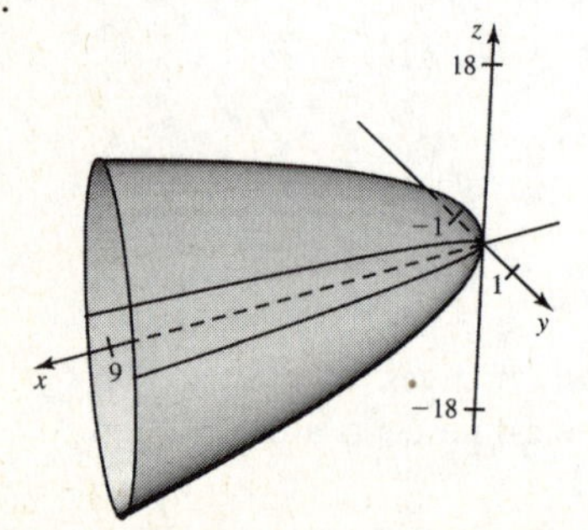

12.1.45

a. The x-intercept is found by setting $y = z = 0$ in the equation of this surface, which gives $x^2 = 25$, so the x-intercepts are $x = \pm 5$. Similarly we see that the y-intercepts are $y = \pm 3$ and there are no z-intercepts.

b. The equations for the xy-, xz- and yz-traces are found be setting $z = 0$, $y = 0$ and $x = 0$ respectively in the equation of the surface, which gives

$$\frac{x^2}{25} + \frac{y^2}{9} = 1, \quad \frac{x^2}{25} - z^2 = 1, \quad \frac{y^2}{9} - z^2 = 1.$$

c.

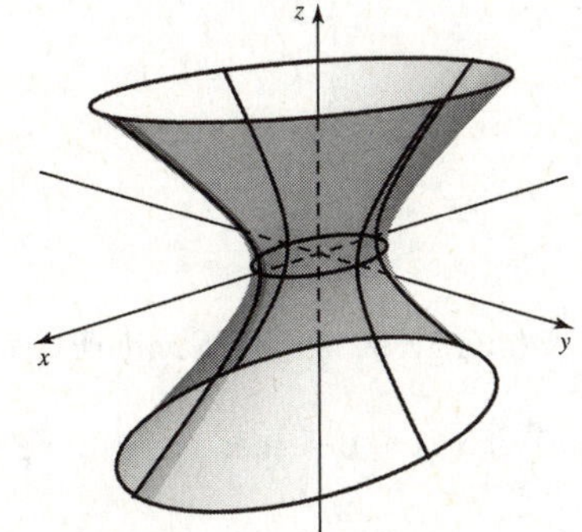

12.1.47

a. The y-intercept is found by setting $x = z = 0$ in the equation of this surface, which gives $y^2 = 144$, so the y-intercepts are $y = \pm 12$. Similarly we see that the z-intercepts are $z = \pm \frac{1}{2}$ and there are no x-intercepts.

b. The equations for the xy-, xz- and yz-traces are found be setting $z = 0$, $y = 0$ and $x = 0$ respectively in the equation of the surface, which gives

$$-\frac{x^2}{4} + \frac{y^2}{16} - 9 = 0, \quad -\frac{x^2}{4} + 36z^2 - 9 = 0, \quad \frac{y^2}{16} + 36z^2 - 9 = 0.$$

c.

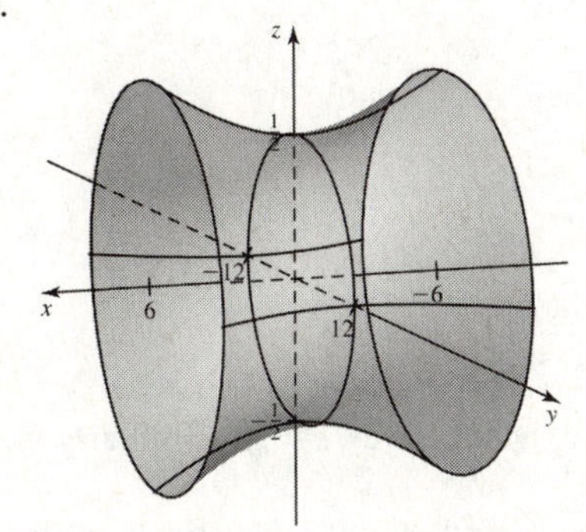

12.1.49

a. The x-intercept is found by setting $y = z = 0$ in the equation of this surface, which gives $x = 0$. Similarly we see that the y-intercept is $y = 0$ and the z-intercept is $z = 0$.

b. The equations for the xy-, xz- and yz-traces are found be setting $z = 0$, $y = 0$ and $x = 0$ respectively in the equation of the surface, which gives

$$\frac{x^2}{9} - y^2 = 0, \quad z = \frac{x^2}{9}, \quad z = -y^2.$$

c.

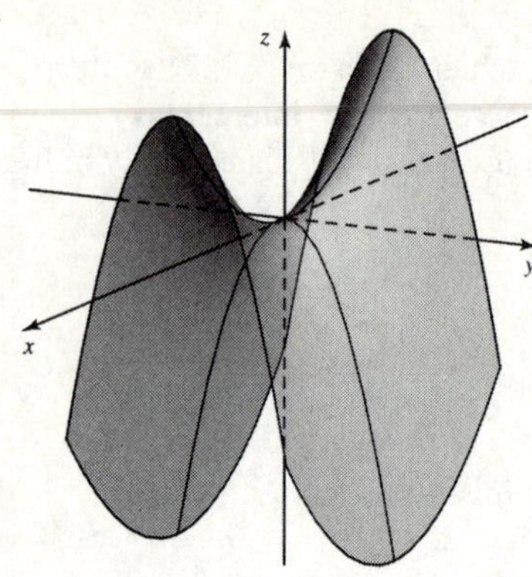

12.1.51

a. The x-intercept is found by setting $y = z = 0$ in the equation of this surface, which gives $x = 0$. Similarly we see that the y-intercept is $y = 0$ and the z-intercept is $z = 0$.

b. The equations for the xy-, xz- and yz-traces are found be setting $z = 0$, $y = 0$ and $x = 0$ respectively in the equation of the surface, which gives

$$5x - \frac{y^2}{25} = 0, \quad 5x + \frac{z^2}{20} = 0, \quad -\frac{y^2}{25} + \frac{z^2}{20} = 0.$$

c.

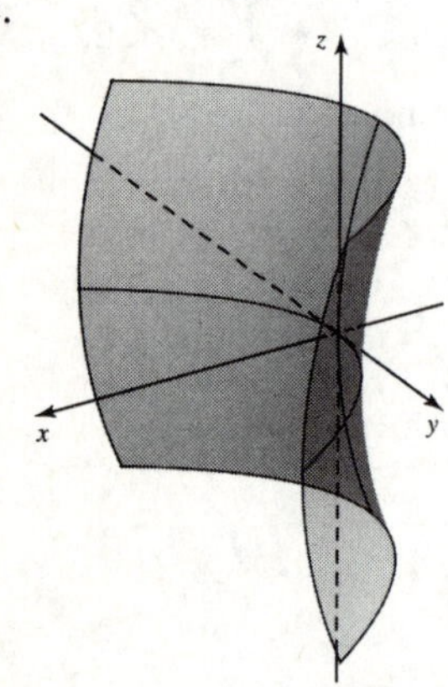

12.1.53

a. The x-intercept is found by setting $y = z = 0$ in the equation of this surface, which gives $x = 0$. Similarly we see that the y-intercept is $y = 0$ and the z-intercept is $z = 0$.

b. The equations for the xy-, xz- and yz-traces are found be setting $z = 0$, $y = 0$ and $x = 0$ respectively in the equation of the surface, which gives

$$x^2 + \frac{y^2}{4} = 0 \iff x = y = z = 0, \quad x^2 = z^2, \quad \frac{y^2}{4} = z^2.$$

c.

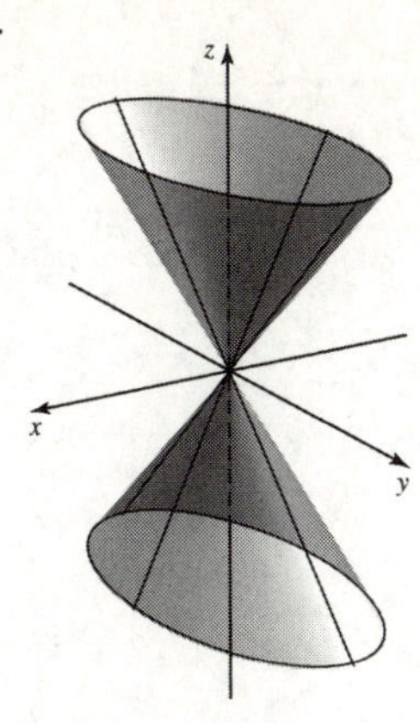

12.1.55

a. The x-intercept is found by setting $y = z = 0$ in the equation of this surface, which gives $x = 0$. Similarly we see that the y-intercept is $y = 0$ and the z-intercept is $z = 0$.

b. The equations for the xy-, xz- and yz-traces are found be setting $z = 0$, $y = 0$ and $x = 0$ respectively in the equation of the surface, which gives

$$\frac{y^2}{18} = 2x^2, \quad \frac{z^2}{32} = 2x^2, \quad \frac{z^2}{32} + \frac{y^2}{18} = 0 \iff x = y = z = 0.$$

c.

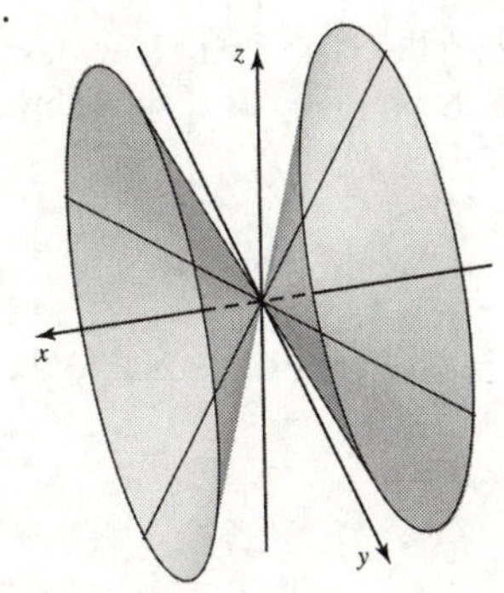

12.1.57

a. The y-intercept is found by setting $x = z = 0$ in the equation of this surface, which gives $y^2 = 4$, so the y-intercepts are $y = \pm 2$. There are no x or z-intercepts.

b. The equations for the xy-, xz- and yz-traces are found be setting $z = 0$, $y = 0$ and $x = 0$ respectively in the equation of the surface, which gives

$$-x^2 + \frac{y^2}{4} = 1, \quad -x^2 - \frac{z^2}{9} = 1 \text{ (no } xz\text{-trace)}, \quad \frac{y^2}{4} - \frac{z^2}{9} = 1.$$

c.

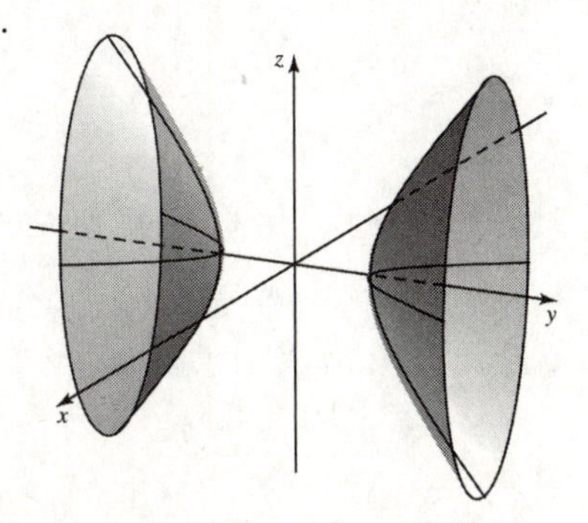

12.1.59

a. The y-intercept is found by setting $x = z = 0$ in the equation of this surface, which gives $3\,y^2 = 1$, so the y-intercepts are $y = \pm\dfrac{\sqrt{3}}{3}$. There are no x or z-intercepts.

b. The equations for the xy-, xz- and yz-traces are found be setting $z = 0$, $y = 0$ and $x = 0$ respectively in the equation of the surface, which gives

$$-\frac{x^2}{3} + 3y^2 = 1, \quad -\frac{x^2}{3} - \frac{z^2}{12} = 1 \text{ (no } xz\text{-trace)}, \quad 3y^2 - \frac{z^2}{12} = 1.$$

c.

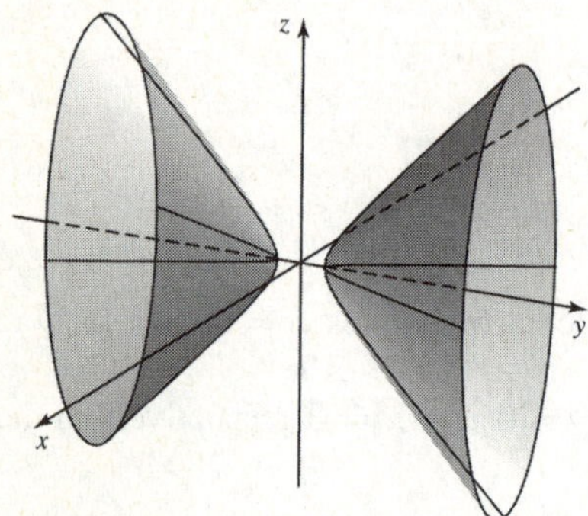
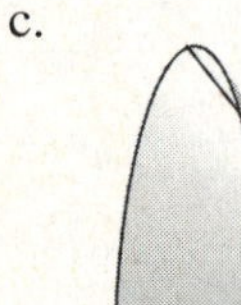

12.1.61

a. True. Observe first that these two planes are parallel since their normal vectors are parallel. The first plane has equation $1 \cdot (x-1) + 2(y-1) - 3(z-1) = 0 \Longleftrightarrow x + 2y - 3z = 0$; the point $(3,\ 0,\ 1)$ is on this plane, so the two planes are identical.

b. False. The point (1, 0, 0) is on the first plane but not the second.

c. False. There are infinite planes orthogonal to the plane Q.

d. True. Any two points on the line ℓ together with P_0 determine the same plane.

e. False. For example, the xz- and yz-coordinate planes both contain the point $(0,\ 0,\ 1)$ and are orthogonal to the xy-coordinate plane.

f. False. Two distinct lines determine a plane only if the lines are parallel or if they intersect.

g. False. Either plane S is plane P or plane S is parallel to plane P.

12.1.63

a. D. This surface is a cylinder parallel to the parabola $y = z^2$ in the yz-plane.

b. A. This surface is a plane.

c. E. This surface is an ellipsoid.

d. F. This surface is a hyperboloid of one sheet.

e. B. This surface is an elliptic cone.

f. C. This surface is a cylinder parallel to the graph $y = |\,x\,|$ in the xy-plane.

12.1.65 This surface is a hyperbolic paraboloid with saddle point at the origin.

12.1.67 This surface is an elliptic paraboloid with axis the y-axis.

12.1.69 Completing the square and rewriting the equation of the surface as $9x^2 + (y+1)^2 - 4z^2 = 1$ shows that this surface is a hyperboloid of one sheet with axis the line $\ell : \mathbf{r} = \langle 0,\ -1,\ t \rangle$.

12.1.71 This surface is a hyperbolic cylinder (the yz-trace is a hyperbola).

12.1.73 Completing the square and rewriting the equation of the surface as $z = \dfrac{(x-4)^2}{4} + (y-5)^2 + 12$ shows that this surface is an elliptic paraboloid with axis the line $\ell : \mathbf{r} = \langle 4,\ 5,\ t \rangle$.

12.1.75 The point $\left(t,\, t^2,\, 3t^2\right)$ lies on the plane $8x+y+z=60$ exactly when

$$8t+t^2+3t^2=60 \quad\Longleftrightarrow\quad t^2+2t-15=0\,,$$

which has solutions $t=-5,\,3$. Therefore the intersection points are $(-5,\,25,\,75)$ and $(3,\,9,\,27)$.

12.1.77 Suppose the point $(x,\,y,\,z)$ lies on both the curve and the plane; then $z=\dfrac{x}{4}$, and substituting this in the equation $2x+3y-12z=0$ gives $y=\dfrac{x}{3}$. We also have $x=4\cos t$ and $\dfrac{x}{3}=4\sin t$ for some t; therefore

$$\left(\frac{x}{4}\right)^2+\left(\frac{x}{12}\right)^2=1 \quad\Longleftrightarrow\quad 10x^2=144$$

which gives $x=\pm\frac{6\sqrt{10}}{5}$, so the intersection points are $\left(\frac{6\sqrt{10}}{5},\,\frac{2\sqrt{10}}{5},\,\frac{3\sqrt{10}}{10}\right)$ and $\left(-\frac{6\sqrt{10}}{5},\,-\frac{2\sqrt{10}}{5},\,-\frac{3\sqrt{10}}{10}\right)$.

12.1.79 The angle between the vectors $\mathbf{n}_1=\langle 5,2,-1\rangle$ and $\mathbf{n}_2=\langle -3,1,2\rangle$ satisfies

$$\cos\theta=\frac{\mathbf{n}_1\cdot\mathbf{n}_2}{|\mathbf{n}_1||\mathbf{n}_2|}=-\frac{15}{\sqrt{30}\sqrt{14}}=-\frac{\sqrt{105}}{14}\,,\ \text{so}\ \theta=\cos^{-1}\left(-\frac{\sqrt{105}}{14}\right)\approx 2.392\ \text{rad}\approx 137^\circ\,.$$

12.1.81 All of the quadric surfaces in Table 12.1 except the hyperbolic paraboloid can have circular cross-sections around a coordinate axis, and so can be generated by revolving a curve in one of the coordinate planes about a coordinate axis.

12.1.83

a.

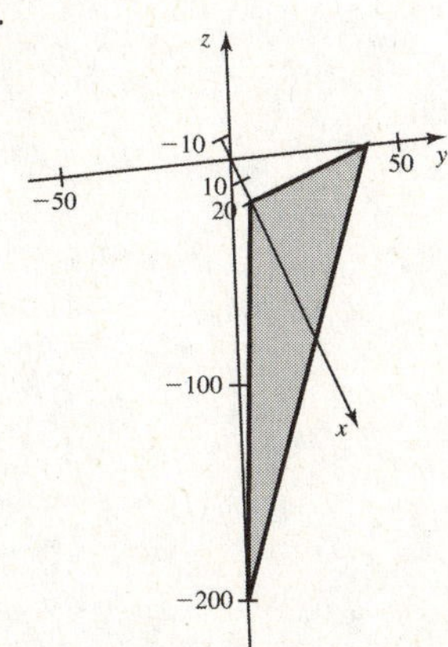

b. The profit is $z=10\cdot 20+5\cdot 10-200=\$50$ which is positive.

c. The profit is 0 when x and y lie on the line $2x+y=40$.

12.1.85

a. Observe that any point $(x,\,y,\,z)$ on this curve satisfies $z=c\,y$, so this gives the equation of the plane P.

b. Plane P has normal vector $\mathbf{n}=\langle 0,-c,\,1\rangle$, so the angle θ that P makes with the xy-plane (which has normal vector $\mathbf{k}$) satisfies $\cos\theta=\dfrac{\mathbf{n}\cdot\mathbf{k}}{|\mathbf{n}||\mathbf{k}|}=\dfrac{1}{\sqrt{1+c^2}}$; hence $\theta=\tan^{-1}c$.

c. The curve can be described as the intersection of the ellipsoid given by $x^2+\dfrac{y^2}{4}+\dfrac{z^2}{4c^2}=1$ with the plane P, which is an ellipse in P.

12.1.87 Let $P(x,\,y,\,z)$ be the projection of the point $P_0(2,\,3,\,-4)$ onto the plane $3x+2y-z=0$; then the vector $\overrightarrow{P_0P}$ is parallel to the normal $\mathbf{n}=\langle 3,\,2,\,-1\rangle$, so we can express $(x,\,y,\,z)=(2,\,3,\,-4)+\lambda(3,\,2,\,-1)$ for some scalar λ. Substituting in the equation of the plane gives

$$3(2+3\lambda)+2(3+2\lambda)-(-4-\lambda)=0 \quad\Longleftrightarrow\quad 16+14\lambda=0\,,$$

so $\lambda=-\frac{8}{7}$ and $(x,\,y,\,z)=\left(-\frac{10}{7},\,\frac{5}{7},\,-\frac{20}{7}\right)$.

12.2 Graphs and Level Curves

12.2.1 The independent variables are x and y and the dependent variable is z.

12.2.3 The domain of g is $\{(x,\, y) : x \neq 0 \text{ or } y \neq 0\}$.

12.2.5 We need three dimensions to plot points $(x,\, y,\, f(x,\, y))$.

12.2.7 The level curves $x^2 + y^2 = z_0$ are circles centered at $(0,\, 0)$ in $\mathbf{R}^2$.

12.2.9 The function f has 6 independent variables, so $n = 6$.

12.2.11 The domain of f is $\mathbf{R}^2$.

12.2.13 The domain of f is $\{(x,\, y) : y \neq 0\}$.

12.2.15 The domain of g is $\{(x,\, y) : y < x^2\}$.

12.2.17 The domain of g is $\{(x,\, y) : x\,y \geq 0,\, (x,\, y) \neq (0,\, 0)\}$.

12.2.19 This surface is a plane; the function's domain is $\mathbf{R}^2$ and its range is $\mathbf{R}$.

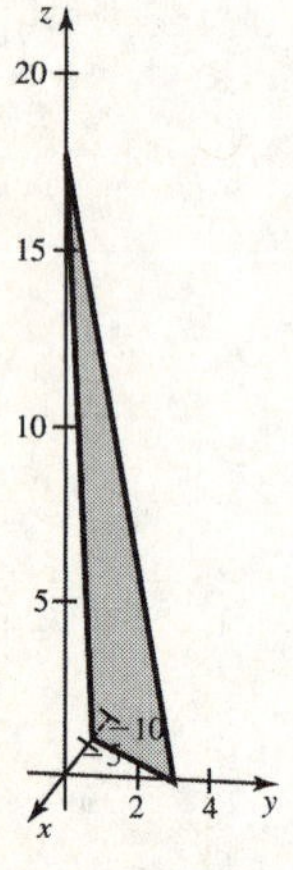

12.2.21 This surface is a hyperbolic paraboloid; the function's domain is $\mathbf{R}^2$ and its range is $\mathbf{R}$.

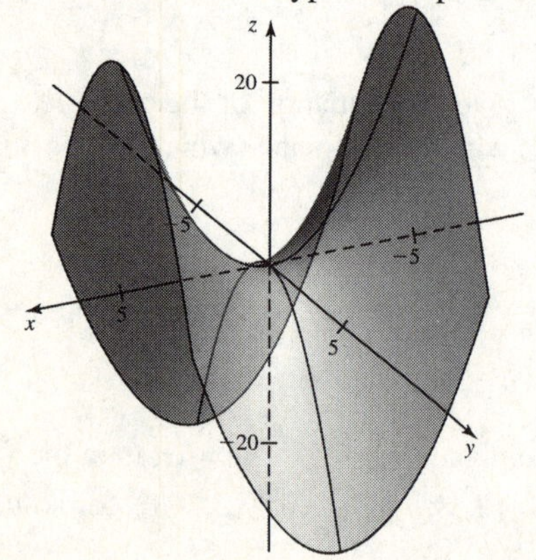

12.2.23 This surface is the lower part of a hyperboloid of two sheets; the function's domain is $\mathbf{R}^2$ and its range is the interval $(-\infty, -1]$.

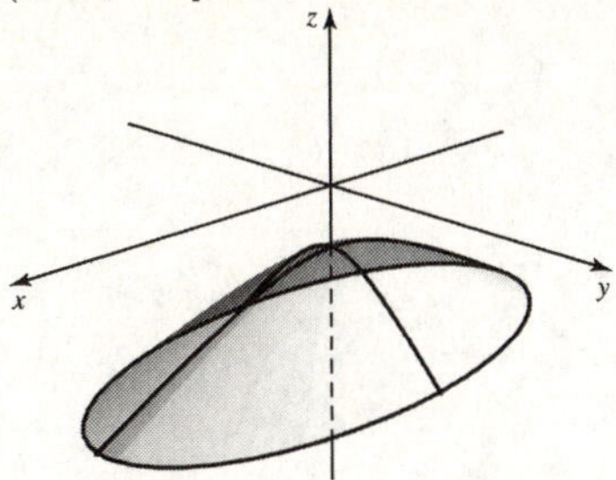

12.2.25 This surface is the upper part of a hyperboloid of one sheet; the function's domain is $\{(x, y) : x^2 + y^2 \geq 1\}$ and its range is the interval $[0, \infty)$.

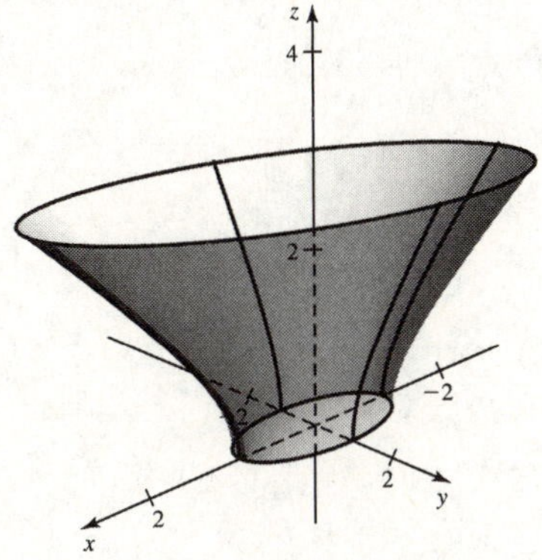

12.2.27

a. A. Notice that the range of the function in (A) is $[-1, 1]$.
b. D. Notice that the function in (D) becomes large and negative for (x, y) near $(0, 0)$.
c. B. Notice that the function in (B) becomes large as you get close to the line $y = x$.
d. C. Notice that the function in (C) is everywhere positive.

12.2.29

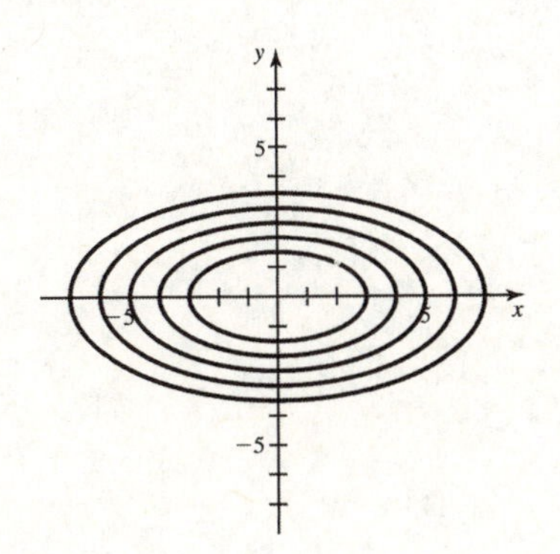

12.2.31

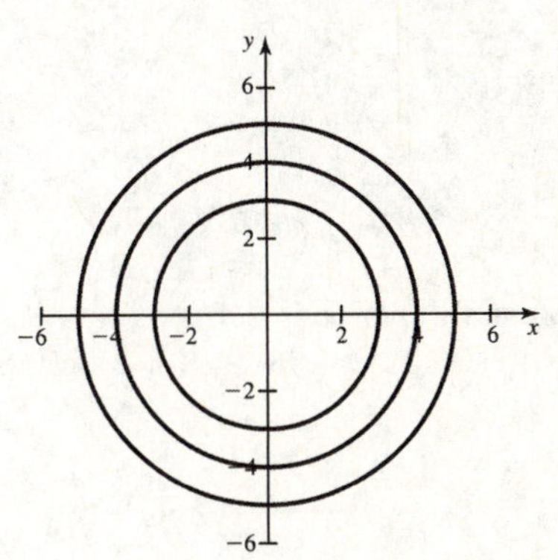

12.2.33

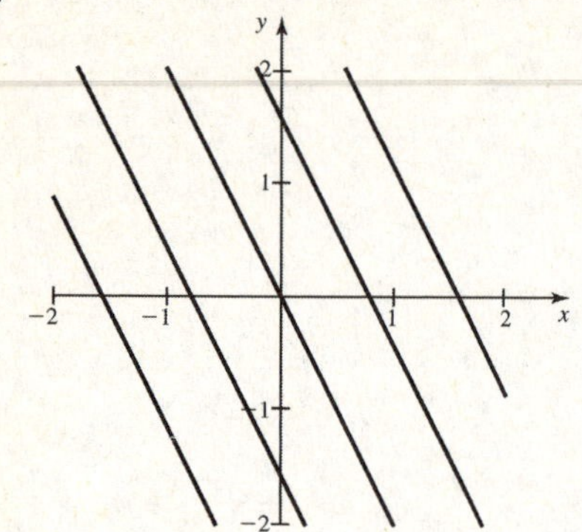

12.2.35

a.

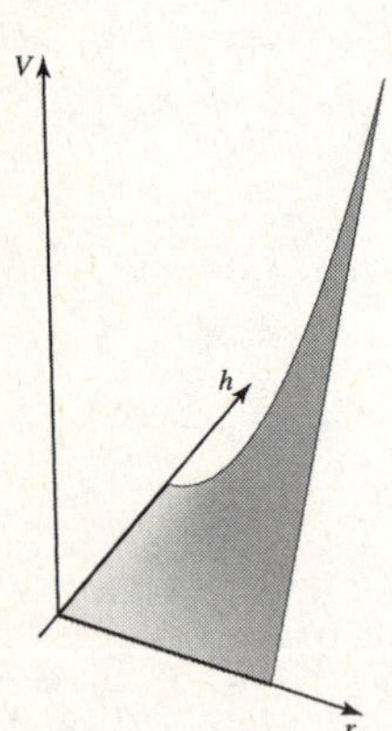

b. The domain is $D = \{(r,\, h) : r > 0,\, h > 0\}$.

c. We have $\pi r^2 h = 300$, so $h = \dfrac{300}{\pi r^2}$.

12.2.37

a.

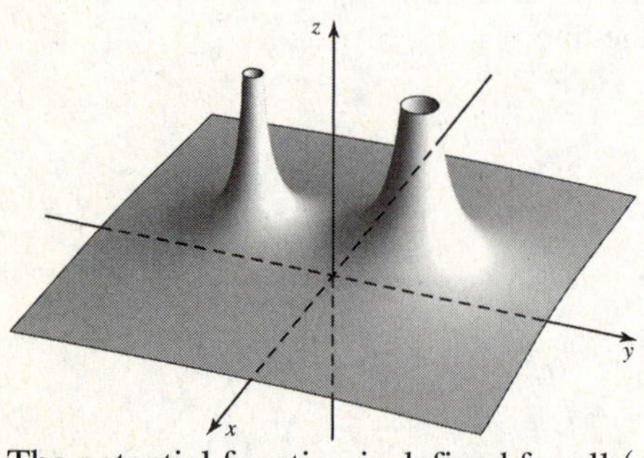

b. The potential function is defined for all $(x,\, y)$ in $\mathbf{R}^2$ except $(0,\, 1)$ and $(0,\, -1)$.

c. We have $\phi(2,\, 3) \approx 0.93 > \phi(3,\, 2) \approx 0.87$.

d.

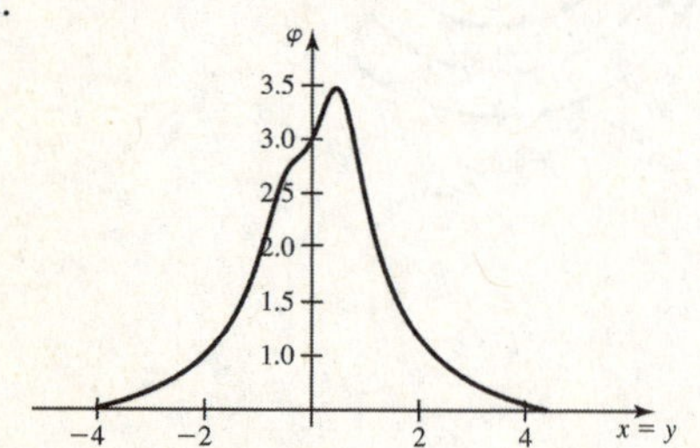

12.2.39

a.

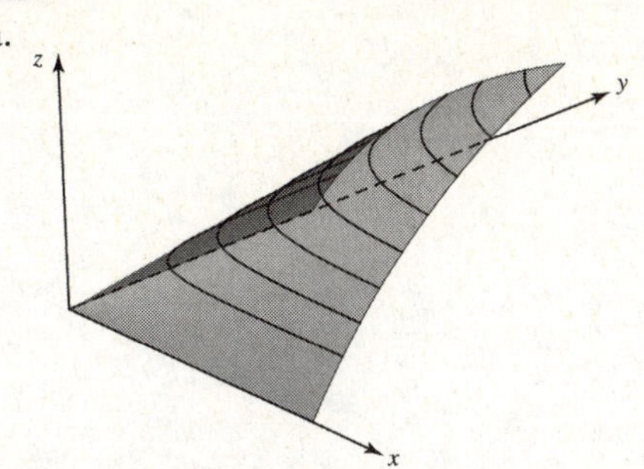

b. The maximum resistance is $R(10, 10) = 5$ ohms.

c. This means $R(x, y) = R(y, x)$.

12.2.41

a.

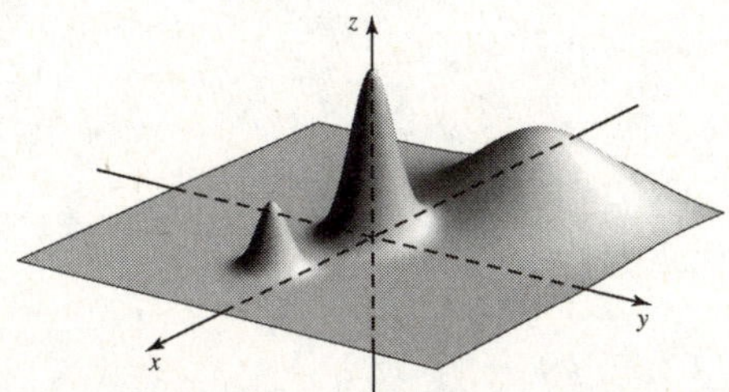

b. The peaks occur near the points $(0, 0)$, $(-5, 3)$ and $(4, -1)$.

c. We have $f(0, 0) \approx 10.17$, $f(-5, 3) \approx 5.00$, $f(4, -1) \approx 4.00$.

12.2.43 The domain of g is $\{(x, y, z) : x \neq z\}$, which is all points in $\mathbf{R}^3$ not on the plane given by $x = z$.

12.2.45 The domain of f is $\{(x, y, z) : y \geq z\}$, which is all points in $\mathbf{R}^3$ on or below the plane given by $z = y$.

12.2.47 The domain of F is $\{(x, y, z) : x^2 \leq y\}$, which is all points on the side of the vertical cylinder $y = x^2$ that contains the positive y-axis.

12.2.49

a. False. This function has domain $\mathbf{R}^2$.

b. False. The domain of a function of 4 variables is a region in $\mathbf{R}^4$.

c. True. The level curves for the function defined by $z = 2x - 3y$ are lines of the form $2x - 3y = c$ for any constant c.

12.2.51

a. The domain is $\mathbf{R}^2$ and the range is the interval $[0, \infty)$.

b.

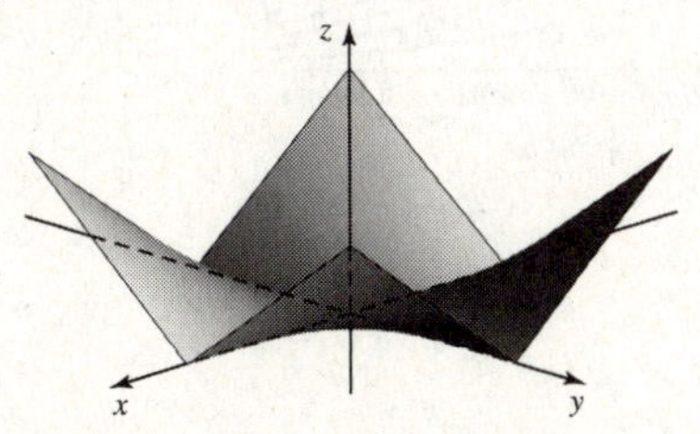

12.2.53

a. The domain is $\{(x, y) : x \neq y\}$ and the range is $\mathbf{R}$.

b.

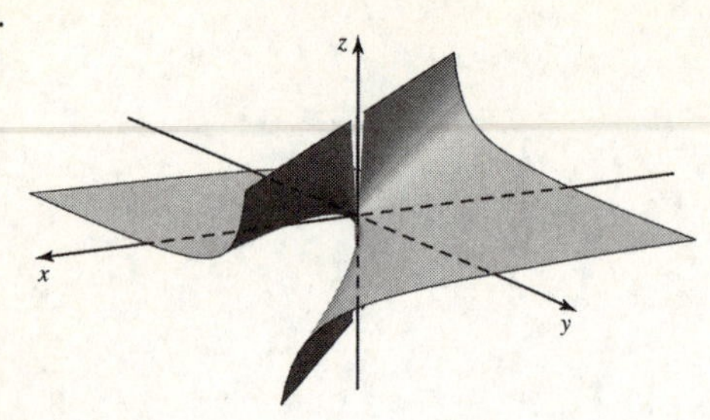

12.2.55

a. The domain is $\{(x, y) : y \neq x + \frac{\pi}{2} + n\pi$ for any integer $n\}$ and the range is the interval $[0, \infty)$.

b.

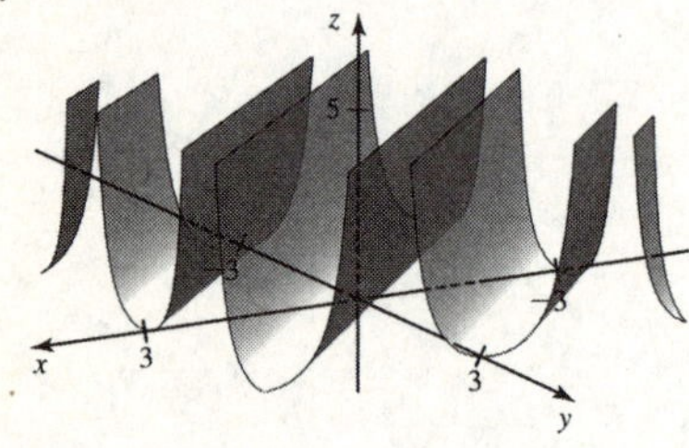

12.2.57 This function has a peak at the origin.

12.2.59 This function has a depression at the point $(1, 0)$.

12.2.61 The level curves are the lines given by $ax + by = d - cz_0$, where z_0 is a constant; these lines all have slope $-\frac{a}{b}$ (in the case $b = 0$ the lines are all vertical).

12.2.63

a. Solving for P in the equation $B(P, r, t) = 20,000$ with $t = 20$ years gives

$$P = \frac{20,000r}{(1+r)^{240} - 1}.$$

b. The level curves are given by

$$P = \frac{Br}{(1+r)^{240} - 1}$$

with $B = 5000, 10,000, 15,000$ and $25,000$.

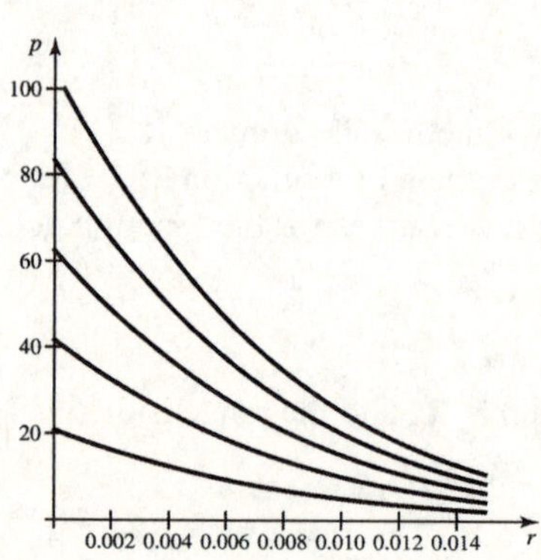

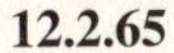

a.

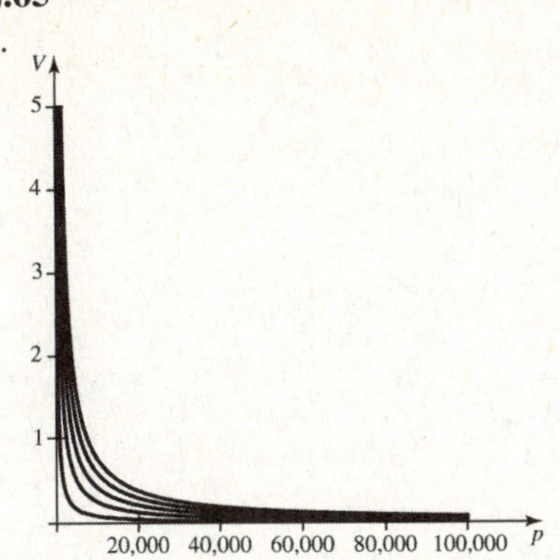

b.

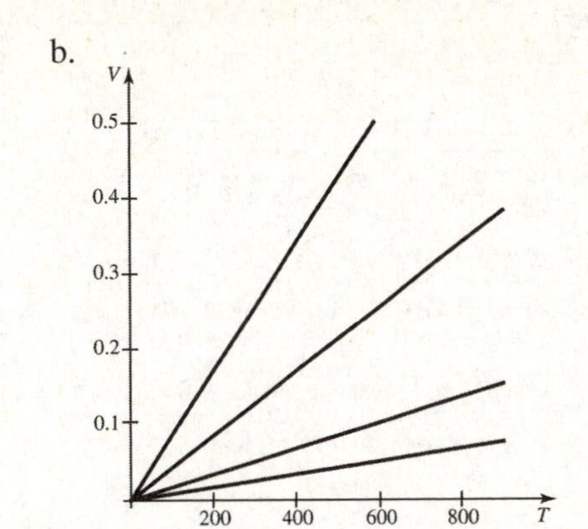

c.

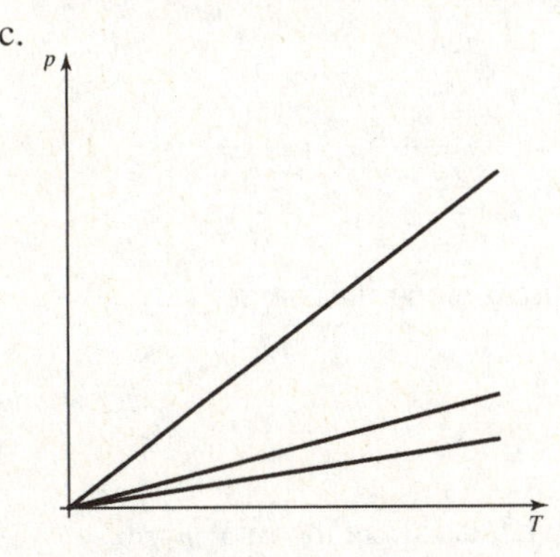

12.2.67 The domain of f is $\{(x, y) : x - 1 \le y \le x + 1\}$. This is the region between the two parallel lines given by $y = x - 1$ and $y = x + 1$.

12.2.69 Factor the equation $z^2 - xz + yz - xy = (z - x)(z + y)$; hence the domain of h is $\{(x, y, z) : (z - x)(z + y) \ge 0\}$, which is equivalent to

$$D = \{(x, y, z) : (x \le z \text{ and } y \ge -z) \text{ or } (x \ge z \text{ and } y \le -z)\}.$$

The domain consists of all points above or below both the planes given by $z = x$ and $z = -y$ as well as the points on either one of these planes.

12.3 Limits and Continuity

12.3.1 The values of $|f(x, y) - L|$ can be made arbitrarily small if (x, y) is sufficiently close to (a, b).

12.3.3 If $f(x, y)$ is a polynomial, then $\lim_{(x,y)\to(a,b)} f(x, y) = f(a, b)$; in other words, the limit can be found by plugging in $x = a$, $y = b$ in $f(x, y)$.

12.3.5 If the limits along different paths do not agree, then the limit does not exist.

12.3.7 The function f must be defined at (a, b), $\lim_{(x,y)\to(a,b)} f(x, y)$ must exist, and the limit must equal $f(a, b)$.

12.3.9 A rational function is continuous at all points where its denominator is nonzero.

12.3.11 $\displaystyle\lim_{(x,y)\to(2,9)} 101 = 101$.

12.3.13 $\displaystyle\lim_{(x,y)\to(-3,3)} (4x^2 - y^2) = 4 \cdot (-3)^2 - (3)^2 = 27$.

12.3.15 $\displaystyle\lim_{(x,y)\to(0,\pi)} \frac{\cos xy+\sin xy}{2y} = \frac{\cos 0+\sin 0}{2\pi} = \frac{1}{2\pi}.$

12.3.17 $\displaystyle\lim_{(x,y)\to(2,0)} \frac{x^2-3xy^2}{x+y} = \frac{2^2-3\cdot 2\cdot 0^2}{2+0} = 2.$

12.3.19 $\displaystyle\lim_{(x,y)\to(6,2)} \frac{x^2-3xy}{x-3y} = \lim_{(x,y)\to(6,2)} \frac{x(x-3y)}{x-3y} = \lim_{(x,y)\to(6,2)} x = 6.$

12.3.21 $\displaystyle\lim_{(x,y)\to(2,2)} \frac{y^2-4}{xy-2x} = \lim_{(x,y)\to(2,2)} \frac{(y+2)(y-2)}{x(y-2)} = \lim_{(x,y)\to(2,2)} \frac{y+2}{x} = \frac{2+2}{2} = 2.$

12.3.23 $\displaystyle\lim_{(x,y)\to(1,2)} \frac{\sqrt{y}-\sqrt{x+1}}{y-x-1} = \lim_{(x,y)\to(1,2)} \frac{\left(\sqrt{y}-\sqrt{x+1}\right)\left(\sqrt{y}+\sqrt{x+1}\right)}{(y-x-1)\left(\sqrt{y}+\sqrt{x+1}\right)}$

$$= \lim_{(x,y)\to(1,2)} \frac{y-x-1}{(y-x-1)\left(\sqrt{y}+\sqrt{x+1}\right)} = \lim_{(x,y)\to(1,2)} \frac{1}{\sqrt{y}+\sqrt{x+1}} = \frac{1}{\sqrt{2}+\sqrt{2}} = \frac{1}{2\sqrt{2}}.$$

12.3.25 Observe that along the line $y=0$, $\displaystyle\lim_{(x,y)\to(0,0)} \frac{x+2y}{x-2y} = \lim_{x\to 0} \frac{x}{x} = 1$, whereas along the line $x=0$, $\displaystyle\lim_{(x,y)\to(0,0)} \frac{x+2y}{x-2y} = \lim_{y\to 0} \frac{2y}{-2y} = -1.$

12.3.27 Observe that along the line $x=0$, $\displaystyle\lim_{(x,y)\to(0,0)} \frac{y^4-2x^2}{y^4+x^2} = \lim_{y\to 0} \frac{y^4}{y^4} = 1$, whereas along the line $y=0$, $\displaystyle\lim_{(x,y)\to(0,0)} \frac{y^4-2x^2}{y^4+x^2} = \lim_{x\to 0} \frac{-2x^2}{x^2} = -2.$

12.3.29 Observe that along the line $y=x$, $\displaystyle\lim_{(x,y)\to(0,0)} \frac{y^3+x^3}{xy^2} = \lim_{x\to 0} \frac{2x^3}{x^3} = 2$, whereas along the line $y=-x$, $\displaystyle\lim_{(x,y)\to(0,0)} \frac{y^3+x^3}{xy^2} = \lim_{x\to 0} \frac{0}{-x^3} = 0.$

12.3.31 The function f is continuous on $\mathbf{R}^2$.

12.3.33 The function p is continuous at all points except the origin, where it is undefined.

12.3.35 The function f is continuous on $\mathbf{R}^2$.

12.3.37 The function h is continuous on $\mathbf{R}^2$.

12.3.39 The function f is continuous on its domain, which is $D=\{(x,y):(x,y)\neq(0,0)\}$.

12.3.41 The function g is continuous on $\mathbf{R}^2$.

12.3.43 $\displaystyle\lim_{(x,y,z)\to(1,\ln 2,3)} z\,e^{xy} = 3\,e^{1\cdot\ln 2} = 6.$

12.3.45 $\displaystyle\lim_{(x,y,z)\to(1,1,1)} \frac{yz-xy-xz-x^2}{yz+xy+xz-y^2} = \frac{1-1-1-1}{1+1+1-1} = -1.$

12.3.47

a. False. The limit may be different or not exist along other paths approaching $(0,0)$.
b. False. We may have $f(a,b)$ undefined, or $f(a,b)\neq L$.

c. True. The limit must exist for f to be continuous at (a, b).
d. False. For example, take $P = (0, 0)$ and the domain of f to be $\{(x, y) : (x, y) \neq (0, 0)\}$.

12.3.49 Observe that $\displaystyle\lim_{(x,y)\to(0,1)} \frac{y\sin x}{x(y+1)} = \left(\lim_{x\to 0}\frac{\sin x}{x}\right)\left(\lim_{y\to 1}\frac{y}{y+1}\right) = 1\cdot\frac{1}{2} = \frac{1}{2}$.

12.3.51 Observe that $\displaystyle\lim_{(x,y)\to(1,0)} \frac{y\ln y}{x} = \left(\lim_{x\to 1}\frac{1}{x}\right)\left(\lim_{y\to 0}\ln y^y\right) = 1\cdot\ln 1 = 0$

12.3.53 Observe that along the line $y = x$, $\displaystyle\lim_{(x,y)\to(0,0)} \frac{|x-y|}{|x+y|} = \lim_{x\to 0}\frac{0}{2|x|} = 0$, whereas along the line $y = 2x$, $\displaystyle\lim_{(x,y)\to(0,0)} \frac{|x-y|}{|x+y|} = \lim_{x\to 0}\frac{|x|}{3|x|} = \frac{1}{3}$, therefore this limit does not exist.

12.3.55 Observe that $\displaystyle\lim_{(x,y)\to(2,0)} \frac{1-\cos y}{xy^2} = \left(\lim_{x\to 2}\frac{1}{x}\right)\left(\lim_{y\to 0}\frac{1-\cos y}{y^2}\right) = \frac{1}{2}\cdot\frac{1}{2} = \frac{1}{4}$ where the y-limit is evaluated by two applications of L'Hôpital's rule.

12.3.57 $\displaystyle\lim_{(x,y)\to(0,0)} \frac{x^2}{x^2+y^2} = \lim_{r\to 0}\frac{r^2\cos^2\theta}{r^2} = \lim_{r\to 0}\cos^2\theta$, which does not exist.

12.3.59 Observe that $\displaystyle\lim_{(x,y)\to(0,0)} \frac{(x-y)^2}{(x^2+y^2)^{3/2}} = \lim_{r\to 0}\frac{r^2(1-2\cos\theta\sin\theta)}{r^3} = \lim_{r\to 0}\frac{1-\sin 2\theta}{r}$. If we approach along any line where $\sin 2\theta < 1$, then this limit is ∞; therefore this limit does not exist.

12.3.61 The limit is 0 along the lines $x = 0$ or $y = 0$; however along the line $x = y$ we have $\displaystyle\frac{a\,x^m y^n}{b\,x^{m+n} + c\,y^{m+n}} = \frac{a}{b+c}$; therefore this limit does not exist unless $a = 0$.

12.3.63 Let $u = xy$; then $u \to 0$ as $(x, y) \to (1, 0)$, so $\displaystyle\lim_{(x,y)\to(1,0)} \frac{\sin xy}{xy} = \lim_{u\to 0}\frac{\sin u}{u} = 1$.

12.3.65 Let $u = xy$; then $u \to 0$ as $(x, y) \to (0, 2)$, so $\displaystyle\lim_{(x,y)\to(0,2)} (2xy)^{xy} = \lim_{u\to 0} 2^u u^u = 1$.

12.3.67 Since $\displaystyle\lim_{(x,y)\to(0,0)} e^{-1/(x^2+y^2)} = 0$, we should define $f(0, 0) = 0$.

12.3.69 For any $\epsilon > 0$, let $\delta = \frac{\epsilon}{2}$; then $|x-a|, |y-b| \le \sqrt{(x-a)^2 + (y-b)^2}$, so

$$0 < \sqrt{(x-a)^2 + (y-b)^2} < \delta \Longrightarrow |x + y - (a+b)| \le |x-a| + |y-b| < \frac{\epsilon}{2} + \frac{\epsilon}{2} = \epsilon .$$

12.3.71 Observe first that this is trivial when $c = 0$, so assume $c \neq 0$ and let $\epsilon > 0$. Then there exists $\delta > 0$ such that

$$0 < \sqrt{(x-a)^2 + (y-b)^2} < \delta \Longrightarrow |f(x, y) - L| < \frac{\epsilon}{|c|};$$

therefore $0 < \sqrt{(x-a)^2 + (y-b)^2} < \delta \Longrightarrow |c\,f(x, y) - c\,L| = |c||f(x, y) - L| < |c|\cdot\frac{\epsilon}{|c|} = \epsilon$.

12.4 Partial Derivatives

12.4.1 The slope parallel to the x-axis is $f_x(a, b)$, and the slope parallel to the y-axis is $f_y(a, b)$.

12.4.3 $f_x(x, y) = \cos(xy) + x(-\sin(xy))y = \cos(x\,y) - x\,y\sin(xy)$, $f_y(x, y) = x(-\sin(xy))x = -x^2\sin(xy)$.

12.4.5 Think of x and y as being fixed, and differentiate with respect to the variable z.

12.4.7 $f_x(x, y) = 6xy$, $f_y(x, y) = 3x^2$.

12.4.9 $g_x(x, y) = (-\sin(2xy))2y = -2y\sin(2xy)$, $g_y(x, y) = (-\sin(2xy))2x = -2x\sin(2xy)$.

12.4.11 $f_w(w, z) = \dfrac{(w^2+z^2)\cdot 1 - w\cdot 2w}{(w^2+z^2)^2} = \dfrac{z^2-w^2}{(w^2+z^2)^2}$, $f_z(w, z) = -w(w^2+z^2)^{-2}\cdot 2z = \dfrac{-2wz}{(w^2+z^2)^2}$.

12.4.13 $s_y(y, z) = z^2(\sec^2 yz)z = z^3\sec^2 yz$, $s_z(y, z) = 2z\tan yz + z^2(\sec^2 yz)y = 2z\tan yz + yz^2\sec^2 yz$

12.4.15 $G_s(s, t) = \dfrac{t}{2\sqrt{st}}\cdot\dfrac{1}{s+t} + \sqrt{st}\cdot\dfrac{-1}{(s+t)^2} = \dfrac{\sqrt{st}(s+t) - 2s\sqrt{st}}{2s(s+t)^2} = \dfrac{\sqrt{st}(t-s)}{2s(s+t)^2}$,

$G_t(s, t) = \dfrac{s}{2\sqrt{st}}\cdot\dfrac{1}{s+t} + \sqrt{st}\cdot\dfrac{-1}{(s+t)^2} = \dfrac{\sqrt{st}(s+t) - 2t\sqrt{st}}{2s(s+t)^2} = \dfrac{\sqrt{st}(s-t)}{2s(s+t)^2}$.

12.4.17 We have $h_x(x, y) = 3x^2 + y^2$, $h_y(x, y) = 2xy$; therefore $h_{xx}(x, y) = 6x$, $h_{yy}(x, y) = 2x$, $h_{xy}(x, y) = h_{yx}(y, x) = 2y$.

12.4.19 We have $f_x(x, y) = 4y^3\cos 4x$, $f_y(x, y) = 3y^2\sin 4x$; therefore $f_{xx}(x, y) = -16y^3\sin 4x$, $f_{yy}(x, y) = 6y\sin 4x$, $f_{xy}(x, y) = f_{yx}(y, x) = 12y^2\cos 4x$.

12.4.21 We have $p_u(u, v) = \dfrac{2u}{u^2+v^2+4}$, $p_v(u, v) = \dfrac{2v}{u^2+v^2+4}$; therefore

$p_{uu}(u, v) = \dfrac{(u^2+v^2+4)\cdot 2 - 2u\cdot 2u}{(u^2+v^2+4)^2} = \dfrac{-2u^2+2v^2+8}{(u^2+v^2+4)^2}$,

$p_{vv}(u, v) = \dfrac{(u^2+v^2+4)\cdot 2 - 2v\cdot 2v}{(u^2+v^2+4)^2} = \dfrac{2u^2-2v^2+8}{(u^2+v^2+4)^2}$,

$p_{uv}(u, v) = p_{vu}(u, v) = -2u(u^2+v^2+4)^{-2}\cdot 2v = \dfrac{-4uv}{(u^2+v^2+4)^2}$.

12.4.23 We have $F_r(r, s) = e^s$, $F_s(r, s) = re^s$; therefore $F_{rr}(r, s) = 0$, $F_{ss}(r, s) = re^s$, $F_{rs}(r, s) = F_{sr}(r, s) = e^s$.

12.4.25 Observe that $f_x(x, y) = 6x^2$, so $f_{xy}(x, y) = 0$; and $f_y(x, y) = 6y$, so $f_{yx}(x, y) = 0$.

12.4.27 Observe that $f_x(x, y) = -y\sin xy$, so $f_{xy}(x, y) = -\sin xy - xy\cos xy$; and $f_y(x, y) = -x\sin xy$, so $f_{yx}(x, y) = -\sin xy - xy\cos xy$.

12.4.29 Observe that $f_x(x, y) = e^{x+y}$, so $f_{xy}(x, y) = e^{x+y}$; and $f_y(x, y) = e^{x+y}$, so $f_{yx}(x, y) = e^{x+y}$.

12.4.31 $f_x(x, y, z) = y + z$; $f_y(x, y, z) = x + z$; $f_z(x, y, z) = x + y$.

12.4.33 $h_x(x, y, z) = h_y(x, y, z) = h_z(x, y, z) = -\sin(x + y + z)$.

12.4.35 $F_u(u, v, w) = \dfrac{1}{v+w}$; $F_v(u, v, w) = -\dfrac{u}{(v+w)^2}$; $F_w(u, v, w) = -\dfrac{u}{(v+w)^2}$.

12.4.37 $f_w(w, x, y, z) = 2w\,x\,y^2$; $f_x(w, x, y, z) = w^2y^2 + y^3z^2$; $f_y(w, x, y, z) = 2w^2x\,y + 3x\,y^2z^2$; $f_z(w, x, y, z) = 2x\,y^3z$.

12.4.39 $h_w(w, x, y, z) = \dfrac{z}{xy}$; $h_x(w, x, y, z) = -\dfrac{wz}{x^2y}$; $h_y(w, x, y, z) = -\dfrac{wz}{xy^2}$; $h_z(w, x, y, z) = \dfrac{w}{xy}$.

12.4.41

a. We have $V = \dfrac{kT}{P}$, so $\dfrac{\partial V}{\partial P} = -\dfrac{kT}{P^2}$. Since this partial derivative is negative, the volume decreases as the pressure increases at a fixed temperature.

b. We have $\dfrac{\partial V}{\partial T} = \dfrac{k}{P}$. Since this partial derivative is positive, the volume increases as the temperature increases at a fixed pressure.

c.

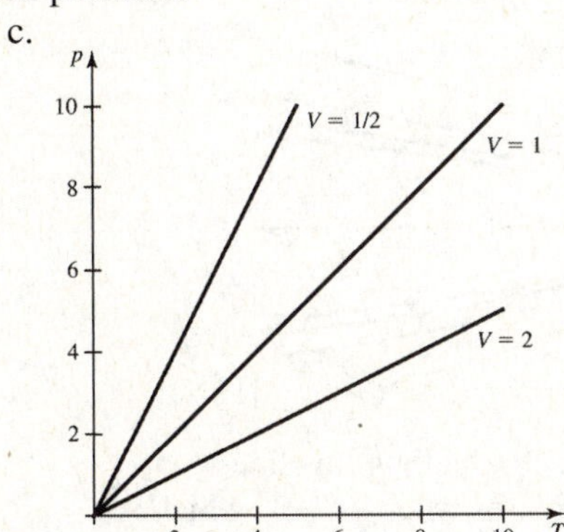

12.4.43

a. Observe that as $f(x, y) = 0$ along either coordinate axis but on the line $y = x$, $f(x, y) = \dfrac{-x^2}{2x^2} = -\dfrac{1}{2}$, so $\lim_{(x,y)\to(0,0)} f(x, y)$ does not exist, and hence f is not continuous at $(0, 0)$.

b. By Theorem 12.6, f is not differentiable at $(0, 0)$.

c. Since f is identically 0 on the coordinate axes, $f_x(0, 0) = f_y(0, 0) = 0$.

d. We have $f_x(x, y) = -\left(\dfrac{(x^2+y^2)y - xy\cdot 2x}{(x^2+y^2)^2}\right) = \dfrac{(x^2-y^2)y}{(x^2+y^2)^2}$. Along the line $x = 2y$, $f_x(x, y) = \dfrac{3y^3}{25y^4} = \dfrac{3}{25}\cdot\dfrac{1}{y}$, which does not converge to 0 as $y \to 0$. Hence f_x is not continuous at $(0, 0)$. A similar argument shows that f_y is also not continuous at $(0, 0)$.

e. Theorem 12.5 does not apply since the partials f_x and f_y are not continuous at $(0, 0)$, and Theorem 12.6 does not apply since f is not differentiable at $(0, 0)$.

12.4.45

a. Observe that $\lim\limits_{(x,y)\to(0,0)} (1 - |xy|) = 1 = f(0, 0)$, so f is continuous at $(0, 0)$.

b. Let $(a, b) = (0, 0)$; then $f(a + \Delta x, b + \Delta y) - f(a, b) = -|\Delta x||\Delta y| = \epsilon_1 \Delta x$ where $\epsilon_1 = \pm\Delta y$ (depending on the sign of x). Since $\epsilon_1 \to 0$ as $\Delta y \to 0$, we see that f is differentiable at $(0, 0)$.

c. Since f is identically equal to 1 on the coordinate axes, $f_x(0, 0) = f_y(0, 0) = 0$.

d. The partial derivative $f_x(0, y)$ does not exist for $y \neq 0$, since the function $|x|$ is not differentiable at $x = 0$. Similarly, the partial derivative $f_y(x, 0)$ does not exist for $x \neq 0$. Hence the partials f_x and f_y are not continuous at $(0, 0)$.

e. Theorem 12.5 does not apply since the partials f_x and f_y are not continuous at $(0, 0)$. Theorem 12.6 implies that f is continuous at $(0, 0)$, which we saw in part (a).

12.4.47

a. False; $\dfrac{\partial}{\partial x} y^{10} = 0$ since x and y are independent variables.

b. False; $\dfrac{\partial^2}{\partial x\,\partial y}(xy)^{1/2} = \dfrac{1}{2} \cdot x^{-1/2} \cdot \dfrac{1}{2} y^{-1/2} = \dfrac{1}{4\sqrt{xy}}$.

c. True; if f has continuous partial derivatives of all orders, then the order of differentiation for mixed partials can be exchanged.

12.4.49 We have $f_x(x, y) = -\dfrac{2x}{1+(x^2+y^2)^2}$ and $f_y(x, y) = -\dfrac{2y}{1+(x^2+y^2)^2}$.

12.4.51 We have $h_x(x, y, z) = z(1+x+2y)^{z-1}$, $h_y(x, y, z) = 2z(1+x+2y)^{z-1}$, and $h_x(x, y, z) = (1+x+2y)^z \ln(1+x+2y)$.

12.4.53

a. We have $z_x = \dfrac{1}{y^2}$ and $z_y = -\dfrac{2x}{y^3}$.

b.

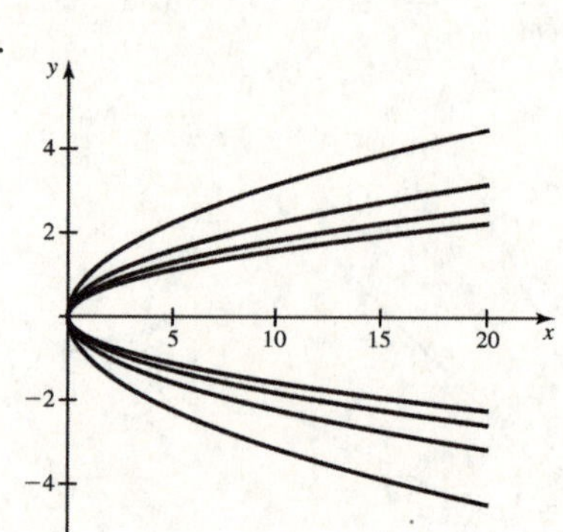

c. We observe that z increases at the same rate as x, which makes sense since $z_x = 1$ along this line.

d. We observe that z increases when $y < 0$, is undefined when $y = 0$ and decreases when $y > 0$, which is consistent with $z_y = -\dfrac{2}{y^3}$ along this line.

12.4.55

a. Since $\cos\left(\dfrac{\pi}{3}\right) = \dfrac{1}{2}$ we have $c = \left(a^2+b^2-ab\right)^{1/2}$, and therefore

$$\frac{\partial c}{\partial a} = \frac{2a-b}{2\sqrt{a^2+b^2-ab}} \quad \text{and} \quad \frac{\partial c}{\partial b} = \frac{2b-a}{2\sqrt{a^2+b^2-ab}}.$$

b. Implicit differentiation gives

$$2c\frac{\partial c}{\partial a} = 2a-b \quad \Longrightarrow \quad \frac{\partial c}{\partial a} = \frac{2a-b}{2c} \quad \text{and} \quad 2c\frac{\partial c}{\partial b} = 2b-a \quad \Longrightarrow \quad \frac{\partial c}{\partial b} = \frac{2b-a}{2c}.$$

c. The necessary relationship is $2a-b>0$ or $a > \dfrac{b}{2}$.

12.4.57

a. We have $\varphi_x(x, y) = -\dfrac{2x}{\left(x^2+(y-1)^2\right)^{3/2}} - \dfrac{x}{\left(x^2+(y+1)^2\right)^{3/2}}$ and

$$\varphi_y(x, y) = -\frac{2(y-1)}{\left(x^2+(y-1)^2\right)^{3/2}} - \frac{y+1}{\left(x^2+(y+1)^2\right)^{3/2}}.$$

b. Observe that $|\varphi_x(x, y)| \le \dfrac{2|x|}{|x|^{3/2}} + \dfrac{|x|}{|x|^{3/2}} = \dfrac{3}{|x|^{1/2}}$ and similarly

$|\varphi_y(x, y)| \le \dfrac{2|y-1|}{|y-1|^{3/2}} + \dfrac{|y+1|}{|y+1|^{3/2}} = \dfrac{2}{|y-1|^{1/2}} + \dfrac{1}{|y+1|^{1/2}}$, which both converge to 0 as $x, y \to \infty$.

c. We see that $\varphi_x(x, y) = 0$ as long as $y \ne \pm 1$. This is consistent with the observation that along horizontal lines $y = y_0$ the potential function takes its maximum at $x = 0$.

d. We see that $\varphi_y(x, 0) = \dfrac{1}{(x^2+1)^{3/2}}$. This implies that if we cross the x-axis at any point from below to above, the potential function is increasing.

12.4.59

a. Solving for R gives $R = \left(R_1^{-1} + R_2^{-1}\right)^{-1}$, so $\dfrac{\partial R}{\partial R_1} = -\left(R_1^{-1} + R_2^{-1}\right)^{-2}\left(-R_1^{-2}\right) = \dfrac{R_2^2}{(R_1+R_2)^2}$ and similarly $\dfrac{\partial R}{\partial R_2} = \dfrac{R_1^2}{(R_1+R_2)^2}$.

b. We have $-R^{-2}\dfrac{\partial R}{\partial R_1} = -R_1^{-2} \implies \dfrac{\partial R}{\partial R_1} = \dfrac{R^2}{R_1^2}$ and similarly $\dfrac{\partial R}{\partial R_1} = \dfrac{R^2}{R_2^2}$.

c. Since $\dfrac{\partial R}{\partial R_1} > 0$, an increase in R_1 causes an increase in R.

d. Since $\dfrac{\partial R}{\partial R_2} > 0$, a decrease in R_2 causes a decrease in R.

12.4.61 Observe that $\dfrac{\partial^2 u}{\partial t^2} = -4c^2\cos(2(x+ct)) = c^2\dfrac{\partial^2 u}{\partial x^2}$.

12.4.63 Observe that $\dfrac{\partial^2 u}{\partial t^2} = Ac^2 f''(x+ct) + Bc^2 g''(x-ct) = c^2\dfrac{\partial^2 u}{\partial x^2}$.

12.4.65 Observe that $u_{xx} + u_{yy} = 6x - 6x = 0$.

12.4.67 Observe that $u_{xx} = \dfrac{2(x-1)y}{\left[(x-1)^2+y^2\right]^2} - \dfrac{2(x+1)y}{\left[(x+1)^2+y^2\right]^2}$ and $u_{yy} = -\dfrac{2(x-1)y}{\left[(x-1)^2+y^2\right]^2} + \dfrac{2(x+1)y}{\left[(x+1)^2+y^2\right]^2}$; so $u_{xx} + u_{yy} = 0$.

12.4.69 We see that $u_t = -16e^{-4t}\cos 2x = u_{xx}$.

12.4.71 We see that $u_t = -a^2Ae^{-a^2t}\cos ax = u_{xx}$.

12.4.73 We have $f(0, 0) = 0$, $f_x(0, 0) = f_y(0, 0) = 0$, and $f(\Delta x, \Delta y) = \Delta x \cdot \Delta y$, so we can take $\epsilon_1 = \Delta y$, $\epsilon_2 = 0$ or $\epsilon_1 = 0$, $\epsilon_2 = \Delta x$.

12.4.75

a. By the fundamental theorem of calculus, $f_x(x, y) = -\dfrac{\partial}{\partial x}\displaystyle\int_x^y h(s)\,ds = -h(x)$ and similarly $f_y(x, y) = h(y)$.

b. Let $H(s)$ be an antiderivative of $h(s)$; then $f(x, y) = H(xy) - H(1) \implies f_x(x, y) = y\,h(xy)$, $f_y(x, y) = x\,h(xy)$.

12.4.77

a. Observe that $u_x = 2x = v_y$ and $u_y = -2y = -v_x$.

b. Observe that $u_x = 3x^2 - 3y^2 = v_y$ and $u_y = -6xy = -v_x$.

c. We have $u_{xx} = v_{yx} = v_{xy} = -u_{yy} \implies u_{xx} + u_{yy} = 0$. The proof that $v_{xx} + v_{yy} = 0$ is similar.

12.5 The Chain Rule

12.5.1 There is one dependent variable (z), two intermediate variables (x and y) and one independent variable (t).

12.5.3 Multiply each of the partial derivatives of w by the t-derivative of the corresponding function, and add all these expressions.

12.5.5

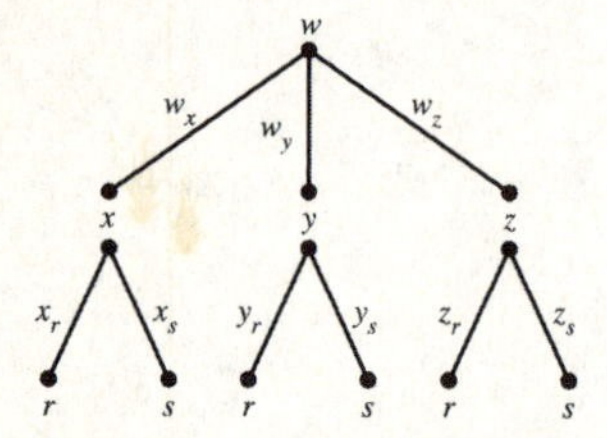

12.5.7 We have $\dfrac{dz}{dt} = \dfrac{\partial z}{\partial x}\dfrac{dx}{dt} + \dfrac{\partial z}{\partial y}\dfrac{dy}{dt} = (\sin y)2t + (x\cos y)12t^2 = 2t\sin\left(4t^3\right) + 12t^4\cos\left(4t^3\right).$

12.5.9 We have

$$\frac{dw}{dt} = \frac{\partial w}{\partial x}\frac{dx}{dt} + \frac{\partial w}{\partial y}\frac{dy}{dt} = (-2\sin 2x\sin 3y)\left(\frac{1}{2}\right) + (3\cos 2x\cos 3y)4t^3 = -\sin t\sin 3t^4 + 12t^3\cos t\cos 3t^4.$$

12.5.11 We have

$$\frac{dw}{dt} = \frac{\partial w}{\partial x}\frac{dx}{dt} + \frac{\partial w}{\partial y}\frac{dy}{dt} + \frac{\partial w}{\partial z}\frac{dz}{dt} = (y\sin z)2t + (x\sin z)12t^2 + (xy\cos z)\cdot 1 = \left(2ty + 12t^2x\right)\sin z + xy\cos z.$$
$$= 20t^4\sin(t+1) + 4t^5\cos(t+1).$$

12.5.13 We have

$$\frac{dU}{dt} = \frac{\partial U}{\partial x}\frac{dx}{dt} + \frac{\partial U}{\partial y}\frac{dy}{dt} + \frac{\partial U}{\partial z}\frac{dz}{dt} = \frac{1}{x+y+z}\cdot 1 + \frac{1}{x+y+z}\cdot 2t + \frac{1}{x+y+z}\cdot 3t^2 = \frac{1+2t+3t^2}{t+t^2+t^3}.$$

12.5.15

a. By the chain rule, $V'(t) = 2\pi r(t)h(t)r'(t) + \pi[r(t)]^2h'(t)$.
b. Substituting $r(t) = e^t$ and $h(t) = e^{-2t}$ gives $V'(t) = 2\pi e^te^{-2t}e^t + \pi e^{2t}\left(-2e^{-2t}\right) = 0$.
c. Since $V'(t) = 0$, the volume remains constant.

12.5.17 We have

$$z_s = \frac{\partial z}{\partial x}\frac{\partial x}{\partial s} + \frac{\partial z}{\partial y}\frac{\partial y}{\partial s} = (y-2xy)\cdot 1 + \left(x-x^2\right)\cdot 1 = s-t-2\left(s^2-t^2\right) + (s+t) - (s+t)^2$$
$$= 2s - 3s^2 - 2st + t^2$$

and

$$z_t = \frac{\partial z}{\partial x}\frac{\partial x}{\partial t} + \frac{\partial z}{\partial y}\frac{\partial y}{\partial t} = (y-2xy)\cdot 1 + \left(x-x^2\right)\cdot(-1) = s-t-2\left(s^2-t^2\right) - (s+t) + (s+t)^2$$
$$= -s^2 - 2t + 2st + 3t^2.$$

12.5.19 We have

$$z_s = \frac{\partial z}{\partial x}\frac{\partial x}{\partial s} + \frac{\partial z}{\partial y}\frac{\partial y}{\partial s} = e^{x+y}\cdot t + e^{x+y}\cdot 1 = (t+1)e^{st+s+t}$$

and

$$z_t = \frac{\partial z}{\partial x}\frac{\partial x}{\partial t} + \frac{\partial z}{\partial y}\frac{\partial y}{\partial t} = e^{x+y}\cdot s + e^{x+y}\cdot 1 = (s+1)e^{st+s+t}.$$

12.5.21 We have

$$w_s = \frac{\partial w}{\partial x}\frac{\partial x}{\partial s} + \frac{\partial w}{\partial y}\frac{\partial y}{\partial s} + \frac{\partial w}{\partial z}\frac{\partial z}{\partial s} = \frac{1}{y+z}\cdot 1 + \frac{z-x}{(y+z)^2}\cdot t - \frac{x+y}{(y+z)^2}\cdot 1 = \frac{(1+t)(z-x)}{(y+z)^2} = \frac{-2t(1+t)}{(st+s-t)^2}$$

and

$$w_t = \frac{\partial w}{\partial x}\frac{\partial x}{\partial t} + \frac{\partial w}{\partial y}\frac{\partial y}{\partial t} + \frac{\partial w}{\partial z}\frac{\partial z}{\partial t} = \frac{1}{y+z}\cdot 1 + \frac{z-x}{(y+z)^2}\cdot s - \frac{x+y}{(y+z)^2}\cdot(-1) = \frac{(1-s)x + 2y + (1+s)z}{(y+z)^2}$$

$$= \frac{2s}{(st+s-t)^2}.$$

12.5.23 $\dfrac{dw}{dt} = \dfrac{dw}{dz}\left(\dfrac{\partial z}{\partial x}\dfrac{dx}{dt} + \dfrac{\partial z}{\partial y}\dfrac{dy}{dt}\right)$

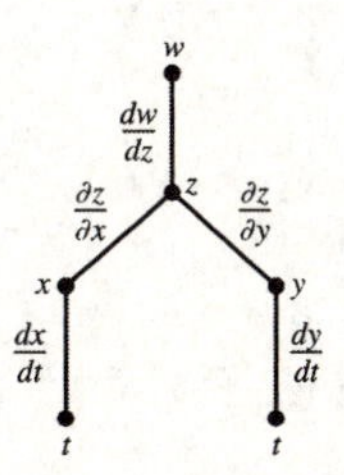

12.5.25 $\dfrac{\partial u}{\partial z} = \dfrac{du}{dv}\left(\dfrac{\partial v}{\partial w}\dfrac{dw}{dz} + \dfrac{\partial v}{\partial x}\dfrac{\partial x}{\partial z} + \dfrac{\partial v}{\partial y}\dfrac{\partial y}{\partial z}\right)$

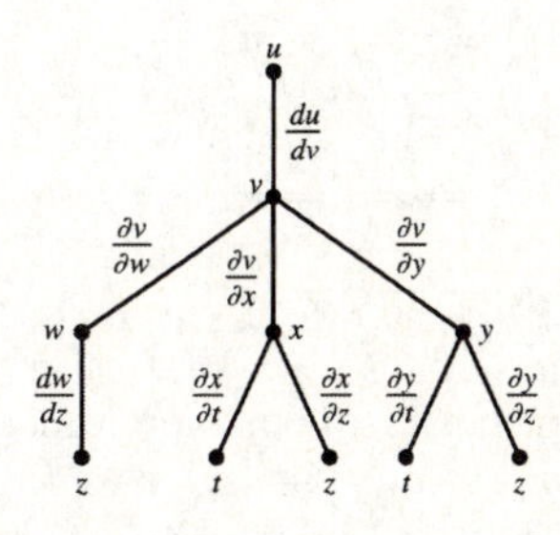

12.5.27 Let $F(x, y) = x^2 - 2y^2 - 1$; then by Theorem 12.9, we have $\dfrac{dy}{dx} = -\dfrac{F_x}{F_y} = -\dfrac{2x}{-4y} = \dfrac{x}{2y}$.

12.5.29 Let $F(x, y) = 2\sin(xy) - 1$; then by Theorem 12.9, we have $\dfrac{dy}{dx} = -\dfrac{F_x}{F_y} = -\dfrac{2y\cos(xy)}{2x\cos(xy)} = -\dfrac{y}{x}$.

12.5.31 Note that we can simplify this equation to $x^2 + 2xy + y^4 = 9$, so let $F(x, y) = x^2 + 2xy + y^4 - 9$; then by Theorem 12.9, we have

$$\frac{dy}{dx} = -\frac{F_x}{F_y} = -\frac{2x+2y}{2x+4y^3} = -\frac{x+y}{x+2y^3}.$$

12.5.33 The chain rule gives

$$\frac{\partial s}{\partial x} = \frac{\partial s}{\partial u}\frac{\partial u}{\partial x} + \frac{\partial s}{\partial v}\frac{\partial v}{\partial x} = \frac{u}{\sqrt{u^2+v^2}}\cdot 0 + \frac{v}{\sqrt{u^2+v^2}}\cdot(-2) = \frac{4x}{\sqrt{4(x^2+y^2)}} = \frac{2x}{\sqrt{x^2+y^2}}$$

and

$$\frac{\partial s}{\partial yx} = \frac{\partial s}{\partial u}\frac{\partial u}{\partial y} + \frac{\partial s}{\partial v}\frac{\partial v}{\partial y} = \frac{u}{\sqrt{u^2+v^2}}\cdot 2 + \frac{v}{\sqrt{u^2+v^2}}\cdot 0 = \frac{4y}{\sqrt{4(x^2+y^2)}} = \frac{2y}{\sqrt{x^2+y^2}}.$$

12.5.35

a. False; the correct equation is $\dfrac{\partial z}{\partial s} = \dfrac{\partial z}{\partial x}\dfrac{dx}{ds} + \dfrac{\partial z}{\partial y}\dfrac{dy}{ds}$.

b. False; w is a function of both s and t, so the rate of change of w with respect to t is the partial derivative $\dfrac{\partial w}{\partial t}$.

12.5.37

a. We have $z = (t^2+2t)^{-1} + (t^3-2)^{-1}$, so $z'(t) = -\dfrac{(2t+2)}{(t^2+2t)^2} - \dfrac{3t^2}{(t^3-2)^2}$.

b. Using the chain rule, $\dfrac{dz}{dt} = \dfrac{\partial z}{\partial x}\dfrac{dx}{dt} + \dfrac{\partial z}{\partial y}\dfrac{dy}{dt} = -\dfrac{(2t+2)}{x^2} - \dfrac{3t^2}{y^2} = -\dfrac{(2t+2)}{(t^2+2t)^2} - \dfrac{3t^2}{(t^3-2)^2}$.

12.5.39 The chain rule gives

$$\begin{aligned}\frac{dw}{dt} &= \frac{\partial w}{\partial x}\frac{dx}{dt} + \frac{\partial w}{\partial y}\frac{dy}{dt} + \frac{\partial w}{\partial z}\frac{dz}{dt} = yz\cdot 8t^3 + xz(-3t^{-2}) + xy(-12t^{-4}) \\ &= \frac{12}{t^4}\cdot 8t^3 + 8t\left(-\frac{3}{t^2}\right) + 6t^3\left(-\frac{12}{t^4}\right) = 0.\end{aligned}$$

This can also be seen by expressing w in terms of t: $w = 2t^4 3t^{-1} 4t^{-3} = 24$, so $\dfrac{dw}{dt} = 0$.

12.5.41 The chain rule gives $-\dfrac{1}{x^2} - \dfrac{1}{z^2}\dfrac{\partial z}{\partial x} = 0 \implies \dfrac{\partial z}{\partial x} = -\dfrac{z^2}{x^2}$.

12.5.43

a. The chain rule gives $w'(t) = aw_x + bw_y + cw_z$.

b. Using part (a), $w'(t) = ayz + bxz + cxy = 3abct^2$.

c. Using part (a), $w'(t) = \dfrac{ax}{\sqrt{x^2+y^2+z^2}} + \dfrac{by}{\sqrt{x^2+y^2+z^2}} + \dfrac{cz}{\sqrt{x^2+y^2+z^2}} = \dfrac{ax+by+cz}{\sqrt{x^2+y^2+z^2}}$

$= \sqrt{a^2+b^2+c^2}\,\dfrac{t}{|t|}$

d. Differentiate the result from part (a) one more time:

$$w''(t) = a(aw_{xx} + bw_{xy} + cw_{xz}) + b(aw_{yx} + bw_{yy} + cw_{yz}) + c(aw_{zx} + bw_{zy} + cw_{zz})$$

which simplifies to $w''(t) = a^2w_{xx} + b^2w_{yy} + c^2w_{zz} + 2abw_{xy} + 2acw_{xz} + 2bcw_{yz}$.

12.5.45 Let $F(x, y, z) = xy + xz + yz - 3$; then the result from Exercise 44 gives

$$\frac{\partial z}{\partial x} = -\frac{F_x}{F_z} = -\frac{y+z}{x+y},$$
$$\frac{\partial z}{\partial y} = -\frac{F_y}{F_z} = -\frac{x+z}{x+y}.$$

12.5.47 Let $F(x, y, z) = xyz + x + y - z$; then the result from Exercise 44 gives

$$\frac{\partial z}{\partial x} = -\frac{F_x}{F_z} = -\frac{yz+1}{xy-1},$$
$$\frac{\partial z}{\partial y} = -\frac{F_y}{F_z} = -\frac{xz+1}{xy-1}.$$

12.5.49

a. The chain rule gives $z'(t) = 2x(-\sin t) + 8y\cos t = 6\sin t\cos t = 3\sin 2t$.

b. Observe that for $0 \le t \le 2\pi$, $z'(t) = 3\sin 2t > 0$ when $0 < t < \dfrac{\pi}{2}$ or $\pi < t < \dfrac{3\pi}{2}$.

12.5.51

a. The chain rule gives $z'(t) = \dfrac{-x}{\sqrt{1-x^2-y^2}}(-e^{-t}) + \dfrac{-y}{\sqrt{1-x^2-y^2}}(-e^{-t}) = \dfrac{2e^{-t}}{\sqrt{1-e^{-2t}}}$.

b. Observe that $z'(t) > 0$ for all t where defined, so the function $z(t)$ is increasing for all $t \ge \dfrac{1}{2}\ln 2$.

12.5.53 The chain rule gives
$$E'(t) = m(uu' + vv') + mgy' = m(x'x'' + y'y'' + gy') = m(u_0 \cdot 0 + y'(y'' + g)) = 0.$$
Therefore, the energy of the projectile remains constant during the motion.

12.5.55

a. If r and R increase at the same rate then $R - r$ is a constant C, so $V = \dfrac{C^2\pi^2}{4}(R + r)$ is increasing.

b. Similarly, if r and R decrease at the same rate then V is decreasing.

12.5.57

a. Consider P as a function of T and V and differentiate with respect to V: $\dfrac{\partial P}{\partial V}V + P \cdot 1 = 0 \implies \dfrac{\partial P}{\partial V} = -\dfrac{P}{V}$. Next, consider T as a function of P and V and differentiate with respect to P: $1 \cdot V = k\dfrac{\partial T}{\partial P} \implies \dfrac{\partial T}{\partial P} = \dfrac{V}{k}$. Lastly, consider V as a function of T and P and differentiate with respect to T: $P\dfrac{\partial V}{\partial T} = k \implies \dfrac{\partial V}{\partial T} = \dfrac{k}{P}$.

b. Observe that $\dfrac{\partial P}{\partial V}\dfrac{\partial T}{\partial P}\dfrac{\partial V}{\partial T} = -\dfrac{P}{V}\dfrac{V}{k}\dfrac{k}{P} = -1$.

12.5.59

a. The chain rule gives
$$w'(t) = \frac{yz}{z^2+1}(-\sin t) + \frac{xz}{z^2+1}(\cos t) + \frac{xy(1-z^2)}{(z^2+1)^2} \cdot 1 = -\frac{(\sin t)t}{t^2+1}\sin t + \frac{(\cos t)t}{t^2+1}\cos t + \frac{(\cos t)(\sin t)(1-t^2)}{(t^2+1)^2}.$$

b. The function $w(t) = \dfrac{t\cos t \sin t}{1+t^2}$ takes its maximum value on $[0, \infty)$ approximately at $t = 0.838$, which gives the point $(0.669, 0.743, 0.838)$ on the spiral.

12.5.61

a. From problem 60 part (d) we have $z_x = \dfrac{x}{r}z_r - \dfrac{y}{r^2}z_\theta$, $z_y = \dfrac{y}{r}z_r + \dfrac{x}{r^2}z_\theta$

b. Differentiating the equation for z_x in part (a) with respect to x gives
$$\begin{aligned}
z_{xx} &= \frac{1}{r}z_r + x\left(-\frac{1}{r^2}\right)r_x z_r + \frac{x}{r}(z_r)_x + \frac{2y}{r^3}r_x z_\theta - \frac{y}{r^2}(z_\theta)_x \\
&= \frac{x}{r}(z_r)_x - \frac{y}{r^2}(z_\theta)_x + \left(\frac{r^2}{r^3} - \frac{x^2}{r^3}\right)z_r + \frac{2xy}{r^4}z_\theta \\
&= \frac{x}{r}\left(\frac{x}{r}z_{rr} - \frac{y}{r^2}z_{r\theta}\right) - \frac{y}{r^2}\left(\frac{x}{r}z_{\theta r} - \frac{y}{r^2}z_{\theta\theta}\right) + \frac{y^2}{r^3}z_r + \frac{2xy}{r^4}z_\theta \\
&= \frac{x^2}{r^2}z_{rr} + \frac{y^2}{r^4}z_{\theta\theta} - \frac{2xy}{r^3}z_{r\theta} + \frac{y^2}{r^3}z_r + \frac{2xy}{r^4}z_\theta.
\end{aligned}$$

c. Differentiating the equation for z_y in part (a) with respect to y gives
$$\begin{aligned}
z_{yy} &= \frac{1}{r}z_r + y\left(-\frac{1}{r^2}\right)r_y z_r + \frac{y}{r}(z_r)_y - \frac{2x}{r^3}r_y z_\theta + \frac{x}{r^2}(z_\theta)_y \\
&= \frac{y}{r}(z_r)_y + \frac{x}{r^2}(z_\theta)_y + \left(\frac{r^2}{r^3} - \frac{y^2}{r^3}\right)z_r - \frac{2xy}{r^4}z_\theta \\
&= \frac{y}{r}\left(\frac{y}{r}z_{rr} + \frac{x}{r^2}z_{r\theta}\right) + \frac{x}{r^2}\left(\frac{y}{r}z_{\theta r} + \frac{x}{r^2}z_{\theta\theta}\right) + \frac{x^2}{r^3}z_r - \frac{2xy}{r^4}z_\theta \\
&= \frac{y^2}{r^2}z_{rr} + \frac{x^2}{r^4}z_{\theta\theta} + \frac{2xy}{r^3}z_{r\theta} + \frac{x^2}{r^3}z_r - \frac{2xy}{r^4}z_\theta.
\end{aligned}$$

d. Adding the results from (b) and (c) gives $z_{xx} + z_{yy} = z_{rr} + \dfrac{1}{r}z_r + \dfrac{1}{r^2}z_{\theta\theta}$.

12.5.63

a. Assuming y is fixed, the chain rule gives $F_x \cdot 1 + F_y \cdot 0 + F_z \cdot \left(\dfrac{\partial z}{\partial x}\right)_y = 0 \implies \left(\dfrac{\partial z}{\partial x}\right)_y = -\dfrac{F_x}{F_z}$.

b. Similarly we find that $\left(\dfrac{\partial y}{\partial z}\right)_x = -\dfrac{F_z}{F_y}$ and $\left(\dfrac{\partial x}{\partial y}\right)_z = -\dfrac{F_y}{F_x}$

c. From (a) and (b) we see that $\left(\dfrac{\partial z}{\partial x}\right)_y \left(\dfrac{\partial y}{\partial z}\right)_x \left(\dfrac{\partial x}{\partial y}\right)_z = -1$.

d. Let $\left(\dfrac{\partial w}{\partial x}\right)_{y,z}$ denote the partial derivative of w with respect to x holding y and z constant, with similar notation for the other possible pairs of variables. A similar derivation as in part (a) and (b) above for $F(w,\, x,\, y,\, z) = 0$ shows that

$$\left(\frac{\partial w}{\partial x}\right)_{y,z}\left(\frac{\partial x}{\partial y}\right)_{w,z}\left(\frac{\partial y}{\partial z}\right)_{w,x}\left(\frac{\partial z}{\partial w}\right)_{x,y} = \left(-\frac{F_x}{F_w}\right)\left(-\frac{F_y}{F_x}\right)\left(-\frac{F_z}{F_y}\right)\left(-\frac{F_w}{F_z}\right) = 1\,.$$

12.5.65

a. We have $\left(\dfrac{\partial w}{\partial x}\right)_y = f_x + f_z\dfrac{dz}{dx} = 2 + 4\cdot 4 = 18\,.$

b. Rewrite $z = 4x - 2y$ as $y = 2x - \dfrac{z}{2}$; therefore $\left(\dfrac{\partial w}{\partial x}\right)_z = f_x + f_y\dfrac{dy}{dx} = 2 + 3\cdot 2 = 8\,.$

c.

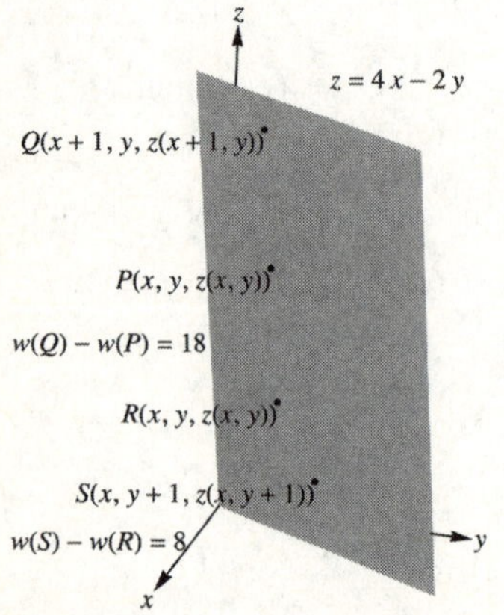

z
z = 4x − 2y
Q(x + 1, y, z(x + 1, y))
P(x, y, z(x, y))
w(Q) − w(P) = 18
R(x, y, z(x, y))
S(x, y + 1, z(x, y + 1))
w(S) − w(R) = 8
y
x

d. Hold x constant; then $\left(\dfrac{\partial w}{\partial y}\right)_x = f_y + f_z\dfrac{dz}{dy} = 3 + 4(-2) = -5$; Hold z constant; then $\left(\dfrac{\partial w}{\partial y}\right)_z = f_x\dfrac{dx}{dy} + f_y = 2\cdot\dfrac{1}{2} + 3 = 4$; Hold x constant; then $\left(\dfrac{\partial w}{\partial z}\right)_x = f_y\dfrac{dy}{dz} + f_z = 3\left(-\dfrac{1}{2}\right) + 4 = \dfrac{5}{2}$;
Hold y constant; then $\left(\dfrac{\partial w}{\partial z}\right)_y = f_x\dfrac{dx}{dz} + f_z = 2\cdot\dfrac{1}{4} + 4 = \dfrac{9}{2}$.

12.6 Directional Derivatives and the Gradient

12.6.1 Take the dot product of the unit direction vector **u** and the gradient of the function.

12.6.3 The direction of the gradient vector is the direction in which the function is increasing the most (steepest ascent).

12.6.5 The gradient is perpendicular to the level curves.

12.6.7

a.

	$(a,\, b) = (2,\, 0)$	$(a,\, b) = (0,\, 2)$	$(a,\, b) = (1,\, 1)$
$\theta = \frac{\pi}{4}$	$-\sqrt{2}$	$-2\sqrt{2}$	$-\frac{3\sqrt{2}}{2}$
$\theta = \frac{3\pi}{4}$	$\sqrt{2}$	$-2\sqrt{2}$	$-\frac{\sqrt{2}}{2}$
$\theta = \frac{5\pi}{4}$	$\sqrt{2}$	$2\sqrt{2}$	$\frac{3\sqrt{2}}{2}$

The gradient of f is $\nabla f = \langle -x, -2y \rangle$; evaluating this at the three points gives $\nabla f(2, 0) = \langle -2, 0 \rangle$, $\nabla f(0, 2) = \langle 0, -4 \rangle$, $\nabla f(1, 1) = \langle -1, -2 \rangle$. The unit vectors with angles $\frac{\pi}{4}$, $\frac{3\pi}{4}$ and $\frac{5\pi}{4}$ are $\left\langle \frac{\sqrt{2}}{2}, \frac{\sqrt{2}}{2} \right\rangle$, $\left\langle -\frac{\sqrt{2}}{2}, \frac{\sqrt{2}}{2} \right\rangle$, $\left\langle -\frac{\sqrt{2}}{2}, -\frac{\sqrt{2}}{2} \right\rangle$ respectively; taking the dot products of these vectors with the gradients above gives the directional derivatives in the table.

b.

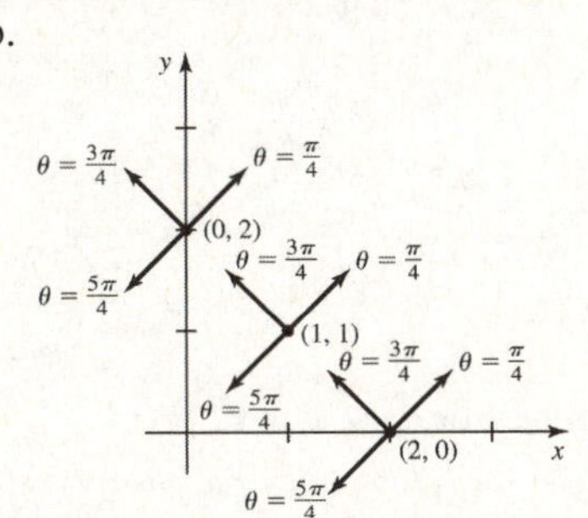

12.6.9 We have $\nabla f(x, y) = \langle 6x, -10y \rangle = 6x\mathbf{i} - 10y\mathbf{j}$; $\nabla f(2, -1) = \langle 12, 10 \rangle = 12\mathbf{i} + 10\mathbf{j}$.

12.6.11 We have $\nabla g(x, y) = \langle 2x - 8xy - 8y^2, -4x^2 - 16xy \rangle = (2x - 8xy - 8y^2)\mathbf{i} + (-4x^2 - 16xy)\mathbf{j}$; $\nabla g(-1, 2) = \langle -18, 28 \rangle = -18\mathbf{i} + 28\mathbf{j}$.

12.6.13 We have $\nabla F(x, y) = \left\langle -2xe^{-x^2-2y^2}, -4ye^{-x^2-2y^2} \right\rangle = -2xe^{-x^2-2y^2}\mathbf{i} - 4ye^{-x^2-2y^2}\mathbf{j}$; $\nabla F(-1, 2) = 2e^{-9}\langle 1, -4 \rangle = 2e^{-9}\mathbf{i} - 8e^{-9}\mathbf{j}$.

12.6.15 Note that the vector $\mathbf{u} = \left\langle \frac{\sqrt{3}}{2}, -\frac{1}{2} \right\rangle$ is a unit vector. We have

$$\nabla f(2, -3) = \langle -6x, y^3 \rangle \Big|_{(2,-3)} = \langle -12, -27 \rangle;$$

therefore, $D_u f(2, -3) = \langle -12, -27 \rangle \cdot \left\langle \frac{\sqrt{3}}{2}, -\frac{1}{2} \right\rangle = \frac{27}{2} - 6\sqrt{3}$.

12.6.17 Note that the vector $\mathbf{u} = \left\langle \frac{1}{\sqrt{5}}, \frac{2}{\sqrt{5}} \right\rangle$ is a unit vector. We have

$$\nabla f(2, -2) = \left\langle -\frac{x}{\sqrt{4 - x^2 - 2y}}, -\frac{1}{\sqrt{4 - x^2 - 2y}} \right\rangle \Bigg|_{(2,-2)} = \left\langle -1, -\frac{1}{2} \right\rangle;$$

therefore, $D_u f(2, -2) = \left\langle -1, -\frac{1}{2} \right\rangle \cdot \left\langle \frac{1}{\sqrt{5}}, \frac{2}{\sqrt{5}} \right\rangle = -\frac{2}{\sqrt{5}}$.

12.6.19 The unit vector in the direction of $\langle 2, 1 \rangle$, is $\mathbf{u} = \left\langle \frac{2}{\sqrt{5}}, \frac{1}{\sqrt{5}} \right\rangle$. We have

$$\nabla P(-1, 2) = \left\langle \frac{2x}{4 + x^2 + y^2}, \frac{2y}{4 + x^2 + y^2} \right\rangle \Bigg|_{(-1,2)} = \left\langle -\frac{2}{9}, \frac{4}{9} \right\rangle;$$

therefore, $D_u P(-1, 2) = \left\langle -\frac{2}{9}, \frac{4}{9} \right\rangle \cdot \left\langle \frac{2}{\sqrt{5}}, \frac{1}{\sqrt{5}} \right\rangle = 0$.

12.6.21

a. At the point $(1, -2)$ the value of the gradient is $\nabla f(1, -2) = \langle 2x, -8y \rangle \big|_{(1,-2)} = \langle 2, 16 \rangle$; therefore, the direction of steepest ascent is $\mathbf{u} = \frac{1}{\sqrt{65}}\langle 1, 8 \rangle$ and the direction of steepest descent is $-\mathbf{u}$.

b. Take any vector perpendicular to **u**; for example, $\mathbf{v} = \langle -8,\ 1\rangle$.

12.6.23

a. At the point $(-1,\ 1)$ the value of the gradient is $\nabla f(-1,\ 1) = \langle 4x^3 - 2xy,\ -x^2 + 2y\rangle\Big|_{(-1,1)} = \langle -2,\ 1\rangle$;

therefore, the direction of steepest ascent is $\mathbf{u} = \dfrac{1}{\sqrt{5}}\langle -2,\ 1\rangle$ and the direction of steepest descent is $-\mathbf{u}$.

b. Take any vector perpendicular to **u**; for example, $\mathbf{v} = \langle 1,\ 2\rangle$.

12.6.25

a. At the point $(-1,\ 1)$ the value of the gradient is

$$\nabla f(-1,\ 1) = \left\langle -xe^{-x^2/2-y^2/2},\ -ye^{-x^2/2-y^2/2}\right\rangle\Big|_{(-1,1)} = e^{-1}\langle 1,\ -1\rangle;$$

therefore, the direction of steepest ascent is $\mathbf{u} = \dfrac{1}{\sqrt{2}}\langle 1,\ -1\rangle$ and the direction of steepest descent is $-\mathbf{u}$.

b. Take any vector perpendicular to **u**; for example, $\mathbf{v} = \langle 1,\ 1\rangle$.

12.6.27

a. The gradient of f at P is $\nabla f(3,\ 2) = \langle -4x,\ -6y\rangle\Big|_{(3,2)} = \langle -12,\ -12\rangle$.

b. The direction of steepest ascent is $\mathbf{u} = \dfrac{1}{\sqrt{2}}\langle -1,\ -1\rangle$ which makes angle $\theta = \dfrac{5\pi}{4}$ with the x-axis; therefore, the angle of maximum decrease is $\theta = \dfrac{\pi}{4}$ and the angles of zero change are $\dfrac{3\pi}{4}$ and $\dfrac{7\pi}{4}$.

c. We have $g(\theta) = \langle -12,\ -12\rangle \cdot \langle \cos\theta,\ \sin\theta\rangle = -12\cos\theta - 12\sin\theta$.

d. The critical points for $g(\theta)$ satisfy $g'(\theta) = 12(\sin\theta - \cos\theta) = 0$, which gives $\theta = \dfrac{\pi}{4}, \dfrac{5\pi}{4}$. By inspection we see that the maximum occurs at $\dfrac{5\pi}{4}$, and we have $g\left(\dfrac{5\pi}{4}\right) = 12\sqrt{2}$.

e. Observe that the maximum value of $g(\theta)$ occurs at the angle found in part (a), and that

$$|\nabla f(3,\ 2)| = 12\sqrt{2} = g\left(\frac{5\pi}{4}\right).$$

12.6.29

a. The gradient of f at P is

$$\nabla f\left(\sqrt{3},\ 1\right) = \left\langle \frac{x}{\sqrt{2+x^2+y^2}},\ \frac{y}{\sqrt{2+x^2+y^2}}\right\rangle\Bigg|_{(\sqrt{3},1)} = \left\langle \frac{\sqrt{3}}{\sqrt{6}},\ \frac{1}{\sqrt{6}}\right\rangle = \frac{\sqrt{6}}{6}\left\langle \sqrt{3},\ 1\right\rangle.$$

b. The direction of steepest ascent is $\mathbf{u} = \left\langle \dfrac{\sqrt{3}}{2},\ \dfrac{1}{2}\right\rangle$ which makes angle $\theta = \dfrac{\pi}{6}$ with the x-axis; therefore, the angle of maximum decrease is $\theta = \dfrac{7\pi}{6}$ and the angles of zero change are $\dfrac{2\pi}{3}$ and $\dfrac{5\pi}{3}$.

c. We have $g(\theta) = \dfrac{\sqrt{6}}{6}\left\langle \sqrt{3},\ 1\right\rangle \cdot \langle \cos\theta,\ \sin\theta\rangle = \dfrac{\sqrt{18}}{6}\cos\theta + \dfrac{\sqrt{6}}{6}\sin\theta$.

d. The critical points for $g(\theta)$ satisfy $g'(\theta) = \dfrac{\sqrt{6}}{6}\left(-\sqrt{3}\sin\theta + \cos\theta\right) = 0$, which gives $\tan\theta = \dfrac{1}{\sqrt{3}}$, so $\theta = \dfrac{\pi}{6}, \dfrac{7\pi}{6}$. By inspection we see that the maximum occurs at $\dfrac{\pi}{6}$, and we have $g\left(\dfrac{\pi}{6}\right) = \dfrac{\sqrt{6}}{3}$.

e. Observe that the maximum value of $g(\theta)$ occurs at the angle found in part (a), and that

$$|\nabla f(3,\ 1)| = \frac{\sqrt{6}}{3} = g\left(\frac{\pi}{6}\right).$$

12.6.31

a. The gradient of f at P is $\nabla f(-1, 0) = \left\langle -2xe^{-x^2-2y^2}, -4ye^{-x^2-2y^2}\right\rangle\Big|_{(-1,0)} = \left\langle 2e^{-1}, 0\right\rangle$.

b. The direction of steepest ascent is $\mathbf{u} = \langle 1, 0\rangle$ which makes angle $\theta = 0$ with the x-axis; therefore the angle of maximum decrease is $\theta = \pi$ and the angles of zero change are $\pm\dfrac{\pi}{2}$.

c. We have $g(\theta) = \dfrac{2}{\text{e}}\langle 1, 0\rangle \cdot \langle \cos\theta, \sin\theta\rangle = \dfrac{2}{\text{e}}\cos\theta$.

d. The maximum value of $g(\theta)$ occurs at $\theta = 0$, and we have $g(0) = \dfrac{2}{\text{e}}$.

e. Observe that the maximum value of $g(\theta)$ occurs at the angle found in part (a), and that

$$|\nabla f(-1, 0)| = \frac{2}{\text{e}} = g(0).$$

12.6.33 The gradient of f at P is $\nabla f(2, -4) = \langle 8x, 4y\rangle\big|_{(2,-4)} = \langle 16, -16\rangle$, which gives the direction of maximum increase.

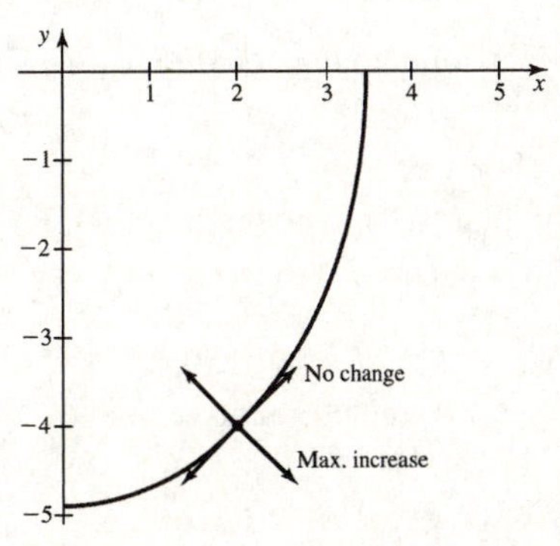

12.6.35 The gradient of f at P is $\nabla f(-3, 3) = \langle 2x + y, x + 2y\rangle\big|_{(-3,3)} = \langle -3, 3\rangle$, which gives the direction of maximum increase.

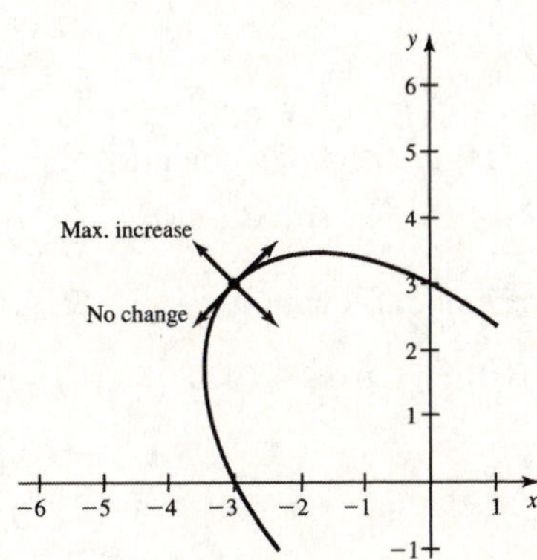

12.6.37 The slope of the level curves for $f(x, y)$ is given by $y'(x) = -\dfrac{f_x}{f_y} = -\dfrac{4x}{y}$, so the tangent line has slope 0 at (0, 16). The gradient of f at this point is $\nabla f(0, 16) = \left\langle -\frac{x}{2}, -\frac{y}{8}\right\rangle\Big|_{(0,16)} = \langle 0, -2\rangle$, which is perpendicular to the tangent line.

12.6.39 The slope of the level curves for $f(x, y)$ is given by $y'(x) = -\dfrac{f_x}{f_y} = -\dfrac{4x}{y}$, which is undefined at (4, 0), so the tangent line is vertical at (4, 0). The gradient of f at this point is $\nabla f(4, 0) = \left\langle -\frac{x}{2}, -\frac{y}{8}\right\rangle\Big|_{(4,0)} = \langle -2, 0\rangle$, which is perpendicular to the tangent line.

12.6.41 Let $z = f(x, y)$; then $z^2 = 1 - \dfrac{x^2}{4} - \dfrac{y^2}{16}$, so

$$2zz_x = -\frac{x}{2} \implies z_x = f_x = -\frac{x}{4z} \text{ and } 2zz_y = -\frac{y}{8} \implies z_y = f_y = -\frac{y}{16z}.$$

The slope of the level curves for $f(x, y)$ is given by $y'(x) = -\frac{f_x}{f_y} = -\frac{4x}{y}$, so the tangent line has slope $-\frac{2}{\sqrt{3}}$ at $\left(\frac{1}{2}, \sqrt{3}\right)$, and its direction is parallel to the vector $\left\langle 1, -\frac{2}{\sqrt{3}}\right\rangle$. The gradient of f at this point is

$$\nabla f\left(\frac{1}{2}, \sqrt{3}\right) = \left\langle -\frac{x}{4z}, -\frac{y}{16z}\right\rangle\Big|_{(1/2, \sqrt{3})} = -\frac{1}{8}\left\langle \frac{2}{\sqrt{3}}, 1\right\rangle,$$

which is perpendicular to the tangent direction.

12.6.43 Let $z = f(x, y)$; then $z^2 = 1 - \frac{x^2}{4} - \frac{y^2}{16}$, so

$$2zz_x = -\frac{x}{2} \Longrightarrow z_x = f_x = -\frac{x}{4z} \text{ and } 2zz_y = -\frac{y}{8} \Longrightarrow z_y = f_y = -\frac{y}{16z}.$$

The slope of the level curves for $f(x, y)$ is given by $y'(x) = -\frac{f_x}{f_y} = -\frac{4x}{y}$, so the tangent line is vertical at $\left(\sqrt{2}, 0\right)$, and its direction is parallel to the vector $\langle 0, 1\rangle$. The gradient of f at this point is

$$\nabla f\left(\sqrt{2}, 0\right) = \left\langle -\frac{x}{4z}, -\frac{y}{16z}\right\rangle\Big|_{(\sqrt{2}, 0)} = \left\langle -\frac{1}{2}, 0\right\rangle,$$

which is perpendicular to the tangent direction.

12.6.45

a. We have $\nabla f = \langle f_x, f_y\rangle = \langle 1, 0\rangle$.

b. Let $(x(t), y(t))$ be the projection into the xy-plane of the path of steepest descent starting at $(x(0), y(0)) = (4, 4)$. We solve $(x'(t), y'(t)) = -\nabla f = \langle -1, 0\rangle$ which together with the initial conditions gives $x = 4 - t$, $y = 4$ for $t \geq 0$.

12.6.47

a. We have $\nabla f = \langle f_x, f_y\rangle = \langle -2x, -4y\rangle$.

b. Let $(x(t), y(t))$ be the projection into the xy-plane of the path of steepest descent starting at $(x(0), y(0)) = \left(\frac{\pi}{2}, 1\right)$. We solve $(x'(t), y'(t)) = -\nabla f = \langle 2x, 4y\rangle$ with the initial conditions $x(0) = 1$ and $y(0) = 1$. The differential equation $\frac{dx}{dt} = -2x$ is separable: we have $\int \frac{1}{x} dx = \int 2\, dt \Longrightarrow \ln|x| = 2t + C$; the initlal condition $x(0) = 1$ gives $C = 0$ so $x(t) = e^{2t}$. Similartly, $y(t) = e^{4t}$. So $y = x^2$ for $x \geq 1$.

12.6.49

a. We have $\nabla f = 2x\mathbf{i} + 4y\mathbf{j} + 8z\mathbf{k}$; $\nabla f(1, 0, 4) = 2\mathbf{i} + 32\mathbf{k}$.

b. The vector $2\mathbf{i} + 32\mathbf{k}$ has length $2\sqrt{257}$, so the unit vector in this direction is $\mathbf{u} = \frac{1}{\sqrt{257}}(\mathbf{i} + 16\mathbf{k})$.

c. The rate of change of f in the direction of maximum increase at P is $||\nabla f(1, 0, 4)|| = 2\sqrt{257}$.

d. The vector $\mathbf{u} = \left\langle \frac{1}{\sqrt{2}}, 0, \frac{1}{\sqrt{2}}\right\rangle$ is a unit vector, so the directional derivative at P in this direction is

$$D_{\mathbf{u}}f(1, 0, 4) = \langle 2, 0, 32\rangle \cdot \left\langle \frac{1}{\sqrt{2}}, 0, \frac{1}{\sqrt{2}}\right\rangle = \frac{34}{\sqrt{2}} = 17\sqrt{2}.$$

12.6.51

a. We have $\nabla f = 4yz\mathbf{i} + 4xz\mathbf{j} + 4xy\mathbf{k}$; $\nabla f(1, -1, -1) = 4\mathbf{i} - 4\mathbf{j} - 4\mathbf{k}$.

b. The vector $4\mathbf{i} - 4\mathbf{j} - 4\mathbf{k}$ has length $4\sqrt{3}$, so the unit vector in this direction is $\mathbf{u} = \frac{1}{\sqrt{3}}(\mathbf{i} - \mathbf{j} - \mathbf{k})$.

c. The rate of change of f in the direction of maximum increase at P is $||\nabla f(1, -1, -1)|| = 4\sqrt{3}$.

d. The vector $\mathbf{u} = \left\langle \frac{1}{\sqrt{3}}, \frac{1}{\sqrt{3}}, -\frac{1}{\sqrt{3}} \right\rangle$ is a unit vector, so the directional derivative at P in this direction is

$$D_u f(1, -1, -1) = \langle 4, -4, -4 \rangle \cdot \left\langle \frac{1}{\sqrt{3}}, \frac{1}{\sqrt{3}}, -\frac{1}{\sqrt{3}} \right\rangle = \frac{4}{\sqrt{3}}.$$

12.6.53

a. We have $\nabla f = \cos(x + 2y - z)(\mathbf{i} + 2\mathbf{j} - \mathbf{k})$; $\nabla f\left(\frac{\pi}{6}, \frac{\pi}{6}, -\frac{\pi}{6}\right) = \cos\left(\frac{2\pi}{3}\right)(\mathbf{i} + 2\mathbf{j} - \mathbf{k}) = -\frac{1}{2}\mathbf{i} - \mathbf{j} + \frac{1}{2}\mathbf{k}$.

b. The vector $-\frac{1}{2}\mathbf{i} - \mathbf{j} + \frac{1}{2}\mathbf{k}$ has length $\frac{\sqrt{3}}{\sqrt{2}} = \frac{\sqrt{6}}{2}$, so the unit vector in this direction is $\mathbf{u} = \frac{1}{\sqrt{6}}(-\mathbf{i} - 2\mathbf{j} + \mathbf{k})$.

c. The rate of change of f in the direction of maximum increase at P is $\left|\nabla f\left(\frac{\pi}{6}, \frac{\pi}{6}, -\frac{\pi}{6}\right)\right| = \frac{\sqrt{6}}{2}$.

d. The vector $\mathbf{u} = \left\langle \frac{1}{3}, \frac{2}{3}, \frac{2}{3} \right\rangle$ is a unit vector, so the directional derivative at P in this direction is

$$D_u f\left(\frac{\pi}{6}, \frac{\pi}{6}, -\frac{\pi}{6}\right) = \left\langle \frac{1}{3}, \frac{2}{3}, \frac{2}{3} \right\rangle \cdot \left\langle -\frac{1}{2}, -1, \frac{1}{2} \right\rangle = -\frac{1}{2}.$$

12.6.55

a. We have $\nabla f = \frac{2}{1 + x^2 + y^2 + z^2}(x\mathbf{i} + y\mathbf{j} + z\mathbf{k})$; $\nabla f(1, 1, -1) = \frac{1}{2}\mathbf{i} + \frac{1}{2}\mathbf{j} - \frac{1}{2}\mathbf{k}$.

b. The vector $\frac{1}{2}\mathbf{i} + \frac{1}{2}\mathbf{j} - \frac{1}{2}\mathbf{k}$ has length $\frac{\sqrt{3}}{2}$, so the unit vector in this direction is $\mathbf{u} = \frac{1}{\sqrt{3}}(\mathbf{i} + \mathbf{j} - \mathbf{k})$.

c. The rate of change of f in the direction of maximum increase at P is $||\nabla f(1, 1, -1)|| = \frac{\sqrt{3}}{2}$.

d. The vector $\mathbf{u} = \left\langle \frac{2}{3}, \frac{2}{3}, -\frac{1}{3} \right\rangle$ is a unit vector, so the directional derivative at P in this direction is

$$D_u f(1, 1, -1) = \left\langle \frac{2}{3}, \frac{2}{3}, -\frac{1}{3} \right\rangle \cdot \left\langle \frac{1}{2}, \frac{1}{2}, -\frac{1}{2} \right\rangle = \frac{5}{6}.$$

12.6.57

a. False; $\nabla f = 2x\mathbf{i} + 2y\mathbf{j}$.

b. False; the gradient is a vector, so it does not make sense to say that it is positive.

c. False; f is a function of three variables, so ∇f has three components.

d. True, because $f_x = f_y = f_z = 0$.

12.6.59 Observe that $\nabla f(x, y) = -8x\mathbf{i} - 2y\mathbf{j} \Rightarrow \nabla f(1, 2) = -8\mathbf{i} - 4\mathbf{j}$. A vector perpendicular to $\nabla f(1, 2)$ is $\mathbf{v} = \mathbf{i} - 2\mathbf{j}$, so the unit vectors perpendicular to $\nabla f(1, 2)$ are $\mathbf{u} = \pm\frac{1}{\sqrt{5}}(\mathbf{i} - 2\mathbf{j})$.

12.6.61 Observe that $\nabla f(x, y) = \frac{1}{\sqrt{3 + 2x^2 + y^2}}(2x\mathbf{i} + y\mathbf{j}) \Rightarrow \nabla f(1, -2) = \frac{2}{3}(\mathbf{i} - \mathbf{j})$. A vector perpendicular to $\nabla f(1, -2)$ is $\mathbf{v} = \mathbf{i} + \mathbf{j}$, so the unit vectors perpendicular to $\nabla f(1, -2)$ are $\mathbf{u} = \pm\frac{1}{\sqrt{2}}(\mathbf{i} + \mathbf{j})$.

12.6.63 The function $f(x, y = ax + by + c)$ has $\nabla f(x, y) = a\mathbf{i} + b\mathbf{j}$, so the path in the xy-plane corresponding to the path of steepest ascent on the plane is given by $x = x_0 + at$, $y = y_0 + bt$.

12.6.65

a. We have $\nabla f(x, y, z) = \langle 2x, 2y, 2z \rangle \Rightarrow \nabla f(1, 1, 1) = \langle 2, 2, 2 \rangle$.

b. Points (x, y, z) on the plane P satisfy $\langle x - 1, y - 1, z - 1 \rangle \cdot \langle 2, 2, 2 \rangle = 0 \Rightarrow x + y + z = 3$ is an equation for P.

12.6.67

a. We have $\nabla f(x, y, z) = e^{x+y-z}\langle 1, 1, -1\rangle \Rightarrow \nabla f(1, 1, 2) = \langle 1, 1, -1\rangle$.

b. Points (x, y, z) on the plane P satisfy $\langle x-1, y-1, z-2\rangle \cdot \langle 1, 1, -1\rangle = 0 \Rightarrow x + y - z = 0$ is an equation for P.

12.6.69

a.

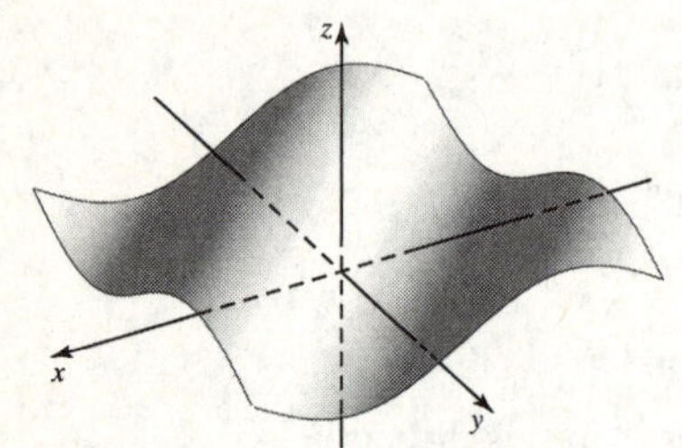

b. The height function has gradient $\nabla z = \cos(x-y)\langle 1, -1\rangle$, so the directions in which the height function has zero change are $\mathbf{v} = \pm\langle 1, 1\rangle$.

c. The direction $\mathbf{v}$ would be the opposite of the direction of ∇z, so $\mathbf{v} = \pm\langle 1, -1\rangle$.

d. These answers are consistent with the graph in part (a).

12.6.71

a. Observe that $V_x = -\dfrac{kQ}{r^2} r_x = -\dfrac{kQ}{r^2} \dfrac{x}{\sqrt{x^2+y^2+z^2}} = -\dfrac{kQx}{r^3}$; similarly $V_y = -\dfrac{kQy}{r^3}$ and $V_z = -\dfrac{kQz}{r^3}$.

Therefore $\mathbf{E}(x, y, z) = -\nabla V(x, y, z) = kQ\left\langle \dfrac{x}{r^3}, \dfrac{y}{r^3}, \dfrac{z}{r^3}\right\rangle$.

b. We have $|\mathbf{E}| = \dfrac{kQ}{r^3}|\langle x, y, z\rangle| = \dfrac{kQ}{r^2}$. Therefore the magnitude of the electric field is inversely proportional to the square of the distance to the point charge.

12.6.73 We have $\langle u, v\rangle = \nabla\varphi = \langle \pi\cos\pi x \sin 2\pi y, 2\pi \sin\pi x \cos 2\pi y\rangle$.

12.6.75 We give the proofs for functions on $\mathbf{R}^2$; the proofs for functions on $\mathbf{R}^3$ are similar.

a. $\nabla(cf) = \langle (cf)_x, (cf)_y\rangle = \langle cf_x, cf_y\rangle = c\nabla f$.

b. $\nabla(f+g) = \langle (f+g)_x, (f+g)_y\rangle = \langle f_x + g_x, f_y + g_y\rangle = \nabla f + \nabla g$.

c. $\nabla(fg) = \langle (fg)_x, (fg)_y\rangle = \langle f_x g + f g_x, f_y g + f g_y\rangle = (\nabla f)g + f\nabla g$.

d. $\nabla\left(\dfrac{f}{g}\right) = \left\langle \left(\dfrac{f}{g}\right)_x, \left(\dfrac{f}{g}\right)_y\right\rangle = \left\langle \dfrac{g f_x - f g_x}{g^2}, \dfrac{g f_y - f g_y}{g^2}\right\rangle = \dfrac{g(\nabla f) - f\nabla g}{g^2}$.

e. $\nabla(f\circ g) = \langle f'(g) g_x, f'(g) g_y\rangle = f'(g)\nabla g$.

12.6.77 Using the quotient rule from Exercise 75 gives

$$\begin{aligned}\nabla\left(\frac{x+y}{x^2+y^2}\right) &= \frac{(x^2+y^2)\nabla(x+y) - (x+y)\nabla(x^2+y^2)}{(x^2+y^2)^2}\\ &= \frac{1}{(x^2+y^2)^2}\left((x^2+y^2)\langle 1, 1\rangle - (x+y)\langle 2x, 2y\rangle\right)\\ &= \frac{1}{(x^2+y^2)^2}\langle y^2 - x^2 - 2xy, x^2 - y^2 - 2xy\rangle.\end{aligned}$$

12.6.79 Using the chain rule from Exercise 75 gives

$$\begin{aligned}\nabla\left(\sqrt{25-x^2-y^2-z^2}\right) &= \frac{1}{2\sqrt{25-x^2-y^2-z^2}}\nabla(25-x^2-y^2-z^2)\\ &= \frac{1}{2\sqrt{25-x^2-y^2-z^2}}\langle -2x,\, -2y,\, -2z\rangle\\ &= -\frac{1}{\sqrt{25-x^2-y^2-z^2}}\langle x,\, y,\, z\rangle.\end{aligned}$$

12.6.81 Using the quotient rule from Exercise 75 gives

$$\begin{aligned}\nabla\left(\frac{x+yz}{y+xz}\right) &= \frac{(y+xz)\nabla(x+yz)-(x+yz)\nabla(y+xz)}{(y+xz)^2}\\ &= \frac{1}{(y+xz)^2}\left((y+xz)\langle 1,\, z,\, y\rangle-(x+yz)\langle z,\, 1,\, x\rangle\right)\\ &= \frac{1}{(y+xz)^2}\langle y(1-z^2),\, x(z^2-1),\, y^2-x^2\rangle.\end{aligned}$$

12.7 Tangent Planes and Linear Approximations

12.7.1 The gradient of F is a multiple of n.

12.7.3 The tangent plane has equation $F_x(a,\, b,\, c)(x-a)+F_y(a,\, b,\, c)(y-b)+F_z(a,\, b,\, c)(z-c)=0$.

12.7.5 Multiply the change in x by f_x and the change in y by f_y and add both terms to $f(a,\, b)$.

12.7.7 In terms of differentials, $dz = f(a,\, b)dx + f_y(a,\, b)dy$.

12.7.9 We have $\nabla F = \langle y+z,\, x+z,\, x+y\rangle$, so $\nabla F(2,\, 2,\, 2) = \langle 4,\, 4,\, 4\rangle$ and the tangent plane at (2, 2, 2) has equation $4(x-2)+4(y-2)+4(z-2)=0$, or $x+y+z=6$. We also have

$$\nabla F\left(-1,\, -2,\, -\frac{10}{3}\right) = \left\langle -\frac{16}{3},\, -\frac{13}{3},\, -3\right\rangle = -\frac{1}{3}\langle 16,\, 13,\, 9\rangle$$

so the tangent plane at $\left(-1,\, -2,\, -\frac{10}{3}\right)$ has equation

$$16(x+1)+13(y+2)+9\left(z+\frac{10}{3}\right)=0$$

or $16x+13y+9z+72=0$.

12.7.11 We have $\nabla F = \langle y\sin z,\, x\sin z,\, yz\cos z\rangle$, so $\nabla F\left(1,\, 2,\, \frac{\pi}{6}\right) = \left\langle 1,\, \frac{1}{2},\, \sqrt{3}\right\rangle$ and the tangent plane at $\left(1,\, 2,\, \frac{\pi}{6}\right)$ has equation $(x-1)+\frac{1}{2}(y-2)+\sqrt{3}\left(z-\frac{\pi}{6}\right)=0$, or $x+\frac{1}{2}y+\sqrt{3}z = 2+\frac{\sqrt{3}\,\pi}{6}$. We also have

$$\nabla F\left(-2,\, -1,\, \frac{5\pi}{6}\right) = \left\langle -\frac{1}{2},\, -1,\, -\sqrt{3}\right\rangle$$

so the tangent plane at $\left(-2,\, -1,\, \frac{5\pi}{6}\right)$ has equation

$$-\frac{1}{2}(x+2)-(y+1)-\sqrt{3}\left(z-\frac{5\pi}{6}\right)=0$$

or $\frac{1}{2}x+y+\sqrt{3}\,z = \frac{5\sqrt{3}\,\pi}{6}-2$.

12.7.13 We have $\nabla F = \left\langle -\frac{x}{8}, -\frac{2y}{9}, 2z \right\rangle$, so $\nabla F\left(4, 3, -\sqrt{3}\right) = \left\langle -\frac{1}{2}, -\frac{2}{3}, -2\sqrt{3} \right\rangle$ and the tangent plane at $\left(4, 3, -\sqrt{3}\right)$ has equation $-\frac{1}{2}(x-4) - \frac{2}{3}(y-3) - 2\sqrt{3}\left(z+\sqrt{3}\right) = 0$, or $\frac{1}{2}x + \frac{2}{3}y + 2\sqrt{3}\,z = -2$. We also have

$$\nabla F\left(-8, 9, \sqrt{14}\right) = \left\langle 1, -2, 2\sqrt{14} \right\rangle$$

so the tangent plane at $\left(-8, 9, \sqrt{14}\right)$ has equation

$$(x+8) - 2(y-9) + 2\sqrt{14}\left(z - \sqrt{14}\right) = 0$$

or $x - 2y + 2\sqrt{14}z = 2$.

12.7.15 We have $f_x = -4x$, $f_y = -2y$, so the tangent plane at $(2, 2, -8)$ has equation

$$z = f(2, 2) + f_x(2, 2)(x-2) + f_y(2, 2)(y-2) = -8 - 8(x-2) - 4(y-2),$$

or $z = -8x - 4y + 16$, and the equation of the tangent plane at $(-1, -1, 1)$ is

$$z = f(-1, -1) + f_x(-1, -1)(x+1) + f_y(-1, -1)(y+1) = 1 + 4(x+1) + 2(y+1),$$

or $z = 4x + 2y + 7$.

12.7.17 We have $f_x = \left(2x + x^2\right)e^{x-y}$, $f_y = -x^2e^{x-y}$, so the tangent plane at $(2, 2, 4)$ has equation

$$z = f(2, 2) + f_x(2, 2)(x-2) + f_y(2, 2)(y-2) = 4 + 8(x-2) - 4(y-2),$$

or $z = 8x - 4y - 4$, and the equation of the tangent plane at $(-1, -1, 1)$ is

$$z = f(-1, -1) + f_x(-1, -1)(x+1) + f_y(-1, -1)(y+1) = 1 - (x+1) - (y+1),$$

or $z = -x - y - 1$.

12.7.19 We have $f_x = \frac{y^2 + 2xy - x^2}{\left(x^2+y^2\right)^2}$, $f_y = \frac{y^2 - 2xy - x^2}{\left(x^2+y^2\right)^2}$, so the tangent plane at $\left(1, 2, -\frac{1}{5}\right)$ has equation

$$z = f(1, 2) + f_x(1, 2)(x-1) + f_y(1, 2)(y-2) = -\frac{1}{5} + \frac{7}{25}(x-1) - \frac{1}{25}(y-2),$$

or $z = \frac{7}{25}x - \frac{1}{25}y - \frac{2}{5}$, and the equation of the tangent plane at $\left(2, -1, \frac{3}{5}\right)$ is

$$z = f(2, -1) + f_x(2, -1)(x-2) + f_y(2, -1)(y+1) = \frac{3}{5} - \frac{7}{25}(x-2) + \frac{1}{25}(y+1),$$

or $z = -\frac{7}{25}x + \frac{1}{25}y + \frac{6}{5}$.

12.7.21

a. We have $f_x = y + 1$, $f_y = x - 1$, so the linear approximation for f at $(2, 3)$ is

$$L(x, y) = f(2, 3) + f_x(2, 3)(x-2) + f_y(2, 3)(y-3) = 5 + 4(x-2) + (y-3),$$

or $L(x, y) = 4x + y - 6$.

b. We have $L(2.1, 2.99) = 5.39$.

12.7.23

a. We have $f_x = -2x$, $f_y = 4y$, so the linear approximation for f at $(3, -1)$ is

$$L(x, y) = f(3, -1) + f_x(3, -1)(x-3) + f_y(3, -1)(y+1) = -7 - 6(x-3) - 4(y+1)$$

or $L(x, y) = -6x - 4y + 7$.

b. We have $L(3.1, -1.04) = -7.44$.

12.7.25

a. We have $f_x = \frac{1}{1+x+y}$, $f_y = \frac{1}{1+x+y}$, so the linear approximation for f at $(0, 0)$ is

$$L(x, y) = f(0, 0) + f_x(0, 0)x + f_y(0, 0)y = x + y.$$

b. We have $L(0.1, -0.2) = -0.1$.

12.7.27 We have $f_x = 2 - 2y$, $f_y = -3 - 2x$, and $dx = 0.1$, $dy = -0.1$, so

$$dz = f_x(1, 4)dx + f_y(1, 4)dy = -6dx - 5dy = -0.1.$$

12.7.29 We have $f_x = e^{x+y}$, $f_y = e^{x+y}$, and $dx = 0.1$, $dy = -0.05$, so

$$dz = f_x(0, 0)dx + f_y(0, 0)dy = dx + dy = 0.05.$$

12.7.31

a. If R increases and r decreases then $R^2 - r^2$ increases, so S increases.
b. If both R and r increase, then it is impossible to say whether $R^2 - r^2$ increases or decreases.
c. We have $dS = 8\pi^2(R\,dR - r\,dr) = 8\pi^2(5.50 \cdot 0.15 - 3 \cdot 0.05) = 5.4\pi^2 \approx 53.296$.
d. We have $dS = 8\pi^2(R\,dR - r\,dr) = 8\pi^2(7 \cdot 0.04 - 3 \cdot (-0.05)) = 3.44\pi^2 \approx 33.951$.
e. The surface area is approximately unchanged when $R\,dR = r\,dr$.

12.7.33 Observe that $dA = \pi(b\,da + a\,db)$ so $\dfrac{dA}{A} = \dfrac{da}{a} + \dfrac{db}{b}$, and hence the percentage increase in the area is approximately 2% + 1.5% = 3.5%.

12.7.35 We have $dw = (y^2 + 2x)dx + (2xy + z^2)dy + (x^2 + 2yz)dz$.

12.7.37 We have $dw = \dfrac{dx}{y+z} - \dfrac{u+x}{(y+z)^2}dy - \dfrac{u+x}{(y+z)^2}dz + \dfrac{du}{y+z}$.

12.7.39

a. Observe that $2c\,dc = (2a - 2b\cos\theta)da + (2b - 2a\cos\theta)db + 2ab\sin\theta\,d\theta$ so

$$dc = \frac{a - b\cos\theta}{c}da + \frac{b - a\cos\theta}{c}db + \frac{ab\sin\theta}{c}d\theta.$$

We have $a = 2$, $b = 4$, $\theta = \dfrac{\pi}{3}$ which gives $c = \sqrt{12}$, and $da = 0.03$, $db = -0.04$, $d\theta = \dfrac{\pi}{90}$; substituting in the equation above gives $dc \approx 0.035$.

b. We have $a = 2$, $b = 4$, $da = 0.03$, $db = -0.04$, $d\theta = 0$, so $dc = -\dfrac{0.01 + 0.04\cos\theta}{c}$; comparing the cases $\theta = \dfrac{\pi}{20}$ and $\theta = \dfrac{9\pi}{20}$ we see that c is smaller and $\cos\theta$ is larger in the first case; therefore the change in c is greater when $\theta = \dfrac{\pi}{20}$.

12.7.41

a. True, because the function $F(x, y, z) = x^2 + z^2$ has $F_y = 0$.
a. True; as $z > 0$ decreases $\dfrac{1}{z}$ increases.
a. False; the gradient $\nabla F(a, b, c)$ is perpendicular to the tangent plane for the surface $F(x, y, z) = 0$ at (a, b, c).

12.7.43 Let $f(x, y) = \tan^{-1}(xy)$; then $f_x(x, y) = \dfrac{y}{1 + (xy)^2}$, $f_y(x, y) = \dfrac{x}{1 + (xy)^2}$ and the tangent plane at $\left(1, 1, \dfrac{\pi}{4}\right)$ has equation $z = f_x(1,1)(x - 1) + f_y(1,1)(y - 1) + f(1,1) = \dfrac{1}{2}x + \dfrac{1}{2}y + \dfrac{\pi}{4} - 1$.

12.7.45 The branch of the surface $\sin(xyz) = \dfrac{1}{2}$ at the point $\left(\pi, 1, \dfrac{1}{6}\right)$ can be described more simply by the equation $F(x, y, z) = xyz = \dfrac{\pi}{6}$. We have $F_x(x, y, z) = yz$, $F_y(x, y, z) = xz$, $Fz(x, y, z) = xy$ so the tangent plane at $\left(\pi, 1, \dfrac{1}{6}\right)$ has equation $F_x\left(\pi, 1, \dfrac{1}{6}\right)(x - \pi) + F_y\left(\pi, 1, \dfrac{1}{6}\right)(y - 1) + Fz\left(\pi, 1, \dfrac{1}{6}\right)\left(z - \dfrac{1}{6}\right) = 0$, or $\dfrac{1}{6}(x - \pi) + \dfrac{\pi}{6}(y - 1) + \pi\left(z - \dfrac{1}{6}\right) = 0$.

12.7.47 Let $F(x, y, z) = x^2 + 2y^2 + z^2 - 2x - 2y + 3$; then $F_x(x, y, z) = 2x - 2$, $F_y(x, y, z) = 2y + 2$. The tangent plane to the surface $F(x, y, z) = 0$ at (a, b, c) is horizontal if and only if $F_x(a, b, c) = F_y(a, b, c) = 0$, which gives $a = 1$, $b = -1$. This implies $c^2 = 1$, so the points are $(1, -1, 1)$ and $(1, -1, -1)$.

12.7.49 Let $f(x, y) = \cos 2x \sin y$; then $f_x(x, y) = -2 \sin 2x \sin y$, $f_y(x, y) = \cos 2x \cos y$; so the tangent plane at $(a, b, \cos 2a \sin b)$ is horizontal at all points where $\sin 2a = \cos b = 0$ or $\sin b = \cos 2a = 0$. In the region $-\pi \le x \le \pi$, $-\pi \le y \le \pi$ the points are $a = 0, \pm\frac{\pi}{2}, \pm\pi$ and $y = \pm\frac{\pi}{2}$, or $a = \pm\frac{\pi}{4}, \pm\frac{3\pi}{4}$ and $y = 0, \pm\pi$.

12.7.51

a. We have $S_r = \pi\sqrt{r^2 + h^2} + \dfrac{\pi r^2}{\sqrt{r^2 + h^2}}$, $S_h = \dfrac{\pi rh}{\sqrt{r^2 + h^2}}$; using the values $r = 2.5$, $h = 0.6$, $dr = 0.05$, $dh = -0.02$ gives $dS = S_r dr + S_h dh = \dfrac{\pi}{\sqrt{r^2 + h^2}}\left((2r^2 + h^2)dr + rh\,dh\right) \approx 0.749$.

b. If $r = 100$, $h = 200$ then $dS = 40\sqrt{5}\pi(3dr + dh)$, so the surface area is more sensitive to small changes in r.

12.7.53

a. The differential of A is given by $dA = \dfrac{dx}{y} - \dfrac{x\,dy}{y^2}$; substituting $x = 60$, $y = 175$, $dx = 2$ and $dy = 5$ gives $dA = \dfrac{2}{1225} \approx 0.00163$.

b. If the batter fails to get a hit, the average decreases by $\dfrac{x}{y} - \dfrac{x}{y+1} = \dfrac{x}{y(y+1)} = \dfrac{A}{y+1}$, whereas if the batter gets a hit, the average increases by $\dfrac{x+1}{y+1} - \dfrac{x}{y} = \dfrac{y-x}{y(y+1)} = \dfrac{1-A}{y+1}$. If $A = 0.350$ the second of these quantities is larger so the answer is no; the batting average changes more if the batter gets a hit than if he fails to get a hit.

c. The answer depends on whether A is less than or greater than 0.500.

12.7.55

a. The centerline velocity is given by $V = \dfrac{R^2}{L}$ so $dV = \dfrac{2R}{L}dR - \dfrac{R^2}{L^2}dL$; evaluate this with $R = 3$, $dR = 0.05$, $L = 50$, $dL = 0.5$ to obtain $dV = \dfrac{21}{5000} = 0.0042\,\text{cm}^3$.

b. Rewrite the formula for dV as $\dfrac{dV}{V} = \dfrac{2\,dR}{R} - \dfrac{dL}{L}$; hence if R decreases 1% and L increases 2%, then V will decrease by approximately 4%.

c. If the radius of the cylinder increases by p%, then the length of the cylinder must decrease by approximately $2p$% in order for the velocity to remain constant.

12.7.57

a. We have $f_r = n(1-r)^{n-1}$ and $f_n = -(1-r)^n \ln(1-r)$.

b. We have $\triangle P \approx f(20, 0.1) \cdot 0.01 = 20 \cdot 0.9^{19} 0.01 \approx 0.027$.

c. We have $\triangle P \approx f(20, 0.9) \cdot 0.01 = 20 \cdot 0.1^{19} 0.01 \approx 2 \times 10^{-20}$.

d. Small changes in the flu rate have a greater effect on the probability of catching the flu when the flu rate is small compared to when the flu rate is large.

12.7.59 From the equation $\dfrac{1}{R} = \dfrac{1}{R_1} + \dfrac{1}{R_2} + \dfrac{1}{R_3}$ we obtain $-\dfrac{1}{R^2}\dfrac{\partial R}{\partial R_1} = -\dfrac{1}{R_1^2} \implies \dfrac{\partial R}{\partial R_1} = \dfrac{R^2}{R_1^2}$, with similar formulas for the other partials. Hence $dR = R^2\left(\dfrac{dR_1}{R_1^2} + \dfrac{dR_2}{R_2^2} + \dfrac{dR_3}{R_3^2}\right)$. Substituting the values $R_1 = 2$, $dR_1 = 0.05$, $R_2 = 3$, $dR_2 = -0.05$, $R_3 = 1.5$, $dR_3 = 0.05$ gives $R = \dfrac{2}{3}$ and $dR = \dfrac{7}{540} \approx 0.0130$ ohms.

12.7.61

a. Suppose f is a function of x and y; then $d(\ln f) = (\ln f)_x\, dx + (\ln f)_y\, dy = \frac{f_x}{f} dx + \frac{f_y}{f} dy = \frac{df}{f}$.

b. The absolute change in ln f is approximately $d(\ln f)$ and the relative change in f is approximately $\frac{df}{f}$; from part (a) these agree.

c. Observe that $df = y\, dx + x\, dy$, so $\frac{df}{f} = \frac{dx}{x} + \frac{dy}{y}$.

d. Observe that $df = \frac{y\, dx - x\, dy}{y^2}$, so $\frac{df}{f} = \frac{dx}{x} - \frac{dy}{y}$.

e. If $f = x_1 x_2 \cdots x_n$ then $\ln f = \ln x_1 + \ln x_2 + \cdots + \ln x_n$ and therefore $\frac{df}{f} = \frac{dx_1}{x_1} + \frac{dx_2}{x_2} + \cdots + \frac{dx_n}{x_n}$.

12.8 Maximum/Minimum Problems

12.8.1 It is locally the highest point on the surface; you cannot get to a higher point in any direction.

12.8.3 The partial derivatives are both zero or one or both do not exist.

12.8.5 The discriminant of f at (a, b) is the determinant given by $D(a, b) = f_{xx}(a, b) f_{yy}(a, b) - f_{xy}(a, b)^2$.

12.8.7 The function f has an absolute minimum value at $(a, b) \in R$ if $f(x, y) \geq f(a, b)$ for all $(x, y) \in R$.

12.8.9 We have $f_x = 2x$, $f_y = 2y$ so $(0, 0)$ is the only critical point of f.

12.8.11 We have $f_x = 6(3x - 2)$, $f_y = 2(y - 4)$ so $\left(\frac{2}{3}, 4\right)$ is the only critical point of f.

12.8.13 We have $f_x = 4x^3 - 16y$, $f_y = 4y^3 - 16x$; solving $f_x = f_y = 0$ gives $y = \frac{x^3}{4}$ and $x = \frac{y^3}{4}$; therefore $x = \frac{x^9}{4^4}$ which gives $x = 0, \pm 2$, and the critical points are $(0, 0)$, $(2, 2)$ and $(-2, -2)$.

12.8.15 We have $f_x = 4x$, $f_y = 6y$; therefore $(0, 0)$ is the only critical point. We also have $f_{xx} = 4$, $f_{yy} = 6$ and $f_{xy} = 0$; hence $D(0, 0) = 4 \cdot 6 - 0^2 = 24 > 0$ and $f_{xx}(0, 0) > 0$, which by the Second Derivative Test implies that f has a local minimum at $(0, 0)$.

12.8.17 We have $f_x = -8x$, $f_y = 16y$; therefore, $(0, 0)$ is the only critical point. We also have $f_{xx} = -8$, $f_{yy} = 16$ and $f_{xy} = 0$; hence $D(0, 0) = -8 \cdot 16 < 0$, which by the Second Derivative Test implies that f has a saddle point at $(0, 0)$.

12.8.19 We have $f_x = 4x^3 - 4y$, $f_y = 4y - 4x$; therefore, the critical points satisfy $y = x$ and $y = x^3$, which gives $x^3 = x$ and therefore $x = 0, \pm 1$, so the critical points are $(0, 0)$, $(1, 1)$ and $(-1, -1)$. We also have $f_{xx} = 12x^2$, $f_{yy} = 4$ and $f_{xy} = -4$; hence $D(x, y) = 48x^2 - 16$. Thus $D(0, 0) = -16 < 0$ so f has a saddle point at $(0, 0)$; and $D(1, 1) = D(-1, -1) = 32 > 0$, $f_{xx}(1, 1) = f_{xx}(-1, -1) = 12 > 0$ so f has a local minimum at $(1, 1)$ and $(-1, -1)$.

12.8.21 Note that $f(x, y)$ has the same critical points as the simpler function

$$g(x, y) = x^2 + y^2 - 4x + 5 = (x - 2)^2 + y^2 + 1;$$

we have $g_x = 2(x - 2)$, $g_y = 2y$; therefore, $(2, 0)$ is the only critical point. We also have $g_{xx} = 2$, $g_{yy} = 2$ and $g_{xy} = 0$; hence $D(2, 0) = 4 > 0$ and $g_{xx}(2, 0) > 0$, which by the Second Derivative Test implies that g (and hence f) has a local minimum at $(2, 0)$.

12.8.23 We have $f_x = 2(1-2x^2)ye^{-x^2-y^2}$, $f_y = 2(1-2y^2)xe^{-x^2-y^2}$; therefore, the critical points are $(0, 0)$, $\pm\left(\frac{1}{\sqrt{2}}, \frac{1}{\sqrt{2}}\right)$ and $\pm\left(\frac{1}{\sqrt{2}}, -\frac{1}{\sqrt{2}}\right)$. We also have

$$f_{xx} = 4(2x^2-3)xye^{-x^2-y^2},\ f_{yy} = 4(2y^2-3)xye^{-x^2-y^2} \text{ and } f_{xy} = 2(1-2x^2)(1-2y^2)e^{-x^2-y^2};$$

hence $D(0, 0) = -4 < 0$, so f has a saddle point at $(0, 0)$ by the Second Derivative Test. We also see that $D(x, y) > 0$ at the four other critical points; also $f_{xx}\left(\pm\left(\frac{1}{\sqrt{2}}, \frac{1}{\sqrt{2}}\right)\right) < 0$ so f has a local maximum at $\pm\left(\frac{1}{\sqrt{2}}, \frac{1}{\sqrt{2}}\right)$, and $f_{xx}\left(\pm\left(\frac{1}{\sqrt{2}}, \frac{1}{\sqrt{2}}\right)\right) > 0$ so f has a local minimum at $\pm\left(\frac{1}{\sqrt{2}}, -\frac{1}{\sqrt{2}}\right)$.

12.8.25 We have $f_x = \dfrac{y^2+2xy-x^2+1}{(1+x^2+y^2)^2}$, $f_y = \dfrac{y^2-2xy-x^2-1}{(1+x^2+y^2)^2}$; therefore, the critical points must satisfy $y^2+2xy-x^2+1 = y^2-2xy-x^2-1 = 0$, which gives $y^2-x^2 = 0$ or $y = \pm x$, and also $2xy+1 = 0$ which gives $y = -\dfrac{1}{2x}$; combining these gives $x = \pm\frac{1}{\sqrt{2}}$ and hence the critical points are $\pm\left(\frac{1}{\sqrt{2}}, -\frac{1}{\sqrt{2}}\right)$. We also have

$$f_{xx} = \frac{(1+x^2+y^2)^2(2y-2x) - (y^2+2xy-x^2+1)\cdot 2(1+x^2+y^2)\cdot 2x}{(1+x^2+y^2)^4},$$

$$f_{yy} = \frac{(1+x^2+y^2)^2(2y-2x) - (y^2-2xy-x^2-1)\cdot 2(1+x^2+y^2)\cdot 2y}{(1+x^2+y^2)^4},$$

$$f_{xy} = \frac{(1+x^2+y^2)^2(2y+2x) - (y^2+2xy-x^2+1)\cdot 2(1+x^2+y^2)\cdot 2y}{(1+x^2+y^2)^4};$$

hence $D\left(\pm\left(\frac{1}{\sqrt{2}}, -\frac{1}{\sqrt{2}}\right)\right) = \frac{1}{2} > 0$. We also have $f_{xx}\left(\frac{1}{\sqrt{2}}, -\frac{1}{\sqrt{2}}\right) < 0$ and $f_{xx}\left(-\frac{1}{\sqrt{2}}, \frac{1}{\sqrt{2}}\right) > 0$; therefore $\left(\frac{1}{\sqrt{2}}, -\frac{1}{\sqrt{2}}\right)$ is a local maximum and $\left(-\frac{1}{\sqrt{2}}, \frac{1}{\sqrt{2}}\right)$ is a local minimum.

12.8.27 We have $f_x = ye^x$, $f_y = e^x - e^y$; therefore, the critical points must satisfy $y = 0$ and $y = x$, so $(0, 0)$ is the only critical point. We also have $f_{xx} = ye^x$, $f_{yy} = -e^y$ and $f_{xy} = e^x$; hence $D(0, 0) = -1 < 0$ so f has a saddle point at $(0, 0)$.

12.8.29 Let $x, y \geq 0$ be the dimensions of the base of the box; then for any given x, y the maximum allowable height is given by $h = 96 - 2x - 2y$, and we must have $h \geq 0$ which implies $x + y \leq 48$. The volume of the box is given by $V = 2xy(48 - x - y)$, which we must maximize over the domain R given by $x, y \geq 0$ and $x + y \leq 48$. The critical points of V satisfy $V_x = 2(48-2x-y)y = 0$, $V_y = 2(48-x-2y)x = 0$; hence the critical points in the interior of R satisfy $2x + y = 2y + x = 48$, which gives $x = y = 16$. Furthermore $V(x, y) = 0$ on the boundary of R, so the maximum volume must occur at the point $(16, 16)$. Therefore the box with largest volume has height 32 in and base 16 in $\times$ 16 in with a volume of 8192 in^3.

12.8.31 Let $x, y \geq 0$ be the dimensions of the base and $h \geq 0$ be the height of the box; then the volume of the box is $xyh = 4$, so $h = \dfrac{4}{xy}$. The four sides of the box plus the base have total area $A = 2xh + 2yh + xy = \dfrac{8}{x} + \dfrac{8}{y} + xy$, which we must maximize over the domain R given by $x, y \geq 0$. The critical points of A satisfy

$$A_x = y - \frac{8}{x^2} = 0,\ A_y = x - \frac{8}{y^2} = 0;$$

hence the critical points in R satisfy $x^2y = xy^2 = 8$, which gives $x = y$ and hence $x, y = 2$; the corresponding height is 1. Furthermore $V(x, y) \to \infty$ as (x, y) approaches the boundary of R or as either $x, y \to \infty$, so the minimum area must occur at the point $(2, 2)$. Therefore the box with smallest area has dimensions 2 m $\times$ 2 m $\times$ 1 m.

12.8.33 Observe that $f_x = 4x^3$, $f_y = 12y^3$ and $f_{xx} = 12x^2$, $f_{yy} = 36y^2$ and $f_{xy} = 0$; therefore $D(0, 0) = 0$ and the Second Derivative Test is inconclusive. Observe that both x^4 and $3y^4$ have absolute minima at 0; therefore, f has an absolute minimum at $(0, 0)$.

12.8.35 Observe that $f_x = 4x^3y^2$, $f_y = 2x^4y$ and $f_{xx} = 12x^2y^2$, $f_{yy} = 2x^4$ and $f_{xy} = 8x^3y$; therefore $D(0,\,0) = 0$ and the Second Derivative Test is inconclusive. Observe that both $x^4 \geq 0$ and $y^2 \geq 0$ have absolute minima at 0; therefore, f has an absolute minimum at $(0,\,0)$ (and in fact along both coordinate axes).

12.8.37 First find the values of f at all critical points in the interior of $R = \{x^2 + y^2 \leq 4\}$; we have $f_x = 2x$, $f_y = 2y - 2$, so $(0,\,1)$ is the only critical point in R, and $f(0,\,1) = 0$. Next, find the minimum and maximum values of f on the boundary of R, which we can parameterize by $x = 2\cos\theta$, $y = 2\sin\theta$ for $0 \leq \theta \leq 2\pi$. Then $f(2\cos\theta,\,2\sin\theta) = 5 - 4\sin\theta$, which has maximum value 9 at $\theta = \frac{3\pi}{2}$ and minimum value 1 at $\theta = \frac{\pi}{2}$. Therefore, the maximum value of f on R is $f(0,\,-2) = 9$ and the minimum value is $f(0,\,1) = 0$.

12.8.39 First find the values of f at all critical points in the interior of R; we have $f_x = 4x$, $f_y = 2y$, so $(0,\,0)$ is the only critical point in R, and $f(0,\,0) = 4$. Next, find the minimum and maximum values of f on the boundary of R, which is a square. On the sides $y = \pm 1$, $-1 \leq x \leq 1$ we have $f(x,\,\pm 1) = 4 + 2x^2 + 1 = 2x^2 + 5$, which has extreme values 5 and 7 on $[-1,\,1]$. On the sides $x = \pm 1$, $-1 \leq y \leq 1$ we have $f(\pm 1,\,y) = 4 + 2 + y^2 = y^2 + 6$, which has extreme values 6 and 7 on $[-1,\,1]$. Therefore, the maximum value of f on R is 7 and the minimum value is 4.

12.8.41 First find the values of f at all critical points in the interior of $R = \{x^2 + y^2 \leq 16\}$; we have $f_x = 2x + 4$, $f_y = 2y - 2$, so $(-2,\,1)$ is the only critical point in R, and $f(-2,\,1) = -5$. Next, find the minimum and maximum values of f on the boundary of R, which we can parameterize by $x = 4\cos\theta$, $y = 4\sin\theta$ for $0 \leq \theta \leq 2\pi$. Then $f(4\cos\theta,\,4\sin\theta) = 16 + 16\cos\theta - 8\sin\theta$, and so we must find the extreme values of the function $g(\theta) = 16 + 16\cos\theta - 8\sin\theta$ on $[0,\,2\pi]$. Let $\mathbf{v} = \langle 16,\,-8\rangle$; then $g(\theta) = 16 + \mathbf{v}\cdot\langle\cos\theta,\,\sin\theta\rangle$, which has extreme values $16 \pm |\mathbf{v}| = 16 \pm 8\sqrt{5}$ on $[0,\,2\pi]$. Therefore, the maximum value of f on R is $16 + 8\sqrt{5} \approx 33.89$ and the minimum value is -5.

12.8.43 First find the values of f at all critical points in the interior of R; we have $f_x = 2x + 2$, $f_y = 8y + 4$, so $\left(-1,\,-\frac{1}{2}\right)$ is the only critical point in R, and $f\left(-1,\,-\frac{1}{2}\right) = -2$. Next, find the minimum and maximum values of f on the boundary of R, which we can parameterize by $x = 4\cos\theta$, $y = \sin\theta$ for $0 \leq \theta \leq 2\pi$. Then

$$f(4\cos\theta,\,\sin\theta) = 16\cos^2\theta + 4\sin^2\theta + 8\cos\theta + 4\sin\theta = 4\left(3\cos^2\theta + 2\cos\theta + \sin\theta + 1\right),$$

and so we must find the extreme values of the function $g(\theta) = 4\left(3\cos^2\theta + 2\cos\theta + \sin\theta + 1\right)$, on $[0,\,2\pi]$. We have $g'(\theta) = 4(-6\cos\theta\sin\theta - 2\sin\theta + \cos\theta)$, which from numerical investigation has extreme values $\approx -1.14,\,24.25$ on $[0,\,2\pi]$. Therefore, the maximum value of f on R is 24.25 and the minimum value is -2.

12.8.45 Observe that $f(0,\,0) = -4$ and $f(x,\,y) \geq -4$ for all points $(x,\,y) \in R$; hence the absolute minimum value of f on R is -4. The function f on R takes on all values in the interval $[-4,\,0)$; therefore f has no absolute maximum on R.

12.8.47 Observe that $f(0,\,0) = 2$ and $f(x,\,y) \leq 2$ for all points $(x,\,y) \in R$; hence, the absolute maximum value of f on R is 2. The function f on R takes on all values in the interval $(0,\,2]$; therefore, f has no absolute minimum on R.

12.8.49 The equation of the plane can be written as $z = 4 - x - y$; so it suffices to minimize the square of the distance from $(x,\,y,\,4 - x - y)$ to $(0,\,3,\,6)$, which is given by $w = x^2 + (y - 3)^2 + (x + y + 2)^2$. We have $w_x = 2(2x + y + 2)$, $w_y = 2(x + 2y - 1)$, so the critical points of w satisfy $2x + y = -2$, $x + 2y = 1$ which gives $(x,\,y,\,z) = \left(-\frac{5}{3},\,\frac{4}{3},\,\frac{13}{3}\right)$. Since we know that there is some point on the plane closest to $(0,\,3,\,6)$, this critical point must be that point.

12.8.51 The distance from a point $(x,\,y,\,x^2 + y^2 + 10)$ on the surface to the plane $x + 2y - z = 0$ is given by

$$d = \frac{|x + 2y - (x^2 + y^2 + 10)|}{\sqrt{6}} = \frac{1}{\sqrt{6}}(x^2 + y^2 - x - 2y + 10)$$

(see Section 12.1 Exercise 86). So it suffices to minimize the function $f(x,\,y) = x^2 + y^2 - x - 2y + 10$ on R^2. We have $f_x = 2x - 1$, $f_y = 2y - 2$, so $\left(\frac{1}{2},\,1\right)$ is the only critical point of f; the corresponding point on the surface is $\left(\frac{1}{2},\,1,\,\frac{45}{4}\right)$, which must be the point on the surface closest to the plane. The corresponding point on the plane must lie on the line through $\left(\frac{1}{2},\,1,\,\frac{45}{4}\right)$ in the direction normal to the plane, and therefore has the form $\left(\frac{1}{2} + t,\,1 + 2t,\,\frac{45}{4} - t\right)$; solving

$$\frac{1}{2}+t+2(1+2t)-\left(\frac{45}{4}-t\right)=6t-\frac{35}{4}=0$$

gives $t=\frac{35}{24}$, so the corresponding point on the plane is $\left(\frac{47}{24},\frac{94}{24},\frac{235}{24}\right)$.

12.8.53

a. True, because our definition of saddle point is a critical point which is neither a local maximum or local minimum.

b. False; a necessary condition for a local maximum at (a, b) is that both $f_x = f_y = 0$ at (a, b), assuming both partials exist.

c. True, because f may take on its absolute maximum or minimum at a point on the boundary of its domain.

d. True; the equation of the tangent plane at a critical point (a, b) is $z = f(a, b)$.

12.8.55 This function has a local minimum near $(0.3, -0.3)$ and a saddle point at $(0, 0)$.

12.8.57 The plane has equation $z = 2 - x + y$; the distance from a point $(x, y, 2-x+y)$ on the plane to the point $(1, 1, 1)$ is given by $d^2 = (x-1)^2 + (y-1)^2 + (x-y-1)^2$. It suffices to minimize the function $f(x, y) = (x-1)^2 + (y-1)^2 + (x-y-1)^2$ on R^2. We have $f_x = 2(x-1+x-y-1) = 2(2x-y-2)$, $f_y = 2(y-1+y-x+1) = 2(-x+2y)$, so the critical point of f satisfies $2x - y = 2$, $x - 2y = 0$ which gives $x=\frac{4}{3}$, $y=\frac{2}{3}$. The corresponding point on the plane is $\left(\frac{4}{3},\frac{2}{3},\frac{4}{3}\right)$. Since there is a point on the plane closest to the point $(1, 1, 1)$, this must be the point we found.

12.8.59

a. Using the relation $z = 200 - x - y$, we see that it suffices to minimize the function $f(x, y) = x^2 + y^2 + (200-x-y)^2$ over the closed bounded region R given by $x, y \geq 0$ and $x + y \leq 200$. We have $f_x = 2(x+x+y-200) = 2(2x+y-200)$, $f_y = 2(y+x+y-200) = 2(x+2y-200)$; therefore, the critical point of f satisfies $2x + y = 200$, $x + 2y = 200$ which gives $x = y = \frac{200}{3}$ (the corresponding value of $z = \frac{200}{3}$ as well), and we have $f\left(\frac{200}{3},\frac{200}{3}\right) = \frac{40{,}000}{3}$. We must also find the extreme values of f on the boundary of R. Along the segment $y = 0, 0 \leq x \leq 200$ we have $f(x, 0) = x^2 + (200-x)^2$ which has range $[20{,}000, 40{,}000]$, and we get the same result along the other two segments that make up the boundary of R. Therefore, the minimum value of $x^2 + y^2 + z^2$ is given by $x = y = z = \frac{200}{3}$.

b. The function $\sqrt{x^2+y^2+z^2}$ takes its minimum at the same point that minimizes $x^2 + y^2 + z^2$, which we saw in part (a) is $x = y = z = \frac{200}{3}$.

c. Using the relation $z = 200 - x - y$, we see that it suffices to minimize the function $f(x, y) = xy(200-x-y) = 200xy - x^2y - xy^2$ over the closed bounded region R given by $x, y \geq 0$ and $x + y \leq 200$. We have $f_x = y(200-2x-y)$, $f_y = x(200-x-2y)$; therefore, the critical points of f in the interior of R satisfy $2x + y = 200$, $x + 2y = 200$ which gives $x = y = \frac{200}{3}$ (the corresponding value of $z = \frac{200}{3}$ as well), and we have $f\left(\frac{200}{3},\frac{200}{3}\right) = \left(\frac{200}{3}\right)^2$. We also observe that $f(x, y) = 0$ on the boundary of R; therefore, the maximum value of xyz is given by $x = y = z = \frac{200}{3}$.

d. The function $x^2y^2z^2$ takes its maximum at the same point that maximizes xyz, which we saw in part (c) is $x = y = z = \frac{200}{3}$.

12.8.61

a. The function to be minimized is

$$f(x, y) = x^2 + y^2 + (x-2)^2 + y^2 + (x-1)^2 + (y-1)^2 = 3x^2 - 6x + 3y^2 - 2y;$$

we have $f_x = 6(x-1)$, $f_y = 2(3y-1)$ so the optimal location is the unique critical point $\left(1, \frac{1}{3}\right)$.

b. The function to be minimized is now

$$f(x, y) = (x-x_1)^2 + (y-y_1)^2 + (x-x_2)^2 + (y-y_2)^2 + (x-x_3)^2 + (y-y_3)^2;$$
$$= 3\left(x^2 - 2\overline{x}x + y^2 - 2\overline{y}y\right) + \text{const}$$

where $\overline{x} = \dfrac{x_1+x_2+x_3}{3}$, $\overline{y} = \dfrac{y_1+y_2+y_3}{3}$. we have $f_x = 6(x-\overline{x})$, $f_y = 6(y-\overline{y})$ so the optimal location is the unique critical point $(\overline{x}, \overline{y})$.

c. The function to be minimized is now

$$f(x, y) = \sum_{i=1}^{n}(x - x_i)^2 + \sum_{i=1}^{n}(y - y_i)^2 = n(x^2 - 2\overline{x}x + y^2 - 2\overline{y}y) + \text{const}$$

where $\overline{x} = \frac{1}{n}\sum_{i=1}^{n} x_i$, $\overline{y} = \frac{1}{n}\sum_{i=1}^{n} y_i$. We have $f_x = n(x - \overline{x})$, $f_y = n(y - \overline{y})$ so the optimal location is the unique critical point $(\overline{x}, \overline{y})$.

d. The actual sum of the distances is given by the function

$$f(x, y) = \sqrt{x^2 + y^2} + \sqrt{(x-2)^2 + y^2} + \sqrt{(x-1)^2 + (y-1)^2}$$

We can minimize this function as follows. First, fix $y = y_0$ and consider the function $g(x) = f(x, y_0)$. Each of the three terms in this function has positive second derivative, and approaches ∞ as $x \to \pm\infty$, so the same is true for their sum; therefore $g(x)$ must have a unique absolute minimum. Observe in addition that g is symmetric in the line $x = 1$, which implies that the absolute minimum must occur at $x = 1$. So to minimize $f(x, y)$, we can set $x = 1$ and reduce to minimizing the function $h(y) = f(1, y) = 2\sqrt{1 + y^2} + |y - 1|$, which has absolute minimum at $y = \frac{1}{\sqrt{3}}$. Therefore, the optimal location is $\left(1, \frac{1}{\sqrt{3}}\right)$, which is different from the point found in part (a).

12.8.63 Given the n data points $(x_1, y_1), \ldots, (x_n, y_n)$, we seek to minimize the function

$$\begin{aligned} E(m, b) &= \sum_{k=1}^{n}(mx_k + b - y_k)^2 \\ &= \left(\sum x_k^2\right)m^2 + 2\left(\sum x_k\right)mb + nb^2 - 2\left(\sum x_k y_k\right)m - 2\left(\sum y_k\right)b + \sum y_k^2. \end{aligned}$$

We have

$$E_m = 2\left(\sum x_k^2\right)m + 2\left(\sum x_k\right)b - 2\left(\sum x_k y_k\right), \quad E_b = 2\left(\sum x_k\right)m + 2nb - 2\left(\sum y_k\right)$$

and solving $E_m = E_b = 0$ gives

$$m = \frac{(\sum x_k)(\sum y_k) - n\sum x_k y_k}{(\sum x_k)^2 - n\sum x_k^2}, \quad b = \frac{1}{n}\left(\sum y_k - m\sum x_k\right).$$

12.8.65 Using the result in Exercise 63, we obtain $m = \frac{2 \cdot 14 - 3 \cdot 24}{2^2 - 3 \cdot 10} = \frac{22}{13}$, $b = \frac{1}{3}\left(14 - \frac{22}{13} \cdot 2\right) = \frac{46}{13}$ so the line has equation $y = \frac{22}{13}x + \frac{46}{13}$.

12.8.67 Let $s = \frac{a+b+c}{2}$ be the semi-perimeter, which we assume is constant. Then we may express $c = 2s - a - b$ and therefore $A^2 = s(s-a)(s-b)(s-(2s-a-b)) = s(s-a)(s-b)(a+b-s)$, so it suffices to maximize the simpler function $f(a, b) = s(s-a)(s-b)(a+b-s)$ over the closed bounded region given by $0 \leq a, b \leq s$ and $a + b \geq s$. We have $f_a = s(s-b)(-(a+b-s) + s - a) = s(s-b)(2s - 2a - b)$ and similarly $f_b = s(s-a)(2s - a - 2b)$, so the critical points of f in the interior of R satisfy $2a + b = 2s$, $a + 2b = 2s$ which gives $a = b = \frac{2}{3}s$, and therefore $c = \frac{2}{3}s$ as well, so we get an equilateral triangle. We also observe that $f(a, b) = 0$ on the boundary of R, so this critical point must give the absolute maximum of f. We conclude that of all triangles with a given perimeter, the maximum area is obtained when all three sides are equal (in the special case of perimeter 9 units, each side length is 3 units).

12.8.69

a. We have $d_1(x, y) = \sqrt{(x - x_1)^2 + (y - y_1)^2}$, so $\nabla d_1(x, y) = \frac{x - x_1}{d_1(x, y)}\mathbf{i} + \frac{y - y_1}{d_1(x, y)}\mathbf{j}$; observe that this is a unit vector in the direction of the vector joining (x_1, y_1) to (x, y).

b. Similarly, we have $\nabla d_2(x, y) = \frac{x - x_2}{d_2(x, y)}\mathbf{i} + \frac{y - y_2}{d_2(x, y)}\mathbf{j}$ and $\nabla d_3(x, y) = \frac{x - x_3}{d_3(x, y)}\mathbf{i} + \frac{y - y_3}{d_3(x, y)}\mathbf{j}$.

c. Since $\nabla f = \nabla d_1 + \nabla d_2 + \nabla d_3$, the condition $f_x = f_y = 0$ is equivalent to $\nabla d_1 + \nabla d_2 + \nabla d_3 = 0$.

d. If three unit vectors add to 0, they must make angles of $\pm\frac{2\pi}{3}$ with each other.

e. In this case the optimal point is the vertex at the large angle.

f. Solving the equations $f_x = f_y = 0$ numerically gives $(0.255457, 0.304504)$.

12.8.71

a. We have $f(x, y) = -2x^4 + 2x^2(e^y + 1) - e^{2y} - 1$; hence $f_x = -8x^3 + 4x(e^y + 1)$, $f_y = 2x^2e^y - 2e^{2y}$, so the critical points must satisfy $x^2 = e^y$, and then the equation $f_x = 0$ reduces to $e^y = 1$, so the critical points are $(\pm 1, 0)$. Next we compute $f_{xx} = -24x^2 + 4(1 + e^y)$, $f_{yy} = 2x^2e^y - 4e^{2y}$, $f_{xy} = 4e^y$ so $f_{xx}f_{yy} - f_{xy}^2 = (-16)(-2) - (4)^2 = 16$ at both critical points; we also have $f_{xx} < 0$ at both critical points, so the critical points are both local maxima.

b. We have $f_x = 8xe^y - 8x^3$, $f_y = 4x^2e^y - 4e^{4y}$, so the critical points must satisfy $x^2 = e^{3y}$, and then the equation $f_x = 0$ reduces to $e^y = 1$, so the critical points are $(\pm 1, 0)$. Next we compute $f_{xx} = 8e^y - 24x^2$, $f_{yy} = 4x^2e^y - 16e^{4y}$, $f_{xy} = 8xe^y$, so $f_{xx}f_{yy} - f_{xy}^2 = (-16)(-12) - (\pm 8)^2 = 128$ at both critical points; we also have $f_{xx} < 0$ at both critical points, so the critical points are both local maxima.

12.9 Lagrange Multipliers

12.9.1 The level curves of f must be tangential to the level curves of g at the optimal point; thus, the gradients are parallel.

12.9.3 We have $\nabla f = \langle 2x, 2y, 2z \rangle$, $\nabla g = \langle 2, 3, -5 \rangle$ so the Lagrange multiplier conditions are $2x = 2\lambda$, $2y = 3\lambda$, $2z = -5\lambda$, $2x + 3y - 5z + 4 = 0$.

12.9.5 We have $\nabla f = \langle 1, 2 \rangle$, $\nabla g = \langle 2x, 2y \rangle$ so the Lagrange multiplier conditions are $1 = 2\lambda x$, $2 = 2\lambda y$, $x^2 + y^2 - 4 = 0$. Hence $\dfrac{1}{x} = 2\lambda = \dfrac{2}{y} \implies y = 2x$; substituting this in the constraint gives $5x^2 = 4$, so $x = \pm \frac{2}{\sqrt{5}}$ and the extreme values of f on the circle $x^2 + y^2 = 4$ must occur at the points $\pm\left(\frac{2}{\sqrt{5}}, \frac{4}{\sqrt{5}}\right)$. We see that $f\left(\frac{2}{\sqrt{5}}, \frac{4}{\sqrt{5}}\right) = 2\sqrt{5}$, $f\left(-\frac{2}{\sqrt{5}}, -\frac{4}{\sqrt{5}}\right) = -2\sqrt{5}$, so these are the maximum and minimum values.

12.9.7 We have $\nabla f = \langle 2ye^{2xy}, 2xe^{2xy} \rangle$, $\nabla g = \langle 3x^2, 3y^2 \rangle$ so the Lagrange multiplier conditions are $2ye^{2xy} = 3\lambda x^2$, $2xe^{2xy} = 3\lambda y^2$, $x^3 + y^3 - 16 = 0$. We can eliminate λ and obtain $x^3 = y^3$, so $y = x$. Substituting $y = x$ in the constraint gives $2x^3 = 16$, so $x = y = 2$. Hence the extreme values (if any) of f along the curve $x^3 + y^3 = 16$ must occur at $(2, 2)$, where $f(2, 2) = e^8$. Observe that as $x \to \infty$ along this curve, $y \to -\infty$ and vice versa; therefore $f(x, y) \to 0$, so f has no minimum along this curve. This shows that f takes its absolute maximum value at $(2, 2)$ along this curve.

12.9.9 We have $\nabla f = \langle -8x, 2y \rangle$, $\nabla g = \langle 2x, 4y \rangle$ so the Lagrange multiplier conditions are $-8x = 2\lambda x$, $2y = 4\lambda y$, $x^2 + 2y^2 - 4 = 0$. If $x, y \neq 0$ the first equation gives $\lambda = -4$, whereas the second gives $\lambda = \frac{1}{2}$ which is a contradiction; hence we must have x or $y = 0$, which gives the points $(\pm 2, 0)$ and $\left(0, \pm\sqrt{2}\right)$, and $f(\pm 2, 0) = = -16$, $f\left(0, \pm\sqrt{2}\right) = 2$. Hence the minimum and maximum values of f on the closed bounded set given by $x^2 + 2y^2 = 4$ are -16 and 2.

12.9.11 We have $\nabla f = \langle 1, 3, -1 \rangle$, $\nabla g = \langle 2x, 2y, 2z \rangle$ so the Lagrange multiplier conditions are $1 = 2\lambda x$, $3 = 2\lambda y$, $-1 = 2\lambda z$, $x^2 + y^2 + z^2 - 4 = 0$. These equations imply $x, y, z \neq 0$, so we can eliminate λ and obtain $y = 3x$, $z = -x$. Then the constraint gives $11x^2 = 4$, so $x = \pm \frac{2}{\sqrt{11}}$ and the solutions are the points $\pm\left(\frac{2}{\sqrt{11}}, \frac{6}{\sqrt{11}}, -\frac{2}{\sqrt{11}}\right)$. We compute $f\left(\frac{2}{\sqrt{11}}, \frac{6}{\sqrt{11}}, -\frac{2}{\sqrt{11}}\right) = 2\sqrt{11}$, $f\left(-\frac{2}{\sqrt{11}}, -\frac{6}{\sqrt{11}}, \frac{2}{\sqrt{11}}\right) = -2\sqrt{11}$, and hence the minimum and maximum values of f on the closed bounded set given by $x^2 + y^2 + z^2 = 4$ are $-2\sqrt{11}$ and $2\sqrt{11}$.

12.9.13 We have $\nabla f = \langle y^2z^3, 2xyz^3, 3xy^2z^2 \rangle$, $\nabla g = \langle 2x, 2y, 4z \rangle$ so the Lagrange multiplier conditions are $y^2z^3 = 2\lambda x$, $2xyz^3 = 2\lambda y$, $3xy^2z^2 = 4\lambda z$, $x^2 + y^2 + 2z^2 - 25 = 0$. Assume first that $x, y, z \neq 0$, so we can eliminate λ

and obtain $\dfrac{xy^2z^3}{\lambda} = 2x^2 = y^2 = \dfrac{4}{3}z^2$ or $y^2 = 2x^2$, $z^2 = \frac{3}{2}x^2$. Then using the constraint we obtain $6x^2 = 25$, so $x = \pm\frac{5}{\sqrt{6}}$, $y = \pm\frac{5}{\sqrt{3}}$ and $z = \pm\frac{5}{2}$. The value of xy^2z^3 at any of these points is $\pm\dfrac{5}{\sqrt{6}}\cdot\dfrac{5^2}{3}\cdot\dfrac{5^3}{2^3} = \pm\dfrac{15,625\sqrt{6}}{144} \approx \pm 265.8$. We also observe that if any of $x, y, z = 0$ then $xy^2z^3 = 0$. Hence, the minimum and maximum values of f on the closed bounded set given by $x^2 + y^2 + 2z^2 = 25$ are $\pm\frac{15,625\sqrt{6}}{144}$.

12.9.15 We have $\nabla f = \langle 2x, 2y, 2z\rangle$, $\nabla g = \langle yz, xz, xy\rangle$, so the Lagrange multiplier conditions are $2x = \lambda yz$, $2y = \lambda xz$, $2z = \lambda xy$, $xyz - 4 = 0$. The first three equations give $\dfrac{\lambda xyz}{2} = x^2 = y^2 = z^2$, so $y = \pm x$, $z = \pm x$. Then using the constraint we obtain $x^3 = \pm 4$, so $x, y, z = \pm\sqrt[3]{4}$. The value of $x^2 + y^2 + z^2$ at any of these points is $f\left(\pm\sqrt[3]{4}, \pm\sqrt[3]{4}, \pm\sqrt[3]{4}\right) = 3\cdot 4^{2/3} = 6\sqrt[3]{2}$. Note that $f(x, y, z)$ is the square of the distance from (x, y, z) to the origin, so this function will have an absolute minimum but no maximum on the surface given by $xyz = 4$; therefore the minimum value of f on the surface is $6\sqrt[3]{2}$.

12.9.17 Let $x, y, z \geq 0$ denote the lengths of the sides of the box, with z the longest side. Then the length plus girth of the box is $2x + 2y + z$ so we must maximize the volume $f(x, y, z) = xyz$ subject to the constraint $g(x, y, z) = 2x + 2y + z - 108 = 0$. We have $\nabla f = \langle yz, xz, xy\rangle$, $\nabla g = \langle 2, 2, 1\rangle$ so the Lagrange multiplier conditions are $yz = 2\lambda$, $xz = 2\lambda$, $xy = \lambda$, $2x + 2y + z - 108 = 0$. Assume that $x, y, z > 0$ (otherwise the volume is 0); then the first three equations give $\dfrac{xyz}{2\lambda} = x = y = \dfrac{z}{2}$, and the constraint gives $6x = 108$, so $x = 18$ and the box has dimensions 18 in $\times$ 18 in $\times$ 36 in. The domain of f is the triangle with vertices $(54, 0, 0)$, $(0, 54, 0)$, $(0, 0, 108)$, which is closed and bounded. We also note that $f = 0$ along any of the edges of the triangle. Therefore, the maximum value of f occurs at the point we found, and the minimum value is 0.

12.9.19 It suffices to find the extreme values of the function $f(x, y) = x^2 + y^2$ subject to the constraint $g(x, y) = x^2 + xy + 2y^2 - 1 = 0$. We have $\nabla f = \langle 2x, 2y\rangle$, $\nabla g = \langle 2x + y, 4y + x\rangle$, so the Lagrange multiplier conditions are $2x = \lambda(2x + y)$, $2y = \lambda(4y + x)$, $x^2 + xy + 2y^2 - 1 = 0$. The first two equations give $2x(4y + x) = \lambda(2x + y)(4y + x) = 2y(2x + y) \Longrightarrow x^2 + 4xy = y^2 + 2xy$, or $x^2 + 2xy - y^2$. This implies that both $x, y \neq 0$, for if, say, $x = 0$ then this condition gives $y = 0$ as well, which violates the constraint (same argument for $y = 0$). Let $r = \dfrac{y}{x}$, and rewrite this equation as $\dfrac{y^2}{x^2} - \dfrac{2y}{x} - 1 = r^2 - 2r - 1 = 0$, which we can solve to obtain $r = 1 \pm \sqrt{2}$. We now use the constraint and the relation $y = rx$ to obtain $x^2(1 + r + 2r^2) = 1$ or $x^2 = \dfrac{1}{2r^2 + r + 1}$. Then the values of the function f are given by

$$f(x, y) = f(x, rx) = x^2(1 + r^2) = \frac{1 + r^2}{2r^2 + r + 1} = \frac{6 \pm 2\sqrt{2}}{7} \approx 0.4531,\ 1.2612,$$

and the corresponding minimum and maximum distances are $\sqrt{\dfrac{6 \pm 2\sqrt{2}}{7}} \approx 0.6731,\ 1.1230$.

12.9.21 Let (x, y) be the vertex of the rectangle in the first quadrant; then the perimeter of the rectangle is $4(x + y)$, so it suffices to maximize the function $f(x, y) = x + y$ subject to the constraint $g(x, y) = 4x^2 + 9y^2 - 36 = 0$ and $x, y \geq 0$. We have $\nabla f = \langle 1, 1\rangle$, $\nabla g = \langle 8x, 18y\rangle$, so the Lagrange multiplier conditions are $1 = 8\lambda x$, $1 = 18\lambda y$, $4x^2 + 9y^2 - 36 = 0$. The first two equations give $4x = \frac{1}{2\lambda} = 9y$; substituting in the constraint gives $13x^2 = 81$, so $x = \frac{9}{\sqrt{13}}$, $y = \frac{4}{\sqrt{13}}$, and we have $f\left(\frac{9}{\sqrt{13}}, \frac{4}{\sqrt{13}}\right) = \sqrt{13} \approx 3.61$. The domain given by the constraint and $x, y \geq 0$ is a closed and bounded arc, and we observe that at the boundary points $f(3, 0) = 3$ and $f(0, 2) = 2$. Therefore, f takes its maximum at $\left(\frac{9}{\sqrt{13}}, \frac{4}{\sqrt{13}}\right)$, and the corresponding rectangle has dimensions $\frac{18}{\sqrt{13}} \times \frac{8}{\sqrt{13}}$.

12.9.23 It suffices to minimize the function $f(x, y, z) = (x - 1)^2 + (y - 2)^2 + (z + 3)^2$ subject to the constraint $g(x, y, z) = x^2 - 2xy + 2y^2 - x + y = 0$. We have $\nabla f = \langle 2x - 2, 2y - 4, 2z + 6\rangle$,

$\nabla g = \langle 2x-2y-1,\, 4y-2x+1,\, 0\rangle$, so the Lagrange multiplier conditions are $2x-2=\lambda(2x-2y-1)$, $2y-4=\lambda(4y-2x+1)$, $2z+6=\lambda\cdot 0$, $x^2-2xy+2y^2-x+y=0$. Therefore, $z=-3$, and

$$(2x-2)(4y-2x+1)=\lambda(2x-2y-1)(4y-2x+1)=(2y-4)(2x-2y-1),$$

which gives the equation $2y^2+(2x-7)y-2x^2+7x-3=0$, and solving for y gives

$y=\dfrac{7-2x\pm\sqrt{20x^2-84x+73}}{4}$. We can also solve the constraint equation for y to obtain

$y=\dfrac{2x-1\pm\sqrt{-4x^2+4x+1}}{4}$; equating these gives

$$\begin{aligned}
4x-8\pm\sqrt{-4x^2+4x+1} &= \pm\sqrt{20x^2-84x+73}\\
16x^2-64x+64-4x^2+4x+1\pm(8x-16)\sqrt{-4x^2+4x+1} &= 20x^2-84x+73\\
\pm(x-2)\sqrt{-4x^2+4x+1} &= x^2-3x+1\\
(x-2)^2\left(-4x^2+4x+1\right) &= \left(x^2-3x+1\right)^2\\
5x^4-26x^3+42x^2-18x-3 &= 0\\
(x-1)\left(5x^3-21x^2+21x+3\right) &= 0
\end{aligned}$$

which has root $x=1$, and using a numerical solver we find that the cubic factor has exactly one root $x\approx -0.1264$. When $x=1$ the two equations for y above give $y=\frac{1}{2}, 1$ and $y=0, \frac{1}{2}$, so $y=\frac{1}{2}$ is the only common solution. Similarly when $x=-0.1264$ the equations for y give $y\approx -0.4772, 4.1036$ and $y\approx -0.4772, -0.1492$ so $y\approx -0.4772$ is the only common solution. Therefore, there are two solutions to the Lagrange multiplier equations: $(x, y, z)=\left(1, \frac{1}{2}, -3\right)$ and $(x, y, z)\approx(-0.1264, -0.4772, -3)$. The first point has distance $\frac{3}{2}$ to the point $(1, 2, -3)$ and the second has distance ≈ 2.7213, so the point on the surface closest to $(1, 2, -3)$ is $\left(1, \frac{1}{2}, -3\right)$.

12.9.25 It suffices to find the extreme values the function $f(x, y, z)=(x-2)^2+(y-3)^2+(z-4)^2$ subject to the constraint $g(x, y, z)=x^2+y^2+z^2-9=0$. We have $\nabla f=\langle 2x-4, 2y-6, 2z-8\rangle$, $\nabla g=\langle 2x, 2y, 2z\rangle$ so the Lagrange multiplier conditions are $2x-4=2\lambda x$, $2y-6=2\lambda y$, $2z-8=2\lambda z$, $x^2+y^2+z^2-9=0$. We can write the first three equations in the form $(1-\lambda)\langle x, y, z\rangle=\langle 2, 3, 4\rangle$ so $\langle x, y, z\rangle=c\langle 2, 3, 4\rangle$ for some scalar c; using the constraint, we find that $c=\pm\frac{3}{\sqrt{29}}$, and hence $\langle x, y, z\rangle=\pm\frac{3}{\sqrt{29}}\langle 2, 3, 4\rangle$; the corresponding values of f are $f\left(\pm\left(\frac{6}{\sqrt{29}}, \frac{9}{\sqrt{29}}, \frac{12}{\sqrt{29}}\right)\right)=38\mp 6\sqrt{29}=\left(\sqrt{29}\mp 3\right)^2$, so the minimum distance is $\sqrt{29}-3$ and the maximum distance is $\sqrt{29}+3$.

12.9.27 Notice that the constraint is equivalent to $\ell+2g=6$. We have $\nabla U=5\langle \ell^{-1/2}g^{1/2}, \ell^{1/2}g^{-1/2}\rangle$ so the Lagrange multiplier conditions are equivalent to $\ell^{-1/2}g^{1/2}=\lambda$, $\ell^{1/2}g^{-1/2}=2\lambda$, $\ell+2g=6$. Eliminating λ from the first two equations gives $\ell^{1/2}g^{-1/2}=2\ell^{-1/2}g^{1/2}$, which simplifies to $g=\frac{\ell}{2}$; substituting in the constraint then gives $\ell=3$ and $g=\frac{3}{2}$. The value of the utility function at this point is $U=15\sqrt{2}$.

12.9.29 Notice that the constraint is equivalent to $5\ell+4g=20$. We have $\nabla U=\frac{8}{5}\langle 4\ell^{-1/5}g^{1/5}, \ell^{4/5}g^{-4/5}\rangle$, so the Lagrange multiplier conditions are equivalent to $4\ell^{-1/5}g^{1/5}=5\lambda$, $\ell^{4/5}g^{-4/5}=4\lambda$, $5\ell+4g=20$. Eliminating λ from the first two equations gives $16\ell^{-1/5}g^{1/5}=5\ell^{4/5}g^{-4/5}$, which simplifies to $g=\frac{5\ell}{16}$; substituting in the constraint then gives $\ell=\frac{16}{5}$, $g=1$. The value of the utility function at this point is $U=8\cdot\left(\frac{16}{5}\right)^{4/5}\approx 20.287$.

12.9.31

a. True, because the tangent plane to a sphere at any point has normal vector in the direction of the line joining the point to the center of the sphere.

b. False in general; in fact the two vectors ∇f and ∇g are in the same direction, so $\nabla f\cdot\nabla g=0$ only if one of these vectors is zero.

12.9.33 Let $x, y>0$ be the dimensions of the base and $h>0$ be the height of the box; then the four sides of the box plus the base have total area $2xh+2yh+xy=2$, which is our constraint. The volume of the box is $V=xyh$, so the Lagrange multiplier conditions are $yh=\lambda(2h+y)$, $xh=\lambda(2h+x)$, $xy=2\lambda(x+y)$, $2xh+2yh+xy=2$. The first two equations give $\dfrac{2h+y}{yh} = \dfrac{2h+x}{xh} \Longrightarrow \dfrac{2}{y} + \dfrac{1}{h} = \dfrac{2}{x} + \dfrac{1}{h}$, so $y = x$. The second and third equations give $\dfrac{2h+x}{xh} = \dfrac{2x+2y}{xy} \Longrightarrow \dfrac{2}{x} + \dfrac{1}{h} = \dfrac{2}{x} + \dfrac{2}{y} = \dfrac{4}{x}$, so $h = \dfrac{x}{2}$. Then substituting in the constraint gives $3x^2 = 2$, so $x = \frac{\sqrt{6}}{3}$. Therefore, the box with largest volume has height $\frac{\sqrt{6}}{6}$ m and base $\frac{\sqrt{6}}{3} \times \frac{\sqrt{6}}{3}$ m.

12.9.35 Let $x, y \geq 0$ be the dimensions of the base and $z \geq 0$ be the height of the box; then $x + 2y + 3z = 6$ is our constraint. The box has volume $V = xyz$, so the Lagrange multiplier conditions are $yz = \lambda$, $xz = 2\lambda$, $xy = 3\lambda$, $x + 2y + 3z = 6$. The first two equations give $y = \dfrac{x}{2}$, the first and third equations give $z = \dfrac{x}{3}$, and substituting in the constraint gives $x = 2$. Therefore, the box with largest volume has dimensions $x = 2$, $y = 1$, $z = \frac{2}{3}$.

12.9.37 The constraint is the equation of the plane $x - y + z = 2$; the distance from a point $(x,\ y,\ z)$ to the point $(1,\ 1,\ 1)$ is given by $d^2 = (x-1)^2 + (y-1)^2 + (x-y-1)^2$, so it suffices to minimize the function $f(x,\ y,\ z) = (x-1)^2 + (y-1)^2 + (z-1)^2$ subject to the constraint $x - y + z = 2$. The Lagrange multiplier conditions are $2x - 2 = \lambda$, $2y - 2 = -\lambda$, $2z - 2 = \lambda$, $x - y + z = 2$. The first two equations give $2(x+y) = 4$, so $y = 2 - x$, the first and third equations give $z = x$, and then substituting in the constraint gives $x = \frac{4}{3}$, so the closest point on the plane to $(1,\ 1,\ 1)$ is $\left(\frac{4}{3},\ \frac{2}{3},\ \frac{4}{3}\right)$.

12.9.39 We have $\nabla f \langle 2x + y,\ -8y + x \rangle$; solving $2x + y = 0$ and $-8y + x = 0$ simultaneously gives unique solution $(0,\ 0)$, and $f(0,\ 0) = 0$. Next we use Lagrange multipliers to find the minimum and maximum values of f on the boundary of R given by $4x^2 + 9y^2 = 36$. The Lagrange multiplier conditions are $2x + y = 8\lambda x$, $-8y + x = 18\lambda y$, $4x^2 + 9y^2 = 36$. The first two equations give $\lambda xy = \dfrac{2xy + y^2}{8} = \dfrac{x^2 - 8xy}{18} \Longrightarrow 9y^2 + 50xy - 4x^2 = 0$; therefore, $y = rx$ where r satisfies the quadratic $9r^2 + 50r - 4 = 0$, which has roots $r = \dfrac{-25 \pm \sqrt{661}}{9}$. Then the constraint gives $(4 + 9r^2)x^2 = 36$, so $x^2 = \dfrac{36}{9r^2 + 4} = \dfrac{18}{8 - 25r}$ (using $9r^2 = 4 - 50r$) and hence

$$f(x,\ y) = \left(1 - 4r^2 + r\right)x^2 = \frac{-7 + 209r}{9} \cdot \frac{18}{4 - 25r} = \frac{418r - 14}{4 - 25r} = \frac{-7 \pm \sqrt{661}}{2},$$

which gives the absolute minimum and maximum values of f on R.

12.9.41 We have $\nabla f = \langle 2(x-1),\ 2(y+1) \rangle$; therefore f has unique critical point $(1,\ -1)$ which is inside R; $f(1,\ -1) = 0$. Next we use Lagrange multipliers to find the minimum and maximum values of f on the boundary of R given by $x^2 + y^2 = 4$. The Lagrange multiplier conditions are equivalent to $x - 1 = \lambda x$, $y + 1 = \lambda y$, $x^2 + y^2 = 4$. The first two equations give $\lambda xy = xy - y = xy + x \Longrightarrow y = -x$; then the constraint gives $2x^2 = 4$, so $x = \pm\sqrt{2}$ and the solutions are $\pm\left(\sqrt{2},\ -\sqrt{2}\right)$. The values of f at these points are $f\left(\sqrt{2},\ -\sqrt{2}\right) = 6 - 4\sqrt{2} \approx 0.343$, $f\left(-\sqrt{2},\ \sqrt{2}\right) = 6 + 4\sqrt{2} \approx 11.657$; therefore, the maximum value of f on R is $6 + 4\sqrt{2}$ and the minimum value is 0.

12.9.43 The maximum and minimum values of f along the curve $g(x,\ y) = 0$ occur at points where the level curves of f are tangent to the curve $g(x,\ y) = 0$; using this, we see that the minimum and maximum values are 1 and 8.

12.9.45 Notice that the constraint is equivalent to $2K + 3L = 30$. We have $\nabla f = \frac{1}{2}\langle K^{-1/2}L^{1/2},\ K^{1/2}L^{-1/2} \rangle$, so the Lagrange multiplier conditions are equivalent to $K^{-1/2}L^{1/2} = 2\lambda$, $K^{1/2}L^{-1/2} = 3\lambda$, $2K + 3L = 30$. Eliminating λ from the first two equations gives $3K^{-1/2}L^{1/2} = 2K^{1/2}L^{-1/2}$, which simplifies to $3L = 2K$; substituting in the constraint then gives $K = 7.5$ and $L = 5$. The domain over which f is to be maximized is a closed line segment; K or L is 0 at the endpoints, and hence so is f. Therefore, the values of K and L found above must maximize f.

12.9.47 We have $\nabla f = \langle aK^{a-1}L^{1-a},\ (1-a)K^aL^{-a} \rangle$, so the Lagrange multiplier conditions are equivalent to $aK^{a-1}L^{1-a} = \lambda p$, $(1-a)K^aL^{-a} = \lambda q$, $pK + qL = B$. Eliminating λ from the first two equations gives $aqK^{a-1}L^{1-a} = (1-a)pK^aL^{-a}$, which simplifies to $aqL = (1-a)pK$; substituting in the constraint then gives

$K = \dfrac{aB}{p}$ and $L = \dfrac{(1-a)B}{q}$. The domain over which f is to be maximized is a closed line segment; K or L is 0 at the endpoints, and hence so is f. Therefore, the values of K and L found above must maximize f.

12.9.49 The function to be maximized is $f(x_1, x_2, x_3, x_4) = x_1 + x_2 + x_3 + x_4$, subject to the constraint $x_1^2 + x_2^2 + x_3^2 + x_4^2 = 16$. The Lagrange multiplier conditions are $1 = 2\lambda x_1$, $1 = 2\lambda x_2$, $1 = 2\lambda x_3$, $1 = 2\lambda x_4$, $x_1^2 + x_2^2 + x_3^2 + x_4^2 = 16$. The first four equations give $x_1 = x_2 = x_3 = x_4$, and then the constraint gives $4x_1^2 = 16$, so $x_1 = \pm 2$. Therefore, the maximum of f on the closed, bounded set given by $x_1^2 + x_2^2 + x_3^2 + x_4^2 = 16$ is $f(2, 2, 2, 2) = 8$ (and the minimum is $f(-2, -2, -2, -2) = -8$).

12.9.51 The function to be maximized is $f(x_1, x_2, \ldots, x_n) = a_1x_1 + a_2x_2 + \cdots + a_nx_n$, subject to the constraint $x_1^2 + x_2^2 + \cdots + x_n^2 = 1$. The Lagrange multiplier conditions are $a_1 = 2\lambda x_1$, $a_2 = 2\lambda x_2$, $\ldots$, $a_n = 2\lambda x_n$, $x_1^2 + x_2^2 + \cdots + x_n^2 = 1$. The first n equations are equivalent to $c(x_1, x_2, \ldots, x_n) = (a_1, a_2, \ldots, a_n)$ for some c, and then the constraint gives $c^2 = a_1^2 + a_2^2 + \cdots + a_n^2$. Therefore, the maximum of f on the closed, bounded set given by $f\left(\frac{a_1}{c}, \frac{a_2}{c}, \ldots, \frac{a_n}{c}\right) = \frac{1}{c}(a_1^2 + a_2^2 + \cdots + a^2) = \sqrt{a_1^2 + a_2^2 + \cdots + a^2}$ (and the minimum is $-\sqrt{a_1^2 + a_2^2 + \cdots + a^2}$).

12.9.53

a. Gradients are perpendicular to level surfaces.

b. If ∇f was not in the plane spanned by ∇g and ∇h, then f could be increased or decreased by moving the point P slightly along the curve C.

c. Since ∇f is in the plane spanned by ∇g and ∇h, we can express ∇f as a linear combination of ∇g and ∇h.

d. The gradient condition from part (c), as well as the constraints, must be satisfied.

12.9.55 We wish to find the extreme values of the function $f(x, y, z) = xyz$ subject to the constraints $g(x, y, z) = x^2 + y^2 - 4 = 0$ and $h(x, y, z) = x + y + z - 1 = 0$. Using the method described in problem 53 above, we solve the equation $\nabla f = \lambda \nabla g + \mu \nabla h$, together with the constraints. This gives the conditions

$$yz = 2\lambda x + \mu,\ xz = 2\lambda y + \mu,\ xy = \mu,\ x^2 + y^2 = 4,\ x + y + z = 1.$$

The first and second equations together give $z(y - x) = 2\lambda(x - y)$, which implies either $x = y$ or $z = -2\lambda$. In the former case the first constraint gives $2x^2 = 4$, so $x = y = \pm\sqrt{2}$, solving for z gives $z = 1 \mp 2\sqrt{2}$ and $f(x, y, z) = 2 \pm 4\sqrt{2}$. In the latter case we obtain $(x + y)z = \mu = xy$ from the first and third equations, and then solving for $z = 1 - x - y$ gives $(x + y)(1 - x - y) - xy = 0 \Longleftrightarrow x + y - 3xy - 4 = 0$, using the relation $x^2 + y^2 = 4$. Therefore,

$$x - 4 = (3x - 1)y$$
$$(x - 4)^2 = (3x - 1)^2(4 - x^2)$$
$$x^2 - 8x + 16 = (9x^2 - 6x + 1)(4 - x^2)$$

which simplifies to $9x^4 - 6x^3 - 34x^2 + 16x + 12 = 0$. This equation has roots $x \approx -0.42, -1.78, 1.96, 0.91$ in the interval $(-2, 2)$; one can check that the corresponding solutions (x, y, z) to the Lagrange conditions do not give values larger or smaller resp. than $2 + 4\sqrt{2}$, $2 - 4\sqrt{2}$, so these are in fact the maximum and minimum values of f subject to the constraints.

12.9.57 We wish to find the extreme values of the function $f(x, y, z) = x^2 + y^2 + z^2$ subject to the constraints $g(x, y, z) = 4x^2 + 4y^2 - z^2 = 0$ and $h(x, y, z) = 2x + 4z - 5 = 0$. Using the method described in problem 53 above, we solve the equation $\nabla f = \lambda \nabla g + \mu \nabla h$, together with the constraints. This gives the conditions

$$2x = 8\lambda x + 2\mu,\ 2y = 8\lambda y,\ 2z = -2\lambda z + 4\mu,\ z^2 = 4x^2 + 4y^2,\ 2x + 4z = 5.$$

The second equation gives $y(1 - 4\lambda) = 0$, so either $y = 0$ or $\lambda = \frac{1}{4}$. Consider first the case $y = 0$; then the first constraint equation gives $z = \pm 2x$. If $z = 2x$ then the second constraint equation gives $x = \frac{1}{2}$, $z = 1$ and we obtain the point $\left(\frac{1}{2}, 0, 1\right)$; similarly if $z = -2x$ then we obtain the point $\left(-\frac{5}{6}, 0, \frac{5}{3}\right)$. In the case $\lambda = \frac{1}{4}$ the first equation gives $\mu = 0$ and then the third equation gives $z = 0$; but then the first of the constraints implies that $x = y = z = 0$, which violates the second constraint. Hence there are no solutions to the Lagrange conditions in this case, and the minimum and maximum values of the function f along this curve are $f\left(\frac{1}{2}, 0, 1\right) = \frac{5}{4}$ and $f\left(-\frac{5}{6}, 0, \frac{5}{3}\right) = \frac{125}{36}$.

12.10 Chapter Twelve Review

12.10.1

a. False; this equation describes a plane in R^3.

b. False; if $2x^2 - 6y^2 > 0$ then $z = \sqrt{2x^2 - 6y^2}$ or $z = -\sqrt{2x^2 - 6y^2}$.

c. False; for example $f(x, y) = x^2 y$ has $f_{xxy} = 2$, $f_{xyy} = 0$.

d. False; ∇f lies in the xy-plane.

e. True; a normal vector for an orthogonal plane can be found by taking the cross product of normal vectors for the two intersecting planes.

12.10.3

a. Let $P = (0, 0, 3)$, $Q = (1, 0, -6)$ and $R = (1, 2, 3)$. Then the vectors $\overrightarrow{PQ} = \langle 1, 0, -9\rangle$ and $\overrightarrow{PR} = \langle 1, 2, 0\rangle$ lie in the plane, so

$$\mathbf{n} = \overrightarrow{PQ} \times \overrightarrow{PR} = \begin{vmatrix} \mathbf{i} & \mathbf{j} & \mathbf{k} \\ 1 & 0 & -9 \\ 1 & 2 & 0 \end{vmatrix} = 18\mathbf{i} - 9\mathbf{j} + 2\mathbf{k}$$

is normal to the plane. The plane has equation $18x - 9y + 2(z - 3) = 0$, which simplifies to $18x - 9y + 2z = 6$.

b. The x-intercept is found by setting $y = z = 0$ and solving $18x = 6$ to obtain $x = \frac{1}{3}$. Similarly, the y and z-intercepts are $y = -\frac{2}{3}$ and $z = 3$.

c.

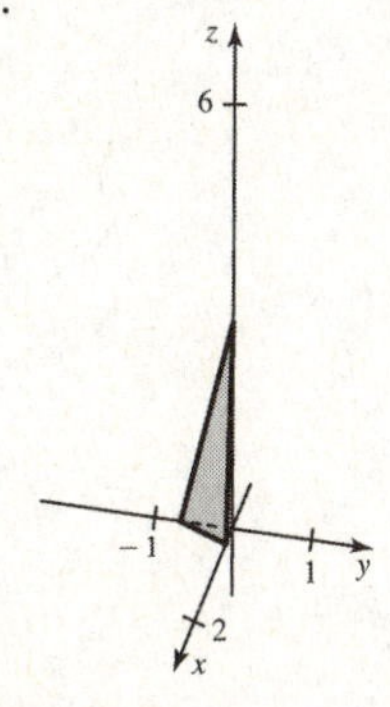

12.10.5 First, note that the vectors normal to the planes, $\mathbf{n}_Q = \langle -3, 1, 2\rangle$ and $\mathbf{n}_R = \langle 3, 3, 4\rangle$ are not multiples of each other; therefore, these planes are not parallel and they intersect in a line ℓ. We need to find a point on ℓ and a vector in the direction of ℓ. Setting $x = 0$ in the equations of the planes gives equations of the lines in which the planes intersect the yz plane:

$$y + 2z = 0$$
$$3y + 4z = 12.$$

Solving these equations simultaneously gives $y = 12$, $z = -6$, so $(0, 12, -6)$ is a point on ℓ. A vector in the direction of ℓ is

$$\mathbf{n}_Q \times \mathbf{n}_R = \begin{vmatrix} \mathbf{i} & \mathbf{j} & \mathbf{k} \\ -3 & 1 & 2 \\ 3 & 3 & 4 \end{vmatrix} = -2\mathbf{i} + 18\mathbf{j} - 12\mathbf{k} = -2\langle 1, -9, 6\rangle .$$

Therefore, ℓ has equation $\mathbf{r}(t) = \langle 0, 12, -6\rangle + t\langle 1, -9, 6\rangle = \langle t, 12 - 9t, -6 + 6t\rangle$, or $x = t$, $y = 12 - 9t$, $z = -6 + 6t$.

12.10.7 Let $P = (-2, 3, 1)$, $Q = (1, 1, 0)$ and $R = (-1, 0, 1)$. Then the vectors $\overrightarrow{PQ} = \langle 3, -2, -1\rangle$ and $\overrightarrow{PR} = \langle 1, -3, 0\rangle$ lie in the plane, so

$$\mathbf{n} = \overrightarrow{PQ} \times \overrightarrow{PR} = \begin{vmatrix} \mathbf{i} & \mathbf{j} & \mathbf{k} \\ 3 & -2 & -1 \\ 1 & -3 & 0 \end{vmatrix} = -(3\mathbf{i} + \mathbf{j} + 7\mathbf{k})$$

is normal to the plane. The plane has equation $3(x + 2) + 1(y - 3) + 7(z - 1) = 0$, which simplifies to $3x + y + 7z = 4$.

12.10.9

a. This surface is a hyperbolic paraboloid.

b. The xy-trace is found by setting $z = 0$ in the equation of the surface, which gives $y = \pm 2x$ (two lines intersecting at the origin). Similarly, we see that the xz-trace is the parabola $z = \frac{x^2}{36}$ and the yz-trace is the parabola $z = -\frac{y^2}{144}$.

c. The x-intercept is found by setting $y = z = 0$ in the equation of the surface, which gives the point $(0, 0, 0)$. Similarly, we see that the y- and z-intercepts are also $(0, 0, 0)$.

d.

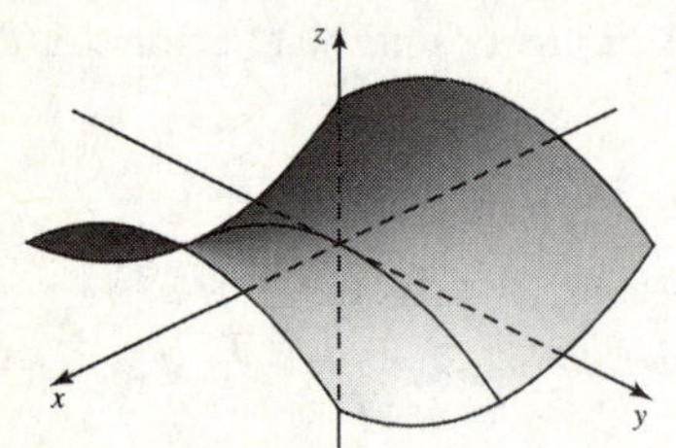

12.10.11

a. This surface is an elliptic cone.

b. The xy-trace is found by setting $z = 0$ in the equation of the surface, which gives $y = \pm 2x$ (two lines intersecting at the origin). Similarly, we see that the xz-trace is $(0, 0, 0)$ and the yz-trace is $z = \pm 5y$.

c. The x-intercept is found by setting $y = z = 0$ in the equation of the surface, which gives the point $(0, 0, 0)$. Similarly, we see that the y- and z-intercepts are also $(0, 0, 0)$.

d.

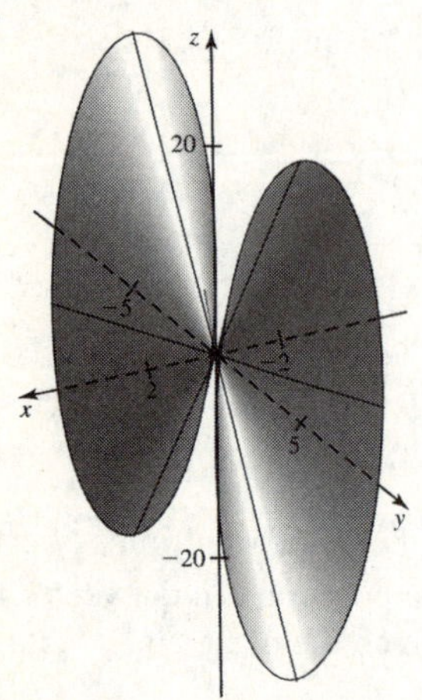

12.10.13

a. This surface is an elliptic paraboloid.

b. The xy-trace is found by setting $z = 0$ in the equation of the surface, which gives $(0, 0, 0)$. Similarly, we see that the xz-trace is the parabola $z = \frac{x^2}{16}$, and the yz-trace is the parabola $z = \frac{y^2}{36}$.

c. The x-intercept is found by setting $y = z = 0$ in the equation of the surface, which gives the point $(0, 0, 0)$. Similarly, we see that the y- and z-intercepts are also $(0, 0, 0)$.

d.

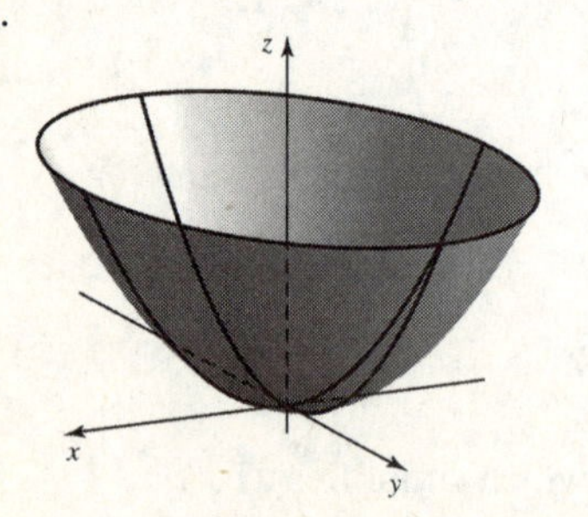

12.10.15

a. This surface is a hyperboloid of one sheet.

b. The xy-trace is found by setting $z = 0$ in the equation of the surface, which gives the hyperbola $y^2 - 2x^2 = 1$. Similarly, we see that the xz-trace is the hyperbola $4z^2 - 2x^2 = 1$, and the yz-trace is the ellipse $y^2 + 4z^2 = 1$.

c. The x-intercept is found by setting $y = z = 0$ in the equation of the surface, which gives no solutions. Similarly, we see that the y-intercepts are $(0, \pm 1, 0)$, and the z-intercepts are $\left(0, 0, \pm \frac{1}{2}\right)$.

d.

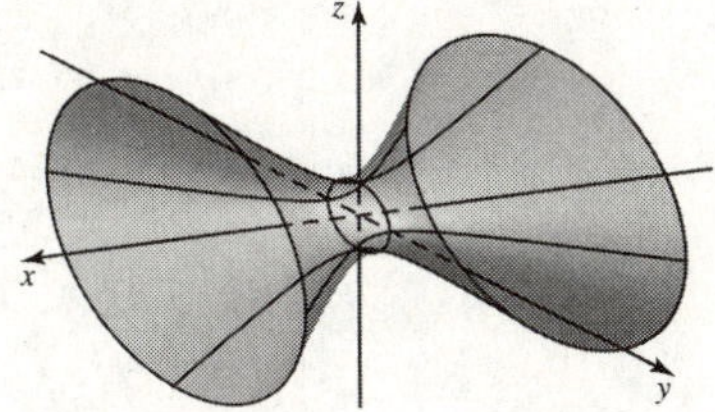

12.10.17

a. This surface is a hyperboloid of one sheet.

b. The xy-trace is found by setting $z = 0$ in the equation of the surface, which gives the ellipse $\dfrac{x^2}{4} + \dfrac{y^2}{16} = 4$. Similarly, we see that the xz-trace is the hyperbola $\dfrac{x^2}{4} - z^2 = 4$, and the yz-trace is the hyperbola $\dfrac{y^2}{16} - z^2 = 4$.

c. The x-intercept is found by setting $y = z = 0$ in the equation of the surface, which gives the points $(\pm 4, 0, 0)$. Similarly, we see that the y-intercepts are $(0, \pm 8, 0)$, and there are no z-intercepts.

d.

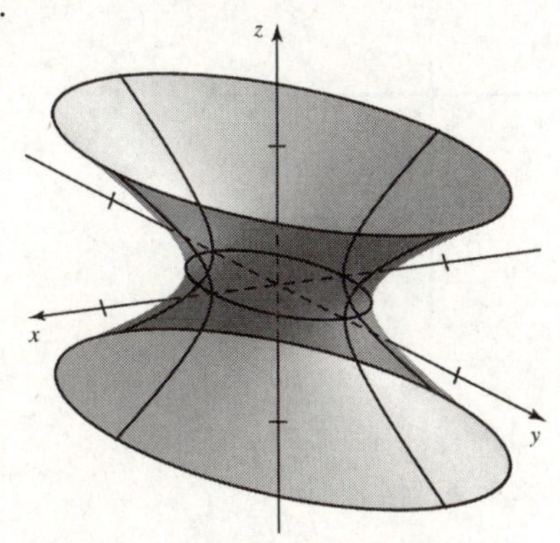

12.10.19

a. This surface is an ellipsoid.

b. The xy-trace is found by setting $z = 0$ in the equation of the surface, which gives the ellipse $\dfrac{x^2}{4} + \dfrac{y^2}{16} = 4$. Similarly, we see that the xz-trace is the ellipse $\dfrac{x^2}{4} + z^2 = 4$, and the yz-trace is the ellipse $\dfrac{y^2}{16} + z^2 = 4$.

c. The x-intercept is found by setting $y = z = 0$ in the equation of the surface, which gives the points $(\pm 4, 0, 0)$. Similarly, we see that the y-intercepts are the points $(0, \pm 8, 0)$, and the z-intercepts are the points $(0, 0, \pm 2)$.

d.

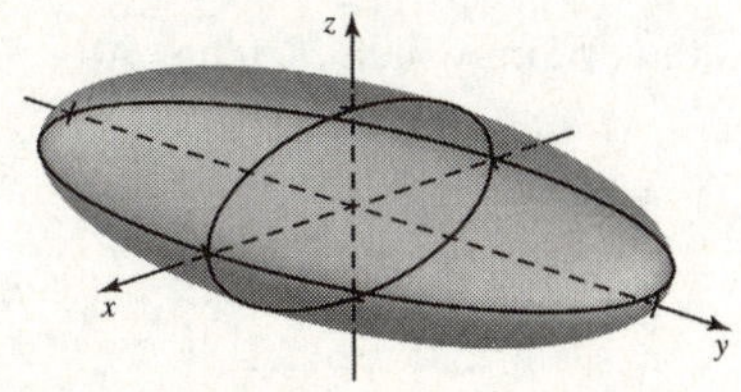

12.10.21

a. This surface is an elliptic cone.

b. The xy-trace is found by setting $z = 0$ in the equation of the surface, which gives the origin $(0, 0, 0)$. Similarly, we see that the xz-trace is $z = \pm\dfrac{8x}{3}$ (two lines intersecting at the origin), and the yz-trace is $y = \pm\dfrac{7z}{8}$.

c. The x-intercept is found by setting $y = z = 0$ in the equation of the surface, which gives the point $(0, 0, 0)$. Similarly, we see that the y and z-intercepts are also $(0, 0, 0)$.

d.

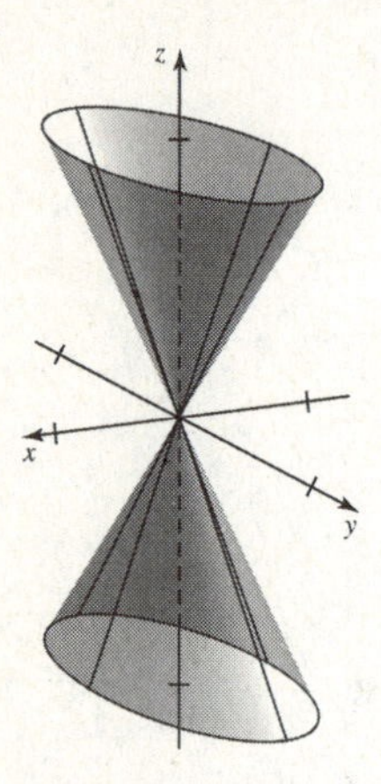

12.10.23 The domain is $D = \{(x, y) : (x, y) \neq (0, 0)\}$.

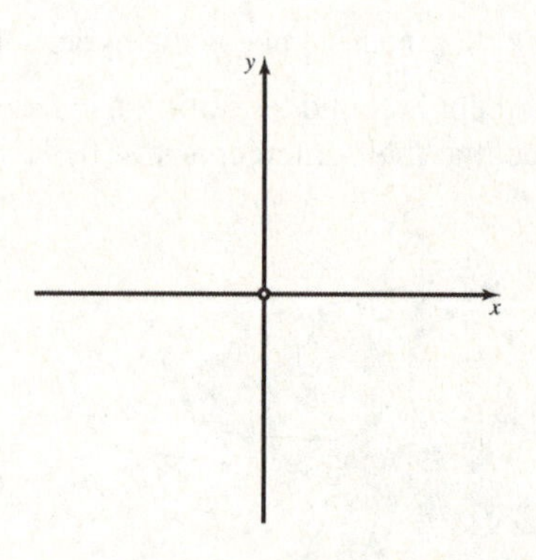

12.10.25 The domain is D=$\{(x, y) : x \geq y^2\}$.

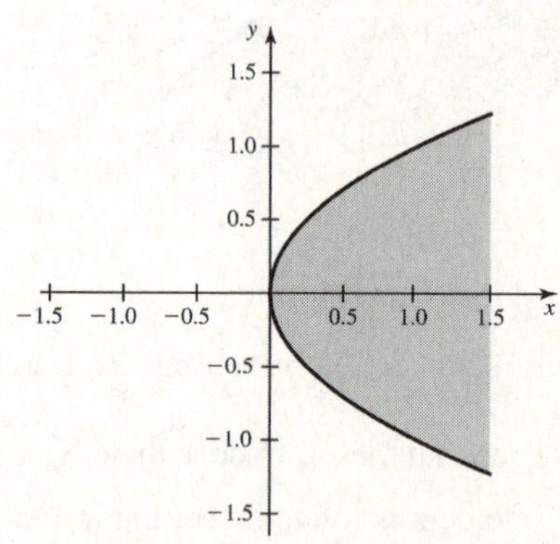

12.10.27

a. The graph of this function is part of a hyperboloid of two sheets and contains the origin, which matches A.
b. The graph of this function is a cylinder, which matches D.
c. The graph of this function is a hyperbolic paraboloid, which matches C.
d. The graph of this function is part of a hyperboloid of one sheet, which matches B.

12.10.29

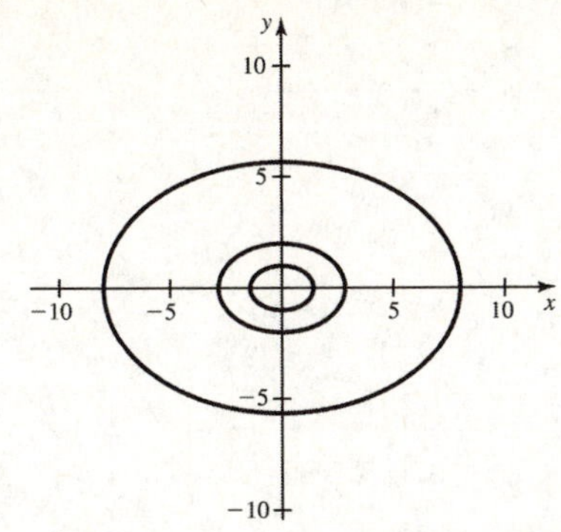

12.10.31 This limit may be evaluated directly by substitution:

$$\lim_{(x,\,y)\to(4,\,-2)} (10x - 5y + 6xy) = 10 \cdot 4 - 5(-2) + 6 \cdot 4(-2) = 2\,.$$

12.10.33 This limit may be evaluated by factoring the numerator and denominator and canceling the common factor $x + y$:

$$\lim_{(x,\,y)\to(-1,\,1)} \frac{x^2 - y^2}{x^2 - xy - 2y^2} = \lim_{(x,\,y)\to(-1,\,1)} \frac{(x-y)(x+y)}{(x-2y)(x+y)} = \lim_{(x,\,y)\to(-1,\,1)} \frac{x-y}{x-2y} = \frac{2}{3}\,.$$

12.10.35 This limit may be evaluated directly by substitution: $\displaystyle\lim_{(x,\,y,\,z)\to(\frac{\pi}{2},\,0,\,\frac{\pi}{2})} 4\cos y \sin\sqrt{xz} = 4\cos 0 \sin\left(\frac{\pi}{2}\right) = 4\,.$

12.10.37 We have $\displaystyle\frac{\partial}{\partial x}\left[xye^{xy}\right] = ye^{xy} + xy^2e^{xy} = y(1+xy)e^{xy}\,,\ \frac{\partial}{\partial y}\left[xye^{xy}\right] = xe^{xy} + x^2ye^{xy} = x(1+xy)e^{xy}\,.$

12.10.39 We have $f_x(x,\,y,\,z) = e^{x+2y+3z}\,,\ f_y(x,\,y,\,z) = 2e^{x+2y+3z}\,,\ f_z(x,\,y,\,z) = 3e^{x+2y+3z}\,.$

12.10.41 Observe that $\displaystyle\frac{\partial^2 u}{\partial x^2} + \frac{\partial^2 u}{\partial y^2} = 6y - 6y = 0\,.$

12.10.43

a. If r is held fixed and R increases then V increases, so $V_R > 0$, whereas, if R is held fixed and r increases then V decreases, so $V_r < 0$.

b. We have $V_R = 4\pi R^2 > 0$ and $V_r = -4\pi r^2 < 0$, consistent with the predictions in part (a).

c. If $R = 3$, $r = 1$ and R is increased by $\triangle R = 0.1$, then $\triangle V \approx 4\pi \cdot 3^2 \cdot 0.1 = 3.6\pi$; if r is decreased by 0.1 then $\triangle V \approx -4\pi \cdot 1^2 \cdot (-0.1) = 0.4\pi$. Therefore, the volume changes more if R is increased.

12.10.45 The chain rule gives

$$\begin{aligned}\frac{dw}{dt} &= \frac{\partial w}{\partial x}\frac{dx}{dt} + \frac{\partial w}{\partial y}\frac{dy}{dt} + \frac{\partial w}{\partial z}\frac{dz}{dt} \\ &= \frac{x}{\sqrt{x^2+y^2+z^2}} \cdot \cos t + \frac{y}{\sqrt{x^2+y^2+z^2}} \cdot (-\sin t) + \frac{z}{\sqrt{x^2+y^2+z^2}} \cdot (-\sin t) \\ &= -\frac{\cos t \sin t}{\sqrt{1+\cos^2 t}}.\end{aligned}$$

12.10.47 The chain rule gives

$$w_r = w_x x_r + w_y y_r = \frac{2x}{x^2+y^2+1} \cdot st + \frac{2y}{x^2+y^2+1} \cdot 1 = \frac{2(rs^2t^2 + r + s + t)}{r^2s^2t^2 + (r+s+t)^2 + 1}$$

and similarly

$$w_s = \frac{2(r^2st^2 + r + s + t)}{r^2s^2t^2 + (r+s+t)^2 + 1}\,,\quad w_t = \frac{2(r^2s^2t + r + s + t)}{r^2s^2t^2 + (r+s+t)^2 + 1}\,.$$

12.10.49 Let $F(x, y) = y\ln(x^2+y^2) - 4$; then if y is determined by $F(x, y) = 0$, we have

$$\frac{dy}{dx} = -\frac{F_x}{F_y} = -\frac{\frac{2xy}{x^2+y^2}}{\ln(x^2+y^2)+\frac{2y^2}{x^2+y^2}} = -\frac{2xy}{2y^2+(x^2+y^2)\ln(x^2+y^2)}.$$

12.10.51

a. The chain rule gives $z'(t) = 2xx'(t) - 4yy'(t) = -24\cos t\sin t = -12\sin 2t$.

b. Walking uphill corresponds to $z'(t) > 0$, which occurs when $\frac{\pi}{2} < t < \pi$ and $\frac{3\pi}{2} < t < 2\pi$.

12.10.53

a.

	$(a,b)=(0,0)$	$(a,b)=(2,0)$	$(a,b)=(1,1)$
$\theta=\frac{\pi}{4}$	0	$4\sqrt{2}$	$-2\sqrt{2}$
$\theta=\frac{3\pi}{4}$	0	$-4\sqrt{2}$	$-6\sqrt{2}$
$\theta=\frac{5\pi}{4}$	0	$-4\sqrt{2}$	$2\sqrt{2}$

b. At each of the points $(a, b) = (0, 0), (2, 0), (1, 1)$ sketch unit vectors which make angles of $\frac{\pi}{4}, \frac{3\pi}{4}, \frac{5\pi}{4}$ with the x-axis.

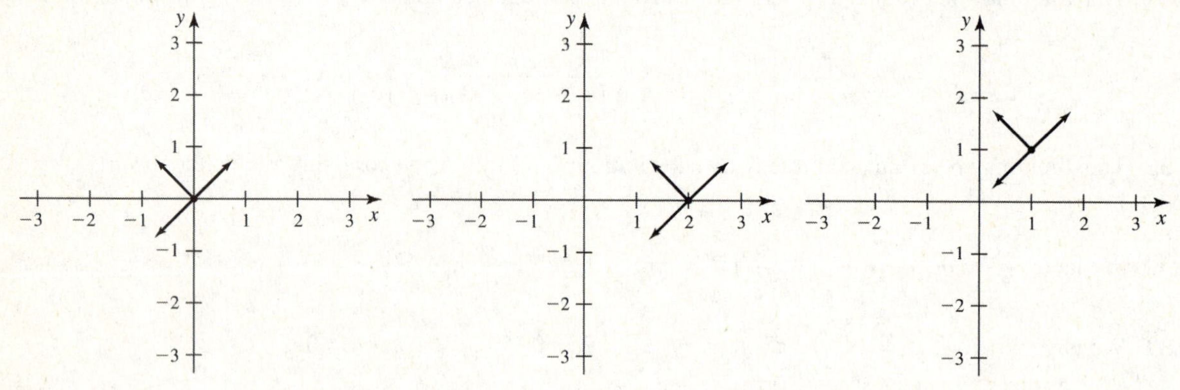

12.10.55 The gradient of h is given by $\nabla h(x, y) = h_x\,\mathbf{i} + h_y\,\mathbf{j} = \dfrac{1}{\sqrt{2+x^2+2y^2}}(x\,\mathbf{i}+2y\,\mathbf{j})$; therefore, $\nabla h(2, 1) = \frac{\sqrt{2}}{2}(\mathbf{i}+\mathbf{j})$ and the directional derivative in the direction $\mathbf{u}$ is given by

$$\nabla h(2,1)\cdot\mathbf{u} = \frac{\sqrt{2}}{10}(\mathbf{i}+\mathbf{j})\cdot(3\mathbf{i}+4\mathbf{j}) = \frac{7\sqrt{2}}{10} \approx 0.9899.$$

12.10.57 The gradient of f is given by $\nabla f(x, y, z) = f_x\,\mathbf{i} + f_y\,\mathbf{j} + f_z\,\mathbf{k} = \cos(x+2y-z)(\mathbf{i}+2\mathbf{j}-\mathbf{k})$; therefore, $\nabla f(\frac{\pi}{6}, \frac{\pi}{6}, -\frac{\pi}{6}) = -\frac{1}{2}(\mathbf{i}+2\mathbf{j}-\mathbf{k})$ and the directional derivative in the direction $\mathbf{u}$ is given by

$$\nabla f\left(\frac{\pi}{6}, \frac{\pi}{6}, -\frac{\pi}{6}\right)\cdot\mathbf{u} = -\frac{1}{6}(\mathbf{i}+2\mathbf{j}-\mathbf{k})\cdot(\mathbf{i}+2\mathbf{j}+2\mathbf{k}) = -\frac{1}{2}.$$

12.10.59

a. The gradient of f is given by $\nabla f(x, y) = f_x\,\mathbf{i} + f_y\,\mathbf{j} = -\dfrac{1}{\sqrt{4-x^2-y^2}}(x\,\mathbf{i}+y\,\mathbf{j})$, so $\nabla f(-1, 1) = \frac{1}{\sqrt{2}}(\mathbf{i}-\mathbf{j})$.

The direction of steepest ascent is the unit vector in this direction, $\mathbf{u} = \frac{\sqrt{2}}{2}\mathbf{i} - \frac{\sqrt{2}}{2}\mathbf{j}$, and the direction of steepest descent is $-\mathbf{u}$.

b. The unit vectors that point in the direction of no change are $\mathbf{v} = \pm\left(\frac{\sqrt{2}}{2}\mathbf{i} + \frac{\sqrt{2}}{2}\mathbf{j}\right)$, since $\mathbf{u}\cdot\mathbf{v} = 0$.

12.10.61 If x is determined by $f(x, y) = C$ we have $\dfrac{dy}{dx} = -\dfrac{f_x}{f_y} = -\dfrac{-2y}{-4x} = \dfrac{y}{2x}$. Therefore, the level curve $f(x, y) = 0$ has a vertical tangent at the point $(0, 0)$, so the tangent line has direction $\mathbf{j}$. The gradient of f at this point is $\nabla f(2, 0) = (-4x\mathbf{i} - 2y\mathbf{j})\Big|_{(2,0)} = -8\mathbf{i}$, which is perpendicular to the tangent direction.

12.10.63 Observe that $V = -\dfrac{k}{2}\left(\ln(x^2 + y^2) - \ln R^2\right)$; therefore

$$\mathbf{E} = -\nabla V = \frac{k}{2}\left(\frac{2x}{x^2 + y^2}\mathbf{i} + \frac{2y}{x^2 + y^2}\mathbf{j}\right) = \frac{kx}{x^2 + y^2}\mathbf{i} + \frac{ky}{x^2 + y^2}\mathbf{j}.$$

12.10.65 Let $F(x, y, z) = yze^{xz} - 8$; then $\nabla F(0, 2, 4) = \left\langle yz^2e^{xz}, ze^{xz}, (y + xyz)e^{xz}\right\rangle\Big|_{(0,2,4)} = \langle 32, 4, 2\rangle$, so the tangent plane at $(0, 2, 4)$ has equation $32x + 4(y - 2) + 2(z - 4) = 0$, or $16x + 2y + z - 8 = 0$. Similarly, $\nabla F(0, -8, -1) = \left\langle yz^2e^{xz}, ze^{xz}, (y + xyz)e^{xz}\right\rangle\Big|_{(0,-8,-1)} = \langle -8, -1, -8\rangle$, so the tangent plane at $(0, -8, -1)$ has equation $-8x - (y + 8) - 8(z + 1) = 0$, or $8x + y + 8z + 16 = 0$.

12.10.67 Let $f(x, y) = \ln(1 + xy)$; then $f_x(x, y) = \dfrac{y}{1 + xy}$ and $f_y(x, y) = \dfrac{x}{1 + xy}$, so the tangent plane at $(1, 2, \ln 3)$ has equation

$$z = f(1, 2) + f_x(1, 2)(x - 1) + f_y(1, 2)(y - 2) = \ln 3 + \tfrac{2}{3}(x - 1) + \tfrac{1}{3}(y - 2) = \tfrac{2}{3}x + \tfrac{1}{3}y + \ln 3 - \tfrac{4}{3}.$$

Similarly, the tangent plane at $(-2, -1, \ln 3)$ has equation

$$z = f(-2, -1) + f_x(-2, -1)(x + 2) + f_y(-2, -1)(y + 1) = \ln 3 - \tfrac{1}{3}(x + 2) - \tfrac{2}{3}(y + 1) = -\tfrac{1}{3}x - \tfrac{2}{3}y + \ln 3 - \tfrac{4}{3}.$$

12.10.69

a. The linear approximation is given by

$$L(x, y) = f(a, b) + f_x(a, b)(x - a) + f_y(a, b)(y - b) = 2 + (x - 2) + 5y = x + 5y.$$

b. This gives the estimate $f(1.95, 0.05) \approx 2.2$ (the actual answer is 2.205 to three decimal places).

12.10.71 We have $dV = \pi(2rh\,dr + r^2dh)$, so $\dfrac{dV}{V} = 2\dfrac{dr}{r} + \dfrac{dh}{h}$. Therefore, if the radius decreases by 3% and the height increases by 2%, the approximate change in volume is -4%.

12.10.73

a. We have $dV = \pi(2rh - h^2)dh = 2\pi(-0.05) = -0.1\pi\ \text{m}^3$ (notice that $r = 1.50$ m is constant, so there is no contribution from the dr term in dV).

b. The surface of the water is a disc with radius $s = \sqrt{2rh - h^2}$, so the surface area is $S = \pi(2rh - h^2)$. Therefore $dS = 2\pi(r - h)dh = -0.05\pi\ \text{m}^2$.

12.10.75 We have $f_x = x^2 + 2y$, $f_y = -y^2 + 2x$; therefore, the critical points satisfy the equations $y = -\dfrac{x^2}{2}$ and $x = \dfrac{y^2}{2}$. Eliminating y gives $x^4 = 8x$ so $x = 0, 2$, and the critical points are $(0, 0)$, $(2, -2)$. We also have $f_{xx} = 2x$, $f_{yy} = -2y$ and $f_{xy} = 2$; hence $D(x, y) = -4(1 + xy)$. We see that $D(0, 0) < 0$ so $(0, 0)$ is a saddle. We have $D(2, -2) > 0$, $f_{xx}(2, -2) > 0$, which by the Second Derivative Test implies that f has a local minimum at $(2, -2)$.

12.10.77 We have $f_x = -3x^2 - 6x$, $f_y = -3y^2 + 6y$; therefore the critical points must have $x = 0, -2$ and $y = 0, 2$. We also have $f_{xx} = -6x - 6$, $f_{yy} = -6y + 6$ and $f_{xy} = 0$; hence $D(x, y) = 36(x + 1)(y - 1)$. We see that $D(0, 0)$, $D(-2, 2) < 0$ so $(0, 0)$ and $(-2, 2)$ are saddles. We have $D(0, 2) > 0$, $f_{xx}(0, 2) < 0$, which by the Second Derivative Test implies that f has a local maximum at $(0, 2)$; $D(-2, 0) > 0$, $f_{xx}(-2, 0) > 0$, which by the Second Derivative Test implies that f has a local minimum at $(-2, 0)$.

12.10.79 First we find the critical points of f inside the square R: we have $f_x = 4x^3 - 4y$, $f_y = 4y^3 - 4x$; the equation $f_x = 0$ gives $y = x^3$, and then substituting this in the equation $f_y = 0$ gives $x^9 = x$, or $x = 0, \pm 1$. Hence, the critical points are $(0, 0)$ and $\pm(1, 1)$, which are all in the interior of R. We observe that the values of f at these points are $f(0, 0) = 1$, $f(1, 1) = f(-1, -1) = -1$. Next we must find the maximum and minimum values of f on the boundary of R, which consists of four segments. On the segment $-2 \le x \le 2$, $y = 2$, let $g(x) = f(x, 2) = x^4 - 8x + 17$: then g has a critical point at $x = \sqrt[3]{2}$, and find that g has extreme values $17 - 6\sqrt[3]{2}$ and 49 on $[0, 3]$. We also note that $f(y, x) = f(-x, -y) = f(x, y)$; therefore f takes the same values on all four segments of the square. Hence, the absolute minimum and maximum values of f on R are $f(1, 1) = f(-1, -1) = -1$ and $f(2, -2) = f(-2, 2) = 49$.

12.10.81 The Lagrange multiplier conditions are $1 = 4\lambda x^3$, $2 = 4\lambda y^3$, $x^4 + y^4 = 1$. Hence, $\dfrac{1}{4x^3} = \lambda = \dfrac{1}{2y^3} \Longrightarrow y = \sqrt[3]{2}x$; substituting this in the constraint gives $(1 + 2^{4/3})x^4 = 1$, so $x = \pm(1 + 2^{4/3})^{-1/4}$, $y = \pm 2^{1/3}(1 + 2^{4/3})^{-1/4}$, and the extreme values of f on the closed bounded set given by $x^4 + y^4 = 1$ are $f\left(\pm(1 + 2^{4/3})^{-1/4}, \pm 2^{1/3}(1 + 2^{4/3})^{-1/4}\right) = \pm\left(1+2\sqrt[3]{2}\right)^{3/4}$.

12.10.83 The Lagrange multiplier conditions are $1 = 2\lambda x$, $2 = 2\lambda y$, $-1 = 2\lambda z$, $x^2 + y^2 + z^2 = 1$. Hence $2\lambda = \dfrac{1}{x} = \dfrac{2}{y} = -\dfrac{1}{z} \Longrightarrow y = 2x$, $z = -x$; substituting these in the constraint gives $6x^2 = 1$ so $x = \pm\frac{\sqrt{6}}{6}$ and we obtain solutions $\pm\left(\frac{\sqrt{6}}{6}, \frac{\sqrt{6}}{3}, -\frac{\sqrt{6}}{6}\right)$. Therefore the extreme values of f on the closed bounded set given by $x^2 + y^2 + z^2 = 1$ are $f\left(\frac{\sqrt{6}}{6}, \frac{\sqrt{6}}{3}, -\frac{\sqrt{6}}{6}\right) = \sqrt{6}$, $f\left(-\frac{\sqrt{6}}{6}, -\frac{\sqrt{6}}{3}, \frac{\sqrt{6}}{6}\right) = -\sqrt{6}$.

12.10.85 Let (x, y) be the corner of the rectangle in the first quadrant; then the perimeter of the rectangle is $4(x + y)$, so it suffices to find the maximum value of $x + y$ subject to the constraint $\dfrac{x^2}{a^2} + \dfrac{y^2}{b^2} = 1$. The Lagrange multiplier conditions are $1 = \dfrac{2\lambda x}{a^2}$, $1 = \dfrac{2\lambda y}{b^2}$, $\dfrac{x^2}{a^2} + \dfrac{y^2}{b^2} = 1$. Hence, $2\lambda = \dfrac{a^2}{x} = \dfrac{b^2}{y} \Longrightarrow y = \dfrac{b^2 x}{a^2}$; substituting in the constraint gives $x^2(a^2 + b^2) = a^4$ so $x = \dfrac{a^2}{\sqrt{a^2 + b^2}}$, $y = \dfrac{b^2}{\sqrt{a^2 + b^2}}$, and the dimensions of the rectangle with greatest perimeter are $\dfrac{2a^2}{\sqrt{a^2 + b^2}}$ by $\dfrac{2b^2}{\sqrt{a^2 + b^2}}$.

12.10.87 It suffices to minimize the function $f(x, y, z) = (x - 1)^2 + (y - 3)^2 + (z - 1)^2$ subject to the constraint $x^2 + y^2 - z^2 = 0$. The Lagrange multiplier conditions are equivalent to $x - 1 = \lambda x$, $y - 3 = \lambda y$, $z - 1 = -\lambda z$, $x^2 + y^2 - z^2 = 0$. The first two equations give $\lambda xy = (x - 1)y = (y - 3)x \Longrightarrow y = 3x$ and similarly, the first and third equations give $\lambda xz = (x - 1)z = -x(z - 1) \Longrightarrow (2x - 1)z = x \Rightarrow z = \sqrt{10}x$. Substituting these equations in the constraint gives $(2x - 1)^2 \cdot 10x^2 = x^2$, so either $x = 0$ (and hence $y = z = 0$ as well) or $10(2x - 1)^2 = 1$, which has solutions $x = \frac{1}{2} \pm \frac{\sqrt{10}}{20}$. Therefore, there are three solutions to the Lagrange conditions: $(0, 0, 0)$, $\left(\frac{1}{2} \pm \frac{\sqrt{10}}{20}, \frac{3}{2} \pm \frac{3\sqrt{10}}{20}, \frac{1}{2} \pm \frac{\sqrt{10}}{2}\right)$. We see that $f(0, 0, 0) = 11$, $f\left(\frac{1}{2} \pm \frac{\sqrt{10}}{20}, \frac{3}{2} \pm \frac{3\sqrt{10}}{20}, \frac{1}{2} \pm \frac{\sqrt{10}}{2}\right) = \frac{11}{2} \mp \sqrt{10}$, so the closest point is $\left(\frac{1}{2} + \frac{\sqrt{10}}{20}, \frac{3}{2} + \frac{3\sqrt{10}}{20}, \frac{1}{2} + \frac{\sqrt{10}}{2}\right)$. (The function $f(x, y, z \to \infty)$ as either x, y or $z \to \infty$ on the cone; therefore, f must have minimum somewhere on the cone, which corresponds to the point we found.)

Chapter 13

13.1 Double Integrals Over Rectangular Regions

13.1.1 $\displaystyle\int_0^2\int_1^3 x\,y\;dy\,dx$

13.1.3 With respect to x first: $\displaystyle\int_1^5\int_{-2}^4 f(x,\,y)\,dx\,dy$; with respect to y first: $\displaystyle\int_{-2}^4\int_1^5 f(x,\,y)\,dy\,dx$

13.1.5 $\displaystyle\int_1^3\int_0^2 x^2\,y\,dx\,dy = \int_1^3 \left(\frac{x^3y}{3}\right)\Bigg|_{x=0}^{x=2} dy = \frac{8}{3}\int_1^3 y\,dy = \frac{8}{3}\left(\frac{y^2}{2}\right)\Bigg|_{y=1}^{y=3} = \frac{32}{3}$

13.1.7 $\displaystyle\int_1^3\int_0^{\pi/2} x\sin y\,dy\,dx = \int_1^3 (-x\cos y)\Bigg|_{y=0}^{y=\pi/2} dx = \int_1^3 x\,dx = \left(\frac{x^2}{2}\right)\Bigg|_{x=1}^{x=3} = 4$

13.1.9 $\displaystyle\int_1^4\int_0^4 \sqrt{u\,v}\,du\,dv = \int_1^4\left(\frac{2}{3}u^{3/2}v^{1/2}\right)\Bigg|_{u=0}^{u=4} dv = \int_1^4\left(\frac{16}{3}v^{1/2}\right)dv = \left(\frac{32}{9}v^{3/2}\right)\Bigg|_{v=1}^{v=4} = \frac{224}{9}$

13.1.11 $\displaystyle\int_1^{\ln 5}\int_0^{\ln 3} e^{x+y}\,dx\,dy = \int_1^{\ln 5}\int_0^{\ln 3}(e^x)\,(e^y)\,dx\,dy = \int_1^{\ln 5}(e^x\cdot e^y\,)\Bigg|_{x=0}^{x=\ln 3} dy = \int_1^{\ln 5}(2e^y\,)\,dy = (2e^y)\Bigg|_{y=1}^{y=\ln 5}$

$\displaystyle = 10 - 2e$

13.1.13 $\displaystyle\iint_R (x+2y)\;dA = \int_1^4\int_0^3 (x+2y)\,dx\,dy = \int_1^4\left(\frac{x^2}{2}+2\,x\,y\right)\Bigg|_{x=0}^{x=3} dy = \int_1^4\left(\frac{9}{2}+6y\right)dy$

$\displaystyle = \left(\frac{9}{2}y+3y^2\right)\Bigg|_{y=1}^{y=4} = \frac{117}{2}$

13.1.15 $\displaystyle\iint_R \sqrt{\frac{x}{y}}\;dA = \int_1^4\int_0^1\left(\frac{x}{y}\right)^{1/2} dx\,dy = \int_1^4\left(\frac{2}{3}\cdot\frac{x^{3/2}}{y^{1/2}}\right)\Bigg|_{x=0}^{x=1} dy = \int_1^4\left(\frac{2}{3\,y^{1/2}}\right)dy = \left(\frac{4}{3}y^{1/2}\right)\Bigg|_{y=1}^{y=4} = \frac{4}{3}$

13.1.17 $\displaystyle\iint_R e^{x+2y}\;dA = \int_1^{\ln 3}\int_0^{\ln 2}(e^x\cdot e^{2y})\,dx\,dy = \int_1^{\ln 3}(e^x\cdot e^{2y})\Bigg|_{x=0}^{x=\ln 2} dy = \int_1^{\ln 3} e^{2y}\,dy = \left(\frac{1}{2}e^{2y}\right)\Bigg|_{y=1}^{y=\ln 3}$

$\displaystyle = \frac{1}{2}(9-e^2)$

13.1.19 $\displaystyle\iint_R (x^5 - y^5)^2\, dA = \int_{-1}^{1}\int_0^1 (x^{10} - 2x^5y^5 + y^{10})\, dx\, dy = \int_{-1}^{1}\left(\frac{x^{11}}{11} - \frac{1}{3}x^6y^5 + x\,y^{10}\right)\Bigg|_{x=0}^{x=1} dy$

$$= \int_{-1}^{1}\left(\frac{1}{11} - \frac{1}{3}y^5 + y^{10}\right)dy = \left(\frac{y}{11} - \frac{y^6}{18} + \frac{y^{11}}{11}\right)\Bigg|_{y=-1}^{y=1} = \frac{4}{11}$$

13.1.21 Integrate first with respect to y.

$$\iint_R x^5 e^{x^3 y}\, dA = \int_0^{\ln 2}\int_0^1 x^5 e^{x^3 y}\, dy\, dx = \int_0^{\ln 2}\left(\frac{x^5}{x^3}e^{x^3 y}\right)\Bigg|_{y=0}^{y=1} dx = \int_0^{\ln 2}\left(x^2 e^{x^3} - x^2\right) dx$$

$$= \left(\frac{1}{3}e^{x^3} - \frac{x^3}{3}\right)\Bigg|_{x=0}^{x=\ln 2} = \frac{1}{3}\left(e^{(\ln 2)^3} - (\ln 2)^3 - 1\right)$$

13.1.23 Integrate first with respect to y.

$$\iint_R \frac{x}{(1+x\,y)^2}\, dA = \int_0^4\int_1^2 \frac{x}{(1+x\,y)^2}\, dy\, dx \quad [\text{Let } u = 1 + x\,y\text{, then } du = x\cdot dy]$$

$$= \int_0^4\left(\frac{-1}{1+x\,y}\right)\Bigg|_{y=1}^{y=2} dx = \int_0^4\left(\frac{1}{1+x} - \frac{1}{1+2x}\right)dx$$

$$= \left(\ln|1+x| - \frac{1}{2}\ln|1+2x|\right)\Bigg|_{x=0}^{x=4} = \ln 5 - \frac{1}{2}\ln 9 = \ln\left(\frac{5}{3}\right)$$

13.1.25 $\displaystyle \overline{f}_{\text{ave}} = \frac{1}{\text{area of } R}\iint_R f(x, y)\, dA = \frac{1}{6\ln 2}\int_0^6\int_0^{\ln 2} e^{-y}\, dy\, dx = \frac{1}{6\ln 2}\int_0^6 (-e^{-y})\Bigg|_{y=0}^{y=\ln 2} dx$

$$= \frac{1}{6\ln 2}\int_0^6\left(\frac{1}{2}\right)dx = \frac{1}{6\ln 2}\left(\frac{x}{2}\right)\Bigg|_{x=0}^{x=6} = \frac{1}{6\ln 2}(3) = \frac{1}{2\ln 2}$$

13.1.27 $\displaystyle \overline{f}_{\text{ave}} = \frac{1}{\text{area of } R}\iint_R f(x, y)\, dA = \frac{1}{8}\int_{-2}^{2}\int_0^2 (x^2 + y^2)\, dy\, dx = \frac{1}{8}\int_{-2}^{2}\left(x^2 y + \frac{1}{3}y^3\right)\Bigg|_{y=0}^{y=2} dx$

$$= \frac{1}{8}\int_{-2}^{2}\left(2x^2 + \frac{8}{3}\right)dx = \frac{1}{8}\left(\frac{2}{3}x^3 + \frac{8}{3}x\right)\Bigg|_{x=-2}^{x=2} = \frac{8}{3}$$

13.1.29

a. True. The region, $R = \{(x, y)\,|\,1 \le x \le 3,\ 4 \le y \le 6\}$ is a rectangle that has width 2 and length 2, thus is a square.

b. False. The region for $\displaystyle\int_4^6\int_1^3 f(x, y)\, dx\, dy$ is $R = \{(x, y)\,|\,1 \le x \le 3,\ 4 \le y \le 6\}$ is not equivalent to the region for $\displaystyle\int_4^6\int_1^3 f(x, y)\, dy\, dx$ which is $R = \{(x, y)\,|\,4 \le x \le 6,\ 1 \le y \le 3\}$ thus Fubini's Theorem does not apply.

c. True. The region, $R = \{(x, y)\,|\,1 \le x \le 3,\ 4 \le y \le 6\}$.

13.1.31

a. total population = (total population region 1) + (total population region 2) + ... + (total population region 9)
= (population density of region 1) × (area of region 1) + (population density of region 2) × (area of region 2)
+ ... + (population density of region 9) × (area of region 9)

$$= (350)\left(\frac{1}{2}\right) + (300)\left(\frac{1}{4}\right) + (150)\left(\frac{3}{4}\right) + (500)\left(\frac{1}{2}\right) + (400)\left(\frac{1}{4}\right) + (250)\left(\frac{3}{4}\right) + (250)(1) + (200)\left(\frac{1}{2}\right) + (150)\left(\frac{3}{2}\right) = 1475$$

b. The calculation above could be expressed as the sum $\sum_{k=1}^{3} f(\overline{x}_k, \overline{y}_k)\Delta A_k$ where $f(\overline{x}_k, \overline{y}_k)$ represents the population density for each sub-region of R with area ΔA_k. This sum can be used as an approximation for a 'continuous' function $f(x, y)$ that represents the population density at each point in R then $\sum_{k=1}^{n} f(\overline{x}_k, \overline{y}_k)\Delta A_k \approx \iint_R f(x, y)\, dA$.

13.1.33 $V = \int_0^6 \int_1^2 10\, dy\, dx = \int_0^6 (10y)\Big|_{y=1}^{y=2} dx$

$$= \int_0^6 10\, dx = (10x)\Big|_{x=0}^{x=6} = 60$$

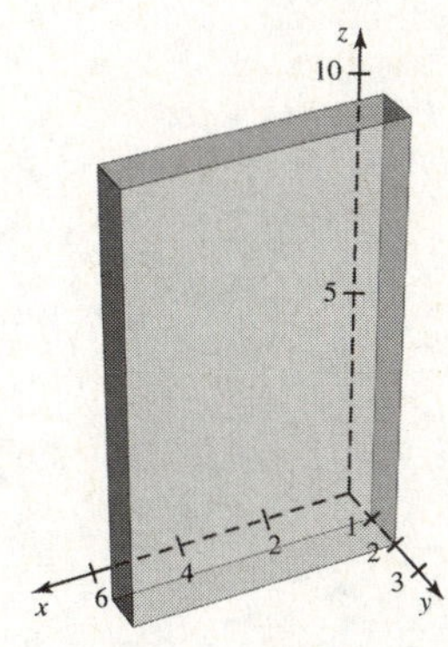

13.1.35 $\int_1^e \int_0^1 \frac{x}{x+y}\, dy\, dx = \int_1^e (x\ln(x+y))\Big|_{y=0}^{y=1} dx = \int_1^e (x\ln(x+1) - x\ln x)\, dx$

$$= \left(\frac{x^2}{2}\ln(x+1) - \frac{1}{4}x^2 + \frac{1}{2}x - \frac{1}{2}\ln(x+1) - \frac{x^2}{2}\ln x + \frac{x^2}{4}\right)\Bigg|_{x=1}^{x=e}$$

$$= \frac{1}{2}(e^2\ln(e+1) - e^2 + e - \ln(e+1) - 1)$$

13.1.37 $\int_0^1 \int_1^4 \frac{3y}{\sqrt{x+y^2}}\, dx\, dy = \int_0^1 \left(6y\sqrt{x+y^2}\right)\Big|_{x=1}^{x=4} dy = 6\int_0^1 \left(y\sqrt{4+y^2} - y\sqrt{1+y^2}\right) dy$

$$= 6\left(\frac{1}{3}(4+y^2)^{3/2} - \frac{1}{3}(1+y^2)^{3/2}\right)\Bigg|_{y=0}^{y=1} = 2\left(5\sqrt{5} - 2\sqrt{2} - 7\right)$$

13.1.39 $\int_{-2}^2 \int_0^{\ln 4} e^{-x}\, dx\, dy = \int_{-2}^2 (-e^{-x})\Big|_{x=0}^{x=\ln 4} dy = \int_{-2}^2 \frac{3}{4} dy = \left(\frac{3}{4}y\right)\Big|_{y=-2}^{y=2} = 3$

13.1.41 $\int_{-1}^3 \int_0^2 (24 - 3x - 4y)\, dy\, dx = \int_{-1}^3 (24y - 3x\,y - 2y^2)\Big|_{y=0}^{y=2} dx = \int_{-1}^3 (40 - 6x) dx = (40x - 3x^2)\Big|_{x=-1}^{x=3}$

$= 136$

13.1.43 $\int_0^a \int_0^\pi \sin(x+y)\,dx\,dy = \int_0^a (-\cos(x+y))\Big|_{x=0}^{x=\pi} dy = \int_0^a (\cos y - \cos(\pi+y))\,dy$

$$= (\sin y - \sin(\pi+y))\Big|_{y=0}^{y=a} = (\sin a - \sin(\pi+a)) = 1$$

Solving for a yields: $(\sin a - \sin(\pi+a)) = 1 \Rightarrow 2\sin a = 1 \Rightarrow \sin a = \frac{1}{2}$. The solutions are $a = \frac{\pi}{6}$ and $a = \frac{5\pi}{6}$.

13.1.45 $\overline{f} = \frac{1}{a^2}\int_0^a \int_0^a (4 - x^2 - y^2)\,dy\,dx = \frac{1}{a^2}\int_0^a \left(4y - x^2 y - \frac{y^3}{3}\right)\Big|_{y=0}^{y=a} dx = \frac{1}{a^2}\int_0^a \left(4a - a\,x^2 - \frac{a^3}{3}\right) dx$

$$= \frac{1}{a^2}\left(4a\,x - \frac{a\,x^3}{3} - \frac{a^3}{3}x\right)\Big|_{x=0}^{x=a} = \frac{1}{a^2}\left(4a^2 - \frac{2a^4}{3}\right) = 4 - \frac{2a^2}{3} = \overline{f}$$

Set $\overline{f} = 0$ and solve for a. $4 - \frac{2a^2}{3} = 0 \Rightarrow a^2 = 6 \Rightarrow a = \pm\sqrt{6}$, choose $a > 0$, thus $a = \sqrt{6}$.

13.1.47

a. $m = \iint\limits_R \rho(x,y)\,dA = \int_0^\pi \int_0^{\pi/2} (1 + \sin x)\,dx\,dy = \int_0^\pi (x - \cos x)\Big|_{x=0}^{x=\pi/2} dy = \int_0^\pi \left(\frac{\pi}{2} + 1\right) dy$

$$= \left(\frac{\pi}{2} + 1\right) y\Big|_{y=0}^{y=\pi} = \frac{\pi^2}{2} + \pi$$

b. $m = \iint\limits_R \rho(x,y)\,dA = \int_0^{\pi/2} \int_0^\pi (1 + \sin y)\,dy\,dx = \int_0^{\pi/2} (y - \cos y)\Big|_{y=0}^{y=\pi} dx = \int_0^\pi (\pi + 2)\,dx$

$$= (\pi + 2)\,x\Big|_{x=0}^{x=\pi/2} = \frac{\pi^2}{2} + \pi$$

c. $m = \iint\limits_R \rho(x,y)\,dA = \int_0^\pi \int_0^{\pi/2} (1 + \sin x \sin y)\,dx\,dy = \int_0^\pi (x - \cos x \sin y)\Big|_{x=0}^{x=\pi/2} dy$

$$= \int_0^\pi \left(\frac{\pi}{2} + \sin y\right) dy = \left(\frac{\pi}{2}y - \cos y\right)\Big|_{y=0}^{y=\pi} = \frac{\pi^2}{2} + 2$$

13.1.49 The area of the constant cross section of S is $A = \int_a^b f(x)\,dx$ for any value of y. The volume of S can be expressed as $\int_c^d \int_a^b f(x)\,dx\,dy = \int_c^d A\,dy = A \cdot y\Big|_{y=c}^{y=d} = A(d-c)$.

13.1.51 $\iint\limits_R \frac{\partial^2 f}{\partial x\,\partial y}\,dA = \int_0^b \int_0^a \left(\frac{\partial^2 f}{\partial x\,\partial y}\right) dx\,dy = \int_0^b \left(\frac{\partial f}{\partial y}\right)\Big|_{x=0}^{x=a} dy = \int_0^b \left(\frac{\partial f}{\partial y}(a,y) - \frac{\partial f}{\partial y}(0,y)\right) dy$

$$= (f(a,y) - f(0,y))\Big|_{y=0}^{y=b} = f(a,b) - f(0,b) - f(a,0) + f(0,0)$$

13.1.53 $\iint\limits_R x^{2r-1} y^{s-1} f(x^r \cdot y^s)\,dA = \int_0^1 \int_0^1 x^{2r-1} y^{s-1} f(x^r \cdot y^s)\,dy\,dx$

Choose $u = x^r \cdot y^s$, then $\frac{du}{s \cdot x^r} = y^{s-1}\,dy$, and

$$\int x^{2r-1} y^{s-1} f(x^r y^s)\,dy = \int \frac{x^{2r-1}}{s \cdot x^r} f(u)\,du = \frac{1}{s}x^{r-1}F(u) + C = \frac{1}{s}x^{r-1}F(x^r y^s) + C.$$

Thus

$$\int_0^1 x^{2r-1}y^{s-1}\,f(x^r y^s)\,dy = \frac{1}{s}x^{r-1}F(x^r y^s)\Big|_{y=0}^{y=1} = \frac{1}{s}x^{r-1}F(x^r)\,dx$$

so

$$\int_0^1\int_0^1 x^{2r-1}y^{s-1}\,f(x^r y^s)\,dy\,dx = \frac{1}{s}\int_0^1 x^{r-1}F(x^r)\,dx$$

Now choose $u = x^r$. Then $\dfrac{du}{r} = x^{r-1}dx$, and

$$\int x^{r-1}F(x^r)\,dx = \frac{1}{r}\int F(u)\,du = \frac{1}{r}G(u) + C = \frac{1}{r}G(x^r) + C$$

Thus

$$\int_0^1\int_0^1 x^{2r-1}y^{s-1}\,f(x^r y^s)\,dy\,dx = \frac{1}{s}\left(\frac{1}{r}G(x^r)\right)\Big|_{x=0}^{x=1} = \frac{G(1)-G(0)}{rs}$$

13.2 Double Integrals Over General Regions

13.2.1 A region is bounded below by $y = f(x)$ and above by $y = g(x)$, on the left by $x = a$ and on the right by $x = b$.
$R = \{(x,\,y)\,|\,a \le x \le b,\ f(x) \le y \le g(x)\}$

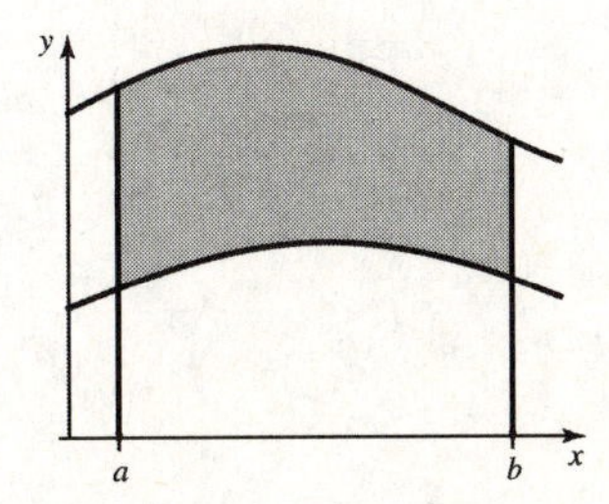

13.2.3 Integrate first with respect to x, then with respect to y. $\displaystyle\iint_R f(x,\,y)\,dA = \int_0^1\int_{y-1}^{1-y} x\,y\;dx\,dy$

13.2.5 $\displaystyle\int_0^1\int_{x^2}^{\sqrt{x}} f(x,\,y)\;dy\,dx$

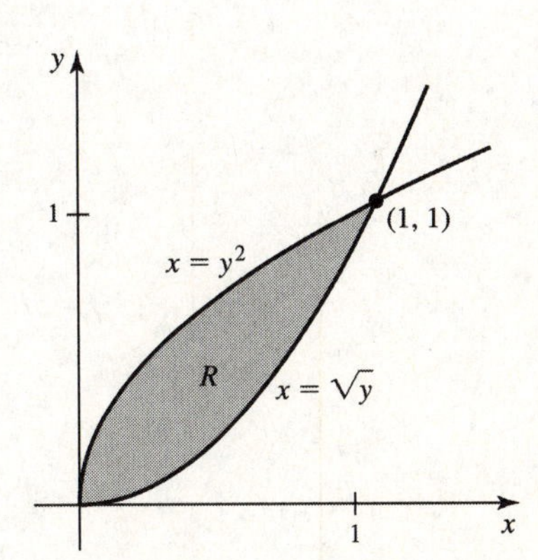

13.2.7 $\displaystyle\iint_R f(x,\,y)\,dA = \int_0^2\int_{x^2}^{4x} f(x,\,y)\;dy\,dx$

13.2.9 $\displaystyle\iint_R f(x, y)\, dA = \int_0^{\frac{\pi}{4}}\int_{\sin x}^{\cos x} f(x, y)\; dy\, dx$

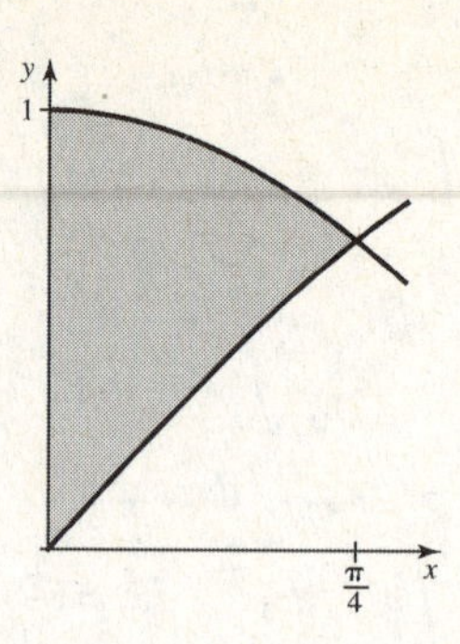

13.2.11 $\displaystyle\iint_R f(x, y)\, dA = \int_1^{2}\int_{x+1}^{2x+4} f(x, y)\; dy\, dx$

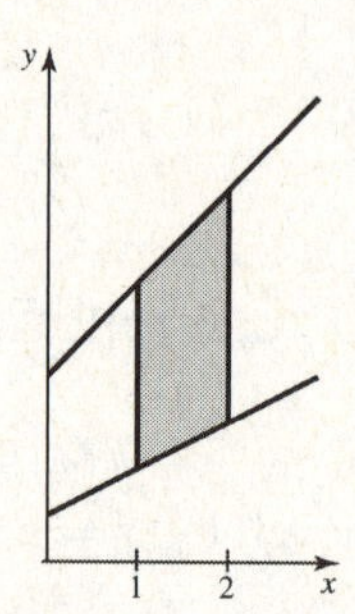

13.2.13 $\displaystyle\int_0^2\int_{x^2}^{2x} x\, y\; dy\, dx = \int_0^2 \left(\frac{x\, y^2}{2}\right)\Bigg|_{y=x^2}^{y=2x} dx = \frac{1}{2}\int_0^2 (4x^3 - x^5)\, dx = \frac{1}{2}\left(x^4 - \frac{x^6}{6}\right)\Bigg|_{x=0}^{x=2} = \frac{8}{3}$

13.2.15 $\displaystyle\int_{-\pi/4}^{\pi/4}\int_{\sin x}^{\cos x} dy\, dx = \int_{-\pi/4}^{\pi/4} (y)\Bigg|_{y=\sin x}^{y=\cos x} dx = \int_{-\pi/4}^{\pi/4} (\cos x - \sin x)\, dx = (\sin x + \cos x)\Bigg|_{x=-\pi/4}^{x=\pi/4} = \sqrt{2}$

13.2.17 $\displaystyle\int_{-2}^2\int_{x^2}^{8-x^2} x\, dy\, dx = \int_{-2}^2 (x\, y)\Bigg|_{y=x^2}^{y=8-x^2} dx = \int_{-2}^2 (8x - 2x^3)\, dx = \left(4x^2 - \frac{x^4}{2}\right)\Bigg|_{x=-2}^{x=2} = 0$

13.2.19 $\displaystyle\iint_R x\, y\, dA = \int_0^1\int_{2x+1}^{-2x+5} (x\, y)\; dy\, dx = \int_0^1 \left(\frac{x\, y^2}{2}\right)\Bigg|_{y=2x+1}^{y=-2x+5} dx = \int_0^1 (12x - 12x^2)\, dx$

$\displaystyle= (6x^2 - 4x^3)\Bigg|_{x=0}^{x=1} = 2$

13.2.21 $\displaystyle\iint_R y^2\, dA = \int_{-1}^1\int_{-x-1}^{2x+2} (y^2)\; dy\, dx = \int_{-1}^1 \left(\frac{y^3}{3}\right)\Bigg|_{y=-x-1}^{y=2x+2} dx = \frac{1}{3}\int_{-1}^1 (8(x+1)^3 + (x+1)^3)\, dx$

$\displaystyle= 3\int_{-1}^1 (x+1)^3\, dx = \frac{3}{4}(x+1)^4\Bigg|_{x=-1}^{x=1} = 12$

13.2.23 $\displaystyle\iint_R f(x, y)\, dA = \int_0^{18}\int_{y/2}^{(y+9)/3} f(x, y)\; dx\, dy$

13.2.25 $\displaystyle\iint_R f(x, y)\, dA = \int_0^{23}\int_{(y-3)/2}^{(y+7)/3} f(x, y)\; dx\, dy$

13.2.27 $\displaystyle\iint_R f(x,\, y)\, dA = \int_1^4 \int_0^{4-y} f(x,\, y)\; dx\, dy = \int_0^3 \int_1^{4-x} f(x,\, y)\, dy\, dx$

13.2.29 $R = \{(x,\, y)\,|\, y \le x \le 4 - y,\ -1 \le y \le 2\}$

$$\int_{-1}^2 \int_y^{4-y} dx\, dy = \int_{-1}^2 (x)\Big|_{x=y}^{x=4-y} dy$$

$$= \int_{-1}^2 (4 - 2y)\, dy = (4 - 2y)\Big|_{y=-1}^{y=2} = 9$$

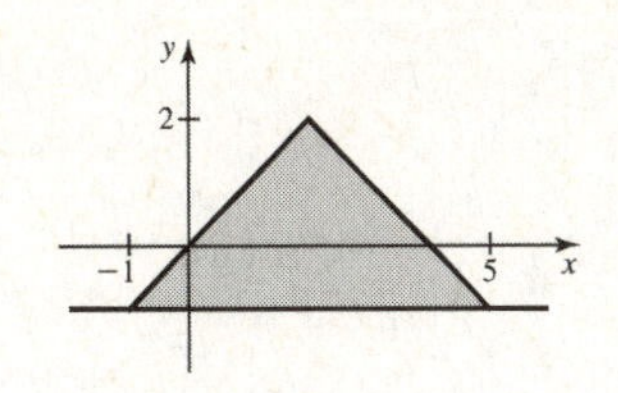

13.2.31 $R = \{(x,\, y)\,|\, -\sqrt{16 - y^2} \le x \le \sqrt{16 - y^2},\ 0 \le y \le 4\}$

$$\int_0^4 \int_{-\sqrt{16-y^2}}^{\sqrt{16-y^2}} 2xy\, dx\, dy$$

$$= \int_0^4 (x^2\, y)\Big|_{x=-\sqrt{16-y^2}}^{x=\sqrt{16-y^2}} dy$$

$$= \int_0^4 0\, dy = 0$$

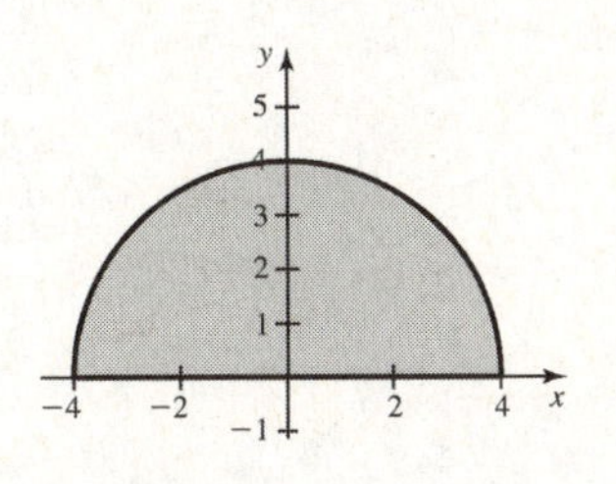

13.2.33 $R = \{(x,\, y)\,|\, e^y \le x \le 2,\ 0 \le y \le \ln 2\}$

$$\int_0^{\ln 2} \int_{e^y}^2 \left(\frac{y}{x}\right) dx\, dy$$

$$= \int_0^{\ln 2} (y \ln|\, x\,|)\Big|_{x=e^y}^{x=2} dy$$

$$= \int_0^{\ln 2} (y \ln 2 - y^2)\, dy$$

$$= \left(\frac{\ln 2}{2} y^2 - \frac{y^3}{3}\right)\Big|_{y=0}^{y=\ln 2} = \frac{(\ln 2)^3}{6}$$

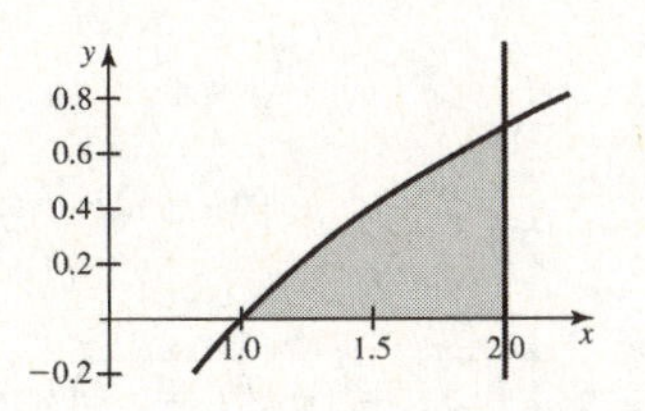

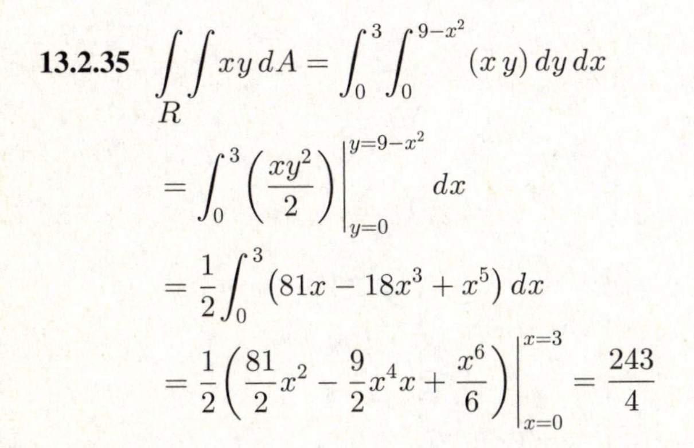

13.2.35 $\displaystyle\iint_R xy\, dA = \int_0^3 \int_0^{9-x^2} (x\, y)\, dy\, dx$

$$= \int_0^3 \left(\frac{xy^2}{2}\right)\Big|_{y=0}^{y=9-x^2} dx$$

$$= \frac{1}{2}\int_0^3 (81x - 18x^3 + x^5)\, dx$$

$$= \frac{1}{2}\left(\frac{81}{2}x^2 - \frac{9}{2}x^4 x + \frac{x^6}{6}\right)\Big|_{x=0}^{x=3} = \frac{243}{4}$$

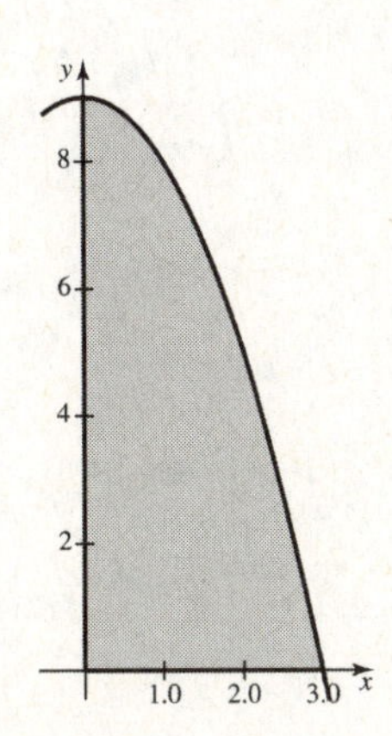

13.2.37 $\displaystyle\iint_R y^2\,dA = \int_0^8 \int_{\frac{y-4}{2}}^{\sqrt[3]{y}} y^2\,dx\,dy$

$\displaystyle= \int_0^8 (xy^2)\Big|_{x=\frac{y-4}{2}}^{x=\sqrt[3]{y}} dy$

$\displaystyle= \int_0^8 \left(y^{7/3} - \frac{y^3}{2} + 2y^2\right) dy$

$\displaystyle= \left(\frac{3}{10}y^{10/3} - \frac{y^4}{8} + \frac{2}{3}y^3\right)\Big|_{y=0}^{y=8} = \frac{2048}{15}$

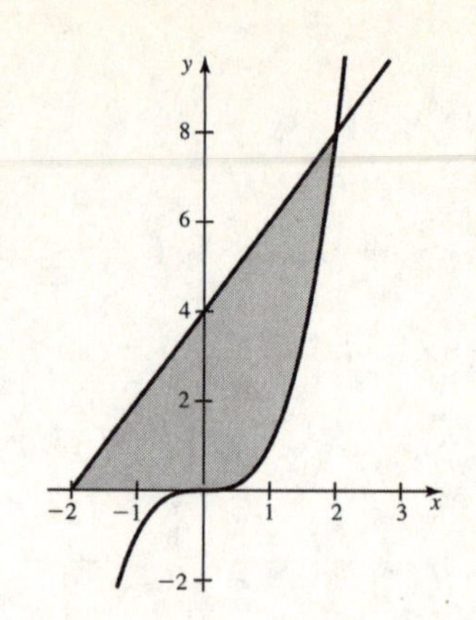

13.2.39 $\displaystyle V = \int_0^4 \int_0^{2-\frac{x}{2}} (8-2x-4y)\,dy\,dx = \int_0^4 (8y - 2xy - 2y^2)\Big|_{y=0}^{y=2-\frac{x}{2}} dx = \int_0^4 \left(8 - 4x - \frac{1}{2}x^2\right) dx$

$\displaystyle= \left(8x - 2x^2 - \frac{1}{6}x^3\right)\Big|_{x=0}^{x=4} = \frac{32}{3},$

or $\displaystyle V = \int_0^2 \int_0^{4-2y} (8-2x-4y)\,dx\,dy = \frac{32}{3}.$

13.2.41 $\displaystyle V = \int_{-1}^1 \int_{-\sqrt{1-x^2}}^{\sqrt{1-x^2}} (12+x+y)\,dy\,dx = \int_{-1}^1 \left(12y + xy + \frac{1}{2}y^2\right)\Big|_{y=-\sqrt{1-x^2}}^{y=\sqrt{1-x^2}} dx$

$\displaystyle= \int_{-1}^1 \left(24\sqrt{1-x^2} + 2x\sqrt{1-x^2}\right) dx = \left(12x\sqrt{1-x^2} + 12\sin^{-1}x - \frac{2}{3}(1-x^2)^{3/2}\right)\Big|_{x=-1}^{x=1}$

$= 12\left(\sin^{-1}(1) - \sin^{-1}(-1)\right) = 12\pi.$

13.2.43 $\displaystyle\int_0^2 \int_{x^2}^{2x} f(x,\,y)\,dy\,dx = \int_0^4 \int_{y/2}^{\sqrt{y}} f(x,\,y)\,dx\,dy$

13.2.45 $\displaystyle\int_{1/2}^1 \int_0^{-\ln y} f(x,\,y)\,dx\,dy = \int_0^{\ln 2} \int_{1/2}^{e^{-x}} f(x,\,y)\,dy\,dx$

13.2.47 $\displaystyle\int_0^1 \int_0^{\cos^{-1}y} f(x,\,y)\,dx\,dy = \int_0^{\pi/2} \int_0^{\cos x} f(x,\,y)\,dy\,dx$

13.2.49 $\displaystyle\int_0^1 \int_y^1 e^{x^2}\,dx\,dy = \int_0^1 \int_0^x e^{x^2}\,dy\,dx$

$\displaystyle= \int_0^1 \left(y\,e^{x^2}\right)\Big|_{y=0}^{y=x} dx = \int_0^1 \left(x\,e^{x^2}\right) dx$

$\displaystyle= \left(\frac{1}{2}e^{x^2}\right)\Big|_{x=0}^{x=1} = \frac{1}{2}(e-1)$

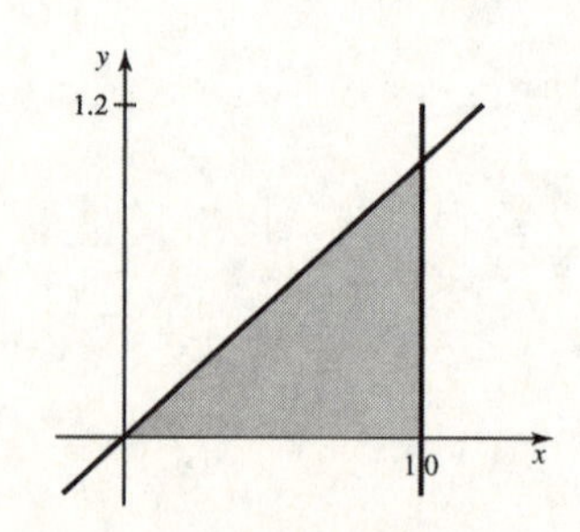

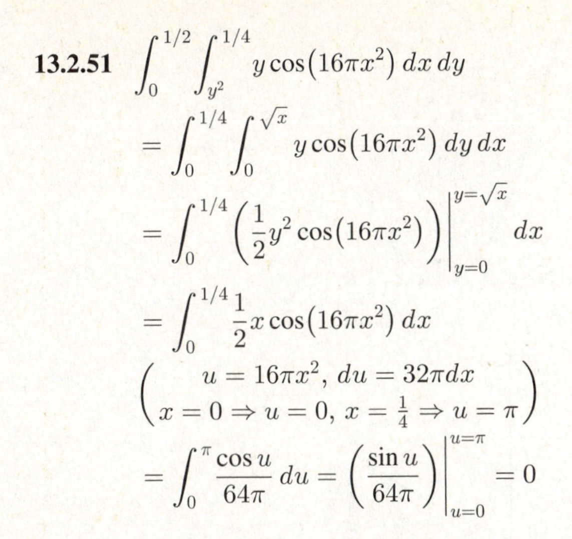

13.2.51 $\displaystyle\int_0^{1/2}\int_{y^2}^{1/4} y\cos(16\pi x^2)\,dx\,dy$

$$= \int_0^{1/4}\int_0^{\sqrt{x}} y\cos(16\pi x^2)\,dy\,dx$$

$$= \int_0^{1/4}\left(\frac{1}{2}y^2\cos(16\pi x^2)\right)\Bigg|_{y=0}^{y=\sqrt{x}}\,dx$$

$$= \int_0^{1/4}\frac{1}{2}x\cos(16\pi x^2)\,dx$$

$$\left(\begin{array}{c} u = 16\pi x^2,\ du = 32\pi dx \\ x = 0 \Rightarrow u = 0,\ x = \frac{1}{4} \Rightarrow u = \pi \end{array}\right)$$

$$= \int_0^{\pi}\frac{\cos u}{64\pi}\,du = \left(\frac{\sin u}{64\pi}\right)\Bigg|_{u=0}^{u=\pi} = 0$$

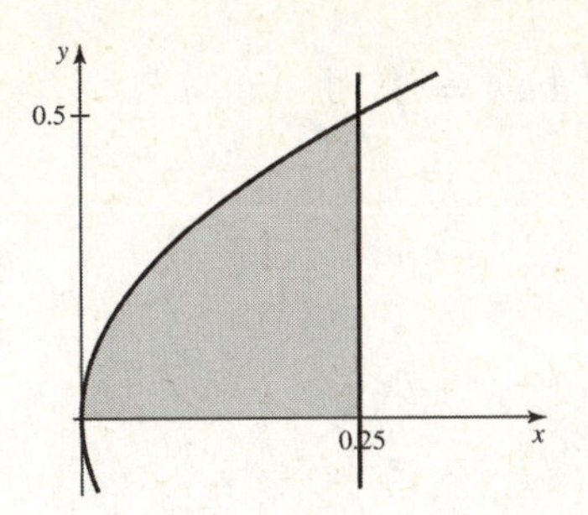

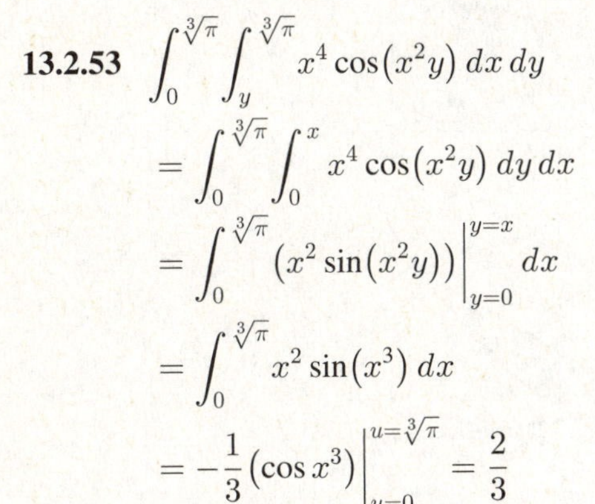

13.2.53 $\displaystyle\int_0^{\sqrt[3]{\pi}}\int_y^{\sqrt[3]{\pi}} x^4\cos(x^2y)\,dx\,dy$

$$= \int_0^{\sqrt[3]{\pi}}\int_0^{x} x^4\cos(x^2y)\,dy\,dx$$

$$= \int_0^{\sqrt[3]{\pi}}\left(x^2\sin(x^2y)\right)\Big|_{y=0}^{y=x}\,dx$$

$$= \int_0^{\sqrt[3]{\pi}} x^2\sin(x^3)\,dx$$

$$= -\frac{1}{3}\left(\cos x^3\right)\Big|_{u=0}^{u=\sqrt[3]{\pi}} = \frac{2}{3}$$

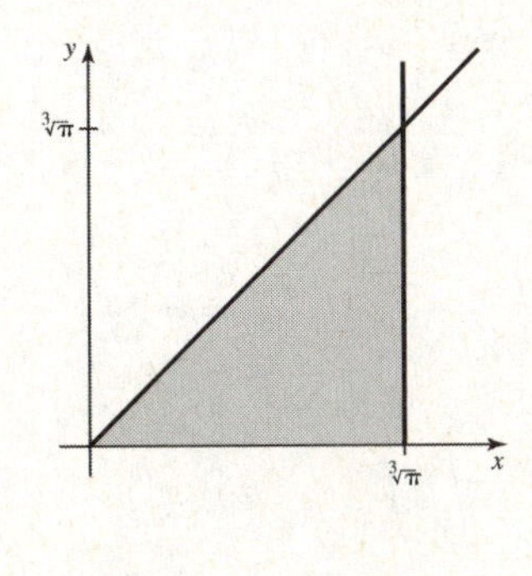

13.2.55 $\displaystyle V = \iint_R (9 - x^2 - y^2)\,dA = \int_{-3}^{3}\int_{-\sqrt{9-x^2}}^{\sqrt{9-x^2}}(9 - x^2 - y^2)\,dy\,dx = \int_{-3}^{3}\left((9 - x^2)y - \frac{1}{3}y^3\right)\Bigg|_{y=-\sqrt{9-x^2}}^{y=\sqrt{9-x^2}}\,dx$

$$= \int_{-3}^{3}\left(12\sqrt{9 - x^2} - \frac{4}{3}x^2\sqrt{9 - x^2}\right)dx = \left(\frac{15}{2}x\sqrt{9 - x^2} - \frac{1}{3}x^2\sqrt{9 - x^2} + \frac{81}{2}\sin^{-1}\left(\frac{x}{3}\right)\right)\Bigg|_{x=-3}^{x=3}$$

$$= \frac{81\pi}{2}.$$

13.2.57 $\displaystyle V = \iint_R (8 - 2x - 3y)\,dA = \int_0^1\int_0^{2-x}(8 - 2x - 3y)\,dy\,dx = \int_0^1\left((8 - 2x)y - \frac{3}{2}y^2\right)\Bigg|_{y=0}^{y=2-x}\,dx$

$$= \int_0^1\left(10 - 6x + \frac{1}{2}x^2\right)dx = \left(10x - 3x^2 + \frac{1}{6}x^3\right)\Bigg|_{x=0}^{x=1} = \frac{43}{6}.$$

13.2.59 $A = \iint_R 1\,dA = \int_{-2}^{2}\int_{x^2}^{4} 1\,dy\,dx$

$$= \int_{-2}^{2} (y)\Big|_{y=x^2}^{y=4}\,dy$$

$$= \int_{-2}^{2} (4 - x^2)\,dx$$

$$= \left(4x - \frac{1}{3}x^3\right)\Big|_{x=-2}^{x=2} = \frac{32}{3}.$$

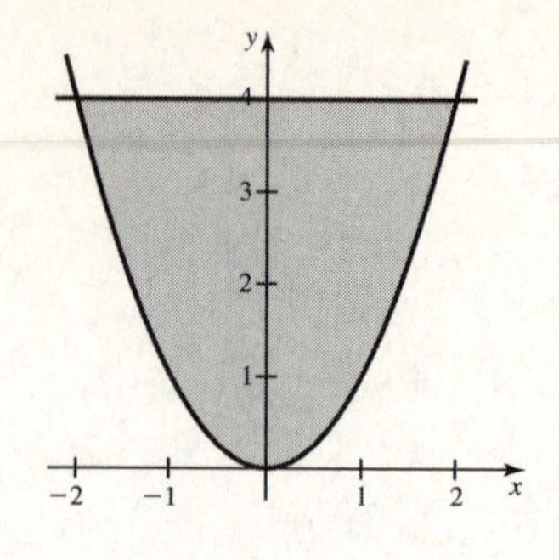

13.2.61 $A = \iint_R 1\,dA = \int_0^{\ln 2}\int_0^{e^x} 1\,dy\,dx$

$$= \int_0^{\ln 2} (y)\Big|_{y=0}^{y=e^x}\,dx$$

$$= \int_0^{\ln 2} e^x\,dx = (e^x)\Big|_{x=0}^{x=\ln 2} = 1$$

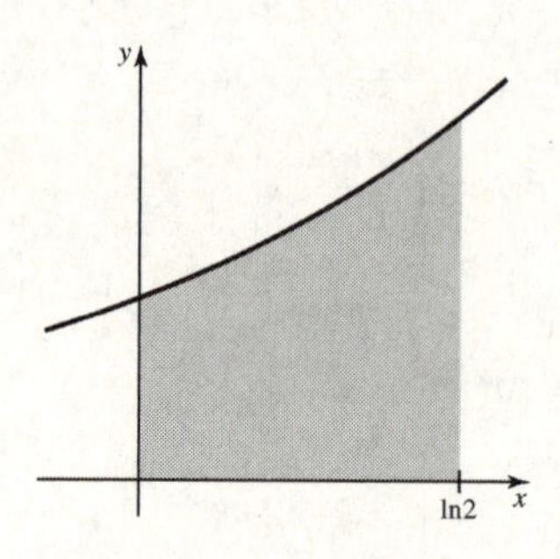

13.2.63 $A = \iint_R 1\,dA$

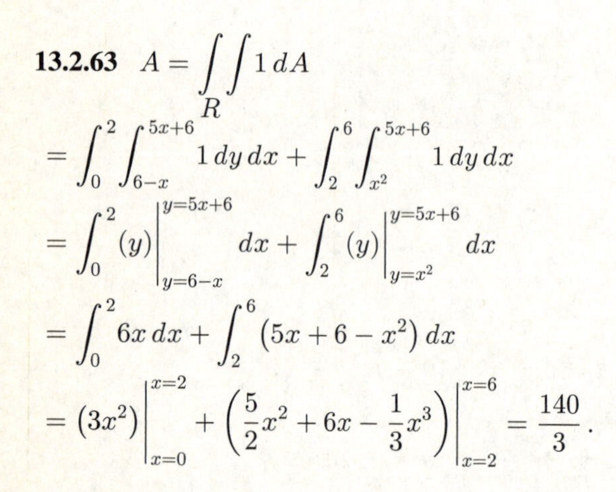

$$= \int_0^2\int_{6-x}^{5x+6} 1\,dy\,dx + \int_2^6\int_{x^2}^{5x+6} 1\,dy\,dx$$

$$= \int_0^2 (y)\Big|_{y=6-x}^{y=5x+6}\,dx + \int_2^6 (y)\Big|_{y=x^2}^{y=5x+6}\,dx$$

$$= \int_0^2 6x\,dx + \int_2^6 (5x + 6 - x^2)\,dx$$

$$= (3x^2)\Big|_{x=0}^{x=2} + \left(\frac{5}{2}x^2 + 6x - \frac{1}{3}x^3\right)\Big|_{x=2}^{x=6} = \frac{140}{3}.$$

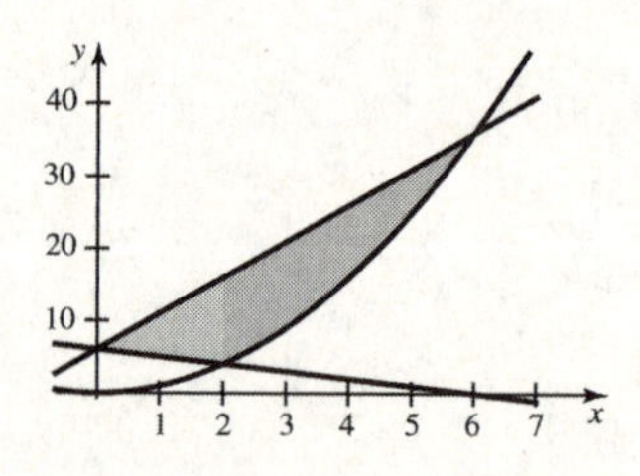

13.2.65

a. False. a and b must be constants or functions of y.
b. False. a and d must be constants.
c. False. Variable limits of integration are not allowed in the outermost integral.

13.2.67 $\iint_R (x+y)\,dA = \int_{1/2}^{2}\int_{1/x}^{5/2-x} (x+y)\,dy\,dx = \int_{1/2}^{2}\left(xy + \frac{1}{2}y^2\right)\Big|_{y=\frac{1}{x}}^{y=\frac{5}{2}-x}\,dx = \int_{1/2}^{2}\left(\frac{17}{8} - \frac{x^2}{2} - \frac{1}{2x^2}\right)dx$

$$= \left(\frac{17}{8}x - \frac{x^3}{6} + \frac{1}{2x}\right)\Big|_{x=1/2}^{x=1} = \frac{9}{8}.$$

13.2.69 $\iint_R x\sec^2 y\, dA = \int_0^{\sqrt{\pi/2}}\int_0^{x^2} x\sec^2 y\, dy\, dx = \int_0^{\sqrt{\pi/2}} x\tan y\Big|_{y=0}^{y=x^2} dx = \int_0^{\sqrt{\pi/2}} x\tan x^2\, dx$

$= \left(\frac{1}{2}\ln|\sec x^2|\right)\Big|_{x=0}^{x=\sqrt{\pi/2}} = \frac{1}{2}\ln\left(\sqrt{2}\right) = \frac{1}{4}\ln 2.$

13.2.71 $\int_0^1\int_{e^y}^e f(x, y)\, dx\, dy + \int_{-1}^0\int_{e^{-y}}^e f(x, y)\, dx\, dy$

$= \int_1^e\int_{-\ln x}^{\ln x} f(x, y)\, dy\, dx$

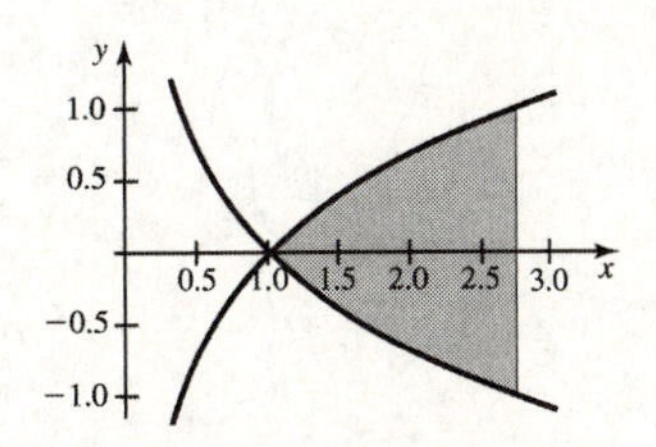

13.2.73 $\overline{f} = \frac{1}{\text{Area of } R}\iint_R f(x, y)\, dA = \frac{1}{\frac{1}{2}a^2}\int_0^a\int_0^{a-x}(a-x-y)\, dy\, dx = \frac{2}{a^2}\int_0^2\left((a-x)y - \frac{1}{2}y^2\right)\Big|_{y=0}^{y=a-x} dx$

$= \frac{2}{a^2}\int_1^9 \frac{1}{2}(a-x)^2\, dx = \frac{1}{a^2}\left(-\frac{1}{3}(a-x)^3\right)\Big|_{x=0}^{x=a} = \frac{1}{a^2}\left(\frac{a^3}{3}\right) = \frac{a}{3}.$

13.2.75

a.

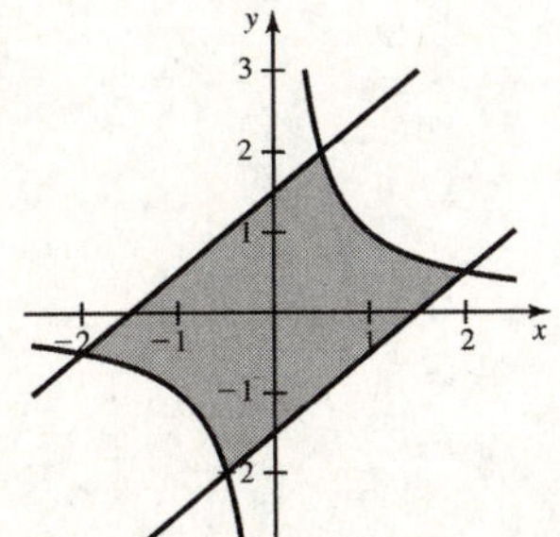

b. $A = \iint_R 1\, dA = \iint_{R_1} 1\, dA + \iint_{R_2} 1\, dA + \iint_{R_3} 1\, dA$

$= \int_{-2}^{-1/2}\int_{1/x}^{x+3/2} 1\, dy\, dx + \int_{-1/2}^{1/2}\int_{x-3/2}^{x+3/2} 1\, dy\, dx + \int_{1/2}^{2}\int_{x-3/2}^{1/x} 1\, dy\, dx$

$= \int_{-2}^{-1/2}\left(x + \frac{3}{2} - \frac{1}{x}\right) dx + \int_{-1/2}^{1/2}(3)\, dx + \int_{1/2}^{2}\left(\frac{1}{x} - x + \frac{3}{2}\right) dx$

$= \left(\frac{1}{2}x^2 + \frac{3}{2}x - \ln|x|\right)\Big|_{x=-2}^{x=-1/2} + (3x)\Big|_{x=-1/2}^{x=1/2} + \left(\ln|x| - \frac{1}{2}x^2 + \frac{3}{2}x\right)\Big|_{x=1/2}^{x=2}$

$= \frac{3}{8} + 2\ln 2 + 3 + \frac{3}{8} + 2\ln 2 = \frac{15}{4} + 4\ln 2$

c. $\displaystyle\iint_R xy\,dA = \int_{-2}^{-1/2}\int_{1/x}^{x+3/2} xy\,dy\,dx + \int_{-1/2}^{1/2}\int_{x-3/2}^{x+3/2} xy\,dy\,dx + \int_{1/2}^{2}\int_{x-3/2}^{1/x} xy\,dy\,dx$

$$= \int_{-2}^{-1/2}\left(\frac{1}{2}xy^2\right)\Big|_{y=1/x}^{y=x+3/2} dx + \int_{-1/2}^{1/2}\left(\frac{1}{2}xy^2\right)\Big|_{y=x-3/2}^{y=x+3/2} dx + \int_{1/2}^{2}\left(\frac{1}{2}xy^2\right)\Big|_{y=x-3/2}^{y=1/x} dx$$

$$= \frac{1}{2}\int_{-2}^{-1/2}\left(x^3+3x^2+\frac{9}{4}x-\frac{1}{x}\right)dx + \frac{1}{2}\int_{-1/2}^{1/2}(6x^2)\,dx + \frac{1}{2}\int_{1/2}^{2}\left(\frac{1}{x}-\frac{9}{4}x+3x^2-x^3\right)dx$$

$$= \frac{1}{2}\left(\frac{1}{4}x^4+x^3+\frac{9}{8}x^2-\ln|x|\right)\Big|_{x=-2}^{x=-1/2} + (x^3)\Big|_{x=-1/2}^{x=1/2} + \frac{1}{2}\left(\ln|x|-\frac{9}{8}x^2+x^3-\frac{1}{4}x^4\right)\Big|_{x=1/2}^{x=2}$$

$$= \frac{1}{2}\left(-\frac{21}{64}+2\ln 2\right)+\frac{1}{4}+\frac{1}{2}\left(-\frac{21}{64}+2\ln 2\right) = -\frac{5}{64}+2\ln 2.$$

13.2.77 $\displaystyle\int_1^\infty\int_0^{e^{-x}} xy\,dy\,dx = \int_1^\infty\left(\frac{1}{2}xy^2\right)\Big|_{y=0}^{y=e^{-x}} dx = \int_1^\infty \frac{1}{2}x\,e^{-2x}\,dx = \lim_{b\to\infty}\int_1^b \frac{1}{2}x\,e^{-2x}\,dx$

$$= \lim_{b\to\infty}\frac{1}{2}\left(e^{-2x}\left(-\frac{1}{2}x-\frac{1}{4}\right)\right)\Big|_{x=1}^{x=b} = \frac{1}{4}\lim_{b\to\infty}\left(e^{-2b}\left(-b-\frac{1}{2}\right)+e^{-2}\left(\frac{3}{2}\right)\right) = \frac{1}{4}\left(0+\frac{3}{2}e^{-2}\right)$$

$$= \frac{3}{8e^2}.$$

13.2.79 $\displaystyle\int_0^\infty\int_0^\infty e^{-x-y}\,dy\,dx = \int_0^\infty\left(\lim_{b\to\infty}\int_0^b e^{-x-y}\,dy\right)dx = \int_0^\infty\left(\lim_{b\to\infty}(-e^{-x-y})\Big|_{y=0}^{y=b}\right)dx$

$$= \int_0^\infty\left(\lim_{b\to\infty}(e^{-x}-e^{-x-b})\right)dx = \int_0^\infty e^{-x}\,dx = \lim_{b\to\infty}\int_0^b -e^{-x}\,dx = \lim_{b\to\infty}(-e^{-x})\Big|_{x=0}^{x=b}$$

$$= \lim_{b\to\infty}(1-e^{-b}) = 1$$

13.2.81 $\displaystyle V = \iint_R f(x,y)\,dA$ (orient y-axis vertically)

$$= \iint_R (y_{\text{top}}-y_{\text{bottom}})\,dz\,dx = \int_0^5\int_0^2\left[(z+1)-\left(\frac{-z-1}{2}\right)\right]dz\,dx = \int_0^5\int_0^2\left(\frac{3}{2}z+\frac{3}{2}\right)dz\,dx$$

$$= \int_0^5\left(\frac{3}{4}z^2+\frac{3}{2}z\right)\Big|_{z=0}^{z=2} dx = \int_0^5 6\,dx = (6x)\Big|_{x=0}^{x=5} = 30.$$

13.2.83 $\displaystyle V = \iint_R (4-x-y)\,dA = \int_{-1}^1\int_{-1}^1(4-x-y)\,dy\,dx = \int_{-1}^1\left((4-x)y-\frac{1}{2}y^2\right)\Big|_{y=-1}^{y=1} dx =$

$$= \int_{-1}^1 2(4-x)\,dx = 2\left(4x-\frac{1}{2}x^2\right)\Big|_{x=-1}^{x=1} = 16.$$

13.2.85 $\displaystyle V = \iint_R (a(2-x)-a(x-2))\,dA = 2a\int_{-1}^1\int_{-\sqrt{1-x^2}}^{\sqrt{1-x^2}}(2-x)\,dy\,dx = 2a\int_{-1}^1\left[(2-x)y\right]\Big|_{y=-\sqrt{1-x^2}}^{y=\sqrt{1-x^2}} dx$

$$= 4a\int_{-1}^1(2-x)\sqrt{1-x^2}\,dx = 4a\int_{-1}^1\left(2\sqrt{1-x^2}-x\sqrt{1-x^2}\right)dx$$

$$= \left[4a\left(x\sqrt{1-x^2}+\sin^{-1}x\right)+\frac{4a}{3}(1-x^2)^{3/2}\right]\Big|_{x=-1}^{x=1} = 4a(\sin^{-1}(1)-\sin^{-1}(-1)) = 4a\pi.$$

13.2.87 $\displaystyle\iint_{R_1} x^{-n}\,dA = \int_1^\infty \int_1^2 x^{-n}\,dy\,dx = \lim_{b\to\infty} \int_1^b \int_1^2 x^{-n}\,dy\,dx = \lim_{b\to\infty} \int_1^b (2-1)x^{-n}dx = \lim_{b\to\infty} \int_1^b x^{-n}dx$

which converges if $n > 1$. $\displaystyle\iint_{R_2} x^{-n}\,dA = \int_1^\infty \int_1^2 x^{-n}\,dx\,dy = \lim_{b\to\infty} \int_1^b \left(\frac{x^{-n+1}}{1-n}\right)\Bigg|_{x=1}^{x=2} dy = \lim_{b\to\infty} \int_1^b \left(\frac{2^{1-n}-1}{1-n}\right) dy$

$\displaystyle= \lim_{b\to\infty} \left(\frac{2^{1-n}-1}{1-n}\right) y \Bigg|_{y=1}^{y=b} = \lim_{b\to\infty} \left(\frac{2^{1-n}-1}{1-n}\right)(b-1)$, which diverges when $n > 1$.

13.3 Double Integrals in Polar Coordinates

13.3.1 A polar rectangle has constant limits on r and θ.

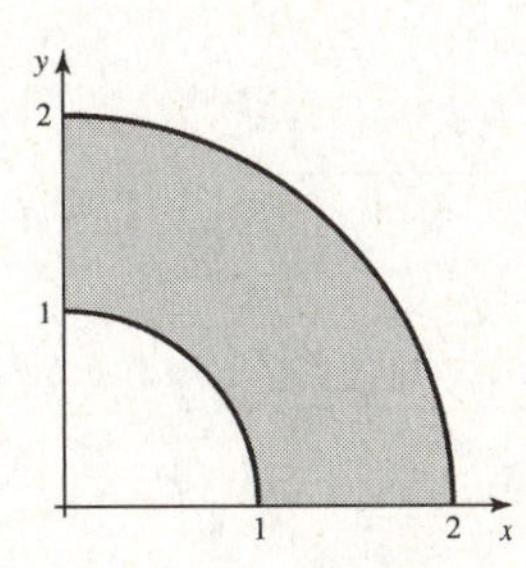

13.3.3 $R = \left\{(r,\,\theta) : \frac{1}{2} \le r \le \cos 2\theta,\ -\frac{\pi}{6} \le \theta \le \frac{\pi}{6}\right\}$

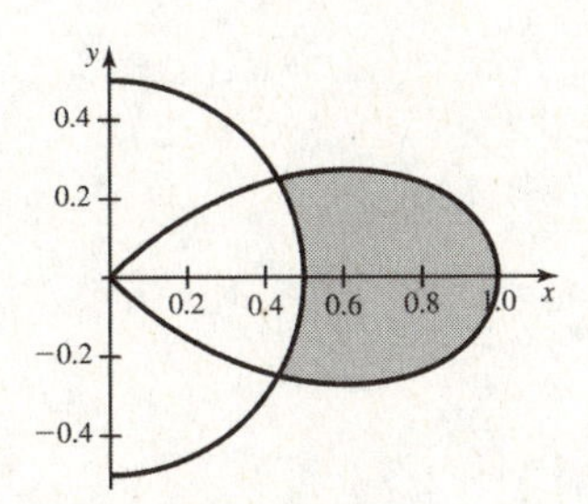

13.3.5 For $R = \{(r,\,\theta) : g(\theta) \le r \le f(\theta),\ \alpha \le \theta \le \beta\}$, the areaof R given by $\displaystyle\iint_R (1)\,dy\,dx$ converts to polar coordinates $\displaystyle\int_\alpha^\beta \int_{g(\theta)}^{f(\theta)} r\,dr\,d\theta$.

13.3.7

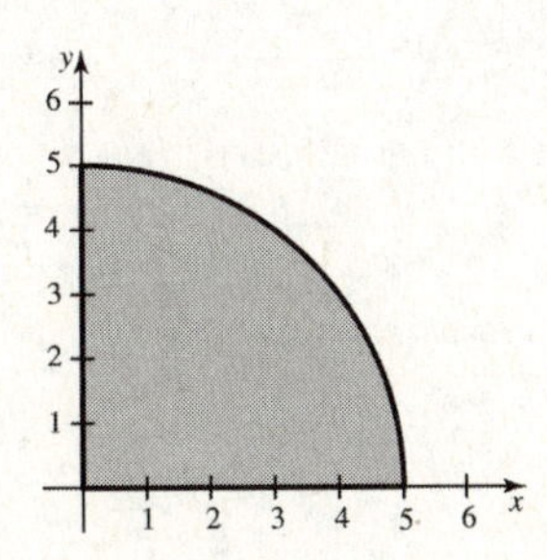

13.3.9

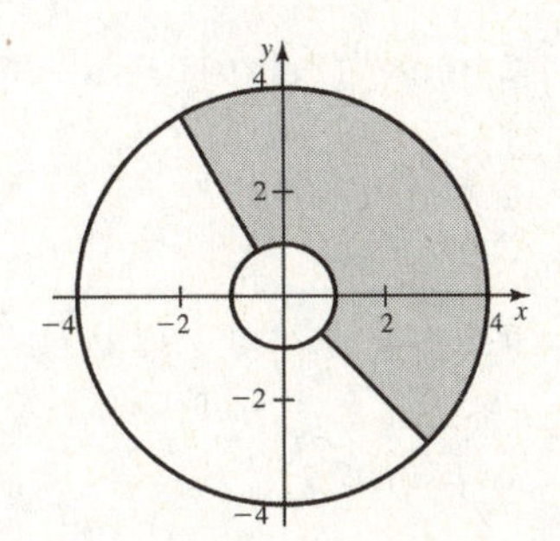

13.3.11 $\displaystyle V = \iint_R f(r\cos\theta,\, r\sin\theta)\,r\,dr\,d\theta = \int_0^{2\pi} \int_0^1 (4-r^2)\,r\,dr\,d\theta = \int_0^{2\pi} \left(2r^2 - \frac{1}{4}r^4\right)\Bigg|_{r=0}^{r=1} d\theta$

$\displaystyle= \int_0^{2\pi} \left(\frac{7}{4}\right) d\theta = \frac{7\pi}{2}.$

13.3.13 $V = \iint\limits_R f(r\cos\theta,\, r\sin\theta)\, r\, dr\, d\theta = \int_0^{2\pi}\int_1^2 (4-r^2)\, r\, dr\, d\theta = \int_0^{2\pi} \left(2r^2 - \frac{1}{4}r^4\right)\Bigg|_{r=1}^{r=2} d\theta$

$= \int_0^{2\pi} \frac{9}{4}\, d\theta = \frac{9\pi}{2}.$

13.3.15 $V = \iint\limits_R \left(5 - \sqrt{1+x^2+y^2}\right) dA = \int_0^{2\pi}\int_0^2 \left(5-\sqrt{1+r^2}\right) r\, dr\, d\theta = \int_0^{2\pi} \left(\frac{5}{2}r^2 - \frac{1}{3}(1+r^2)^{3/2}\right)\Bigg|_{r=0}^{r=2} d\theta$

$= \int_0^{2\pi} \left(10 - \frac{5\sqrt{5}-1}{3}\right) d\theta = \left(\frac{31-5\sqrt{5}}{3}\right)\cdot \theta\Bigg|_{\theta=0}^{\theta=2\pi} = 2\pi\left(\frac{31-5\sqrt{5}}{3}\right) = \frac{\left(62-10\sqrt{5}\right)\pi}{3}$

13.3.17 $V = \iint\limits_R \left(5 - \sqrt{1+x^2+y^2}\right) dA = \int_0^{2\pi}\int_1^2 \left(5-\sqrt{1+r^2}\right) r\, dr\, d\theta = \int_0^{2\pi} \left(\frac{5}{2}r^2 - \frac{1}{3}(1+r^2)^{3/2}\right)\Bigg|_{r=1}^{r=2} d\theta$

$= \int_0^{2\pi} \left(\frac{15}{2} + \frac{2\sqrt{2}-5\sqrt{5}}{3}\right) d\theta = \left(\frac{15}{2} + \frac{2\sqrt{2}-5\sqrt{5}}{3}\right)\cdot \theta\Bigg|_{\theta=0}^{\theta=2\pi} = \pi\left(15 + \frac{4\sqrt{2}-10\sqrt{5}}{3}\right)$

13.3.19 $\iint\limits_R (x^2+y^2)\, dA$

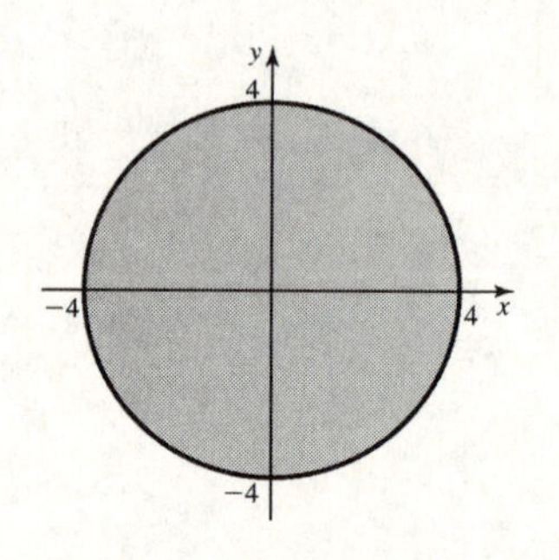

$= \int_0^{2\pi}\int_0^4 (r^2)\, r\, dr\, d\theta$

$= \int_0^{2\pi}\int_0^4 r^3\, dr\, d\theta$

$= \int_0^{2\pi} \left(\frac{1}{4}r^4\right)\Bigg|_{r=0}^{r=4} d\theta$

$= \int_0^{2\pi} 64\, d\theta = 128\pi$

13.3.21 $\iint\limits_R (2xy)\, dA$

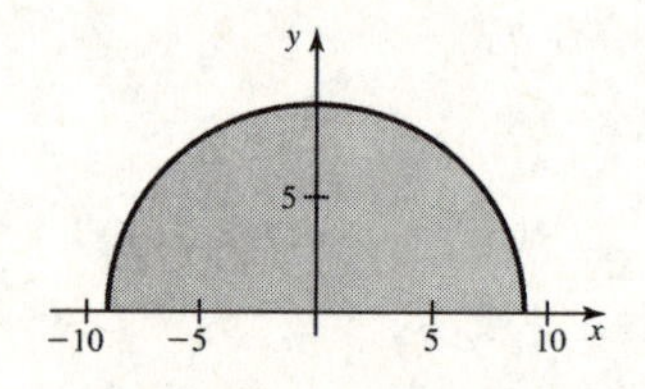

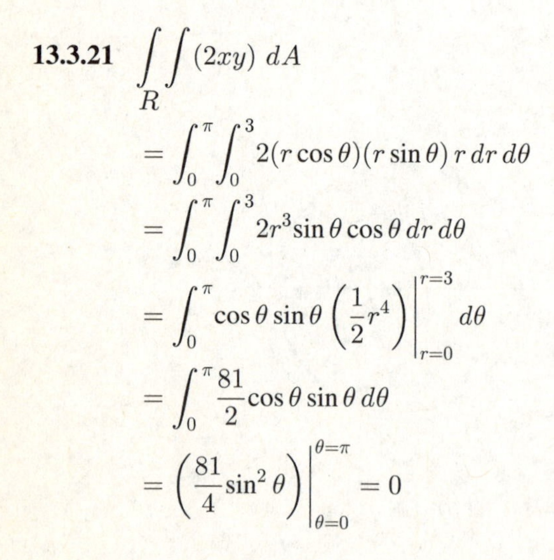

$= \int_0^{\pi}\int_0^3 2(r\cos\theta)(r\sin\theta)\, r\, dr\, d\theta$

$= \int_0^{\pi}\int_0^3 2r^3 \sin\theta\cos\theta\, dr\, d\theta$

$= \int_0^{\pi} \cos\theta\sin\theta \left(\frac{1}{2}r^4\right)\Bigg|_{r=0}^{r=3} d\theta$

$= \int_0^{\pi} \frac{81}{2}\cos\theta\sin\theta\, d\theta$

$= \left(\frac{81}{4}\sin^2\theta\right)\Bigg|_{\theta=0}^{\theta=\pi} = 0$

13.3.23 $\displaystyle\iint_R \frac{1}{\sqrt{16-x^2-y^2}}\,dA$

$\displaystyle = \int_0^{\pi/2}\int_0^2 \frac{1}{\sqrt{16-r^2}}\,r\,dr\,d\theta$

$\displaystyle = \int_0^{\pi/2}\int_0^2 \frac{r}{\sqrt{16-r^2}}\,dr\,d\theta$

$\displaystyle = \int_0^{\pi/2}\left(-\sqrt{16-r^2}\right)\Bigg|_{r=0}^{r=2} d\theta$

$\displaystyle = \int_0^{\pi/2}\left(-2\sqrt{3}+4\right)d\theta = \pi\left(2-\sqrt{3}\right)$

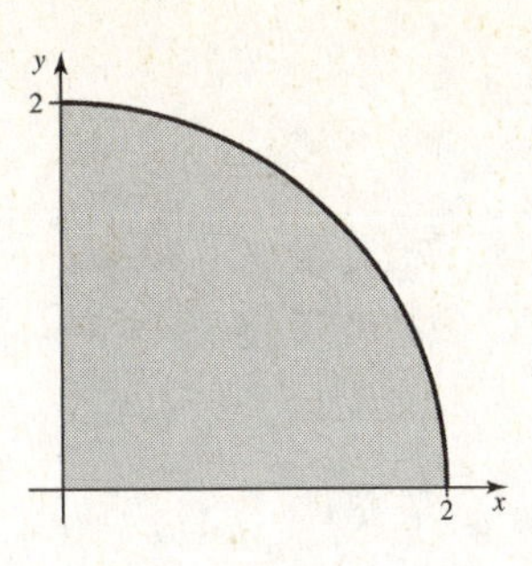

13.3.25 $z \geq 0 \Longrightarrow e^{-(x^2+y^2)/8} - e^{-2} \geq 0 \Longrightarrow e^{-(x^2+y^2)/8} \geq e^{-2}$ if $\dfrac{x^2+y^2}{8} \leq 2$, when R: $x^2+y^2 \leq 16$ thus $0 \leq r \leq 4$ and $0 \leq \theta \leq 2\pi$. $\displaystyle V = \int_0^{2\pi}\int_0^4 \left(e^{-r^2/8} - e^{-2}\right) r\,dr\,d\theta = \int_0^{2\pi}\int_0^4 \left(r\,e^{-r^2/8} - r\,e^{-2}\right)dr\,d\theta$

$\displaystyle = \int_0^{2\pi}\left(-4\,e^{-r^2/8} - \frac{1}{2}r^2\,e^{-2}\right)\Bigg|_{r=0}^{r=4} d\theta = \int_0^{2\pi}\left(-12e^{-2}+4\right)d\theta = 8\pi\left(1-3e^{-2}\right)$

13.3.27 $z \geq 0 \Longrightarrow 25 - \sqrt{x^2+y^2} \geq 0 \Longrightarrow \sqrt{x^2+y^2} \leq 25$ when R: $x^2+y^2 \leq 625$ thus $0 \leq r \leq 25$ and $0 \leq \theta \leq 2\pi$. $\displaystyle V = \int_0^{2\pi}\int_0^{25}\left(25-\sqrt{r^2}\right) r\,dr\,d\theta = \int_0^{2\pi}\int_0^{25}\left(25r - r^2\right)dr\,d\theta = \int_0^{2\pi}\left(\frac{25}{2}r^2 - \frac{1}{3}r^3\right)\Bigg|_{r=0}^{r=25} d\theta$

$\displaystyle = \int_0^{2\pi}\frac{15625}{6}\,d\theta = \frac{15625\pi}{3}$

13.3.29 $\displaystyle\iint_R f(r,\theta)\,dA =$

$\displaystyle = \int_0^{2\pi}\int_0^{1+\frac{1}{2}\cos\theta} f(r,\theta)\;r\,dr\,d\theta$

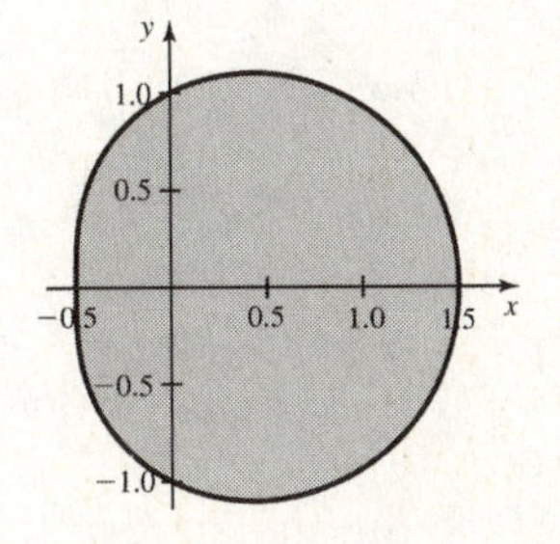

13.3.31 $\displaystyle\iint_R f(r,\theta)\,dA =$

$\displaystyle = \int_0^{\pi/2}\int_0^{\sqrt{2\sin(2\theta)}} f(r,\theta)\;r\,dr\,d\theta$

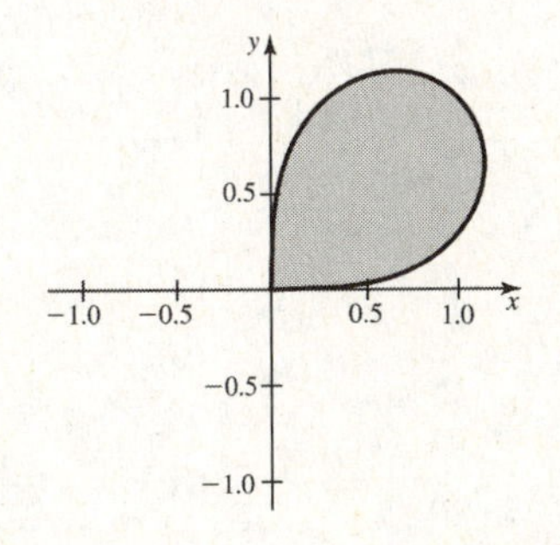

13.3.33 $\displaystyle\iint_R f(r,\theta)\,dA =$

$$= \int_{\pi/18}^{5\pi/18}\int_1^{2\sin(3\theta)} f(r,\theta)\;r\,dr\,d\theta$$

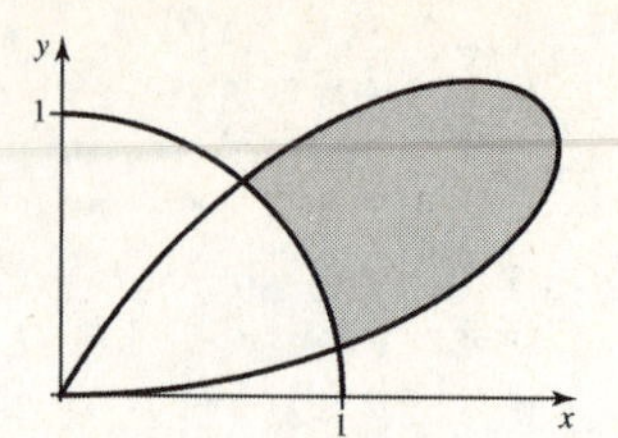

13.3.35 $\displaystyle A = \iint_R 1\,dA = \int_0^{\pi}\int_1^2 r\,dr\,d\theta$

$$= \int_0^{\pi}\left(\frac{1}{2}r^2\right)\Bigg|_{r=1}^{r=2} d\theta = \int_0^{\pi}\left(\frac{3}{2}\right)d\theta$$

$$= \left(\frac{3}{2}\theta\right)\Bigg|_{\theta=0}^{\theta=\pi} = \frac{3\pi}{2}$$

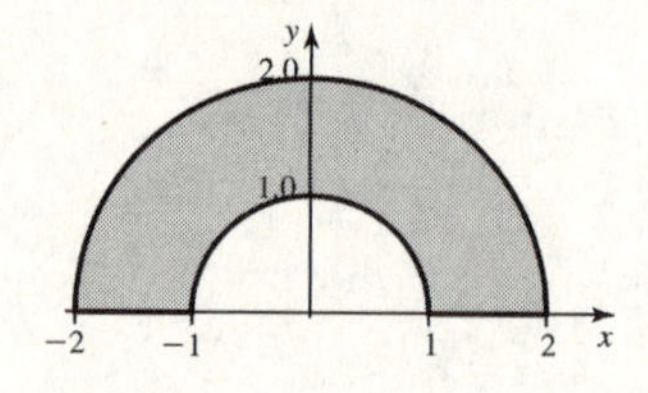

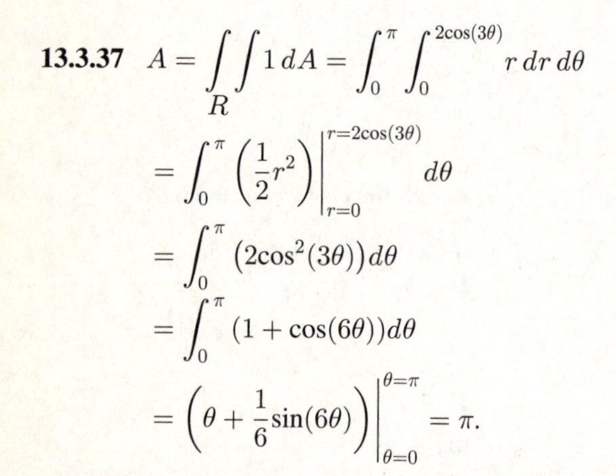

13.3.37 $\displaystyle A = \iint_R 1\,dA = \int_0^{\pi}\int_0^{2\cos(3\theta)} r\,dr\,d\theta$

$$= \int_0^{\pi}\left(\frac{1}{2}r^2\right)\Bigg|_{r=0}^{r=2\cos(3\theta)} d\theta$$

$$= \int_0^{\pi}\left(2\cos^2(3\theta)\right)d\theta$$

$$= \int_0^{\pi}(1+\cos(6\theta))d\theta$$

$$= \left(\theta + \frac{1}{6}\sin(6\theta)\right)\Bigg|_{\theta=0}^{\theta=\pi} = \pi.$$

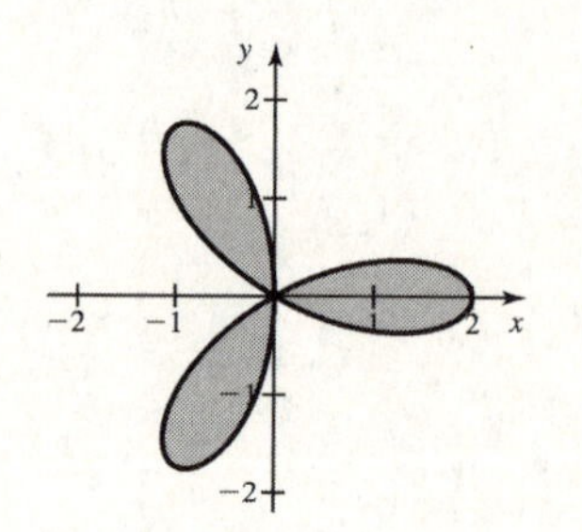

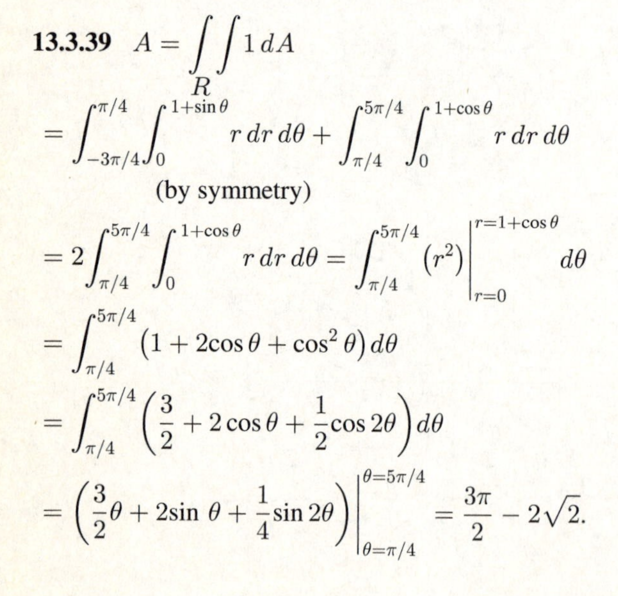

13.3.39 $\displaystyle A = \iint_R 1\,dA$

$$= \int_{-3\pi/4}^{\pi/4}\int_0^{1+\sin\theta} r\,dr\,d\theta + \int_{\pi/4}^{5\pi/4}\int_0^{1+\cos\theta} r\,dr\,d\theta$$

(by symmetry)

$$= 2\int_{\pi/4}^{5\pi/4}\int_0^{1+\cos\theta} r\,dr\,d\theta = \int_{\pi/4}^{5\pi/4}(r^2)\Bigg|_{r=0}^{r=1+\cos\theta} d\theta$$

$$= \int_{\pi/4}^{5\pi/4}\left(1 + 2\cos\theta + \cos^2\theta\right)d\theta$$

$$= \int_{\pi/4}^{5\pi/4}\left(\frac{3}{2} + 2\cos\theta + \frac{1}{2}\cos 2\theta\right)d\theta$$

$$= \left(\frac{3}{2}\theta + 2\sin\theta + \frac{1}{4}\sin 2\theta\right)\Bigg|_{\theta=\pi/4}^{\theta=5\pi/4} = \frac{3\pi}{2} - 2\sqrt{2}.$$

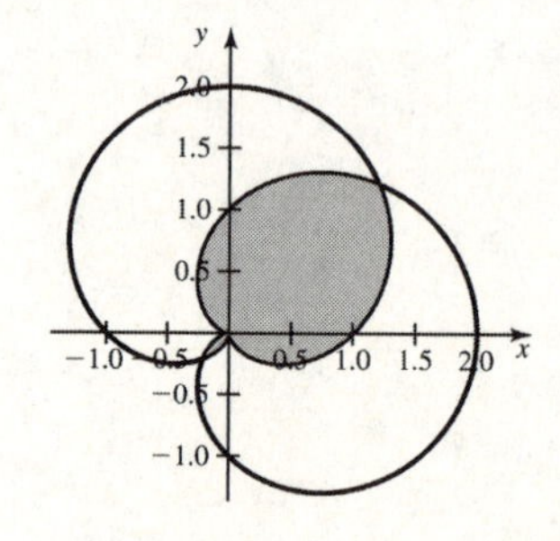

13.3.41 $\overline{f} = \dfrac{1}{\text{area of } R}\iint\limits_R f(r,\theta)\,dA = \dfrac{1}{\pi a^2}\int_0^{2\pi}\int_0^a (r)\,r\,dr\,d\theta = \dfrac{1}{\pi a^2}\int_0^{2\pi}\left(\dfrac{1}{3}r^3\right)\Big|_{r=0}^{r=a} d\theta = \dfrac{1}{\pi a^2}\int_0^{2\pi}\left(\dfrac{1}{3}a^3\right)d\theta$

$$= \frac{a}{3\pi}(\theta)\Big|_{\theta=0}^{\theta=2\pi} = \frac{2a}{3}$$

13.3.43 The square of the distance from a point to $(1,\,1)$ is

$$(x-1)^2+(y-1)^2 = x^2-2x+y^2-2y+2 = r^2-2r\cos\theta-2r\sin\theta+2$$

$$\overline{f} = \frac{1}{\text{area of } R}\iint\limits_R (x^2-2x+y^2-2y+2)\,dA$$

$$= \frac{1}{\pi}\int_0^{2\pi}\int_0^1 (r^2-2r\cos\theta-2r\sin\theta+2)\,r\,dr\,d\theta = \frac{1}{\pi}\int_0^{2\pi}\left(\frac{1}{4}r^4-\frac{2}{3}r^3\cos\theta-\frac{2}{3}r^3\sin\theta+r^2\right)\Big|_{r=0}^{r=1} d\theta$$

$$= \frac{1}{\pi}\int_0^{2\pi}\left(\frac{5}{4}-\frac{2}{3}\cos\theta-\frac{2}{3}\sin\theta\right)d\theta = \frac{1}{\pi}\left(\frac{5}{4}\theta-\frac{2}{3}\sin\theta+\frac{2}{3}\cos\theta\right)\Big|_{\theta=0}^{\theta=2\pi} = \frac{5}{2}$$

13.3.45

a. False. $\iint\limits_R (x^2+y^2)\,dA = \int_0^{2\pi}\int_0^1 (r^2)\,r dr\,d\theta\,.$

b. True. The distance from every point on a hemisphere with radius 2 to the origin is 2.

c. True. $\int_0^1\int_0^{\sqrt{1-y^2}} e^{x^2+y^2}\,dx\,dy$ converts to $\int_0^{\pi/2}\int_0^r r\,e^{r^2}\,dr\,d\theta\,.$

13.3.47 $\displaystyle\int_{-1}^1\int_{-\sqrt{1-x^2}}^{\sqrt{1-x^2}} (x^2+y^2)^{3/2}dy\,dxy$

$$= \int_0^{2\pi}\int_0^1 (r^2)^{3/2}\,r\,dr\,d\theta$$

$$= \int_0^{2\pi}\left(\frac{1}{5}r^5\right)\Big|_{r=0}^{r=1} d\theta = \int_0^{2\pi}\left(\frac{1}{5}\right)d\theta$$

$$= \left(\frac{1}{5}\theta\right)\Big|_{\theta=0}^{\theta=2\pi} = \frac{2\pi}{5}$$

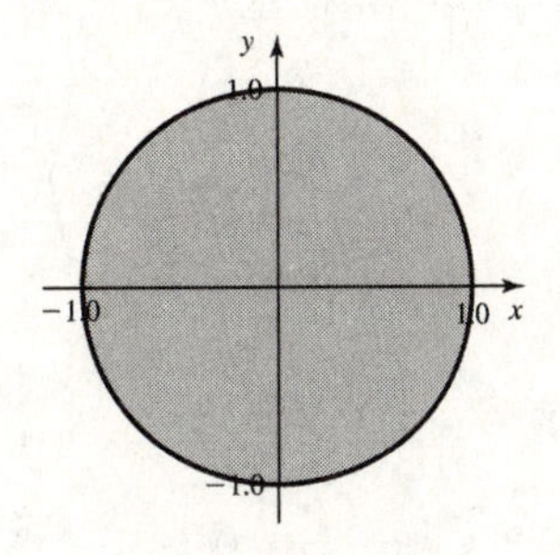

13.3.49 $\displaystyle\int_0^{\pi/4}\int_0^{\sec\theta} r^3\,dr\,d\theta$

$$= \int_0^1\int_0^x (x^2+y^2)\,dy\,dx$$

$$= \int_0^1\left(x^2y+\frac{1}{3}y^3\right)\Big|_{y=0}^{y=x} dx$$

$$= \int_0^1\left(\frac{4}{3}x^3\right)dx = \left(\frac{1}{3}x^4\right)\Big|_{x=0}^{x=1} = \frac{1}{3}.$$

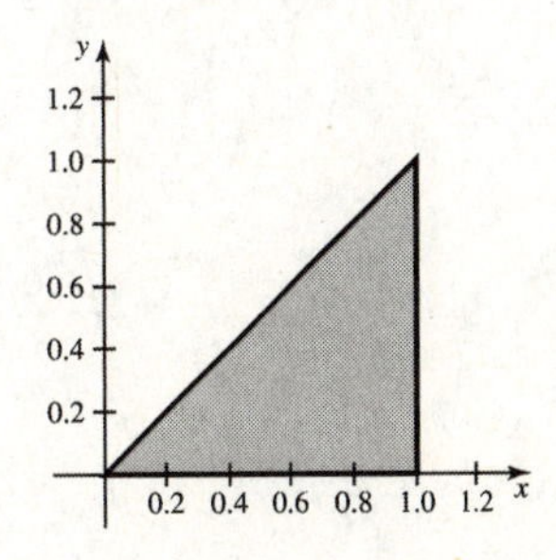

13.3.51

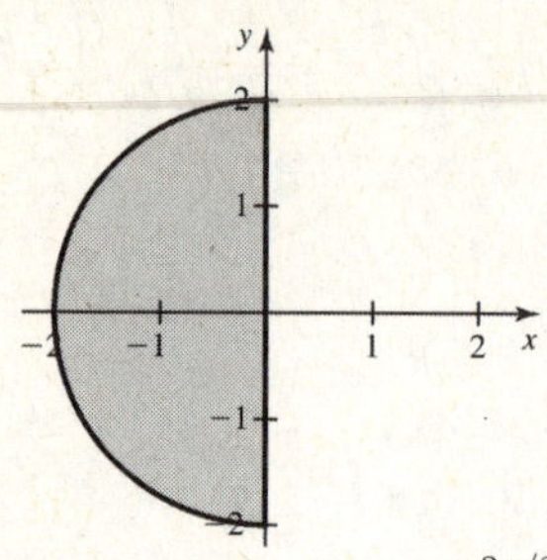

$$\iint_R \frac{1}{4+\sqrt{x^2+y^2}}\, dA = \int_{\pi/2}^{3\pi/2}\int_0^2 \frac{1}{4+r}\, r\, dr\, d\theta = \int_{\pi/2}^{3\pi/2}\int_0^2 \left(1-\frac{4}{r+4}\right) dr\, d\theta$$

$$= \int_{\pi/2}^{3\pi/2} (r - 4\ln|r+4|)\Big|_{r=0}^{r=2} d\theta = 2\int_{\pi/2}^{3\pi/2}\left(1-2\ln\left(\frac{3}{2}\right)\right) d\theta = 2\left(\theta\left(1-2\ln\left(\frac{3}{2}\right)\right)\right)\Big|_{\theta=\pi/2}^{\theta=3\pi/2} = 2\pi\left(1-2\ln\left(\frac{3}{2}\right)\right)$$

13.3.53 Paraboloid:

$$V = \iint_R (x^2+y^2)\, dA = \int_0^{2\pi}\int_0^2 (r^2)\, r\, dr\, d\theta = \int_0^{2\pi}\int_0^2 r^3\, dr\, d\theta = \int_0^{2\pi}\left(\frac{1}{4}r^4\right)\Big|_{r=0}^{r=2} d\theta = \int_0^{2\pi} 4\, d\theta = 8\pi$$

Cone:

$$V = \iint_R \sqrt{x^2+y^2}\, dA = \int_0^{2\pi}\int_0^4 \sqrt{r^2}\, r\, dr\, d\theta = \int_0^{2\pi}\int_0^4 r^2\, dr\, d\theta = \int_0^{2\pi}\left(\frac{1}{3}r^3\right)\Big|_{r=0}^{r=4} d\theta = \int_0^{2\pi}\frac{64}{3}\, d\theta = \frac{128\pi}{3}$$

Hyperboloid:

$$V = \iint_R \left(\sqrt{1+x^2+y^2}-1\right) dA = \int_0^{2\pi}\int_0^{\sqrt{24}}\left(\sqrt{1+r^2}-1\right) r\, dr\, d\theta = \int_0^{2\pi}\int_0^{\sqrt{24}}\left(r\sqrt{1+r^2}-r\right) dr\, d\theta$$

$$= \int_0^{2\pi}\left(\frac{1}{3}(1+r^2)^{3/2} - \frac{1}{2}r^2\right)\Big|_{r=0}^{r=\sqrt{24}} d\theta = \int_0^{2\pi}\frac{88}{3}\, d\theta = \frac{176\pi}{3} \Longrightarrow \text{largest}$$

13.3.55

a. $z \geq 0 \Longrightarrow x^2+y^2 \geq 0 \Longrightarrow x^2 \geq y^2 \Longrightarrow x \geq |y|$ or $x \leq -|y|$ so $R = \{(r,\theta)| -\frac{\pi}{4} \leq \theta \leq \frac{\pi}{4} \text{ or } \frac{3\pi}{4} \leq \theta \leq \frac{5\pi}{4}\}$.

b. $$V = \iint_R (x^2-y^2)\, dA = \int_{-\pi/4}^{\pi/4}\int_0^a (r^2\cos^2\theta - r^2\sin^2\theta)\, r\, dr\, d\theta = \int_{-\pi/4}^{\pi/4}\int_0^a (\cos^2\theta - \sin^2\theta)\, r^3\, dr\, d\theta$$

$$= \int_{-\pi/4}^{\pi/4} (\cos^2\theta - \sin^2\theta)\left(\frac{1}{4}r^4\right)\Big|_{r=0}^{r=a} d\theta = \left(\frac{1}{4}a^4\right)\int_{-\pi/4}^{\pi/4}\cos 2\theta\, d\theta = \left(\frac{1}{4}a^4\right)\left(\frac{1}{2}\sin 2\theta\right)\Big|_{\theta=-\pi/4}^{\theta=\pi/4} = \frac{a^4}{4}$$

13.3.57 $$\int_0^{\pi/2}\int_1^{\infty}\frac{\cos\theta}{r^3}\, r\, dr d\theta = \lim_{b\to\infty}\int_0^{\pi/2}\int_1^b \frac{\cos\theta}{r^2} dr d\theta = \lim_{b\to\infty}\int_0^{\pi/2}\left(-\frac{\cos\theta}{r}\right)\Big|_{r=1}^{r=b} d\theta$$

$$= \lim_{b\to\infty}\int_0^{\pi/2}\cos\theta\left(1-\frac{1}{b}\right) d\theta = \lim_{b\to\infty}\left(\sin\theta\left(1-\frac{1}{b}\right)\right)\Big|_{\theta=0}^{\theta=\pi/2} = \lim_{b\to\infty}\left(1-\frac{1}{b}\right) = 1.$$

13.3.59 $\iint\limits_R e^{-x^2-y^2}\,dA = \int_0^{\pi/2}\int_0^{\infty} e^{-r^2}\,r\,dr\,d\theta = \int_0^{\pi/2}\left(\lim_{b\to\infty}\int_0^b e^{-r^2}\,r\,dr\right)d\theta$

$$= \int_0^{\pi/2}\left(\lim_{b\to\infty}\left(-\frac{1}{2}e^{-r^2}\right)\Bigg|_{r=0}^{r=b}\right)d\theta = \int_0^{\pi/2}\left(\lim_{b\to\infty}\frac{1}{2}\left(1-e^{-b^2}\right)\right)d\theta = \int_0^{\pi/2}\left(\frac{1}{2}(1-0)\right)d\theta$$

$$= \int_0^{\pi/2}\frac{1}{2}\,d\theta = \frac{\pi}{4}$$

13.3.61

a. $A = \iint\limits_R 1\,dA = \int_0^{2\pi}\int_0^{2+\cos\theta} r\,dr\,d\theta = \int_0^{2\pi}\left(\frac{1}{2}r^2\right)\Bigg|_{r=0}^{r=2+\cos\theta} d\theta = \frac{1}{2}\int_0^{2\pi}(2+\cos\theta)^2\,d\theta$

$$= \frac{1}{2}\int_0^{2\pi}\left(4+4\cos\theta+\cos^2\theta\right)d\theta = \frac{1}{2}\left(4\theta+4\sin\theta+\frac{1}{2}\theta+\frac{1}{4}\sin 2\theta\right)\Bigg|_{\theta=0}^{\theta=2\pi} = \frac{9\pi}{2}$$

b. $r = 0 \Longrightarrow 0 = 1+2\cos\theta \Longrightarrow \theta = \frac{2\pi}{3}, \frac{4\pi}{3}$.

$$A = \iint\limits_R 1\,dA = \int_0^{2\pi}\int_0^{1+2\cos\theta} r\,dr\,d\theta - \int_{2\pi/3}^{4\pi/3}\int_0^{1+2\cos\theta}(1)\,r\,dr\,d\theta$$

$$= \frac{1}{2}\int_0^{2\pi}(1+2\cos\theta)^2 d\theta - \int_{2\pi/3}^{4\pi/3}(1+2\cos\theta)^2 d\theta$$

$$= \frac{1}{2}(3\theta+4\sin\theta+\sin 2\theta)\Bigg|_{\theta=0}^{\theta=2\pi} - (3\theta+4\sin\theta+\sin 2\theta)\Bigg|_{\theta=2\pi/3}^{\theta=4\pi/3} = \left(3\pi-\left(\pi-\frac{3\sqrt{3}}{2}\right)\right) = 2\pi+\frac{3\sqrt{3}}{2}.$$

c. $A = \iint\limits_R 1\,dA = \int_{2\pi/3}^{4\pi/3}\int_0^{1+2\cos\theta} r\,dr\,d\theta = \int_{2\pi/3}^{4\pi/3}\left(\frac{1}{2}r^2\right)\Bigg|_{r=0}^{r=1+2\cos\theta} d\theta$

$$= \frac{1}{2}\int_{2\pi/3}^{4\pi/3}\left(1+4\cos\theta+4\cos^2\theta\right)d\theta = \frac{1}{2}(3\theta+4\sin\theta+\sin 2\theta)\Bigg|_{\theta=2\pi/3}^{\theta=4\pi/3} = \frac{1}{2}\left(2\pi-4\sqrt{3}+\sqrt{3}\right)$$

$$= \pi - \frac{3\sqrt{3}}{2}$$

13.3.63 $m = \iint\limits_R \rho(r,\theta)\,dA = \int_0^{\pi}\int_1^4(4+r\sin\theta)\,r\,dr\,d\theta = \int_0^{\pi}\left(2r^2+\frac{1}{3}r^3\sin\theta\right)\Bigg|_{r=1}^{r=4} d\theta = \int_0^{\pi}(30+21\sin\theta)\,d\theta$

$$= (30\theta-21\cos\theta)\Bigg|_{\theta=0}^{\theta=\pi} = 30\pi+42$$

13.3.65

a. $\int_{-\infty}^{\infty}\int_{-\infty}^{\infty} e^{-x^2-y^2}\,dx\,dy = \int_0^{2\pi}\int_0^{\infty} e^{-r^2}\,r\,dr\,d\theta = \int_0^{2\pi}\left(\lim_{b\to\infty}\int_0^b e^{-r^2}\,r\,dr\right)d\theta$

$$= \int_0^{2\pi}\left(\lim_{b\to\infty}\left(-\frac{1}{2}e^{-r^2}\right)\Bigg|_{r=0}^{r=b}\right)d\theta = \int_0^{2\pi}\left(\lim_{b\to\infty}\frac{1}{2}\left(1-e^{-b^2}\right)\right)d\theta = \int_0^{2\pi}\frac{1}{2}\,d\theta = \pi$$

Thus $\int_{-\infty}^{\infty}\int_{-\infty}^{\infty} e^{-x^2-y^2}\,dxdy = \pi$.

$$\int_{-\infty}^{\infty}\int_{-\infty}^{\infty} e^{-x^2-y^2}\,dxdy = \int_{-\infty}^{\infty}\int_{-\infty}^{\infty}\left(e^{-x^2}\cdot e^{-y^2}\right)dxdy = \left(\int_{-\infty}^{\infty} e^{-x^2}dx\right)\cdot\left(\int_{-\infty}^{\infty} e^{-y^2}dy\right)$$

$$= \left(\int_{-\infty}^{\infty} e^{-x^2}dx\right)^2 = \pi$$

Thus $\int_{-\infty}^{\infty} e^{-x^2}dx = \sqrt{\pi}$.

b. (i). $\int_{0}^{\infty} e^{-x^2}dx = \frac{1}{2}\int_{-\infty}^{\infty} e^{-x^2}dx = \frac{\sqrt{\pi}}{2}$.

(ii). Let $u = -x^2$. Then $-\frac{du}{2} = x\,dx$, and the integral becomes

$$\frac{1}{2}\int_{-\infty}^{0} e^u du = \left(\frac{1}{2}\right)\lim_{b\to-\infty}\int_{b}^{0} e^u du = \left(\frac{1}{2}\right)\lim_{b\to-\infty}(e^u)\Big|_{u=b}^{u=0} = \left(\frac{1}{2}\right)\lim_{b\to-\infty}(1-e^b) = \frac{1}{2}(1-0) = \frac{1}{2}$$

(iii). $\int_{0}^{\infty} x^2 e^{-x^2}dx = \lim_{b\to\infty}\int_{0}^{b} x^2 e^{-x^2}dx$ By parts: let $u = x$, $dv = xe^{-x^2}$, then $du = dx$, $v = -\frac{1}{2}e^{-x^2}$

$$= \lim_{b\to\infty}\left(-\frac{1}{2}x\,e^{-x^2}\Big|_{u=0}^{x=b} + \frac{1}{2}\int_{0}^{b} e^{-x^2}dx\right) = 0 + \frac{1}{2}\cdot\frac{\sqrt{\pi}}{2} = \frac{\sqrt{\pi}}{4} \text{ from (i).}$$

13.3.67

a. $$\int_0^1\int_0^1 \frac{1}{(1+x^2+y^2)^2}\,dydx = \int_0^{\pi/4}\int_0^{\sec\theta}\frac{1}{(1+r^2)^2}\,r\,drd\theta + \int_{\pi/4}^{\pi/2}\int_0^{\csc\theta}\frac{1}{(1+r^2)^2}\,r\,drd\theta$$

$$= \frac{1}{2}\int_0^{\pi/4}\frac{\sec^2\theta}{1+\sec^2\theta}d\theta + \frac{1}{2}\int_{\pi/4}^{\pi/2}\frac{\csc^2\theta}{1+\csc^2\theta}d\theta = \frac{1}{2}\int_0^{\pi/4}\frac{\sec^2\theta}{2+\tan^2\theta}d\theta + \frac{1}{2}\int_{\pi/4}^{\pi/2}\frac{\csc^2\theta}{2+\cot^2\theta}d\theta$$

$$= \frac{1}{2}\int_0^1\frac{du}{2+u^2} + \frac{1}{2}\int_0^1\frac{du}{2+u^2} = \int_0^1\frac{du}{2+u^2} = \frac{1}{\sqrt{2}}\tan^{-1}\left(\frac{u}{\sqrt{2}}\right)\Big|_{u=0}^{u=1} = \frac{1}{\sqrt{2}}\tan^{-1}\left(\frac{1}{\sqrt{2}}\right).$$

b. $$\int_0^1\int_0^a \frac{1}{(1+x^2+y^2)^2}\,dydx$$

$$= \int_0^{\tan^{-1}(a)}\int_0^{\sec\theta}\frac{1}{(1+r^2)^2}\,r\,drd\theta + \int_{\tan^{-1}(a)}^{\pi/2}\int_0^{a\csc\theta}\frac{1}{(1+r^2)^2}\,r\,drd\theta$$

$$= \frac{1}{2}\int_0^{\tan^{-1}(a)}\frac{\sec^2\theta}{1+\sec^2\theta}d\theta + \frac{1}{2}\int_{\tan^{-1}(a)}^{\pi/2}\frac{a^2\csc^2\theta}{1+a^2\csc^2\theta}d\theta$$

$$= \frac{1}{2}\int_0^{\tan^{-1}(a)}\frac{\sec^2\theta}{2+\tan^2\theta}d\theta + \frac{1}{2}\int_{\tan^{-1}(a)}^{\pi/2}\frac{\csc^2\theta}{\left(\frac{1+a^2}{a^2}\right)+\cot^2\theta}d\theta$$

$$= \frac{1}{2}\int_0^1\frac{du}{2+u^2} + \frac{1}{2}\int_0^{1/a}\frac{du}{\left(\frac{1+a^2}{a^2}\right)+u^2} = \frac{1}{2\sqrt{2}}\tan^{-1}\left(\frac{u}{\sqrt{2}}\right)\Big|_{u=0}^{u=1} + \frac{a}{2\sqrt{1+a^2}}\tan^{-1}\left(\frac{a-u}{\sqrt{1+a^2}}\right)\Big|_{u=0}^{u=1/a}$$

$$= \frac{1}{2\sqrt{2}}\tan^{-1}\left(\frac{a}{\sqrt{2}}\right) + \frac{a}{2\sqrt{1+a^2}}\tan^{-1}\left(\frac{1}{\sqrt{1+a^2}}\right)$$

c. $$\lim_{a\to\infty}\int_0^1\int_0^a \frac{1}{(1+x^2+y^2)^2}\,dydx = \lim_{a\to\infty}\left[\frac{1}{2\sqrt{2}}\tan^{-1}\left(\frac{a}{\sqrt{2}}\right) + \frac{a}{2\sqrt{1+a^2}}\tan^{-1}\left(\frac{1}{\sqrt{1+a^2}}\right)\right]$$

$$= \frac{1}{2\sqrt{2}}\left(\frac{\pi}{2}\right) + \frac{1}{2}(1)(0) = \frac{\pi\sqrt{2}}{8}$$

13.4 Triple Integrals

13.4.1

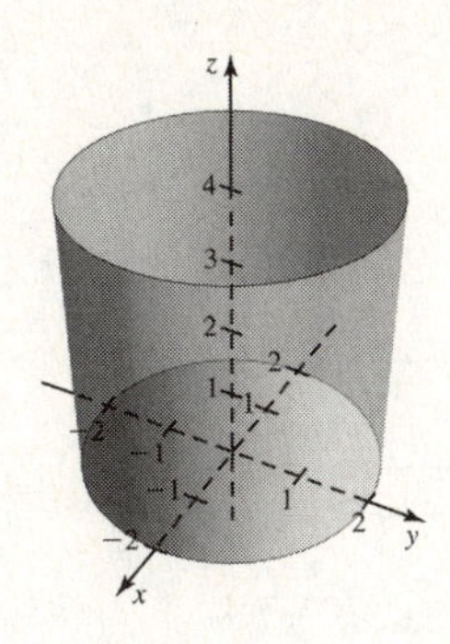

13.4.3 $\displaystyle\iiint_D f(x, y, z)\,dV = \int_{-3}^{3}\int_{-\sqrt{9-x^2}}^{\sqrt{9-x^2}}\int_{-\sqrt{9-x^2-y^2}}^{\sqrt{9-x^2-y^2}} f(x, y, z)\,dz\,dy\,dx$

13.4.5 $\displaystyle\int_0^1\int_0^{\sqrt{1-z^2}}\int_0^{\sqrt{1-x^2-z^2}} f(x, y, z)\,dy\,dx\,dz$

13.4.7 $\displaystyle\int_{-2}^{2}\int_3^6\int_0^2 dx dy dz = \int_{-2}^{2}\int_3^6 (x)\Big|_{x=0}^{x=2} dy dz = \int_{-2}^{2}\int_3^6 2\,dy dz = \int_{-2}^{2}(2y)\Big|_{y=3}^{y=6} dz = \int_{-2}^{2} 6\,dz$

$\displaystyle= (6z)\Big|_{z=-2}^{z=2} = 24$

13.4.9 $\displaystyle\int_{-2}^{2}\int_1^2\int_1^e \frac{xy^2}{z}dz\,dx\,dy = \int_{-2}^{2}\int_1^2 xy^2\left(xy^2\ln|z|\Big|_{z=1}^{z=e}\right)dx\,dy = \int_{-2}^{2}\int_1^2 xy^2\,dx\,dy = \int_{-2}^{2}\left(\frac{1}{2}x^2y^2\right)\Big|_{x=1}^{x=2} dy$

$\displaystyle= \int_{-2}^{2}\frac{3}{2}y^2\,dy = \left(\frac{1}{2}y^3\right)\Big|_{y=-2}^{y=2} = 8$

13.4.11 $\displaystyle\int_0^{\pi/2}\int_0^1\int_0^{\pi/2} \sin\pi x\cdot\cos y\cdot\sin 2z\,dy\,dx\,dz = \int_0^{\pi/2}\int_0^1 \sin\pi x\sin 2z\left((\sin y)\Big|_{y=0}^{y=\pi/2}\right)dx\,dz$

$\displaystyle= \int_0^{\pi/2}\int_0^1 \sin\pi x\sin 2z\,(1-0)\,dx\,dz = \int_0^{\pi/2}\int_0^1 \sin\pi x\sin 2z\,dx\,dz = \int_0^{\pi/2}\sin 2z\left(\left(-\frac{1}{\pi}\cos\pi x\right)\Big|_{x=0}^{x=1}\right)dz$

$\displaystyle= \int_0^{\pi/2}\sin 2z\left(\frac{1}{\pi}+\frac{1}{\pi}\right)dz = \frac{2}{\pi}\int_0^{\pi/2}\sin 2z\,dz = \frac{2}{\pi}\left(-\frac{1}{2}\cos 2z\right)\Big|_{z=0}^{z=\pi/2} = \frac{2}{\pi}\left(\frac{1}{2}+\frac{1}{2}\right) = \frac{2}{\pi}$

13.4.13 $\displaystyle\iiint_D (xy+xz+yz)\,dV = \int_{-1}^{1}\int_{-2}^{2}\int_{-3}^{3}(xy+xz+yz)\,dz\,dy\,dx$

$\displaystyle= \int_{-1}^{1}\int_{-2}^{2}\left(xyz+\frac{1}{2}xz^2+\frac{1}{2}yz^2\right)\Big|_{z=-3}^{z=3} dy\,dx = \int_{-1}^{1}\int_{-2}^{2}(6xy)\,dy\,dx = \int_{-1}^{1}(3xy^2)\Big|_{y=-2}^{y=2} dx = \int_{-1}^{1}(0)\,dx = 0$

13.4.15 $V = \iiint_D 1\,dV = \int_0^6 \int_0^{4-2x/3} \int_0^{2-x/3-y/2} 1\,dz\,dy\,dx = \int_0^6 \int_0^{4-2x/3} (z)\Big|_{z=0}^{z=2-x/3-y/2} dy\,dx$

$$= \int_0^6 \int_0^{4-2x/3} \left(2 - \frac{1}{3}x - \frac{1}{2}y\right) dy\,dx = \int_0^6 \left(2y - \frac{1}{3}xy - \frac{1}{4}y^2\right)\Big|_{y=0}^{y=4-2x/3} dx$$

$$= \int_0^6 \left(\frac{1}{9}x^2 - \frac{4}{3}x + 4\right) dx = \left(\frac{1}{27}x^3 - \frac{2}{3}x^2 + 4x\right)\Big|_{x=0}^{x=6} = 8$$

13.4.17 $V = \iiint_D 1\,dV = \int_{-2}^2 \int_{-\sqrt{4-x^2}}^{\sqrt{4-x^2}} \int_{\sqrt{x^2+y^2}}^{\sqrt{8-x^2-y^2}} 1\,dz\,dy\,dx = \int_{-2}^2 \int_{-\sqrt{4-x^2}}^{\sqrt{4-x^2}} (z)\Big|_{z=\sqrt{x^2+y^2}}^{z=\sqrt{8-x^2-y^2}} dy\,dx$

$$= \int_{-2}^2 \int_{-\sqrt{4-x^2}}^{\sqrt{4-x^2}} \left(\sqrt{8-x^2-y^2} - \sqrt{x^2+y^2}\right) dy\,dx \quad \text{(Convert to polar coordinates)}$$

$$= \int_0^{2\pi} \int_0^2 \left(\sqrt{8-r^2} - \sqrt{r^2}\right) r\,dr d\theta = \int_0^{2\pi} \int_0^2 \left(r\sqrt{8-r^2} - r^2\right) dr d\theta$$

$$= \int_0^{2\pi} \left(-\frac{1}{3}(8-r^2)^{3/2} - \frac{1}{3}r^3\right)\Big|_{r=0}^{r=2} d\theta = \int_0^{2\pi} \left(\frac{-16+16\sqrt{2}}{3}\right) d\theta = \frac{32\pi}{3}\left(\sqrt{2}-1\right)$$

13.4.19 $V = \iiint_D 1\,dV = \int_{-2}^2 \int_{-\sqrt{4-x^2}}^0 \int_0^{-y} 1\,dz\,dy\,dx = \int_{-2}^2 \int_{-\sqrt{4-x^2}}^0 (z)\Big|_{z=0}^{z=-y} dy\,dx = \int_{-2}^2 \int_{-\sqrt{4-x^2}}^0 (-y)\,dy\,dx$

$$= \int_{-2}^2 \left(-\frac{1}{2}y^2\right)\Big|_{y=-\sqrt{4-x^2}}^{y=0} dx = \frac{1}{2}\int_{-2}^2 (4-x^2)\,dx = \frac{1}{2}\left(4x - \frac{1}{3}x^3\right)\Big|_{x=-2}^{x=2} = \frac{16}{3}$$

13.4.21 $V = \iiint_D 1\,dV = \int_{-3}^3 \int_{-\sqrt{9-x^2}}^{\sqrt{9-x^2}} \int_{\sqrt{1+x^2+y^2}}^{\sqrt{19-x^2-y^2}} 1\,dz\,dy\,dx = \int_{-3}^3 \int_{-\sqrt{9-x^2}}^{\sqrt{9-x^2}} (z)\Big|_{z=\sqrt{1+x^2+y^2}}^{z=\sqrt{19-x^2-y^2}} dy\,dx$

$$= \int_{-3}^3 \int_{-\sqrt{9-x^2}}^{\sqrt{9-x^2}} \left(\sqrt{19-x^2-y^2} - \sqrt{1+x^2+y^2}\right) dy\,dx \quad \text{(Convert to polar coordinates)}$$

$$= \int_0^{2\pi} \int_0^3 \left(\sqrt{19-r^2} - \sqrt{1+r^2}\right) r\,dr\,d\theta = \int_0^{2\pi} \int_0^3 \left(r\sqrt{19-r^2} - r\sqrt{1+r^2}\right) dr\,d\theta$$

$$= \int_0^{2\pi} \left(-\frac{1}{3}\right)\left((19-r^2)^{3/2} + (1+r^2)^{3/2}\right)\Big|_{r=0}^{r=3} d\theta = \left(-\frac{1}{3}\right)\int_0^{2\pi} \left(20\sqrt{10} - 19\sqrt{19} - 1\right) d\theta$$

$$= \frac{1+19\sqrt{19}-20\sqrt{10}}{3} \int_0^{2\pi} d\theta = \frac{2\pi}{3}\left(1 + 19\sqrt{19} - 20\sqrt{10}\right)$$

13.4.23 $V = \iiint_D 1\,dV = \int_{-2}^2 \int_{-\frac{1}{2}\sqrt{4-x^2}}^{\frac{1}{2}\sqrt{4-x^2}} \int_{x-3}^{3-x} 1\,dz\,dy\,dx = \int_{-2}^2 \int_{-\frac{1}{2}\sqrt{4-x^2}}^{\frac{1}{2}\sqrt{4-x^2}} (z)\Big|_{z=x-3}^{z=3-x} dy\,dx$

$$= \int_{-2}^2 \int_{-\frac{1}{2}\sqrt{4-x^2}}^{\frac{1}{2}\sqrt{4-x^2}} (6-2x)\,dy\,dx = \int_{-2}^2 (6-2x)y\Big|_{y=-\frac{1}{2}\sqrt{4-x^2}}^{y=\frac{1}{2}\sqrt{4-x^2}} dx = \int_{-2}^2 \left(6\sqrt{4-x^2} - 2x\sqrt{4-x^2}\right) dx$$

$$= \left(3x\sqrt{4-x^2} + 12\sin^{-1}\left(\frac{x}{2}\right) + \frac{2}{3}\sqrt{4-x^2}\right)\Big|_{x=-2}^{x=2} = 12\pi$$

13.4.25 $\int_0^1\int_0^{\sqrt{1-x^2}}\int_0^{\sqrt{1-x^2}} dz\,dy\,dx = \int_0^1\int_0^{\sqrt{1-x^2}} (z)\Big|_{z=0}^{z=\sqrt{1-x^2}} dy\,dx = \int_0^1\int_0^{\sqrt{1-x^2}} \sqrt{1-x^2}\,dy\,dx$

$= \int_0^1 y\sqrt{1-x^2}\Big|_{y=0}^{y=\sqrt{1-x^2}} dx = \int_0^1 (1-x^2)\,dx = \left(x - \frac{1}{3}x^3\right)\Big|_{x=0}^{x=1} = \frac{2}{3}$

13.4.27 $\int_0^4\int_{-2\sqrt{16-y^2}}^{2\sqrt{16-y^2}}\int_0^{16-(x^2/4)-y^2} dz\,dx\,dy = \int_0^4\int_{-2\sqrt{16-y^2}}^{2\sqrt{16-y^2}} (z)\Big|_{z=0}^{z=16-(x^2/4)-y^2} dx\,dy$

$= \int_0^4\int_{-2\sqrt{16-y^2}}^{2\sqrt{16-y^2}} \left(16 - \frac{x^2}{4} - y^2\right) dx\,dy = \int_0^4 \left(16x - \frac{1}{12}x^3 - xy^2\right)\Big|_{x=-2\sqrt{16-y^2}}^{x=2\sqrt{16-y^2}} dy$

$= \int_0^4 \left(64\sqrt{16-y^2} - \frac{4}{3}(16-y^2)^{3/2} - 4y^2\sqrt{16-y^2}\right) dy$

$= \left(32y\sqrt{16-y^2} + 512\sin^{-1}\left(\frac{y}{4}\right) - \frac{40}{3}\sqrt{16-y^2} + \frac{y^3}{3}\sqrt{16-y^2} - 128\sin^{-1}\left(\frac{y}{4}\right) + 8y\sqrt{16-y^2}\right.$

$\left. - y^3\sqrt{16-y^2} - 128\sin^{-1}\left(\frac{y}{4}\right)\right)\Big|_{y=0}^{y=4} = \left(\frac{80y}{3}\sqrt{16-y^2} - \frac{2}{3}y^2\sqrt{16-y^2} + 256\sin^{-1}\left(\frac{y}{4}\right)\right)\Big|_{y=0}^{y=4}$

$= 128\pi.$

13.4.29 $\int_0^3\int_0^{\sqrt{9-z^2}}\int_0^{\sqrt{1+x^2+z^2}} dy\,dx\,dz = \int_0^3\int_0^{\sqrt{9-z^2}} \sqrt{1+x^2+z^2}\,dz\,dy$ (Use polar with $x^2+z^2=r^2$)

$= \int_0^{\pi/2}\int_0^3 \sqrt{1+r^2}\,r\,dr\,d\theta = \int_0^{\pi/2}\left(\frac{1}{3}(1+r^2)^{3/2}\right)\Big|_{r=0}^{r=3} d\theta = \int_0^{\pi/2}\frac{1}{3}(10^{3/2}-1)\,d\theta = \frac{\pi}{6}\left(10\sqrt{10}-1\right)$

13.4.31 $\int_1^{\ln 8}\int_1^{\sqrt{z}}\int_{\ln y}^{\ln(2y)} e^{x+y^2-z}\,dx\,dy\,dz = \int_1^{\ln 8}\int_1^{\sqrt{z}} e^{x+y^2-z}\Big|_{x=\ln y}^{x=\ln(2y)} dy\,dz$

$= \int_1^{\ln 8}\int_1^{\sqrt{z}}\left(2ye^{y^2-z} - ye^{y^2-z}\right) dy\,dz = \int_1^{\ln 8}\int_1^{\sqrt{z}} ye^{y^2-z}\,dydz = \int_1^{\ln 8}\left(\frac{1}{2}e^{y^2-z}\right)\Big|_{y=1}^{y=\sqrt{z}} dz$

$= \frac{1}{2}\int_1^{\ln 8}(1-e^{1-z})\,dz = \frac{1}{2}(z+e^{1-z})\Big|_{z=1}^{z=\ln 8} = \frac{1}{2}\left(\ln 8 + \frac{e}{8} - (1+1)\right) = \frac{1}{2}\ln 8 + \frac{e}{16} - 1$

13.4.33 $\int_0^2\int_0^4\int_{y^2}^4 \sqrt{x}\,dz\,dx\,dy = \int_0^2\int_0^4 z\sqrt{x}\Big|_{z=y^2}^{z=4} dx\,dy = \int_0^2\int_0^4 (4-y^2)\sqrt{x}\,dx\,dy$

$= \int_0^2 (4-y^2)\left(\frac{2}{3}x^{3/2}\right)\Big|_{x=0}^{x=4} dy = \frac{16}{3}\int_0^2 (4-y^2)dy = \frac{16}{3}\left(4y - \frac{1}{3}y^3\right)\Big|_{y=0}^{y=2} = \frac{256}{9}$

13.4.35 $\int_0^5\int_{-1}^0\int_0^{4x+4} dydxdz$

$0 \le z \le 5, -1 \le x \le 0, 0 \le y \le 4x+4$, thus $y \le 4x+4 \Rightarrow \frac{y-4}{4} \le x$ so $x \le 0 \Rightarrow y \le 4$. Switch

$= \int_0^4\int_{(y-4)/4}^0\int_0^5 dz\,dx\,dy = \int_0^4\int_{(y-4)/4}^0 5\,dx\,dy = \int_0^4 -\frac{5}{4}(y-4)\,dy = -\frac{5}{4}\left(\frac{1}{2}y^2 - 4y\right)\Big|_{y=0}^{y=4} = 10$

13.4.37 $\int_0^1\int_0^{\sqrt{1-x^2}}\int_0^{\sqrt{1-x^2}} dy\,dz\,dx$

$0 \le x \le 1, 0 \le y \le \sqrt{1-x^2}, 0 \le z \le \sqrt{1-x^2}$. Switch

$$= \int_0^1 \int_0^{\sqrt{1-x^2}} \int_0^{\sqrt{1-x^2}} dz\,dy\,dx = \int_0^1 \int_0^{\sqrt{1-x^2}} \sqrt{1-x^2}\,dy\,dx = \int_0^1 \left(\sqrt{1-x^2}\right)\left(\sqrt{1-x^2}\right) dx$$

$$= \int_0^1 (1-x^2)\,dx = \left(x - \frac{1}{3}x^3\right)\Big|_{x=0}^{x=1} = \frac{2}{3}$$

13.4.39 average value $= \dfrac{1}{\text{volume of } D}\displaystyle\iiint_D T(x,\,y,\,z)\,dV = \dfrac{1}{\ln 2 \cdot \ln 4 \cdot \ln 8}\int_0^{\ln 2}\int_0^{\ln 4}\int_0^{\ln 8} 100\,e^{-x-y-z}\,dz\,dy\,dx$

$$= \frac{100}{6(\ln 2)^3}\int_0^{\ln 2}\int_0^{\ln 4}\int_0^{\ln 8} e^{-x}e^{-y}e^{-z}\,dz\,dy\,dx = \frac{50}{3(\ln 2)^3}\int_0^{\ln 2}\int_0^{\ln 4} e^{-x}e^{-y}(-e^{-z})\Big|_{z=0}^{z=\ln 8} dy\,dx$$

$$= \frac{50}{3(\ln 2)^3}\int_0^{\ln 2}\int_0^{\ln 4} e^{-x}e^{-y}\left(1-\frac{1}{8}\right)dy\,dx = \frac{175}{12(\ln 2)^3}\int_0^{\ln 2}\int_0^{\ln 4} e^{-x}e^{-y}dy\,dx$$

$$= \frac{175}{12(\ln 2)^3}\int_0^{\ln 2} e^{-x}(-e^{-y})\Big|_{y=0}^{y=\ln 4} dx = \frac{175}{12(\ln 2)^3}\int_0^{\ln 2} e^{-x}\left(1-\frac{1}{4}\right)dx$$

$$= \frac{175}{16(\ln 2)^3}\int_0^{\ln 2} e^{-x}dx = \frac{175}{16(\ln 2)^3}(-e^{-x})\Big|_{x=0}^{x=\ln 2} = \frac{175}{16(\ln 2)^3}\left(1-\frac{1}{2}\right) = \frac{175}{32(\ln 2)^3}$$

13.4.41 average value $= \dfrac{1}{\text{volume of } D}\displaystyle\iiint_D f(x,\,y,\,z)\,dV = \dfrac{1}{(\pi\cdot 2^2\cdot 2)}\int_{-2}^{2}\int_{-\sqrt{4-x^2}}^{\sqrt{4-x^2}}\int_0^2 (x^2+y^2+z^2)\,dz\,dy\,dx$

$$= \frac{1}{8\pi}\int_{-2}^{2}\int_{-\sqrt{4-x^2}}^{\sqrt{4-x^2}}\left((x^2+y^2)z + \frac{1}{3}z^3\right)\Big|_{z=0}^{z=2} dy\,dx = \frac{1}{8\pi}\int_{-2}^{2}\int_{-\sqrt{4-x^2}}^{\sqrt{4-x^2}}\left(2(x^2+y^2)+\frac{8}{3}\right)dy\,dx$$

$$= \frac{1}{8\pi}\int_{-2}^{2}\left(2x^2y + \frac{2}{3}y^3 + \frac{8}{3}y\right)\Big|_{y=-\sqrt{4-x^2}}^{y=\sqrt{4-x^2}} dx = \frac{1}{2\pi}\int_{-2}^{2}\left(x^2\sqrt{4-x^2} + \frac{1}{3}(4-x^2)^{3/2} + \frac{4}{3}\sqrt{4-x^2}\right)dx$$

$$= \frac{1}{2\pi}\left(-\frac{x}{4}(x^2-2)\sqrt{4-x^2} + 2\sin^{-1}\left(\frac{x}{2}\right) + \frac{x}{12}(x^2-10) + 2\sin^{-1}\left(\frac{x}{2}\right)\right.$$

$$\left. + \frac{x}{6}\sqrt{4-x^2} + \frac{8}{3}\sin^{-1}\left(\frac{x}{2}\right)\right)\Big|_{x=-2}^{x=2} = \frac{1}{2\pi}\left(\frac{10\pi}{3}+\frac{10\pi}{3}\right) = \frac{10}{3}.$$

13.4.43 average value $= \dfrac{1}{\text{volume of } D}\displaystyle\iiint_D f(x,\,y,\,z)\,dV = \dfrac{1}{\frac{1}{2}\left(\frac{4}{3}\pi\cdot 4^3\right)}\int_{-4}^{4}\int_{-\sqrt{16-x^2}}^{\sqrt{16-x^2}}\int_0^{\sqrt{16-x^2-y^2}} (z)\,dz\,dy\,dx$

$$= \frac{3}{128\pi}\int_{-4}^{4}\int_{-\sqrt{16-x^2}}^{\sqrt{16-x^2}}\left(\frac{1}{2}z^2\right)\Big|_{z=0}^{z=\sqrt{16-x^2-y^2}} dy\,dx = \frac{3}{256\pi}\int_{-4}^{4}\int_{-\sqrt{16-x^2}}^{\sqrt{16-x^2}}(16-x^2-y^2)\,dy\,dx$$

(Convert to polar)

$$= \frac{3}{256\pi}\int_0^{2\pi}\int_0^4 (16-r^2)r\,dr\,d\theta = \frac{3}{256\pi}\int_0^{2\pi}\int_0^4 (16r-r^3)dr\,d\theta$$

$$= \frac{3}{256\pi}\int_0^{2\pi}\left(8r^2-\frac{1}{4}r^4\right)\Big|_{r=0}^{r=4} d\theta = \frac{3}{256\pi}\int_0^{2\pi}(128-64)\,d\theta = \frac{3}{4\pi}\int_0^{2\pi} d\theta = \frac{3}{2}$$

13.4.45

a. False. Only six iterations are possible.
b. False. The outermost limits of integration must be constants.
c. False. D is intersection of two cylinders in first octant.

13.4.47 $V = \displaystyle\int_0^1\int_z^{z+1}\int_0^1 dx\,dy\,dz = \int_0^1\int_z^{z+1} 1\,dy\,dz = \int_0^1 1\,dz = 1$

13.4.49 $V = \int_0^2 \int_0^{4-2y} \int_0^{(4-z)/2} 1\,dx\,dz\,dy = \int_0^2 \int_0^{4-2y} \left(\frac{4-z}{2}\right) dz\,dy$
$= \int_0^2 \left(2z - \frac{1}{4}z^2\right)\Big|_{z=0}^{z=4-2y} dy = \int_0^2 (4-y^2)\,dy = \left(4y - \frac{1}{3}y^3\right)\Big|_{y=0}^{y=2} = \frac{16}{3}$

13.4.51 $V = \int_{-1}^0 \int_{-x-1}^{x+1} \int_0^{1-x-y} 1\,dz\,dy\,dx + \int_0^1 \int_{x-1}^{-x+1} \int_0^{1-x-y} 1\,dz\,dy\,dx$
$= \int_{-1}^0 \int_{-x-1}^{x+1} (1-x-y)\,dy\,dx + \int_0^1 \int_{x-1}^{-x+1} (1-x-y)\,dy\,dx$
$= \int_{-1}^0 \left((1-x)y - \frac{1}{2}y^2\right)\Big|_{y=-x-1}^{y=x+1} dx + \int_0^1 \left((1-x)y - \frac{1}{2}y^2\right)\Big|_{y=x-1}^{y=-x+1} dx$
$= \int_{-1}^0 (2-2x^2)\,dx + \int_0^1 (2-4x+2x^2)dx = \left(2x - \frac{2}{3}x^3\right)\Big|_{x=-1}^{x=0} + \left(2x - 2x^2 + \frac{2}{3}x^3\right)\Big|_{x=0}^{x=1}$
$= \left(2 - \frac{2}{3}\right) + \left(2 - 2 + \frac{2}{3}\right) = 2$

13.4.53 mass $= \iiint_D \rho(x, y, z)\,dV$,

$m_1 = \int_0^4 \int_0^{4-x} \int_0^{4-x-y} (8-z)\,dz\,dy\,dx = \int_0^4 \int_0^{4-x} \left(8z - \frac{1}{2}z^2\right)\Big|_{z=0}^{z=4-x-y} dy\,dx$
$= \int_0^4 \int_0^{4-x} \left(24 - 4x - \frac{x^2}{2} - 4y - xy - \frac{y^2}{2}\right) dy\,dx = \int_0^4 \left(24y - 4xy - \frac{x^2}{2}y - 2y^2 - \frac{1}{2}xy^2 - \frac{y^3}{6}\right)\Big|_{y=0}^{y=4-x} dx$
$= \int_0^4 \left(\frac{160}{3} - 24x + 2x^2 + \frac{x^3}{6}\right) dx = \left(\frac{160x}{3} - 12x^2 + \frac{2}{3}x^3 + \frac{x^4}{24}\right)\Big|_{x=0}^{x=4} = \frac{224}{3}$

$m_2 = \int_0^4 \int_0^{4-x} \int_0^{4-x-y} (4+z)\,dzdydx = \int_0^4 \int_0^{4-x} \left(4z + \frac{1}{2}z^2\right)\Big|_{z=0}^{z=4-x-y} dy\,dx$
$= \int_0^4 \int_0^{4-x} \left(24 - 8x - \frac{x^2}{2} - 8y + xy - \frac{y^2}{2}\right) dydx = \int_0^4 \left(24y - 8xy - \frac{x^2y}{2} - 4y^2 + \frac{1}{2}xy^2 - \frac{y^3}{6}\right)\Big|_{y=0}^{y=4-x} dx$
$= \int_0^4 \left(\frac{128}{3} - 24x + 4x^2 - \frac{x^3}{6}\right) dx = \left(\frac{128x}{3} - 12x^2 + \frac{4}{3}x^3 - \frac{x^4}{24}\right)\Big|_{x=0}^{x=4} = \frac{160}{3}$

Solid 1 has greater mass since the density is greater near the bottom where the tetrahedron is wider.

13.4.55 Equation of a cone with height h and whose base is centered at the origin with radius r in xy-plane is $z = h - \frac{h}{r}\sqrt{x^2+y^2}$.

$V = \int_{-r}^r \int_{-\sqrt{r^2-x^2}}^{\sqrt{r^2-x^2}} \int_0^{h-\frac{h}{r}\sqrt{x^2+y^2}} 1\,dz\,dy\,dx = \int_{-r}^r \int_{-\sqrt{r^2-x^2}}^{\sqrt{r^2-x^2}} \left(h - \frac{h}{r}\sqrt{x^2+y^2}\right) dy\,dx$

(Use polar coordinates. Let $x^2 + y^2 = a^2$ to avoid confusion with constant r.)

$= \int_0^{2\pi} \int_0^r \left(h - \frac{h}{r}a\right) a\,da\,d\theta = \int_0^{2\pi} \int_0^r \left(ha - \frac{h}{r}a^2\right) da\,d\theta = \int_0^{2\pi} \left(\frac{1}{2}ha^2 - \frac{h}{3r}a^3\right)\Big|_{a=0}^{a=r} d\theta$
$= \int_0^{2\pi} \left(\frac{1}{2}r^2h - \frac{hr^3}{3r}\right) d\theta = \int_0^{2\pi} \frac{1}{6}r^2h\,d\theta = \frac{1}{3}\pi r^2 h$

13.4.57 Equation of a sphere with radius R is $x^2+y^2+z^2=R^2$.

$$V=\int_{-\sqrt{2Rh-h^2}}^{\sqrt{2Rh-h^2}}\int_{-\sqrt{2Rh-h^2-x^2}}^{\sqrt{2Rh-h^2-x^2}}\int_{R-h}^{\sqrt{R^2-x^2-y^2}} 1\,dz\,dy\,dx$$

$$=\int_{-\sqrt{2Rh-h^2}}^{\sqrt{2Rh-h^2}}\int_{-\sqrt{2Rh-h^2-x^2}}^{\sqrt{2Rh-h^2-x^2}}\left(\sqrt{R^2-x^2-y^2}-(R-h)\right)dy\,dx \qquad \text{(Convert to polar.)}$$

$$=\int_0^{2\pi}\int_0^{\sqrt{2Rh-h^2}}\left(\sqrt{R^2-r^2}-(R-h)\right)r\,dr\,d\theta=\int_0^{2\pi}\int_0^{\sqrt{2Rh-h^2}}\left(r\sqrt{R^2-r^2}-r(R-h)\right)dr\,d\theta$$

$$=\int_0^{2\pi}\left(\left(-\frac{1}{3}(R^2-r^2)^{3/2}-\frac{1}{2}r^2(R-h)\right)\right)\Bigg|_{r=0}^{r=\sqrt{2Rh-h^2}}d\theta$$

$$=\int_0^{2\pi}\left[-\frac{1}{3}(R-h)^3-\frac{1}{2}(2Rh-h^2)(R-h)+\frac{1}{3}R^3\right]d\theta=\int_0^{2\pi}\left(\frac{h^2}{6}(3R-h)\right)d\theta=\frac{1}{3}\pi h^2(3R-h)$$

13.4.59 The equation of the given ellipsoid is $\dfrac{x^2}{a^2}+\dfrac{y^2}{b^2}+\dfrac{z^2}{c^2}=1$. Using symmetry, the volume V of the ellipsoid is eight times the volume of the portion of the ellipsoid in the first octant. Thus

$$V=8\int_0^a\int_0^{b\sqrt{1-x^2/a^2}}\int_0^{c\sqrt{1-x^2/a^2-y^2/b^2}}1\,dz\,dy\,dx=8c\int_0^a\int_0^{b\sqrt{1-x^2/a^2}}\sqrt{1-\frac{x^2}{a^2}-\frac{y^2}{b^2}}\,dy\,dx$$

$$=8c\int_0^a\left(\frac{y}{2}\sqrt{1-\frac{x^2}{a^2}-\frac{y^2}{b^2}}+\frac{b(a^2-x^2)}{2a^2}\sin^{-1}\left(\frac{ay}{b\sqrt{a^2-x^2}}\right)\right)\Bigg|_{y=0}^{y=b\sqrt{1-x^2/a^2}}dx$$

$$=8c\int_0^a\frac{b\pi(a^2-x^2)}{4a^2}dx=\frac{2\pi bc}{a^2}\left(a^2x-\frac{x^3}{3}\right)\Bigg|_{x=0}^{x=a}=\frac{2\pi bc}{a^2}\cdot\frac{2a^3}{3}=\frac{4}{3}\pi abc$$

13.4.61 Hypervolume $=\displaystyle\int_0^1\int_0^{1-x}\int_0^{1-x-y}\int_0^{1-x-y-z}1\,dwdzdydx$

$$=\int_0^1\int_0^{1-x}\int_0^{1-x-y}(1-x-y-z)\,dz\,dy\,dx$$

$$=\int_0^1\int_0^{1-x}\left((1-x-y)z-\frac{1}{2}z^2\right)\Bigg|_{z=0}^{z=-x-y}dy\,dx$$

$$=\int_0^1\int_0^{1-x}\left((1-x-y)^2-\frac{1}{2}(1-x-y)^2\right)dy\,dx=\frac{1}{2}\int_0^1\int_0^{1-x}(1-x-y)^2\,dy\,dx$$

$$=\frac{1}{2}\int_0^1-\frac{1}{3}(1-x-y)^3\Bigg|_{y=0}^{y=1-x}dx=\frac{1}{6}\int_0^1(1-x)^3\,dx=\frac{1}{6}\left(-\frac{1}{4}(1-x)^4\right)\Bigg|_{x=0}^{x=1}=\frac{1}{24}$$

13.5 Triple Integrals in Cylindrical and Spherical Coordinates

13.5.1 The triple $(r,\,\theta,\,z)$ describes the point (written in Cartesian coordinates) $(r\cos\theta,\,r\sin\theta,\,z)$.

13.5.3 A double cone (opening both upwards and downwards), where the radius is always 4 times the distance to the xy-plane.

13.5.5 Looking back at the definition of the integral using Riemann sums, one sees that the volume of a volume element gets larger as the volume element gets further from the origin, and the factor is measured by r

13.5.7 $\displaystyle\iiint_D f(r,\,\theta,\,z)dV=\int_\alpha^\beta\int_{g(\theta)}^{h(\theta)}\int_{G(r,\theta)}^{H(r,\theta)}f(r,\,\theta,\,z)dz\,r\,dr\,d\theta$

13.5.9 Cylindrical coordinates, since in cylindrical coordinates $x^2 + y^2$ simplifies to r^2.

13.5.11 This is a wedge of a cylinder of radius 3 from $z = 1$ to $z = 4$, where the wedge angle is $\frac{\pi}{3}$.

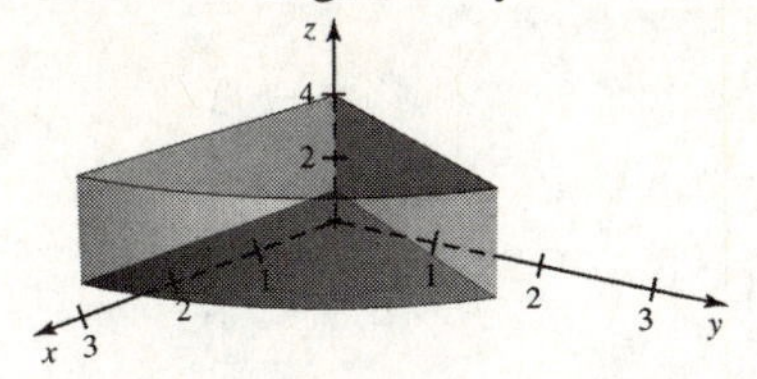

13.5.13 This is the solid upward pointing cone given by $z = 2r$ from $z = 0$ to $z = 4$.

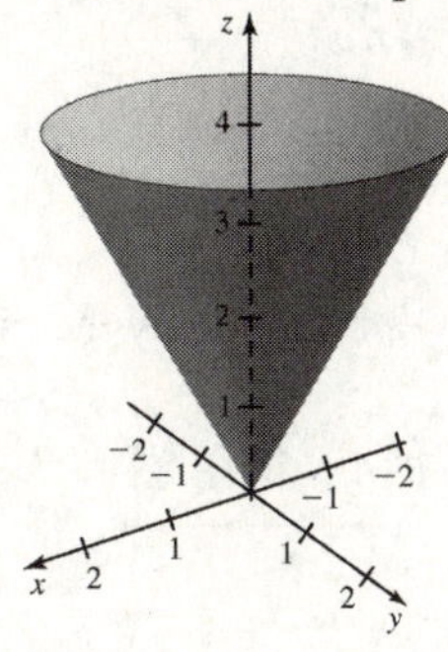

13.5.15 $$\int_0^{2\pi}\int_0^1\int_{-1}^1 dz\,r\,dr\,d\theta = \int_0^{2\pi}\int_0^1 r\,z\Big|_{z=-1}^{z=1} dr\,d\theta = \int_0^{2\pi}\int_0^1 2r\,dr\,d\theta = \int_0^{2\pi} 1\,d\theta = 2\pi.$$

13.5.17 This is the integral of $(x^2 + y^2)^{3/2}$ over a cylinder with radius 1 from $z = -1$ to $z = 1$. We convert to cylindrical coordinates, so the integrand becomes r^3 since $x^2 + y^2 = r^2$. The integral is thus

$$\int_0^{2\pi}\int_0^1\int_{-1}^1 r^3\,dz\,r\,dr\,d\theta = \int_0^{2\pi}\int_0^1 2\,r^4\,dr\,d\theta = \frac{2}{5}\int_0^{2\pi} r^5\Big|_{r=0}^{r=1} = \frac{4\pi}{5}.$$

13.5.19 The region of integration is a wedge from $\theta = \frac{\pi}{4}$ to $\theta = \frac{\pi}{2}$ with radius 1, between $z = 0$ and $z = 4$. This can be seen by noting that the limits of integration for y range from x to the boundary of the unit circle (y-coordinate $\sqrt{1 - x^2}$) and that $x = \frac{\sqrt{2}}{2}$corresponds to $\frac{\pi}{4}$. Thus the integral is

$$\int_{\pi/4}^{\pi/2}\int_0^1\int_0^4 e^{-r^2}\,dz\,r\,dr\,d\theta = \int_{\pi/4}^{\pi/2}\int_0^1 4r\,e^{-r^2}\,dr\,d\theta = -2\int_0^{\pi} e^{-r^2}\Big|_{r=0}^{r=1} = -2\int_0^{\pi}\left(e^{-1} - 1\right)d\theta = \frac{\pi}{2}\left(1 - e^{-1}\right).$$

13.5.21 The region of integration is below the cone $x^2 + y^2 = z^2$ in the first octant, so the integral is

$$\int_0^{\pi/2}\int_0^3\int_0^r \frac{1}{r}\,dz\,r\,dr\,d\theta = \int_0^{\pi/2}\int_0^3 r\,dr\,d\theta = \int_0^{\pi/2}\frac{9}{2}\,d\theta = \frac{9\pi}{4}.$$

13.5.23 $$\int_0^{2\pi}\int_0^4\int_0^{10}\left(1 + \frac{z}{2}\right)dz\,r\,dr\,d\theta = \int_0^{2\pi}\int_0^4\left(z + \frac{z^2}{4}\right)\Big|_{z=0}^{z=10} r\,dr\,d\theta = \int_0^{2\pi}\int_0^4 35r\,dr\,d\theta = 560\pi$$

13.5.25 $$\int_0^{2\pi}\int_0^6\int_0^{6-r}(7 - z)\,dz\,r\,dr\,d\theta = \int_0^{2\pi}\int_0^6\left(7z - \frac{1}{2}z^2\right)\Big|_{z=0}^{z=6-r} r\,dr\,d\theta$$
$$= \int_0^{2\pi}\int_0^4\left(7r(6 - r) - \frac{r}{2}(6 - r)^2\right)dr\,d\theta = 396\pi$$

13.5.27 The base of both surfaces in the xy-plane is the area bounded by the unit circle $r = 1$. The mass of the solid bounded by the xy-plane and $z = 4 - 4r$ is given by

$$\int_0^{2\pi}\int_0^1\int_0^{4-4r} (10-2z)\,dz\,r\,dr\,d\theta = \int_0^{2\pi}\int_0^1 \left(10z - z^2\right)\Big|_{z=0}^{z=4-4r} r\,dr\,d\theta$$
$$= \int_0^{2\pi}\int_0^1 \left(10r(4-4r) - r(4-4r)^2\right) dr\,d\theta = \frac{32\pi}{3}$$

The mass of the solid bounded by the xy-plane and $z = 4 - 4r^2$ is given by

$$\int_0^{2\pi}\int_0^1\int_0^{4-4r^2} (10-2z)\,dz\,r\,d\,rd\theta = \int_0^{2\pi}\int_0^1 \left(10z - z^2\right)\Big|_{z=0}^{z=4-4r^2} r\,dr\,d\theta$$
$$= \int_0^{2\pi}\int_0^1 \left(10r\left(4-4r^2\right) - r\left(4-4r^2\right)^2\right) drd\theta = \frac{44\pi}{3}$$

Thus the mass of the solid bounded by the paraboloid is larger. This can also be seen by noting that for $r \le 1$, $r^2 \le r$ so that $4 - 4r^2 \ge 4 - 4r$, so that the cone is contained in the paraboloid.

13.5.29 The base of this solid in the xy-plane is given by $\sqrt{17} = \sqrt{1 + x^2 + y^2}$ so is the area bounded by the circle $x^2 + y^2 = 16$ or $r = 4$. Thus the volume is

$$\int_0^{2\pi}\int_0^4\int_0^{\sqrt{17}-\sqrt{1+r^2}} 1\,dz\,r\,dr\,d\theta = \int_0^{2\pi}\int_0^4 \left(r\sqrt{17} - r\sqrt{1+r^2}\right) dr\,d\theta = \frac{\pi\left(14\sqrt{17}+2\right)}{3}.$$

13.5.31 This solid sits over the circle $r = 5$ in the xy-plane, so the volume is

$$\int_0^{2\pi}\int_0^5\int_{\sqrt{4+r^2}}^{\sqrt{29}} 1\,dz\,r\,d\,rd\theta = \int_0^{2\pi}\int_0^5 \left(r\sqrt{29} - r\sqrt{4+r^2}\right) drd\theta = \frac{\pi\left(17\sqrt{29}+16\right)}{3}.$$

13.5.33 The first octant is determined by $0 \le \theta \le \frac{\pi}{2}$, $0 \le z$, and the condition $z = x$ in cylindrical coordinates becomes $z = r\cos\theta$, so the volume of the solid is

$$\int_0^{\pi/2}\int_0^1\int_0^{r\cos\theta} 1\,dz\,r\,dr\,d\theta = \int_0^{\pi/2}\int_0^1 r^2\cos\theta\,dr\,d\theta = \frac{1}{3}.$$

13.5.35 This is a spherical shell centered at the origin with outer radius 3 and inner radius 1.

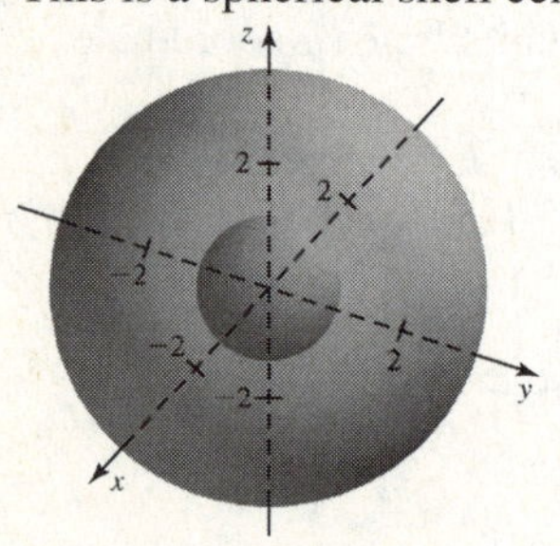

13.5.37 This is a sphere of radius 2 centered at $(0, 0, 2)$. To see this, note that $\rho = 4\cos\varphi$ implies that $\rho^2 = 4\rho\cos\varphi$. Converting to rectangular coordinates gives $x^2 + y^2 + z^2 = 4z$, and completing the square yields $x^2 + y^2 + (z-2)^2 = 4$.

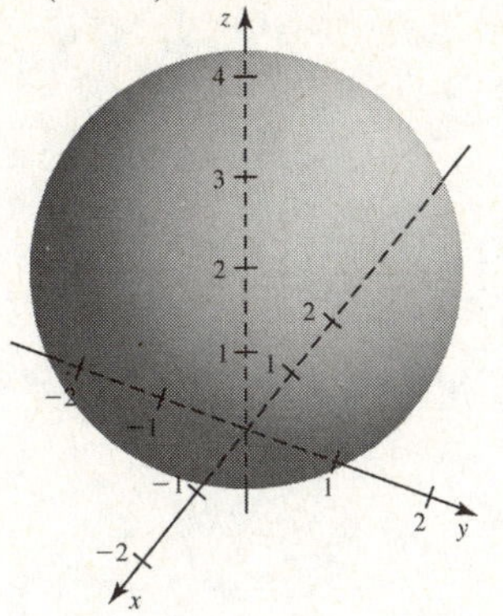

13.5.39 $$\iiint_D (x^2+y^2+z^2)^{5/2}\,dV = \int_0^1\int_0^{2\pi}\int_0^{\pi} \rho^5\,\rho^2 \sin\varphi\; d\varphi\, d\theta\, d\rho = \int_0^1\int_0^{2\pi}\int_0^{\pi} \rho^7 \sin\varphi\; d\varphi\, d\theta\, d\rho$$
$$= \int_0^1\int_0^{2\pi} 2\rho^7\, d\theta\, d\rho = 4\pi\int_0^1 \rho^7\, d\rho = \frac{\pi}{2}$$

13.5.41 $$\iiint_D (x^2+y^2+z^2)^{-3/2}\,dV = \int_0^{2\pi}\int_0^{\pi}\int_1^2 \rho^{-3}\,\rho^2 \sin\varphi\; d\rho\, d\varphi\, d\theta = \int_0^{2\pi}\int_0^{\pi}\int_1^2 \rho^{-1} \sin\varphi\, d\rho d\varphi\, d\theta$$
$$= \ln 2\int_0^{2\pi}\int_0^{\pi} \sin\varphi\; d\varphi\, d\theta = 4\pi\ln 2\,.$$

13.5.43 $$\int_0^{\pi}\int_0^{\pi/6}\int_{2\sec\varphi}^{4} \rho^2 \sin\varphi\; d\rho\, d\varphi\, d\theta = \frac{1}{3}\int_0^{\pi}\int_0^{\pi/6} (64 - 8\sec^3\varphi)\, \sin\varphi\; d\varphi\, d\theta = \left(\frac{188}{9} - \frac{32}{3}\sqrt{3}\right)\pi$$

13.5.45 $$\int_0^{2\pi}\int_{\pi/6}^{\pi/3}\int_0^{2\csc\varphi} \rho^2 \sin\varphi\; d\rho\, d\varphi\, d\theta = \frac{8}{3}\int_0^{2\pi}\int_{\pi/6}^{\pi/3} \csc^3\varphi\, \sin\varphi\; d\varphi\, d\theta = \frac{32}{9}\pi\sqrt{3}$$

13.5.47 The two spheres intersect where $2\cos\varphi = 1$, i.e. when $\varphi = \frac{\pi}{3}$. That circle is then at $z = \frac{1}{2}$ (again looking at $\cos\varphi = \frac{1}{2}$ and using the fact that the lower sphere has radius 1). The volume in question is the sum of the volumes above and below that circle of intersection; since the circle is at height $\frac{1}{2}$ and both spheres have radius 1, the volumes above and below are identical. Thus we need only compute the upper volume and double it:
$$2\int_0^{\pi/3}\int_{(\sec\varphi)/2}^{1}\int_0^{2\pi} \rho^2 \sin\varphi\; d\theta\, d\rho\, d\varphi = \frac{4\pi}{3}\int_0^{\pi/3}\left(1 - \frac{\sec^3\varphi}{8}\right) \sin\varphi\; d\theta = \frac{5}{12}\pi$$

13.5.49 The portion of the sphere is determined by $\frac{\pi}{4} \le \varphi \le \frac{\pi}{2}$, so the integral is
$$\int_0^{2\pi}\int_{\pi/4}^{\pi/2}\int_0^{4\cos\varphi} \rho^2 \sin\varphi\; d\rho\, d\varphi\, d\theta = \frac{64}{3}\int_0^{\pi/3}\int_{\pi/4}^{\pi/2} \cos^3\varphi\, \sin\varphi d\varphi\, d\theta = \frac{8}{3}\pi$$

13.5.51 The value of φ corresponding to the circle of intersection of the sphere ρ and the plane $z = 2\sqrt{3}$ is found by noting that $\cos\varphi = \frac{2\sqrt{3}}{4} = \frac{\sqrt{3}}{2}$, so that $\varphi = \frac{\pi}{6}$. Similarly, the intersection with the plane $z = 2$ is where $\cos\varphi = \frac{2}{4} = \frac{1}{2}$ so that $\varphi = \frac{\pi}{3}$. We can find the volume of the required region by determining the volume of the regions above $z = 2$ and $z = 2\sqrt{3}$ and subtracting. The volume of the region above the plane $z = 2$ is
$$\int_0^{2\pi}\int_0^{\pi/3}\int_{2\sec\varphi}^{4} \rho^2 \sin\varphi\, d\rho\, d\varphi\, d\theta = \frac{1}{3}\int_0^{2\pi}\int_0^{\pi/3} (64 - 8\sec^3\varphi) \sin\varphi\, d\varphi\, d\theta = \frac{40}{3}\pi$$

and the volume of the region above $z = 2\sqrt{3}$ is

$$\int_0^{2\pi}\int_0^{\pi/6}\int_{2\sqrt{3}\sec\varphi}^{4} \rho^2 \sin\varphi\, d\rho\, d\varphi\, d\theta = \frac{1}{3}\int_0^{2\pi}\int_0^{\pi/3}\left(64 - 24\sqrt{3}\sec^3\varphi\right)\sin\varphi\, d\varphi\, d\theta = \frac{128}{3}\pi - 24\sqrt{3}\pi$$

so that the volume of the region between the two planes is $\left(24\sqrt{3} - \frac{88}{3}\right)\pi$

13.5.53

a. True. In either set of coordinates, any value of θ may be chosen.
b. True. Note that $r = z$ if and only if $\rho\sin\varphi = \rho\cos\varphi$, which happens if and only if $\varphi = \frac{\pi}{4}$.

13.5.55 $\rho^2 = -\sec(2\varphi)$ means $-1 = \rho^2\cos(2\varphi) = \rho^2(\cos^2\varphi - \sin^2\varphi) = z^2 - x^2 - y^2$ so that $x^2 + y^2 - z^2 = 1$, which is a hyperboloid of one sheet. Since $\frac{\pi}{4} < \varphi < \frac{3\pi}{4}$, we get only the upper sheet.

13.5.57

$$\int_0^{2\pi}\int_0^{\pi}\int_0^{8} 2e^{-\rho^3}\rho^2\sin\varphi\, d\rho\, d\varphi\, d\theta = -\frac{2}{3}\int_0^{2\pi}\int_0^{\pi} e^{-\rho^3}\Big|_{\rho=0}^{\rho=8} \sin\varphi\, d\varphi\, d\theta$$

$$= -\frac{2}{3}\left(e^{-512} - 1\right)\int_0^{2\pi}\int_0^{\pi/3} \sin\varphi\, d\varphi\, d\theta = \frac{8}{3}\left(1 - e^{-512}\right)\pi$$

13.5.59 Note that because of the absolute value sign, the mass is symmetric around the xy-plane, so we can compute the mass for $0 \le z \le 1$ and double it. Using cylindrical coordinates, the mass is then

$$2\int_0^1\int_0^{2\pi}\int_0^2 (2-z)(4-r)r\, dr\, d\theta\, dz = 2\int_0^1\int_0^{2\pi}\int_0^2 (2-z)\left(4r - r^2\right) dr\, d\theta\, dz$$

$$= 2\int_0^1\int_0^{2\pi} (2-z)\left(2r^2 - \frac{1}{3}r^3\right)\Big|_{r=0}^{r=2} d\theta\, dz$$

$$= \frac{64\pi}{3}\int_0^1 (2-z)dz = 32\pi$$

13.5.61

$$\int_0^{2\pi}\int_0^{\sqrt{2}}\int_r^{\sqrt{4-r^2}} f(r,\theta,z)r\, dz\, dr\, d\theta$$

$$\int_0^{2\pi}\int_0^{\sqrt{2}}\int_0^{z} f(r,\theta,z)r\, dr\, dz\, d\theta + \int_0^{2\pi}\int_{\sqrt{2}}^{2}\int_0^{\sqrt{4-z^2}} f(r,\theta,z)r\, dr\, dz\, d\theta$$

$$\int_0^{\sqrt{2}}\int_r^{\sqrt{4-r^2}}\int_0^{2\pi} f(r,\theta,z)r\, d\theta\, dz\, dr$$

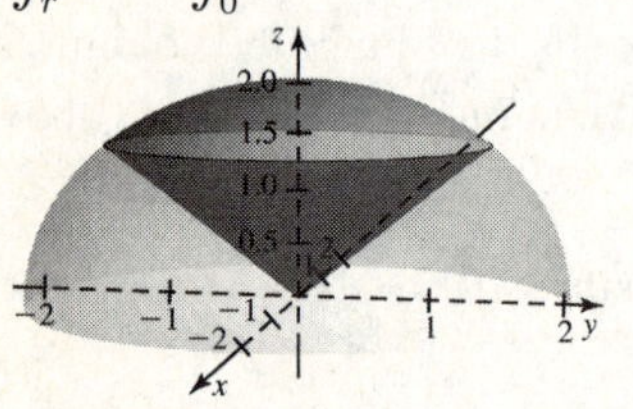

13.5.63 The region of integration is the solid between the upper half-sphere of radius 2 centered at the origin and a cylinder of radius 1 oriented along the z-axis.

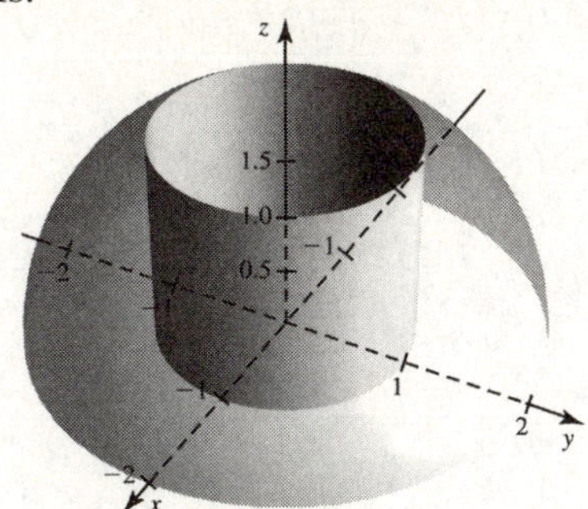

We have

$$\int_{\pi/6}^{\pi/2}\int_0^{2\pi}\int_{\csc\varphi}^{2} f(\rho,\varphi,\theta)\,\rho^2\sin\varphi\,d\rho\,d\theta\,d\varphi$$
$$\int_{\pi/6}^{\pi/2}\int_{\csc\varphi}^{2}\int_0^{2\pi} f(\rho,\varphi,\theta)\,\rho^2\sin\varphi\,d\theta\,d\rho\,d\varphi$$

13.5.65 This region is symmetric about the xy-plane, so we compute the volume of the region inside the solid cylinder for $z \ge 0$ that is below the cone $\varphi = \frac{\pi}{3}$ and double it. Use spherical coordinates. For a given value of φ, we have $0 \le r \le 2\csc\varphi$, so the volume is

$$2\int_{\pi/3}^{\pi/2}\int_0^{2\csc\varphi}\int_0^{2\pi}\rho^2\sin\varphi\,d\theta\,d\rho\,d\varphi = \frac{32\pi}{3}\int_{\pi/3}^{\pi/2}\csc^3\varphi\sin\varphi\,d\varphi = \frac{32}{9}\pi\sqrt{3}$$

13.5.67 Use cylindrical coordinates. Note that $x + y \ge 0$ for $-\frac{\pi}{4} \le \theta \le \frac{3\pi}{4}$, so this is the range of integration for θ.

$$\int_0^1\int_{-\pi/4}^{3\pi/4}\int_0^{r\cos\theta+r\sin\theta} r\,dz\,d\theta\,dr = \int_0^1\int_{-\pi/4}^{3\pi/4} r^2(\cos\theta+\sin\theta)d\theta\,dr = \int_0^1 r^2(\sin\theta-\cos\theta)\Big|_{\theta=-\pi/4}^{\theta=3\pi/4}$$
$$= 2\sqrt{2}\int_0^1 r^2\,dr = \frac{2\sqrt{2}}{3}$$

13.5.69 The planes $z = x - 2$ and $z = 2 - x$ intersect when $x = 2$, which is at the boundary of the cardioid, so we can simply integrate between those two planes. Thus the volume is

$$\int_0^{2\pi}\int_0^{1+\cos\theta}\int_{r\cos\theta-2}^{2-r\cos\theta} r\,dz\,dr\,d\theta = \int_0^{2\pi}\int_0^{1+\cos\theta} r(4-2r\cos\theta)\,d\theta\,dr = \frac{7\pi}{2}$$

13.5.71 Due to symmetry, this region is made up of eight identical pieces, one in each octant. Consider the piece in the first octant. A particle moving through this region parallel to the positive y-axis would start on the xz-plane ($y = 0$) within the unit circle $x^2 + z^2 = 1$ and would end on the cylinder that runs parallel to the x-axis ($y^2 + z^2 = 1$). The total volume is thus

$$V = 8\int_0^1\int_0^{\sqrt{1-z^2}}\int_0^{\sqrt{1-z^2}} 1\,dy\,dx\,dz = 8\int_0^1\int_0^{\sqrt{1-z^2}}\sqrt{1-z^2}\,dx\,dz = 8\int_0^1\left(1-z^2\right)dz = \frac{16}{3}$$

13.5.73 $\displaystyle\int_0^{2\pi}\int_0^2\int_0^8 r\left(1-0.05e^{-0.01r^2}\right)dz\,dr\,d\theta \approx 95.60362$

13.5.75

a. With $x = \cos\varphi$ we have $\sin\varphi = \sqrt{1-x^2}$ and $dx = -\sin\varphi\,d\varphi$ so that $d\varphi = -\dfrac{1}{\sqrt{1-x^2}}dx$. The limits of integration for φ become 1 to -1, so the integral then becomes

$$F(d) = -\frac{GMm}{4\pi}\int_0^{2\pi}\int_{-1}^{1}\frac{(d-Rx)\sqrt{1-x^2}}{(R^2+d^2-2Rdx)^{3/2}\left(-\sqrt{1-x^2}\right)}\,dx\,d\theta$$
$$= \frac{GMm}{4\pi}\int_{-1}^{1}\int_0^{2\pi}\frac{d-Rx}{(R^2+d^2-2Rdx)^{3/2}}\,d\theta\,dx$$
$$= \frac{GMm}{2}\int_{-1}^{1}\left[\frac{d}{(R^2+d^2-2Rdx)^{3/2}} - \frac{Rx}{(R^2+d^2-2Rdx)^{3/2}}\right]dx$$
$$= \frac{GMm}{2}\left(\frac{1}{R\sqrt{R^2+d^2-2Rdx}} - \frac{R^2+d^2-Rdx}{Rd^2\sqrt{R^2+d^2-2Rdx}}\right)\Bigg|_{x=-1}^{x=1}$$
$$= \frac{GMm}{2}\frac{Rdx-R^2}{Rd^2\sqrt{R^2+d^2-2Rdx}}\Bigg|_{x=-1}^{x=1}$$

Assuming $d > R$, this simplifies to

$$F(d) = \frac{GMm}{2}\left(\frac{Rd-R^2}{Rd^2(d-R)} + \frac{Rd+R^2}{Rd^2(d+R)}\right) = \frac{GMm}{2}\left(\frac{1}{d^2}+\frac{1}{d^2}\right) = \frac{GMm}{d^2}$$

b. If $d < R$, then

$$F(d) = \frac{GMm}{2}\frac{Rdx-R^2}{Rd^2\sqrt{R^2+d^2-2Rdx}}\Bigg|_{x=-1}^{x=1}$$
$$= \frac{GMm}{2}\left(\frac{Rd-R^2}{Rd^2(R-d)} + \frac{Rd+R^2}{Rd^2(R+d)}\right) = \frac{GMm}{2}\left(-\frac{1}{d^2}+\frac{1}{d^2}\right) = 0$$

13.5.77 Assume the base of the cone lies in the xy-plane and that the center of the base is at the origin, with the vertex of the cone on the positive z-axis. Then the equation of the cone in cylindrical coordinates $(a,\,\theta,\,z)$ is $z = h - \frac{h}{r}a$ where h and r are the given constants. Thus the volume is

$$\int_0^r\int_0^{h-\frac{h}{r}a}\int_0^{2\pi} a\,d\theta\,dz\,da = 2\pi\int_0^r a\left(h-\frac{h}{r}a\right)da = 2\pi\left(\frac{hr^2}{2}-\frac{hr^2}{3}\right) = \frac{1}{3}\pi hr^2$$

13.5.79 Using similar triangles, if the frustum is extended to a complete cone, the height of the cone is $\frac{Rh}{R-r}$. Thus the equation of the cone (in cylindrical coordinates$(a,\,\theta,\,z)$ is $z = \frac{Rh}{R-r} - \frac{h}{R-r}a$ so that $a = R - z\frac{R-r}{h}$ and the volume of the frustum is

$$\int_0^h\int_0^{R-z\frac{R-r}{h}}\int_0^{2\pi} a\,d\theta\,da\,dz = \pi\int_0^r\left(R - z\frac{R-r}{h}\right)dz = \frac{\pi}{3}(R^2+rR+r^2)h$$

13.5.81 The two spheres are $x^2+y^2+z^2 = R^2$, $x^2+y^2+(z-r)^2 = r^2$. The equation of the second sphere simplifies to $x^2+y^2+z^2-2zr = 0$, so the two spheres meet when $R^2 - 2zr = 0$ or $z = \frac{R^2}{2r}$. This is the plane of intersection of the spheres. The volume in question now consists of two spherical caps, one on either side of this plane. The upper one is a spherical cap of the sphere of radius R with height $R - \frac{R^2}{2r} = \frac{2Rr-R^2}{2r}$, so by problem 78 has volume

$$\frac{\pi}{3}\left(\frac{2Rr-R^2}{2r}\right)^2\left(3R - \frac{2Rr-R^2}{2r}\right)$$

The lower one is a spherical cap of the sphere of radius r with height $\frac{R^2}{2r}$, so again by problem 78 it has volume

$$\frac{\pi}{3}\frac{R^4}{4r^2}\left(3r - \frac{R^2}{2r}\right)$$

Adding these two and simplifying gives for the volume $\dfrac{\pi R^3(8r-3R)}{12r}$.

13.6 Integrals for Mass Calculations

13.6.1 By definition, the system will balance when the pivot is located at the center of mass of the two people.

13.6.3 Integrate the density function over the region to find the mass; then integrate x times the density function and divide by the mass to get the y-coordinate and integrate y times the density function and divide by the mass to get the x-coordinate.

13.6.5 Integrate the density function over the region to find the mass. To find M_{yz}, integrate x times the density function over the region; the x-coordinate is then $\frac{M_{yz}}{M}$. Similarly for the other coordinates.

13.6.7 The center of mass is $\frac{1}{13}(10 \cdot 3 + 3(-1)) = \frac{27}{13}$.

13.6.9 The mass is $\displaystyle\int_0^{\pi} (1 + \sin x)\, dx = \pi + 2$, and the center of mass is then

$$\frac{1}{\pi + 2}\int_0^{\pi} x(1 + \sin x)\, dx = \frac{1}{\pi + 2}\left(\frac{1}{2}\pi^2 + \pi\right) = \frac{\pi}{2}$$

13.6.11 The mass is $\displaystyle\int_0^4 \left(2 - \frac{x^2}{16}\right) dx = 8 - \frac{4}{3} = \frac{20}{3}$, so the center of mass is

$$\frac{1}{20/3}\int_0^4 x\left(2 - \frac{x^2}{16}\right) dx = \frac{3}{20}(16 - 4) = \frac{9}{5}$$

13.6.13 The mass is $1 \cdot 2 + \displaystyle\int_2^4 (1 + x)\, dx = 10$ so that the center of mass is $\displaystyle\frac{1}{10}\left(\int_0^2 x\, dx + \int_2^4 x(1 + x)\, dx\right) = \frac{8}{3}$.

13.6.15 The region is symmetric with respect to $x = \frac{\pi}{2}$ and $y = \frac{1}{2}$, so its center of mass is at $\left(\frac{\pi}{2}, \frac{1}{2}\right)$.

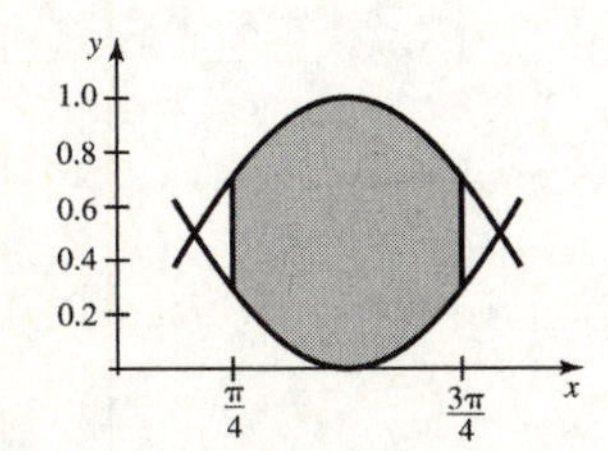

13.6.17 Assume density 1. By symmetry, $\overline{x} = 0$. The mass of the region is 1, since the two triangles together make a unit square. Thus

$$\overline{y} = \frac{M_x}{M} = \int_{-1}^{0}\int_0^{1+x} y\, dy\, dx + \int_0^1\int_0^{1-x} y\, dy\, dx = \frac{1}{3}$$

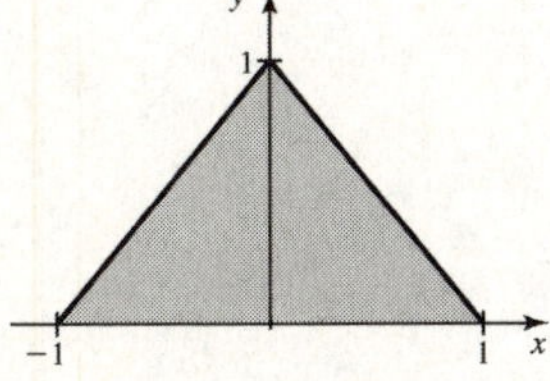

13.6.19 Assume density 1. The mass is $\displaystyle\int_1^e\int_0^{\ln x} 1\, dy\, dx = \int_1^e \ln x\, dx = 1$. Then

$$\overline{x} = \frac{M_y}{M} = \frac{1}{1}\int_1^e\int_0^{\ln x} x\, dy\, dx = \int_1^e x\ln x\, dx = \frac{1}{4}(e^2 + 1),$$

$$\overline{y} = \frac{M_x}{M} = \frac{1}{1}\int_1^e\int_0^{\ln x} x\, dy\, dx = \frac{1}{2}\int_1^e (\ln x)^2\, dx = \frac{1}{2}e - 1,$$

so that the center of mass is $\left(\frac{1}{4}(e^2+1), \frac{1}{2}e-1\right)$.

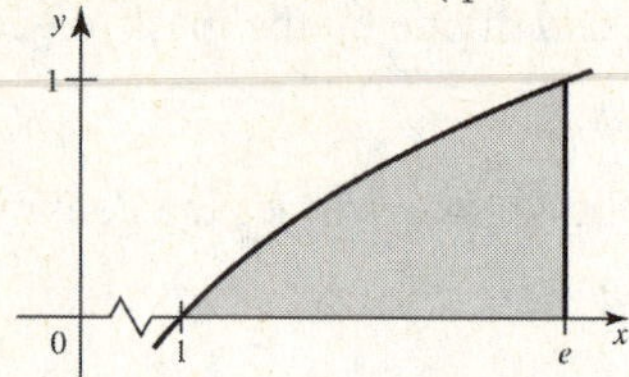

13.6.21 The mass is $\int_0^4\int_0^2\left(1+\frac{x}{2}\right)dy\,dx = 2\int_0^4\left(1+\frac{x}{2}\right)dx = 16$. Then

$$\overline{x} = \frac{M_y}{M} = \frac{1}{16}\int_0^4\int_0^2\left(x+\frac{x^2}{2}\right)dy\,dx = \frac{1}{8}\int_0^4\left(x+\frac{x^2}{2}\right)dx = \frac{7}{3}$$

$$\overline{y} = \frac{M_x}{M} = \frac{1}{16}\int_0^4\int_0^2 y\left(1+\frac{x}{2}\right)dy\,dx = \frac{1}{8}\int_0^4\left(1+\frac{x}{2}\right)dx = 1$$

so that the center of mass is at $\left(\frac{7}{3}, 1\right)$. The density of the plate increases as you move toward the right

13.6.23 The mass is $\int_0^4\int_0^{4-x}(1+x+y)\,dy\,dx = \int_0^4\left(4-x+x(4-x)+\frac{(4-x)^2}{2}\right)dx = \frac{88}{3}$.

By symmetry, $\overline{x}=\overline{y}$ (both the region and the density function are symmetric around $x=y$), and

$$\overline{x} = \frac{M_y}{M} = \frac{1}{88/3}\int_0^4\int_0^{4-x}(x+x^2+xy)\,dy\,dx = \frac{3}{88}\int_0^4\left(x(4-x)+x^2(4-x)+x\frac{(4-x)^2}{2}\right)dx = \frac{16}{11},$$

so the center of mass is $\left(\frac{16}{11}, \frac{16}{11}\right)$. The density of the plate increases as you move right and/or up.

13.6.25 The mass is $\int_{-3}^3\int_0^{\sqrt{9-x^2}/3}(1+y)\,dy\,dx = \int_{-3}^3\left(\frac{\sqrt{9-x^2}}{3}+\frac{9-x^2}{18}\right)dx = \frac{3}{2}\pi+2 = \frac{3\pi+4}{2}$.

The density function does not depend on x, and the region is symmetric about the y-axis, so $\overline{x}=0$. Then

$$\overline{y} = \frac{M_x}{M} = \frac{1}{(3\pi+4)/2}\int_{-3}^3\int_0^{\sqrt{9-x^2}/3}(y+y^2)\,dy\,dx = \frac{3\pi+16}{12\pi+16},$$

so the center of mass is $\left(0, \frac{3\pi+16}{12\pi+16}\right)$. The density increases as you move up the plate.

13.6.27 Assuming density 1, the mass is the volume of a half-sphere of radius 4, which is $\frac{1}{2}\cdot\frac{4}{3}\pi\cdot 4^3 = \frac{128\pi}{3}$. By symmetry, $\overline{x}=\overline{y}=0$, and

$$\overline{z} = \frac{1}{(128\pi)/3}\int_0^4\int_0^{2\pi}\int_0^{\pi/2}\rho\cos\varphi\cdot\rho^2\sin\varphi\,d\varphi\,d\theta\,d\rho = \frac{3}{128\pi}\int_0^4\int_0^{2\pi}\int_0^{\pi/2}\rho^3\cos\varphi\sin\varphi\,d\varphi\,d\theta\,d\rho$$

$$= \frac{3}{128}\int_0^4\rho^3\,d\rho = \frac{3}{2}$$

so the center of mass is at $\left(0, 0, \frac{3}{2}\right)$.

13.6.29 Assuming density 1, the mass is the volume of a pyramid with height 1 and base area $\frac{1}{2}$, so the volume is $\frac{1}{6}$. The region is symmetric with respect to the line $x=y=z$, $\overline{x}=\overline{y}=\overline{z}$, and

$$\overline{z} = \frac{1}{1/6}\int_0^1\int_0^{1-x}\int_0^{1-x-y}z\,dz\,dy\,dx = 3\int_0^1\int_0^{1-x}(1-x-y)^2\,dy\,dx = \frac{1}{4},$$

so the center of mass is $\left(\frac{1}{4}, \frac{1}{4}, \frac{1}{4}\right)$.

13.6.31 Assuming density 1, the mass is $\int_0^1\int_0^{2\pi}\int_0^{1-r\sin\theta} r\,dz\,d\theta\,dr = \int_0^1\int_0^{2\pi} r(1-r\sin\theta)\,d\theta\,dr = \pi$

The region is symmetric around the yz-plane, so $\overline{x}=0$.

$$\overline{y} = \frac{1}{\pi}\int_0^1\int_0^{2\pi}\int_0^{1-r\sin\theta} r^2\sin\theta\,dz\,d\theta\,dr = \frac{1}{\pi}\int_0^1\int_0^{2\pi} r^2(1-r\sin\theta)\sin\theta\,d\theta\,dr = -\frac{1}{4}$$

$$\overline{z} = \frac{1}{\pi}\int_0^1\int_0^{2\pi}\int_0^{1-r\sin\theta} rz\,dz\,d\theta\,dr = \frac{1}{2\pi}\int_0^1\int_0^{2\pi} r(1-r\sin\theta)^2\,d\theta\,dr = \frac{5}{8}$$

so that the center of mass is $\left(0, -\frac{1}{4}, \frac{5}{8}\right)$.

13.6.33 The mass is $\int_0^4\int_0^1\int_0^1 \left(1+\frac{x}{2}\right)dz\,dy\,dx = \int_0^4\left(1+\frac{x}{2}\right)dx = 8$.

By symmetry, since the density depends only on x, $\overline{y}=\overline{z}=\frac{1}{2}$, while

$$\overline{x} = \frac{1}{8}\int_0^4\int_0^1\int_0^1\left(x+\frac{x^2}{2}\right)dz\,dy\,dx = \frac{1}{8}\int_0^4\left(x+\frac{x^2}{2}\right)dx = \frac{7}{3}.$$

so the center of mass is $\left(\frac{7}{3}, \frac{1}{2}, \frac{1}{2}\right)$.

13.6.35 The mass of the sphere is

$$\int_0^{16}\int_0^{\pi/2}\int_0^{2\pi}\left(1+\frac{\rho}{4}\right)\rho^2\sin\varphi\,d\theta\,d\varphi\,d\rho = 2\pi\int_0^{16}\left(1+\frac{\rho}{4}\right)\rho^2\,d\rho = \frac{32768\pi}{3}.$$

By symmetry, since the density function depends only on ρ, $\overline{x}=\overline{y}=0$, and

$$\begin{aligned}\overline{z} &= \frac{1}{(32768\pi)/3}\int_0^{16}\int_0^{\pi/2}\int_0^{2\pi}\left(1+\frac{\rho}{4}\right)\rho^2\sin\varphi\cdot\rho\cos\varphi\,d\theta\,d\varphi\,d\rho\\ &= \frac{3}{16384}\int_0^{16}\int_0^{\pi/2}\rho^3\left(1+\frac{\rho}{4}\right)\sin\varphi\cos\varphi\,d\varphi\,d\rho\\ &= \frac{3}{16384}\int_0^{16}\rho^3\left(1+\frac{\rho}{4}\right)d\rho = \frac{63}{10}.\end{aligned}$$

so the center of mass is $\left(0, 0, \frac{63}{10}\right)$.

13.6.37 The mass is $\int_0^1\int_0^4\int_0^x (2+y)\,dz\,dy\,dx = \int_0^1\int_0^4 (2x+xy)\,dy\,dx = \int_0^1 16x\,dx = 8$. Then

$$\overline{x} = \frac{1}{8}\int_0^1\int_0^4\int_0^x (2x+xy)\,dz\,dy\,dx = \frac{1}{8}\int_0^1\int_0^4 x^2(y+2)dy\,dx = 2\int_0^1 x^2\,dx = \frac{2}{3}$$

$$\overline{y} = \frac{1}{8}\int_0^1\int_0^4\int_0^x (2y+y^2)\,dz\,dy\,dx = \frac{1}{8}\int_0^1\int_0^4 x(y+2y^2)dy\,dx = \frac{14}{3}\int_0^1 x\,dx = \frac{7}{3}$$

$$\overline{z} = \frac{1}{8}\int_0^1\int_0^4\int_0^x (2z+yz)\,dz\,dy\,dx = \frac{1}{8}\int_0^1\int_0^4\left(x^2+\frac{x^2y}{2}\right)dy\,dx = \frac{1}{8}\int_0^1 8x^2\,dx = \frac{1}{3}$$

so that the center of mass is $\left(\frac{2}{3}, \frac{7}{3}, \frac{1}{3}\right)$.

13.6.39

a. False. It has a center of mass with a y-coordinate of zero.

b. True, since every point is balanced by the corresponding point on the other side of the origin.

c. False. For example, the annulus $1\le r\le 3$ has center of mass at the origin.

d. False. For example, the solid reulting from revolving the annulus in part (c) about the x-axis is connected, but its center of mass is at the origin.

13.6.41 The mass of the rod is $\int_0^L \frac{10}{1+x^2}\,dx = 10\arctan(L)$, so its center of mass is

$$\frac{1}{\arctan(L)}\int_0^L \frac{10x}{1+x^2}\,dx = \frac{\ln(1+L^2)}{\arctan(L)}$$

As $L\to\infty$, the numerator grows without bound while the denominator approaches $\frac{\pi}{2}$, so $x\to\infty$ as $L\to\infty$.

13.6.43 The mass of the plate, assuming density 1, is $8+4=12$ since the area of the rectangle is 8 and the two triangles together to form a 2×2 square. By symmetry, $\overline{x}=0$. Compute $\overline{y}$ by computing M_x for each of the three pieces. The moments around the x-axis are equal for the two triangles. So we need only compute M_x for one of the triangles. This is

$$M_x=\int_2^4\int_0^{4-x} y\,dy\,dx=\int_2^4 \frac{1}{2}(4-x)^2\,dx=\frac{4}{3}$$

The moment around the x-axis for the rectangle is $M_x=\int_{-2}^{2}\int_0^{2} y\,dy\,dx=8$, so that $\overline{y}=\frac{1}{12}\left(\frac{4}{3}+8+\frac{4}{3}\right)=\frac{8}{9}$. Thus the center of mass is at $\left(0,\,\frac{8}{9}\right)$.

13.6.45 The mass of the region (assuming density 1) is $\frac{1}{2}\pi r^2=2\pi$, by symmetry, $\overline{x}=0$, and

$$\overline{y}=\frac{1}{2\pi}\int_0^{\pi}\int_0^{2} r^2\sin\theta\,dr\,d\theta=\frac{1}{2\pi}\int_0^{\pi}\frac{8}{3}\sin\theta\,d\theta=\frac{8}{3\pi},$$

so the center of mass is at $\left(0,\,\frac{8}{3\pi}\right)$.

13.6.47 The mass of the cardioid is $\int_0^{2\pi}\int_0^{1+\cos\theta} r\,dr\,d\theta=\frac{1}{2}\int_0^{2\pi}(1+\cos\theta)^2\,d\theta=\frac{3\pi}{2}$. By symmetry, $\overline{y}=0$, and

$$\overline{x}=\frac{1}{(3\pi)/2}\int_0^{2\pi}\int_0^{1+\cos\theta} r^2\cos\theta\,dr\,d\theta=\frac{2}{9\pi}\int_0^{2\pi}(1+\cos\theta)^3\cos\theta\,d\theta=\frac{5}{6},$$

so that the center of mass is at $\left(\frac{5}{6},\,0\right)$.

13.6.49 The mass of the leaf is $\int_0^{\pi/2}\int_0^{\sin 2\theta} r\,dr\,d\theta=\frac{1}{2}\int_0^{\pi/2}\sin^2 2\theta\,d\theta=\frac{\pi}{8}$. By symmetry, $\overline{x}=\overline{y}$, and

$$\overline{x}=\frac{1}{\pi/8}\int_0^{\pi/2}\int_0^{\sin 2\theta} r^2\cos\theta\,dr\,d\theta=\frac{8}{3\pi}\int_0^{\pi/2}\sin^3 2\theta\cos\theta\,d\theta=\frac{128}{105\pi},$$

so that the center of mass is at $\left(\frac{128}{105\pi},\,\frac{128}{105\pi}\right)$.

13.6.51 We compute the center of mass of a very thin half-annulus whose outer edge is the semicircle of radius r and whose width is $\triangle r$. The area of such an annulus is the difference of the areas of the outer and inner semicircles, so is $\frac{1}{2}\pi\left(r^2-(r-\triangle r)^2\right)$. By symmetry, $\overline{x}=0$, and

$$\overline{y}=\frac{1}{\pi\left(r^2-(r-\triangle r)^2\right)/2}\int_0^{\pi}\int_{r-\triangle r}^{r} a^2\sin\theta\,da\,d\theta=\frac{2}{3\pi\left(2r\triangle r-(\triangle r)^2\right)}\int_0^{\pi}\left(r^3-(r-\triangle r)^3\right)^2\sin\theta\,d\theta$$

$$=\frac{4\left(3r^2\triangle r-3r(\triangle r)^2+(\triangle r)^3\right)}{3\pi\left(2r\triangle r-(\triangle r)^2\right)}=\frac{12r^2-12r\triangle r+4(\triangle r)^2}{\pi(6r-3\triangle r)}$$

Taking the limit as $\triangle r\to 0$ gives $\overline{y}=\frac{2r}{\pi}$. Thus the center of mass is at $\left(0,\,\frac{2r}{\pi}\right)$.

13.6.53 The mass of the region, assuming density 1, is $a^2-\frac{\pi}{4}a^2=\frac{4-\pi}{4}a^2$. By symmetry, $\overline{x}=\overline{y}$, and

$$\overline{x}=\frac{1}{a^2(4-\pi)/4}\int_0^{a}\int_{\sqrt{a^2-x^2}}^{a} x\,dy\,dx=\frac{4}{a^2(4-\pi)}\int_0^{a} x\left(a-\sqrt{a^2-x^2}\right)dx=\frac{2a}{3(4-\pi)},$$

so that the center of mass is at $\left(\frac{2a}{3(4-\pi)},\,\frac{2a}{3(4-\pi)}\right)$.

13.6.55 Place the origin at the vertex of the cone and let the z-axis be the axis of the cone. Then the cone has equation (in cylindrical coordinates $(a,\,\theta,\,z)$) $z=\frac{h}{r}a$. The mass of the cone is $\frac{1}{3}\pi r^2 h$, and by symmetry $\overline{x}=\overline{y}=0$. Now

$$\overline{z}=\frac{1}{\pi r^2 h/3}\int_0^{r}\int_0^{(h/r)a}\int_0^{2\pi} a z\,d\theta\,dz\,da=\frac{3h}{r^4}\int_0^{r} a^3\,da=\frac{3h}{4},$$

so that the center of mass is one quarter of the way from the base to the vertex.

13.6.57 Place the origin at the middle of the base of the triangle. Then the y-coordinate of the center of mass can be determined. If h is the height of the triangle, its area is $\frac{bh}{2}$, $\overline{x} = 0$ by symmetry, and

$$\overline{y} = \frac{1}{(bh)/2}\int_0^h \int_{b(y-h)/(2h)}^{-b(y-h)/(2h)} y\,dx\,dy = \frac{2}{bh}\int_0^h \frac{-b(y-h)}{h}\,dy = -\frac{2}{h^2}\int_0^h (y^2 - hy)\,dy = \frac{h}{3},$$

so that the center of mass is $\frac{1}{3}$ of the way from the base to the vertex.

13.6.59 Place the origin at the center of the ellipse with the circular base of radius r in the xy-plane, so that the top of the ellipsoid is on the positive z-axis, at $(0, 0, a)$. Use cylindrical coordinates (ρ, θ, z); then the equation for the top half of the ellipsoid is $z = a\sqrt{1 - \frac{\rho^2}{r^2}}$. We know (Problem 80, Section 13.5) that the volume of the top half of this ellipsoid is $\frac{2\pi r^2 a}{3}$, so

$$\overline{z} = \frac{1}{(2\pi r^2 a)/3}\int_0^r \int_0^{a\sqrt{1-\rho^2/r^2}} \int_0^{2\pi} \rho z\,d\theta\,dz\,d\rho = \frac{3a}{2r^2}\int_0^r \rho\left(1 - \frac{\rho^2}{r^2}\right)d\rho = \frac{3}{8}a,$$

so the center of mass is $\frac{3}{8}$ of the way from the base to the top of the ellipsoid.

13.6.61

a. The mass of the plate is the difference of the area of the two semicircles, so is $\frac{1}{2}\pi(1 - a^2)$. The y-coordinate of the center of mass is then

$$\overline{y} = \frac{1}{\pi(1-a^2)/2}\int_a^1 \int_0^\pi r^2 \sin\theta\,d\theta\,dr = \frac{4}{3\pi(1-a^2)}\int_a^1 r^2\,dr = \frac{4(1-a^3)}{3\pi(1-a^2)}.$$

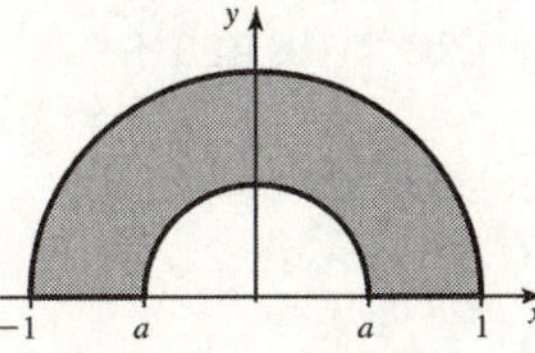

b. The center of mass always has x coordinate 0, so it lies on the edge of the plate exactly when

$$\frac{4(1-a^3)}{3\pi(1-a^2)} = a \text{ or } 1.$$

Solving for equality with 1 gives no solutions in the range $0 \le a \le 1$. Solving for equality with a gives

$$a = -\frac{1}{2}\left(1 - \sqrt{\frac{3(\pi+4)}{3\pi - 4}}\right) \approx 0.49366$$

while the other solution is outside of the range $0 \le a \le 1$.

13.6.63 Place the origin at the center of the bottom of the soda can. If the height of soda in the can is h, for $0 \le h \le 12$, the mass of the can is $16\pi h + \frac{16}{1000}\pi(12 - h) = \frac{6}{125}\pi(333h + 4)$. To compute $\overline{z}$, compute the moments around the x-axis separately for the soda and the air:

$$\begin{aligned}\overline{z} &= \frac{1}{6\pi(333h+4)/125}\int_0^4 \int_0^{2\pi}\left(\frac{1}{1000}\int_h^{12} zr\,dz + \int_0^h zr\,dz\right)d\theta\,dr \\ &= \frac{125}{6\pi(333h+4)}\int_0^4\int_0^{2\pi}\left(\frac{1}{2000}(144 - h^2)r + \frac{1}{2}h^2 r\right)d\theta\,dr \\ &= \frac{125}{6\pi(333h+4)} \cdot \frac{\pi(144 + 999h^2)}{125} = \frac{3}{2}\cdot\frac{111h^2 + 16}{333h + 4}\end{aligned}$$

The center of mass is at its lowest point when the derivative of this function is zero, i.e. when

$$\frac{333h}{333h+4} - \frac{999(16 + 111h^2)}{2(333h+4)^2} = 0$$

Placing over a common denominator, setting the numerator to zero and solving gives $h = \frac{40\sqrt{10}-4}{333} \approx 0.3678$ cm.

13.6.65

a. Place the origin at Q, with the circles to the right of the y-axis. Then using polar coordinates $(a,\,\theta)$, the equation of the circles are $a = 2R\cos\theta$ and $a = 2r\cos\theta$ for $0 \le \theta \le \pi$. The mass of the earring is $\pi(R^2 - r^2)$, and $\overline{y} = 0$.

$$\overline{x} = \frac{1}{\pi(R^2 - r^2)}\int_0^{\pi}\int_{2r\cos\theta}^{2R\cos\theta} a^2\cos\theta\,da\,d\theta = \frac{8(R^3 - r^3)}{3\pi(R^2 - r^2)}\int_0^{\pi}\cos^4\theta\,d\theta = \frac{R^2 + Rr + r^2}{R + r}.$$

With the origin instead at the center of the large circle, the equations of the circles are $x^2 + y^2 = R^2$ and $(x - (R - r))^2 + y^2 = r^2$. Then the moment around the y-axis of the large circle is zero, so to compute $\overline{x}$ for the earring, we compute

$$\begin{aligned}\overline{x} &= \frac{1}{\pi(R^2 - r^2)}\left(0 - \int_{R-2r}^{R}\int_{-\sqrt{r^2-(x-(R-r))^2}}^{\sqrt{r^2-(x-(R-r))^2}} x\,dy\,dx\right)\\ &= \frac{-1}{\pi(R^2 - r^2)}\int_{R-2r}^{R} x\sqrt{r^2 - (x - (R - r))^2}\,dx = \frac{-r^2}{R + r}.\end{aligned}$$

b. With the origin at the center of the large circle, point P is $(R - 2r,\,0)$, so we want $R - 2r = \dfrac{-r^2}{R + r}$. Multiplying both sides by $R + r$, dividing by r^2 and letting $x = \dfrac{R}{r}$, we find that

$$x = \frac{1}{x - 1} \quad\text{or}\quad x^2 - x - 1 = 0$$

which has roots $\frac{1\pm\sqrt{5}}{2}$. Since $R,\,r > 0$, it must be the positive value, so $x = \frac{1+\sqrt{5}}{2}$ satisfies the condition.

13.7 Change of Variables in Multiple Integrals

13.7.1 It is the square with vertices $(0,\,0)$, $(2,\,0)$, $(2,\,2)$, and $(0,\,2)$.

13.7.3 The Jacobian is $\begin{vmatrix} 1 & 1 \\ 1 & -1 \end{vmatrix} = -2$, so

$$\iint_R f(x,\,y)\,dA = \iint_S f(u + v,\,u - v)|J(u,\,v)|dA = \iint_S 2f(u + v,\,u - v)dA = \int_0^1\int_0^1 2f(u + v,\,u - v)dv\,du$$

13.7.5 $u = \frac{x}{2}$, so $0 \le u \le 1$ means $0 \le \frac{x}{2} \le 1$ or $0 \le x \le 2$. $v = 2y$, so $0 \le v \le 1$ means $0 \le 2y \le 1$, or $0 \le y \le \frac{1}{2}$. Thus the image is the region $\left\{0 \le x \le 2,\,0 \le y \le \frac{1}{2}\right\}$, which is a rectangle in the first quadrant with vertices $(0,\,0)$, $(2,\,0)$, $\left(2,\,\frac{1}{2}\right)$, $\left(0,\,\frac{1}{2}\right)$.

13.7.7 Solving for u and v gives $u = x + y$ and $v = x - y$. The region is thus the square bounded by the lines $x + y = 0$, $x + y = 1$, $x - y = 0$, and $x - y = 1$.

13.7.9 From $(0,\,0)$ to $(1,\,0)$, $x = u^2$ and $y = 0$, so this traces out the segment from $(0,\,0)$ to $(1,\,0)$. Similarly, from $(0,\,1)$ to $(0,\,0)$, $x = -v^2$ and $y = 0$, so this traces out the segment from $(-1,\,0)$ to $(0,\,0)$. From $(1,\,0)$ to $(1,\,1)$, $(x,\,y) = (1 - v^2,\,2v)$, so that $(x,\,y)$ satisfies the equation $y^2 = 4 - 4x$. From $(1,\,1)$ to $(0,\,1)$, $(x,\,y) = (u^2 - 1,\,2u)$, so that $(x,\,y)$ satisfies the equation $y^2 = 4 + 4x$. So the result is the region enclosed by the x-axis and the parabolas $y^2 = 4 - 4x$ and $y^2 = 4 + 4x$.

13.7.11 As $(u,\,v)$ goes from $(0,\,0)$ to $(1,\,0)$, $(x,\,y)$ goes from $(0,\,0)$ to $(1,\,0)$ along the x-axis. As $(u,\,v)$ goes from $(1,\,0)$ to $(1,\,1)$, $(x,\,y)$ traces out the upper half of the unit circle. As $(u,\,v)$ goes from $(1,\,1)$ to $(0,\,1)$, $(x,\,y)$ goes from $(-1,\,0)$ to $(0,\,0)$ along the x-axis. Finally, as $(u,\,v)$ goes from $(0,\,1)$ to $(0,\,0)$, $(x,\,y)$ is stationary at the origin. Thus the region swept out is the upper half of the unit circle.

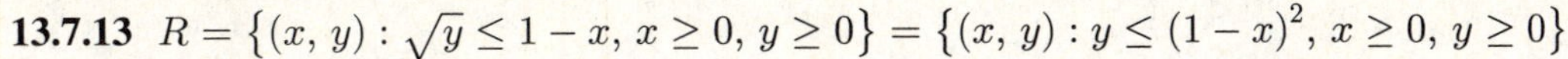

13.7.13 $R = \left\{(x,\, y) : \sqrt{y} \le 1 - x,\, x \ge 0,\, y \ge 0\right\} = \left\{(x,\, y) : y \le (1-x)^2,\, x \ge 0,\, y \ge 0\right\}$

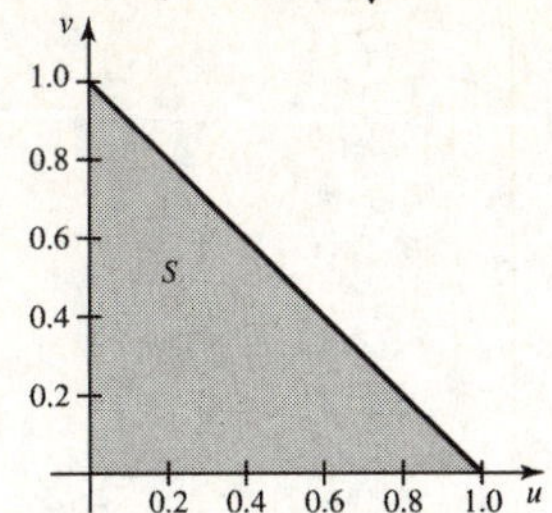

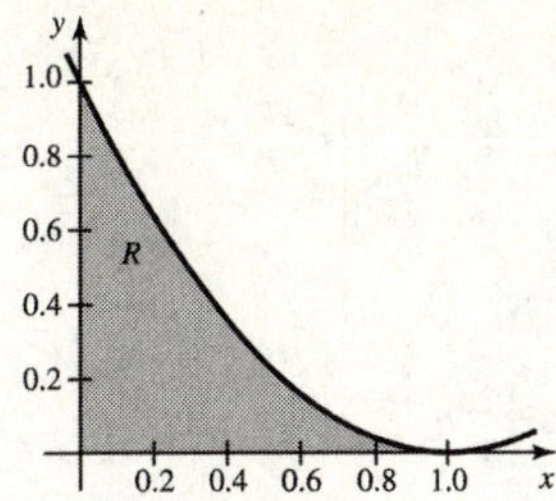

13.7.15 $R = \{(x,\, y) : 1 \le xy \le 3,\, 2 \le y \le 4\} = \left\{(x,\, y) : 2 \le y \le 4,\, \frac{1}{y} \le x \le \frac{3}{y}\right\}$

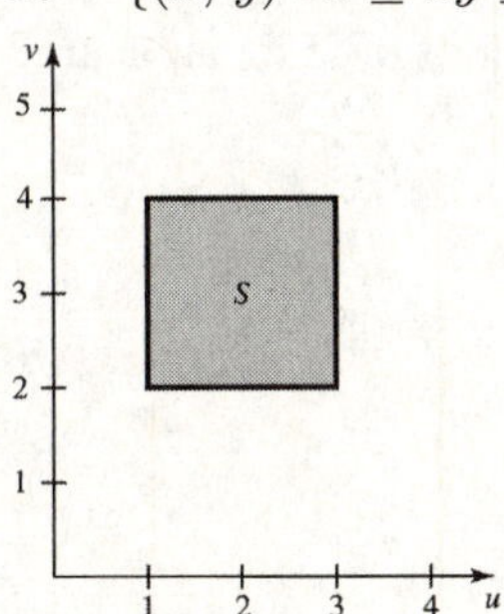

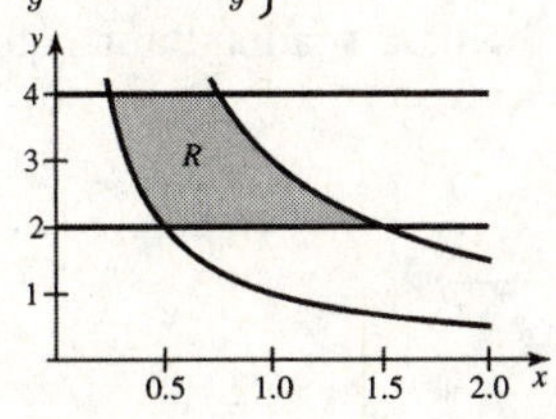

13.7.17 $J = \begin{vmatrix} 3 & 0 \\ 0 & -3 \end{vmatrix} = -9$

13.7.19 $J = \begin{vmatrix} 2v & 2u \\ 2u & -2v \end{vmatrix} = -4(u^2 + v^2)$

13.7.21 $J = \begin{vmatrix} \frac{1}{\sqrt{2}} & \frac{1}{\sqrt{2}} \\ \frac{1}{\sqrt{2}} & -\frac{1}{\sqrt{2}} \end{vmatrix} = -1$

13.7.23 Add the two equations to get $3x = u + v$, so $x = \dfrac{u+v}{3}$, and $y = u - \dfrac{u+v}{3} = \dfrac{2u-v}{3}$. The Jacobian is then

$$J = \begin{vmatrix} \frac{1}{3} & \frac{1}{3} \\ \frac{2}{3} & -\frac{1}{3} \end{vmatrix} = -\frac{1}{3}$$

13.7.25 Add twice the second equation to the first to get $-y = u + 2v$ so that $y = -u - 2v$, and then $x = -u - 3v$. The Jacobian is $J = \begin{vmatrix} -1 & -3 \\ -1 & -2 \end{vmatrix} = -1.$

13.7.27

a.

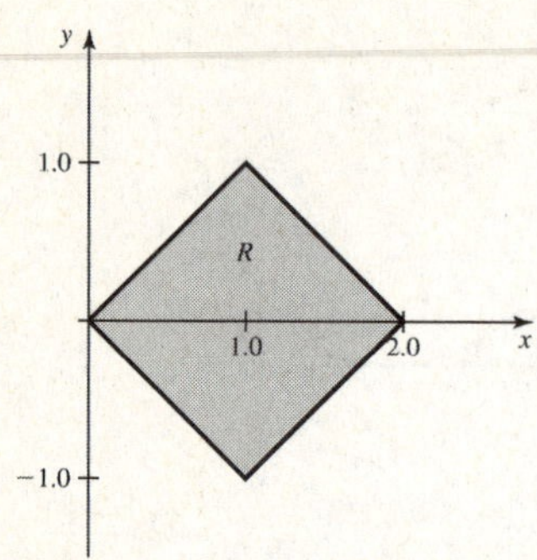

b. The region S is the first-quadrant unit square with one vertex at the origin, so the limits of integration are $0 \le u,\, v \le 1$.

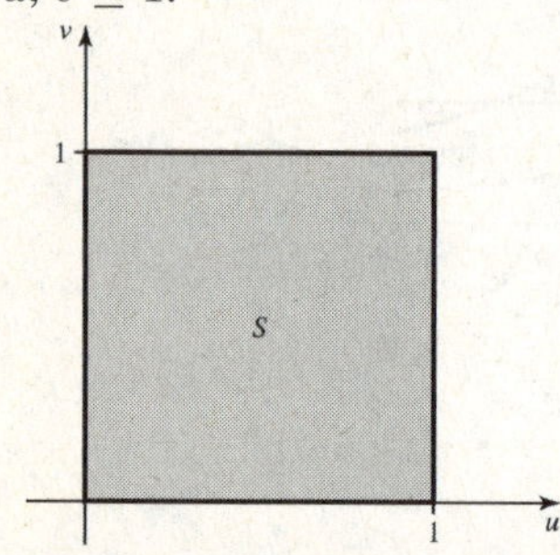

c. The Jacobian of the transformation is $\begin{vmatrix} 1 & 1 \\ 1 & -1 \end{vmatrix} = -2$.

d. $\displaystyle \iint_R xy\, dA = \int_0^1 \int_0^1 \frac{u+v}{2} \cdot \frac{u-v}{2} |-2|\, dv\, du = \frac{1}{2} \int_0^1 \int_0^1 (u^2 - v^2)\, dv\, du = \frac{1}{2} \int_0^1 \left(u^2 - \frac{1}{3}\right) du = 0\,.$

13.7.29

a. $S = \{(u,\, v) : 0 \le 2u \le 2,\ -u \le v - u \le 1 - 2u\} = \{(u,\, v) : 0 \le u \le 1,\ 0 \le v \le 1 - u\}$

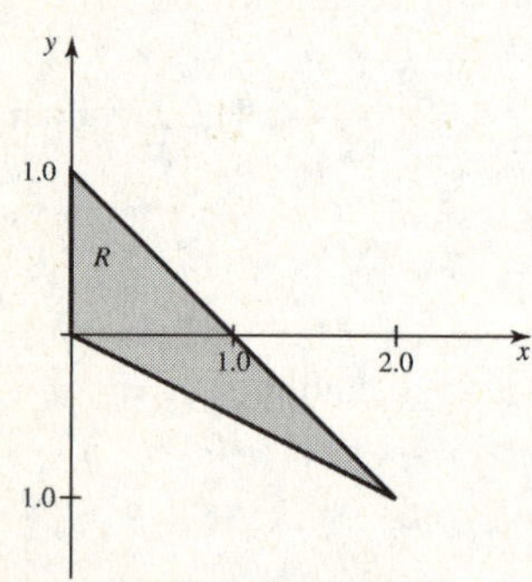

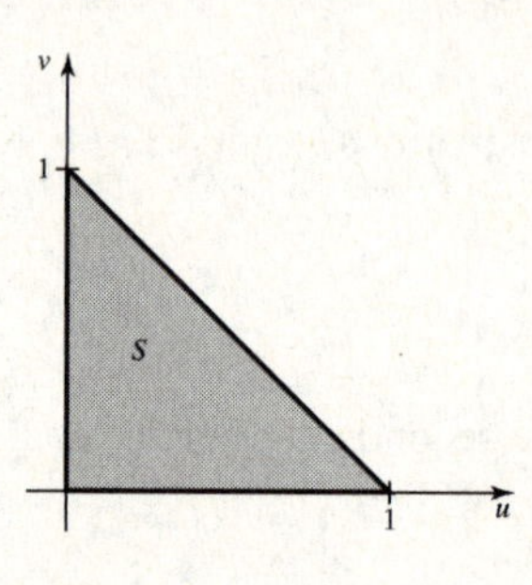

b. From the above, the new limits of integration are $0 \le u \le 1,\ 0 \le v \le 1 - u$.

c. The Jacobian is $\begin{vmatrix} 2 & 0 \\ -1 & 1 \end{vmatrix} = 2$.

d. $\displaystyle \iint_R x^2 \sqrt{x + 2y}\, dA = 8 \int_0^1 \int_0^{1-u} u^2 \sqrt{2v}\, dv\, du = \frac{16\sqrt{2}}{3} \int_0^1 u^2 (1-u)^{3/2}\, du = \frac{256\sqrt{2}}{945}\,.$

13.7.31 Use the substitution $y = v$, $x = u + v$. Then

$$S = \{(u, v) : 0 \le v \le 1,\ v \le u + v \le v + 2\} = \{(u, v) : 0 \le v \le 1,\ 0 \le u \le 2\}$$

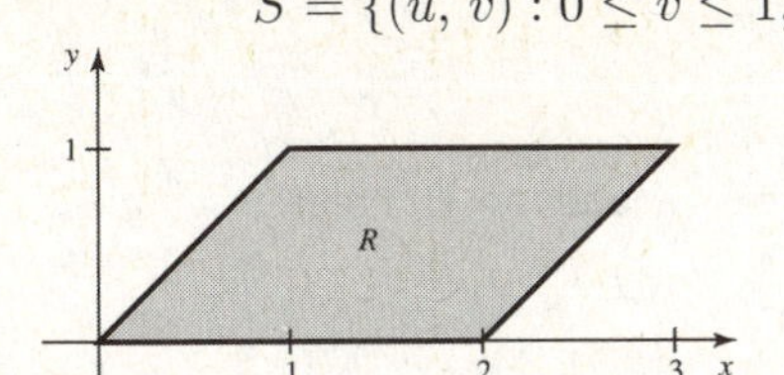

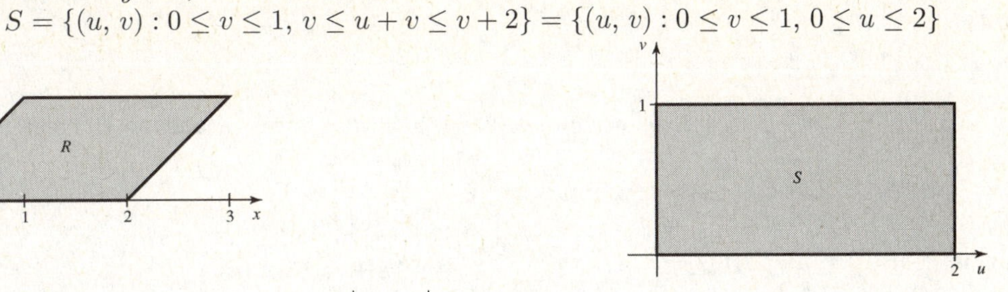

The Jacobian of this transformation is $J(u, v) = \begin{vmatrix} 1 & 1 \\ 0 & 1 \end{vmatrix} = 1$ so that

$$\int_0^1 \int_y^{y+2} \sqrt{x - y}\, dx\, dy = \int_0^1 \int_0^2 \sqrt{u}\, du\, dv = \frac{2}{3} 2^{3/2} = \frac{4\sqrt{2}}{3}.$$

13.7.33 The points of intersection of the given lines are $\left(-\frac{1}{3}, \frac{2}{3}\right)$, $\left(-\frac{2}{3}, \frac{4}{3}\right)$, $\left(\frac{2}{3}, \frac{8}{3}\right)$, and $(1, 2)$. Setting $u = y - x$, $v = y + 2x$ sends these points into the rectangle with vertices $(1, 0)$, $(2, 0)$, $(2, 4)$, and $(1, 4)$.

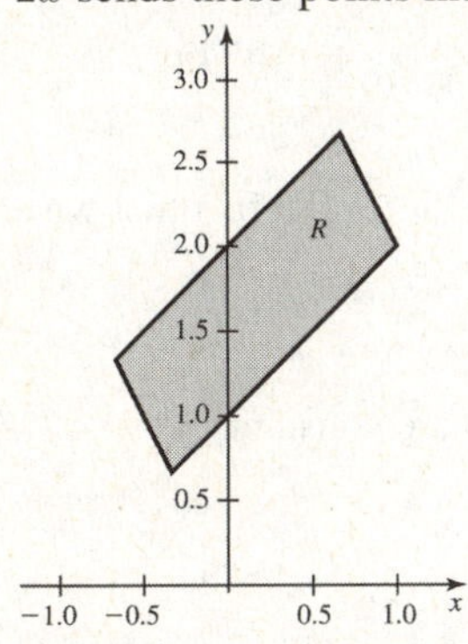

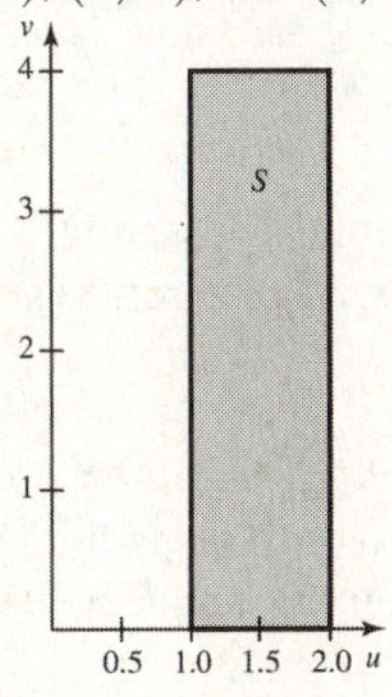

So use the transformation (solving for x, y) $x = \dfrac{v - u}{3}$, $y = \dfrac{v + 2u}{3}$, $J(u, v) = \begin{vmatrix} -\frac{1}{3} & \frac{1}{3} \\ \frac{2}{3} & \frac{1}{3} \end{vmatrix} = -\frac{1}{3}$, so that

$$\iint\limits_R \left(\frac{y - x}{y + 2x + 1}\right)^4 dA = \frac{1}{3} \int_1^2 \int_0^4 \left(\frac{u}{v + 1}\right)^4 dv\, du = \frac{1}{3} \int_1^2 \frac{124}{375} u^4\, du = \frac{3844}{5625}.$$

13.7.35 Use the transformation $u = xy$, $v = y$, so that $x = \frac{u}{v}$, $y = v$. The new region is the square with vertices $(1, 1)$, $(4, 1)$, $(4, 3)$, and $(1, 3)$. The Jacobian of the transformation is $\begin{vmatrix} \frac{1}{v} & -\frac{u}{v^2} \\ 0 & 1 \end{vmatrix} = \frac{1}{v}$.

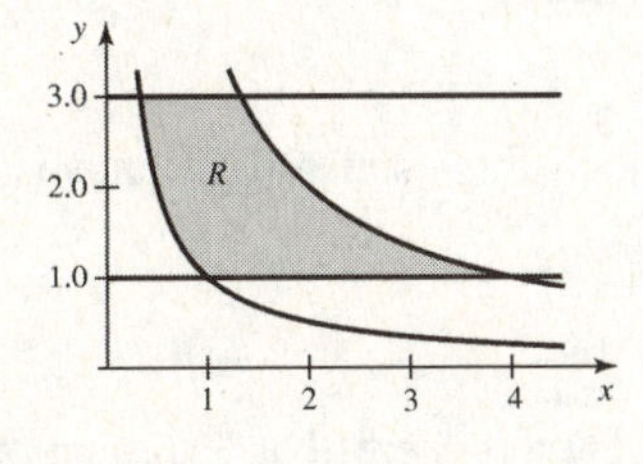

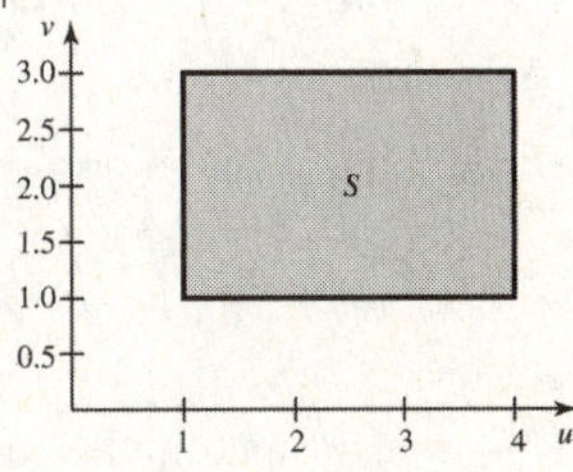

Thus $\displaystyle\iint\limits_R xy\, dA = \int_1^4 \int_1^3 \frac{u}{v}\, dv\, du = \int_1^4 u \ln 3\, du = \frac{15 \ln 3}{2}$.

13.7.37 $J(u, v, w) = \begin{vmatrix} 0 & 1 & 1 \\ 1 & 0 & 1 \\ 1 & 1 & 0 \end{vmatrix} = 2$

13.7.39 $J(u, v, w) = \begin{vmatrix} 0 & w & v \\ w & 0 & u \\ 2u & -2v & 0 \end{vmatrix} = 2w(u^2 - v^2)$

13.7.41 Let $u = y - x$, $v = z - y$, $w = z$; then $x = w - v - u$, $y = w - v$, $z = w$ and the Jacobian is

$$J(u, v, w) = \begin{vmatrix} -1 & -1 & 1 \\ 0 & -1 & 1 \\ 0 & 0 & 1 \end{vmatrix} = 1$$

The new region is clearly $0 \le u \le 2, 0 \le v \le 1, 0 \le w \le 3$, so we get

$$\iiint_D xy\,dV = \int_0^2 \int_0^1 \int_0^3 (w - v - u)(w - v)\,dw\,dv\,du = 5$$

13.7.43 Using the given change of variables, the Jacobian is $\begin{vmatrix} 4\cos v & -4u\sin v & 0 \\ 2\sin v & 2u\cos v & 0 \\ 0 & 0 & 1 \end{vmatrix} = 8u$, and the new range of integration is $0 \le u \le 1, 0 \le v \le 2\pi, 0 \le w \le 16 - 16u^2$. Thus

$$\iiint_D z\,dV = \int_0^1 \int_0^{2\pi} \int_0^{16-16u^2} 8uw\,dw\,dv\,du = \int_0^1 2048\pi\,u\left(u^2 - 1\right)^2 du = \frac{1024\pi}{3}.$$

(Note that this change of variables is essentially cylindrical coordinates, with an adjustment for the fact that we are integrating over a paraboloid with differently sized axes.)

13.7.45

a. True, since $g(u, v)$ and $h(u, v)$ are of the form $au + bv$ so their partial derivatives are constants.
b. True, since the transformation maps lines to lines.
c. True; it simply halves lengths and reflects in the x axis.

13.7.47 Expand the determinant about the third row.

$$\begin{aligned} J(\rho, \varphi, \theta) &= \begin{vmatrix} \sin\varphi\cos\theta & \rho\cos\varphi\cos\theta & -\rho\sin\varphi\sin\theta \\ \sin\varphi\sin\theta & \rho\cos\varphi\sin\theta & \rho\sin\varphi\cos\theta \\ \cos\varphi & -\rho\sin\varphi & 0 \end{vmatrix} \\ &= \cos\varphi\left(\rho^2\sin\varphi\cos\varphi\cos^2\theta + \rho^2\sin\varphi\cos\varphi\sin^2\theta\right) + \rho\sin\varphi\left(\rho\sin^2\varphi\cos^2\theta + \rho\sin^2\varphi\sin^2\theta\right) \\ &= \rho^2\sin\varphi\cos^2\varphi + \rho^2\sin^3\varphi = \rho^2\sin\varphi \end{aligned}$$

13.7.49 This integral is four times the integral over the first quadrant. This is

$$4ab\int_0^1 \int_0^{\sqrt{1-u^2}} abuv\,dv\,du = 2a^2b^2\int_0^1 u\left(1 - u^2\right)du = \frac{a^2b^2}{2}.$$

13.7.51 The distance of a point from the origin is $\sqrt{x^2 + y^2} = \sqrt{a^2u^2 + b^2v^2}$, so the average squared distance is

$$\frac{1}{\pi ab}\int_{-1}^1 \int_{-\sqrt{1-u^2}}^{\sqrt{1-u^2}} ab\left(a^2u^2 + b^2v^2\right)dv\,du = \frac{a^2 + b^2}{4}.$$

13.7.53 Under the given transformation, the equation becomes $u^2 + v^2 + w^2 = 1$, the unit sphere. The Jacobian of the transformation is abc, so the volume of D is abc times the volume of the unit sphere, or $\frac{4}{3}\pi abc$.

13.7.55 The mass of the upper half is $\frac{2\pi abc}{3}$ by Problem 53, and $\overline{x} = \overline{y} = 0$ by symmetry.

$$\begin{aligned} \overline{z} &= \frac{1}{(2\pi abc)/3}\int_{-1}^1 \int_{-\sqrt{1-u^2}}^{\sqrt{1-u^2}} \int_0^{\sqrt{1-u^2-v^2}} abc^2w\,dw\,dv\,du \\ &= \frac{3c}{2\pi}\int_{-1}^1 \int_{-\sqrt{1-u^2}}^{\sqrt{1-u^2}} \left(1 - u^2 - v^2\right)dv\,du = \frac{3c}{8} \end{aligned}$$

13.7.57

a. The line $u = a$ is the set $\{(a, v)\}$, which maps under T to $\{(a^2 - v^2, 2av)\}$. But points of this form satisfy the equation $a^2 - x = \frac{1}{4a^2}y^2$, or $x = -\frac{1}{4a^2}y^2 + a^2$, which is a parabola opening in the negative x direction. The vertex of the parabola $x = Ay^2 + By + C$ is at $\left(C - \frac{B^2}{4A}, -\frac{B}{2A}\right)$, which for this parabola is $(a^2, 0)$, which lies on the positive x-axis.

b. The line $v = b$ is the set $\{(u, b)\}$, which maps under T to $\{(u^2 - b^2, 2ub)\}$. But points of this form satisfy the equation $b^2 + x = \frac{1}{4b^2}y^2$, or $x = \frac{1}{4b^2}y^2 - b^2$, which is a parabola opening in the positive x direction. The vertex of the parabola $x = Ay^2 + By + C$ is at $\left(C - \frac{B^2}{4A}, -\frac{B}{2A}\right)$, which for this parabola is $(-b^2, 0)$, which lies on the negative x-axis.

c. $J(u, v) = \begin{vmatrix} 2u & -2v \\ 2v & 2u \end{vmatrix} = 4(u^2 + v^2)$.

d. Use the transformation $x = v^2 - u^2$, $y = 2uv$. The Jacobian of this transformation is $\begin{vmatrix} -2u & 2v \\ 2v & 2u \end{vmatrix} = -4(u^2 + v^2)$. $x = 4 - \frac{1}{16}y^2$ corresponds to the lines $u = \pm 2$ (by an analysis similar to part (a)), while $x = \frac{1}{4}y^2 - 1$ corresponds to the line $v = \pm 1$, so that the rectangle with vertices $(-2, -1)$, $(-2, 1)$, $(2, 1)$, $(2, -1)$ is mapped to the region bounded by the parabolas. However, note that the area of that rectangle to the left of the v-axis and the area to the right of the v-axis are each mapped onto that region (that is, the map is 2 : 1; it is a simple computation to see that (a, b) and $(-a, -b)$ map to the same xy point). Thus to determine the area of the original region, we want to integrate over only (say) the right half of the rectangle. So the area is

$$\int_0^2 \int_{-1}^1 4(u^2 + v^2)\, dv\, du = \frac{80}{3}.$$

e. As in part (d), use the transformation $x = v^2 - u^2$, $y = 2uv$, with Jacobian $-4(u^2 + v^2)$. The correspondences are:

$$x = \frac{y^2}{4} - 1 \Leftrightarrow u = 1$$

$$x = \frac{y^2}{64} - 16 \Leftrightarrow u = 4$$

$$x = 9 - \frac{y^2}{36} \Leftrightarrow v = \pm 3$$

$$x = 4 - \frac{y^2}{16} \Leftrightarrow v = \pm 2$$

Since we are looking at the positive portion of the bounded piece, the new range of integration is $1 \le u \le 4$, $2 \le v \le 3$. Thus the area is

$$\int_1^4 \int_2^3 4(u^2 + v^2)\, dv\, du = 160$$

f. This simply reverses the roles of x and y in parts (a) and (b). Thus lines $u = a$ in the uv plane map to parabolas in the xy plane that open in the negative y direction with vertices on the positive y-axis, while lines $v = b$ in the uv plane map to parabolas in the xy plane that open in the positive y direction with vertices on the positive y-axis.

13.7.59

a. The z coordinate remains constant, while the other coordinates are stretched by an amount depending on all three coordinates. The result is thus a parallelepiped with its base in the xy plane.

b. $J(u, v, w) = \begin{vmatrix} a & b & c \\ 0 & d & e \\ 0 & 0 & 1 \end{vmatrix} = ad$

c. D has a height of 1 since S and the z coordinate remains unchanged. Its base is in the xy plane, and is the parallelogram that is the result of the shear transformation $x = au + by$, $y = dv$ (set $w = 0$ in the original equations to see this). The area of this parallelogram, from Problem 58, is ad, so the total volume is ad (again what we would expect from the Jacobian).

d. By symmetry, the center of mass is $\left(\frac{a+b+c}{2}, \frac{d+e}{2}, \frac{1}{2}\right)$.

13.7.61

a. This is just the definition of the transformation T, which is given by $x = g(u, v)$, $y = h(u, v)$, so that the image of (x, y) is $T((x, y)) = (g(u, v), h(u, v))$. Apply this to the coordinates of the points O, P, Q.

b. The Taylor expansions of g and h at $(0, 0)$ are

$$g(u, v) = g(0, 0) + u\, g_u(0, 0) + v\, g_v(0, 0) + \text{ terms involving higher derivatives}$$
$$h(u, v) = h(0, 0) + u\, h_u(0, 0) + v\, h_v(0, 0) + \text{terms involving higher derivatives}$$

Substituting the two points $(\triangle u, 0)$ and $(0, \triangle v)$ into these equations and considering only the terms up through the first derivative gives the desired result.

c. From part (b),

$$P' \approx (g(0, 0){+}g_u(0, 0)\triangle u,\ h(0, 0) + h_u(0, 0)\triangle u)$$
$$Q' \approx (g(0, 0){+}g_v(0, 0)\triangle v,\ h(0, 0) + h_v(0, 0)\triangle v)$$

so that $\overrightarrow{O'P'} \approx \triangle u \langle g_u(0, 0), h_u(0, 0)\rangle$ and $\overrightarrow{O'Q'} \approx \triangle v \langle g_v(0, 0), h_v(0, 0)\rangle$. The area of the parallelogram determined by $\overrightarrow{O'P'}$ and $O'Q'$ is the magnitude of the cross product of these vectors (considered as vectors in 3-space with zero z coordinate), so the area of the resulting region is approximately

$$\begin{aligned} |\triangle u \langle g_u(0, 0), h_u(0, 0), 0\rangle \times \triangle v \langle g_v(0, 0), h_v(0, 0), 0\rangle| &= \triangle u \triangle v |\langle g_u(0, 0), h_u(0, 0), 0\rangle \times \langle g_v(0, 0), h_v(0, 0), 0\rangle| \\ &= \triangle u \triangle v |g_u(0, 0)\, h_v(0, 0) - g_v(0, 0) h_u(0, 0)| \\ &= |J(u, v)| \triangle u \triangle v \end{aligned}$$

d. Since the area of R is $\triangle u \triangle v$, the ratio is approximately $|J(u, v)|$.

13.8 Chapter Thirteen Review

13.8.1

a. False. For example, if $g(x, y) = 2$, then $\displaystyle\int_c^d \int_a^b 2\, dx\, dy = 2(b-a)(d-c)$ while $\displaystyle\left(\int_c^d 2\, dy\right)\left(\int_a^b 2\, dx\right) = b(b-a)(d-c)$.

b. True. The first set is the set whose φ coordinate is $\frac{\pi}{2}$; φ is the angle the line to the point makes with the z-axis, so this is the set of points in the xyplane.

c. False. For example, it maps the standard unit square into the square with vertices $(0, 0)$, $(0, -1)$, $(1, -1)$, $(1, 0)$.

13.8.3 $\displaystyle\int_1^3 \int_1^{e^x} \frac{x}{y} dy\, dx = \int_1^3 x \ln y \Big|_{y=1}^{y=e^x} dx = \int_1^3 x^2 dx = \frac{26}{3}.$

13.8.5 $\displaystyle\int_0^1 \int_{-\sqrt{y}}^{\sqrt{y}} f(x, y)\, dx\, dy$

13.8.7 $\displaystyle\int_0^1 \int_0^{\sqrt{1-x^2}} f(x, y)\, dy\, dx$

13.8.9

$$\begin{aligned} &2\int_0^4 \int_x^{20-x^2} 1\, dy\, dx \\ &= 2\int_0^4 (20 - x^2 - x)\, dx \\ &= \frac{304}{3} \end{aligned}$$

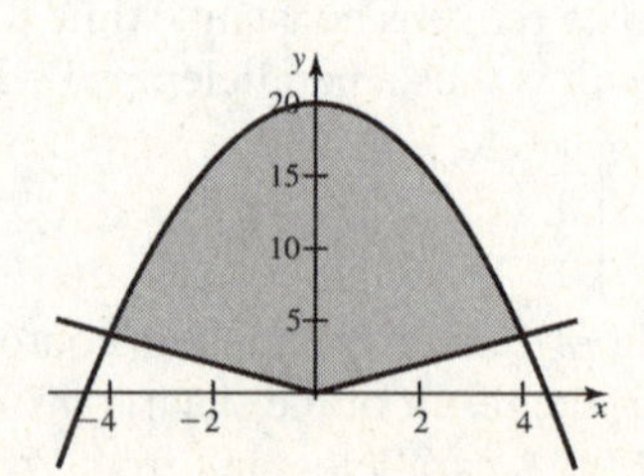

13.8.11 $\displaystyle\int_1^2 \int_0^{x^{3/2}} \frac{2y}{\sqrt{x^4+1}} dy\, dx = \int_1^2 \frac{x^3}{\sqrt{x^4+1}} dx = \frac{\sqrt{17} - \sqrt{2}}{2}$

13.8.13 $\displaystyle\int_0^{\pi}\int_0^{4\sin\theta} r^2(\cos\theta+\sin\theta)\,dr\,d\theta = \frac{64}{3}\int_0^{\pi}\sin^3\theta\,(\cos\theta+\sin\theta)\,d\theta = 8\pi$

13.8.15 $\displaystyle\int_0^1\int_{y^{1/3}}^1 x^{10}\cos(\pi x^4 y)\,dx\,dy = \int_0^1\int_0^{x^3} x^{10}\cos(\pi x^4 y)\,dy\,dx = \int_0^1 \frac{x^{10}}{\pi x^4}\sin(\pi x^7)\,dx$

$$= \frac{1}{\pi}\int_0^1 x^6\sin(\pi x^7)\,dx = \frac{2}{7\pi^2}$$

13.8.17 $\displaystyle\int_0^1\int_0^{\pi/2} 3(r\cos\theta)^2\, r\sin\theta\, r\,d\theta\,dr = \int_0^1\int_0^{\pi/2} 3r^4\cos^2\theta\sin\theta\,d\theta\,dr = \int_0^1 r^4\,dr = \frac{1}{5}$

13.8.19 The area is four times the area of one leaf, so is

$$4\int_{-\pi/4}^{\pi/4}\int_0^{3\cos 2\theta} r\,dr\,d\theta$$

$$= 18\int_{-\pi/4}^{\pi/4}\cos^2 2\theta\,d\theta = \frac{9\pi}{2}$$

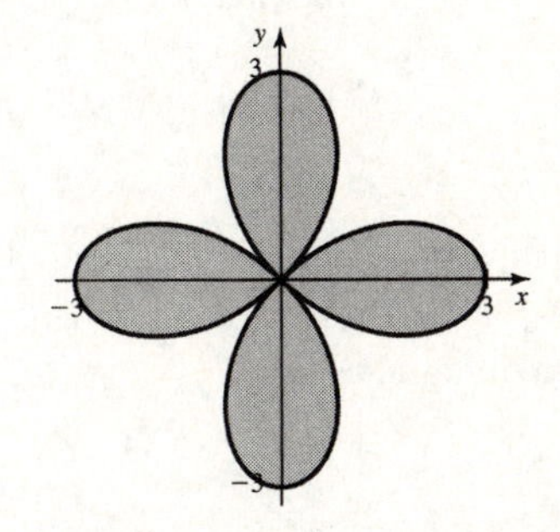

13.8.21 The area is four times the area bounded by the cardioid $2-2\cos\theta$ and the y-axis for positive x (i.e. for $0\le\theta\le\frac{\pi}{2}$), since all four portions of the area are congruent. Thus it is

$$4\int_0^{\pi/2}\int_0^{2-2\cos\theta} r\,dr\,d\theta = 8\int_0^{\pi/2}(1-\cos\theta)^2\,d\theta = 6\pi-16$$

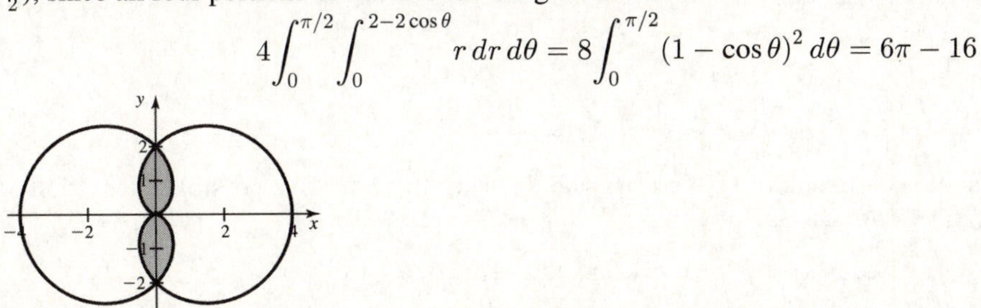

13.8.23 The volume of the cone is $\frac{1}{3}$ times the area of the base times the height. The base (at $z=8$) is a circle of radius 4, so the volume is $\frac{1}{3}\pi\cdot 4^2\cdot 8 = \frac{128\pi}{3}$. Use cylindrical coordinates to integrate; then the distance to the z-axis is r and the average is

$$\frac{1}{(128\pi)/3}\int_0^8\int_0^{z/2}\int_0^{2\pi} r^2\,d\theta\,dr\,dz = \frac{1}{64}\int_0^8 \frac{z^3}{8}\,dz = 2$$

13.8.25 $\displaystyle\int_0^4\int_0^{\sqrt{16-z^2}}\int_0^{\sqrt{16-y^2-z^2}} f(x,y,z)\,dx\,dy\,dz$

13.8.27 $\displaystyle\int_0^1\int_{-z}^{z}\int_{-\sqrt{1-x^2}}^{\sqrt{1-x^2}} dy\,dx\,dz = 2\int_0^1\int_{-z}^{z}\sqrt{1-x^2}\,dx\,dz = 2\int_0^1\left(z\sqrt{1-z^2}+\arcsin(z)\right)dz = \pi-\frac{4}{3}$

13.8.29 The region in the xy plane can be restated as $0\le x\le 2$, $0\le y\le\frac{x}{2}$. Reordering the integral gives

$$\int_0^9\int_0^2\int_0^{x/2}\frac{4\sin(x^2)}{\sqrt{z}}\,dy\,dx\,dz = 2\int_0^9\int_0^2\frac{x\sin(x^2)}{\sqrt{z}}\,dx\,dz = (1-\cos 4)\int_0^9 z^{-1/2}\,dz = 6-6\cos 4$$

13.8.31 Reorder to integrate with respect to x last. Since $0 \le y \le 2$, we have $0 \le x \le 4$; then the integral becomes

$$\int_0^4 \int_{\sqrt{x}}^2 \int_0^{y^{1/3}} yz^5\left(1+x+y^2+z^6\right)^2 dz\,dy\,dx = \frac{1}{18}\int_0^4 \int_{\sqrt{x}}^2 (7y^7+9y^5+9xy^5+3y^3+6xy^3+3x^2y^3)\,dy\,dx$$

$$= \frac{1}{18}\int_0^4 \left(332 - \frac{25}{8}x^4 - 3x^3 + \frac{45}{4}x^2 + 120x\right) dx = \frac{848}{9}$$

13.8.33 $\displaystyle\int_0^2 \int_0^\pi \int_0^{r\sin\theta} r\,dz\,d\theta\,dr = \int_0^2 \int_0^\pi r^2 \sin\theta\,d\theta\,dr = \int_0^2 2r^2\,dr = \frac{16}{3}$

13.8.35 Look at the intersection of the cylinders from the positive z-axis. The vertical sides of the region lie on the cylinder $x^2+y^2=4$, and the top and bottom lie on $x^2+z^2=4$. Thus the region is

$$\left\{(x,\,y,\,z) : x^2+y^2 \le 4,\, -\sqrt{4-x^2} \le z \le \sqrt{4-x^2}\right\}$$

Thus the volume is

$$\int_{-2}^2 \int_{-\sqrt{4-x^2}}^{\sqrt{4-x^2}} \int_{-\sqrt{4-x^2}}^{\sqrt{4-x^2}} 1\,dz\,dy\,dx = \int_{-2}^2 \int_{-\sqrt{4-x^2}}^{\sqrt{4-x^2}} 2\sqrt{4-x^2}\,dy\,dx = \int_{-2}^2 4(4-x^2)\,dx = \frac{128}{3}$$

13.8.37 Rewrite the integral as

$$\int_0^{1/2} \int_{\sin^{-1}x}^{\sin^{-1}2x} 1\,dy\,dx$$

and then change the order of integration. This results in the integral breaking up into two integrals, and we get

$$\int_0^{\pi/6} \int_{(\sin y)/2}^{\sin y} 1\,dx\,dy + \int_{\pi/6}^{\pi/2} \int_{(\sin y)/2}^{1/2} 1\,dx\,dy = \int_0^{\pi/6} \frac{\sin y}{2}dy + \int_{\pi/6}^{\pi/2} \left(\frac{1}{2} - \frac{\sin y}{2}\right) dy = \frac{1-\sqrt{3}}{2} + \frac{\pi}{6}$$

13.8.39

a. $\displaystyle\int_0^2 \int_0^{z^3} \int_0^{y^2} 1\,dx\,dy\,dz = \int_0^2 \int_0^{z^3} y^2\,dy\,dz = \frac{1}{3}\int_0^2 z^9\,dz = \frac{512}{15}$

b. In theory, there are a total of six arrangements of dx, dy and dz. Thus there are five possible integration orders other than this one. For example, let's use $dx\,dz\,dy$:

$$\int_0^8 \int_{\sqrt[3]{y}}^2 \int_0^{y^2} 1\,dx\,dz\,dy = \int_0^8 \int_{\sqrt[3]{y}}^2 y^2\,dz\,dy = \int_0^8 y^2(2-\sqrt[3]{y})\,dy = \frac{512}{15}$$

c. $\displaystyle\int_0^2 \int_0^{z^q} \int_0^{y^p} 1\,dx\,dy\,dz = \int_0^2 \int_0^{z^q} y^p\,dy\,dz = \frac{1}{p+1}\int_0^2 z^{q(p+1)}\,dz = \frac{2^{q(p+1)+1}}{(p+1)(q(p+1)+1)}$

13.8.41 The volume of the prism is $\displaystyle\int_0^1 \int_0^{3-3x} \int_0^2 1\,dz\,dy\,dx = \int_0^1 (6-6x)\,dx = 3$. Thus the average x coordinate is $\displaystyle\frac{1}{3}\int_0^1 \int_0^{3-3x} \int_0^2 x\,dz\,dy\,dx = \frac{1}{3}\int_0^1 2x(3-3x)\,dx = \frac{1}{3}$

13.8.43 Use cylindrical coordinates where the cylinder is oriented along the y-axis, so that $r^2 = x^2+z^2$. The integral is then

$$\int_{-2}^2 \int_0^1 \int_0^\pi \frac{r}{(1+r^2)^2}\,d\theta\,dr\,dy = \int_{-2}^2 \frac{\pi}{4}\,dy = \pi$$

13.8.45 $\displaystyle\int_0^4 \int_0^\pi \int_0^{2\cos\theta} r\,dr\,d\theta\,dz = 2\int_0^4 \int_0^\pi \cos^2\theta\,d\theta\,dz = 4\pi$

13.8.47 $\displaystyle\int_0^\pi \int_0^{\pi/4} \int_{2\sec\varphi}^{4\sec\varphi} \rho^2 \sin\varphi\,d\rho\,d\varphi\,d\theta = \frac{56}{3}\int_0^\pi \int_0^{\pi/4} \sec^3\varphi \sin\varphi\,d\varphi\,d\theta = \frac{28\pi}{3}$

13.8.49 $\displaystyle\int_0^{\pi/2}\int_0^{4\sin 2\varphi}\int_0^{2\pi} \rho^2 \sin\varphi \, d\theta \, d\rho \, d\varphi = \frac{128\pi}{3}\int_0^{\pi/2} \sin^3 2\varphi \sin\varphi \, d\varphi = \frac{2048\pi}{105}$

13.8.51 .

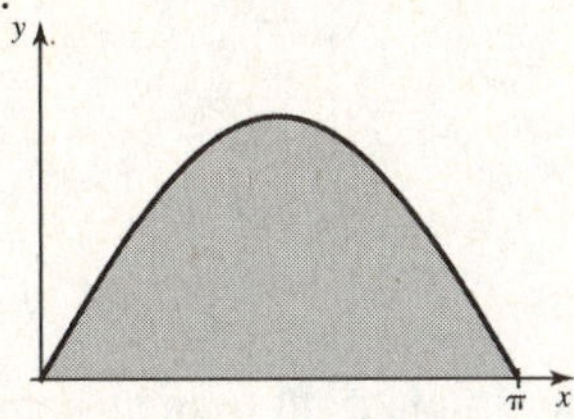

The mass of the plate is $\displaystyle\int_0^{\pi} \sin x \, dx = -\cos\pi + \cos 0 = 2$. By symmetry, $\overline{x} = \frac{\pi}{2}$, and

$$\overline{y} = \frac{1}{2}\int_0^{\pi}\int_0^{\sin x} y \, dy \, dx = \frac{1}{4}\int_0^{4} \sin^2 x \, dx = \frac{\pi}{8},$$

so that the center of mass is $\left(\frac{\pi}{2}, \frac{\pi}{8}\right)$.

13.8.53

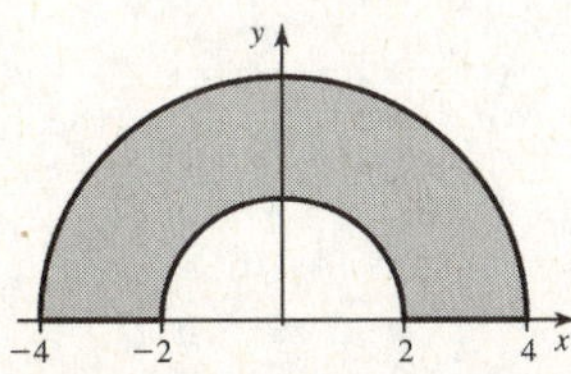

By symmetry, $\overline{x} = 0$. The mass of the region is $8\pi - 2\pi = 6\pi$, so

$$\overline{y} = \frac{1}{6\pi}\int_0^{\pi}\int_2^{4} r^2 \sin\theta \, dr \, d\theta = \frac{28}{9\pi}\int_0^{\pi} \sin\theta \, d\theta = \frac{56}{9\pi}$$

and the center of mass is at $\left(0, \frac{56}{9\pi}\right)$.

13.8.55 By symmetry, $\overline{x} = \overline{y} = 0$. The mass of the bowl is (using cylindrical coordinates)

$$\int_0^{6}\int_{r^2}^{36}\int_0^{2\pi} r \, d\theta \, dz \, dr = 2\pi\int_0^{6} r(36 - r^2) \, dr = 648\pi$$

so

$$\overline{z} = \frac{1}{648\pi}\int_0^{6}\int_{r^2}^{36}\int_0^{2\pi} rz \, d\theta \, dz \, dr = \frac{1}{648}\int_0^{6} r(1296 - r^4) \, dr = 24$$

so the center of mass is $(0, 0, 24)$.

13.8.57 Use spherical coordinates. The mass is

$$\int_0^{\pi/2}\int_0^{16}\int_0^{2\pi}\left(1 + \frac{\rho}{4}\right)\rho^2 \sin\varphi \, d\theta \, d\rho \, d\varphi = 2\pi\int_0^{\pi/2}\int_0^{16}\left(\rho^2 + \frac{\rho^3}{4}\right)\sin\varphi \, d\rho \, d\varphi$$

$$= \frac{32768\pi}{3}\int_0^{\pi/2} \sin\varphi \, d\varphi = \frac{32768\pi}{3}$$

By symmetry of the region and of the density function around the z axis, we have $\overline{x} = \overline{y} = 0$, and

$$\overline{z} = \frac{1}{32768\pi/3}\int_0^{\pi/2}\int_0^{16}\int_0^{2\pi} \rho\cos\varphi\left(1 + \frac{\rho}{4}\right)\rho^2 \sin\varphi \, d\theta \, d\rho \, d\varphi$$

$$= \frac{3}{32768}\int_0^{\pi/2}\int_0^{16}\left(\rho^3 + \frac{\rho^4}{4}\right)\cos\varphi \sin\varphi \, d\rho \, d\varphi = \frac{3}{32768}\int_0^{\pi/2} \frac{16384 \cdot 21}{5}\cos\varphi \sin\varphi \, d\varphi$$

$$= \frac{63}{10}$$

so the center of mass is $\left(0, 0, \frac{63}{10}\right)$

13.8.59 Place the vertex at the origin with the paraboloid opening upwards along the positive z-axis. Then the equation of the paraboloid is $z = \frac{h}{R^2}r^2$. Its mass is

$$\int_0^R \int_0^{2\pi} \int_{hr^2/R^2}^{h} r\,dz\,d\theta\,dr = 2\pi \int_0^R \left(r - \frac{h}{R^2}r^2\right) dr = \frac{1}{2}\pi h R^2$$

Then

$$\overline{z} = \frac{1}{\pi h R^2/2} \int_0^R \int_0^{2\pi} \int_{hr^2/R^2}^{h} rz\,dz\,d\theta\,dr = \frac{2}{hR^2}\int_0^R rh^2 \frac{R^4 - r^4}{R^4} dr = \frac{2}{3}h$$

so that the center of mass is $\frac{1}{3}$ of the way from the base to the vertex.

13.8.61 Place one of the vertices of the base at the origin and the other at $(b, 0)$. Some simple right triangle analysis shows that the third vertex is at $\left(\frac{b}{2}, \sqrt{s^2 - \frac{b^2}{4}}\right)$. The area of the triangle is half the base times the height, or

$$\frac{b}{2}\cdot\frac{1}{2}\sqrt{4s^2 - b^2} = \frac{b\sqrt{4s^2 - b^2}}{4}$$

Then

$$\begin{aligned}
\overline{y} &= \frac{1}{b\sqrt{4s^2 - b^2}/4}\left(\int_0^{b/2}\int_0^{x\sqrt{4s^2-b^2}/b} y\,dy\,dx + \int_{b/2}^{b}\int_0^{-(x-b)\sqrt{4s^2-b^2}/b} y\,dy\,dx\right) \\
&= \frac{2}{b\sqrt{4s^2 - b^2}}\left(\int_0^{b/2} \frac{4s^2 - b^2}{b^2}x^2\,dx + \int_{b/2}^{b} \frac{4s^2 - b^2}{b^2}(x-b)^2\,dx\right) \\
&= \frac{2\sqrt{4s^2 - b^2}}{b^3}\left(\frac{b^3}{24} + \frac{b^3}{24}\right) = \frac{\sqrt{4s^2 - b^2}}{6}
\end{aligned}$$

which is one third the height of the triangle.

13.8.63 The equation of the cone in cylindrical coordinates is $z = 2 - \frac{1}{2}r$. The volume of the cone is one third the area of the base times the height, or $\frac{32\pi}{3}$.

a. The volume of a slice of $\frac{\pi}{4}$ radians is

$$\int_0^4 \int_0^{\pi/4} \int_0^{2-r/2} r\,dz\,d\theta\,dr = \frac{\pi}{4}\int_0^4 r\left(2 - \frac{r}{2}\right) dr = \frac{4\pi}{3}$$

The volume of the wedge is in fact one eighth the volume of the entire cone ($\frac{\pi}{4}$ radians is an eighth-circle).

b. The volume of a slice of Q radians is

$$\int_0^4 \int_0^{Q} \int_0^{2-r/2} r\,dz\,d\theta\,dr = Q\int_0^4 r\left(2 - \frac{r}{2}\right) dr = \frac{16}{3}Q$$

Geometrically, Q radians is $\frac{Q}{2\pi}$ of a circle, and indeed the volume of the slice is $\frac{Q}{2\pi}$ times the volume of the cone.

13.8.65 The transformation just switches the coordinates, so the image is again the unit square.

13.8.67 As (u, v) goes from $(0, 0)$ to $(1, 0)$, (x, y) goes from $(0, 0)$ to $\left(\frac{1}{2}, \frac{1}{2}\right)$; as (u, v) goes from $(1, 0)$ to $(1, 1)$, (x, y) goes from $\left(\frac{1}{2}, -\frac{1}{2}\right)$ to $(1, 0)$; as (u, v) goes from $(1, 1)$ to $(0, 1)$, (x, y) goes from $(1, 0)$ to $\left(\frac{1}{2}, \frac{1}{2}\right)$. Thus the image of S is the diamond with vertices $(0, 0)$, $\left(\frac{1}{2}, \frac{1}{2}\right)$, $(1, 0)$, and $\left(\frac{1}{2}, -\frac{1}{2}\right)$.

13.8.69 $J(u, v) = \begin{vmatrix} 4 & -1 \\ -2 & 3 \end{vmatrix} = 10$

13.8.71 $J(u, v) = \begin{vmatrix} 3 & 0 \\ 0 & 2 \end{vmatrix} = 6$

13.8.73

a.

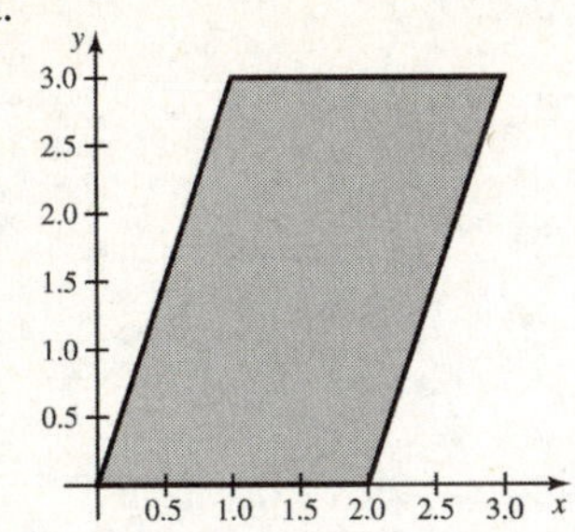

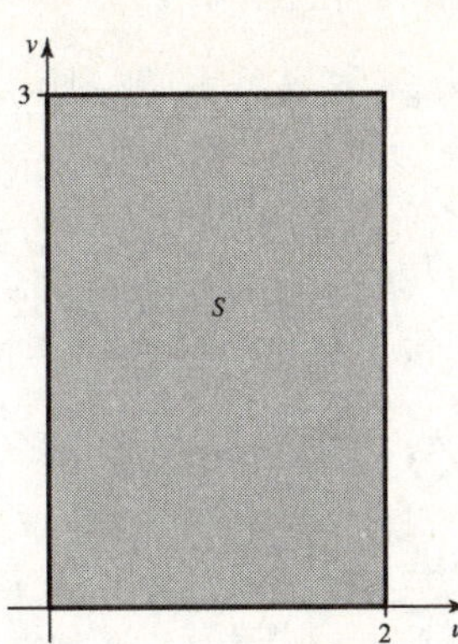

b. $S = \left\{(u, v) : \frac{v}{3} \le u + \frac{v}{3} \le \frac{v+6}{3},\ 0 \le v \le 3\right\} = \{(u, v) : 0 \le u \le 2,\ 0 \le v \le 3\}$.

c. $J(u, v) = \begin{vmatrix} 1 & \frac{1}{3} \\ 0 & 1 \end{vmatrix} = 1$

d. $\displaystyle\iint_R xy^2\, dA = \int_0^2 \int_0^3 \left(u + \frac{v}{3}\right) v^2\, dv\, du = \frac{63}{2}$

13.8.75

a.

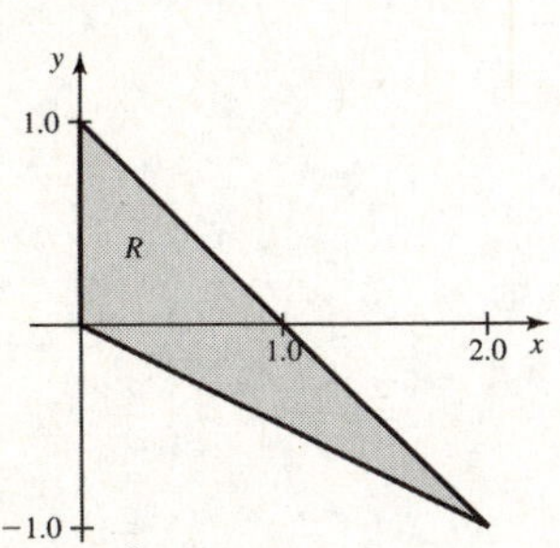

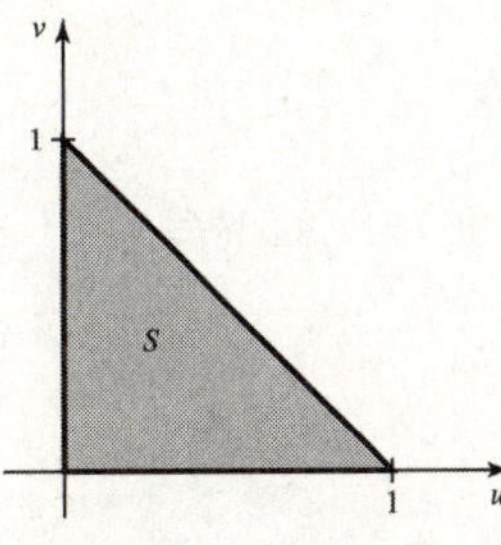

b. $S = \left\{(u, v) : 0 \le 2u \le 2,\ -\frac{2u}{2} \le v - u \le 1 - 2u\right\} = \{(u, v) : 0 \le u \le 1,\ 0 \le v \le 1 - u\}$.

c. $J(u, v) = \begin{vmatrix} 2 & 0 \\ -1 & 1 \end{vmatrix} = 2$

d. Switch the order of integration to get

$$\iint_R x^2 \sqrt{x + 2y}\, dA = 2\int_0^1 \int_0^{1-v} (2u)^2 \sqrt{2v}\, du\, dv = \frac{8\sqrt{2}}{3} \int_0^1 (1 - v)^3 \sqrt{v}\, dv = \frac{256\sqrt{2}}{945}$$

13.8.77 Use the transformation $u = xy$; $v = \frac{y}{x}$ so that $x = \sqrt{\frac{u}{v}}$, $y = \sqrt{uv}$. Then $xy = u$ and $\frac{y}{x} = v$, so the new region S is $\{(u, v) : 1 \le u \le 4,\ 1 \le v \le 3\}$. The Jacobian of this transformation is

$$J(u, v) = \begin{vmatrix} \frac{u^{-1/2}v^{-1/2}}{2} & -\frac{u^{1/2}v^{-3/2}}{2} \\ \frac{v^{1/2}u^{-1/2}}{2} & \frac{u^{1/2}v^{-1/2}}{2} \end{vmatrix} = \frac{1}{2v}$$

so that $\displaystyle\iint_R y^4\, dA = \frac{1}{2}\int_1^4 \int_1^3 u^2 v\, dv\, du = 42$

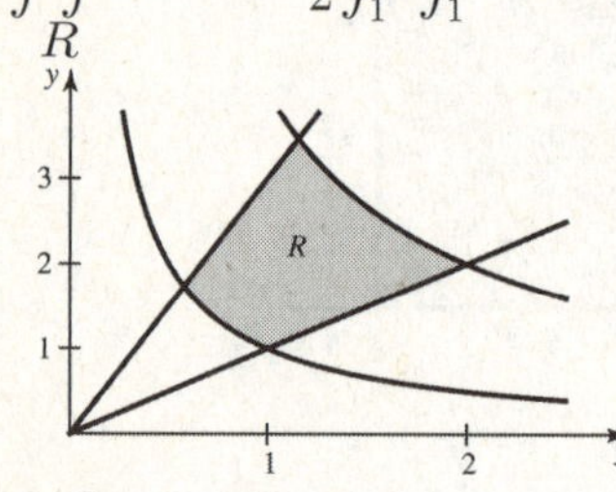

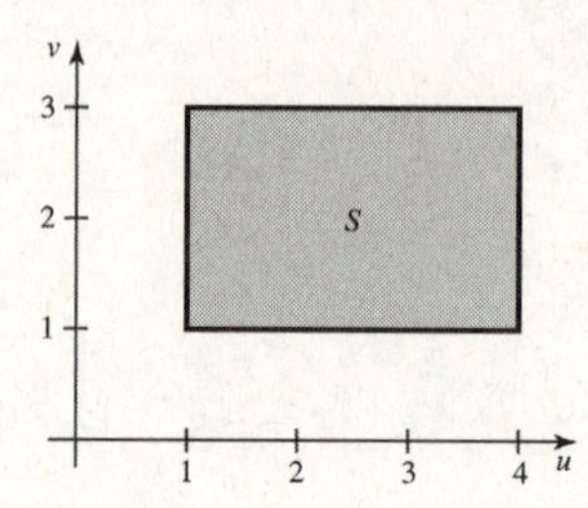

13.8.79 Use the transformation $u = x + 2y$, $v = x - z$, $w = 2y - z$; solving for x, y, z gives

$$x = \frac{u + v - w}{2},\quad y = \frac{u - v + w}{4},\quad z = \frac{u - v - w}{2}.$$

The new range of integration becomes $1 \le u \le 2$, $0 \le v \le 2$, $0 \le w \le 3$. The Jacobian is

$$J(u, v, w) = \begin{vmatrix} \frac{1}{2} & \frac{1}{2} & -\frac{1}{2} \\ \frac{1}{4} & -\frac{1}{4} & \frac{1}{4} \\ \frac{1}{2} & -\frac{1}{2} & -\frac{1}{2} \end{vmatrix} = \frac{1}{4}$$

so that

$$\iiint_D yz\, dV = \frac{1}{32}\int_1^2 \int_0^2 \int_0^3 (u - v + w)(u - v - w)\, dw\, dv\, du = -\frac{7}{16}$$

Chapter 14

14.1 Vector Fields

14.1.1 A vector field describes the motion of the air as a vector at each point in the room.

14.1.3 At each point (x, y), the x-coordinate of the vector is given by $f(x, y)$; the y-coordinate by $g(x, y)$.

14.1.5 The gradient field gives, at each point, the direction in which the temperature is increasing most rapidly and the amount of increase.

14.1.7

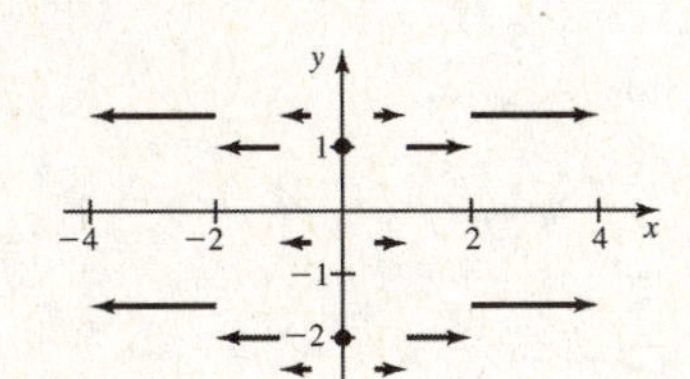

14.1.9

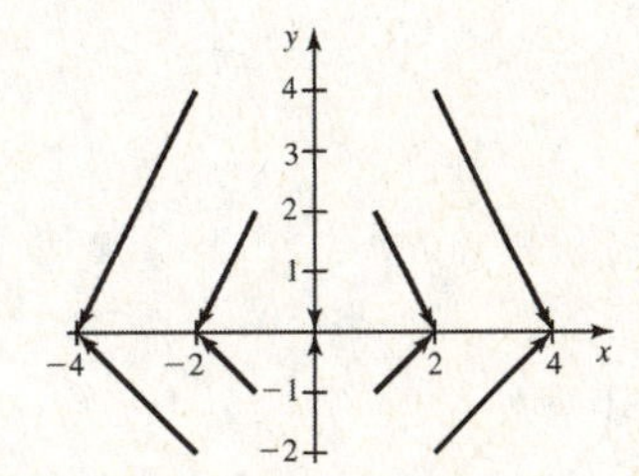

14.1.11

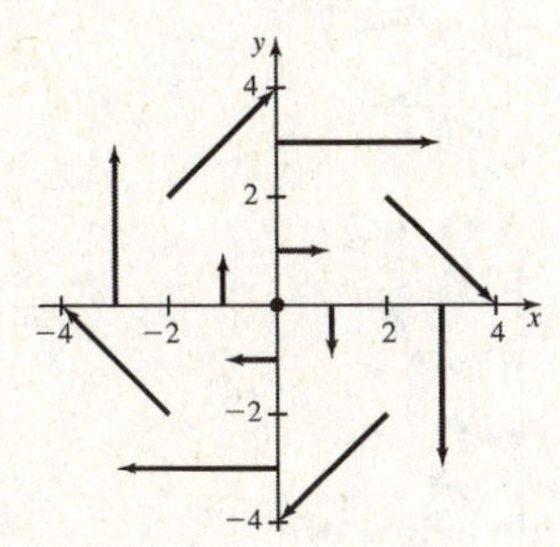

14.1.13

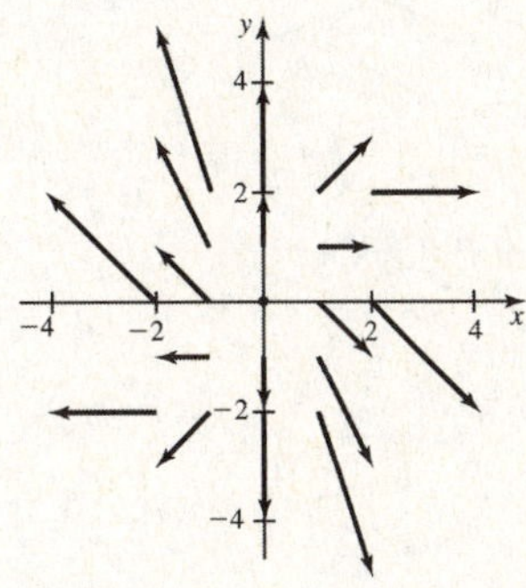

14.1.15

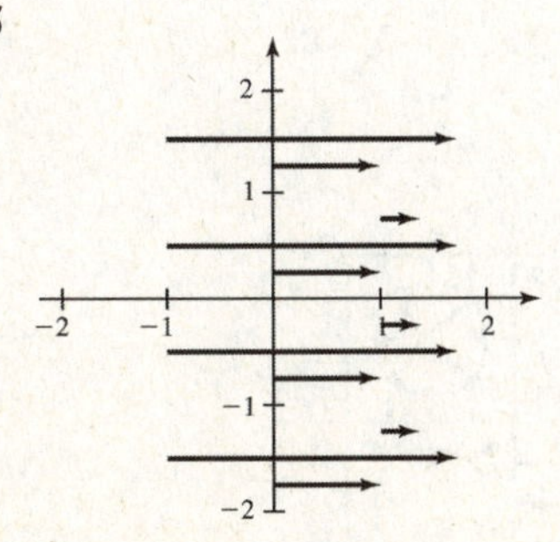

14.1.17 Here C is the circle of radius 2, so a tangent vector to the circle is $\langle -y, x \rangle$. So **F** is normal to C at (x, y) since $\langle -y, x \rangle \cdot \langle x, y \rangle = 0$.

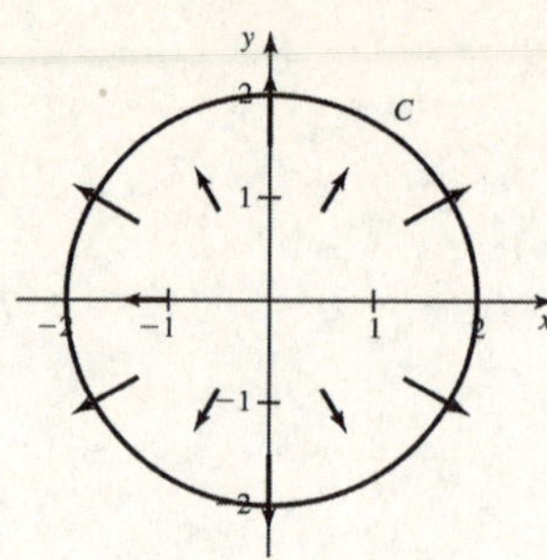

14.1.19 C is the vertical line at $x = 1$, so the tangent vector is a multiple of $\langle 0, y \rangle$ at all points. Then $\langle x, y \rangle$ is never a multiple of $(0, y)$ for any point on C (since $x = 1$ there), and $\langle x, y \rangle \cdot \langle 0, y \rangle = y^2$ is zero for $y = 0$, so that **F** is normal to C at $(1, 0)$.

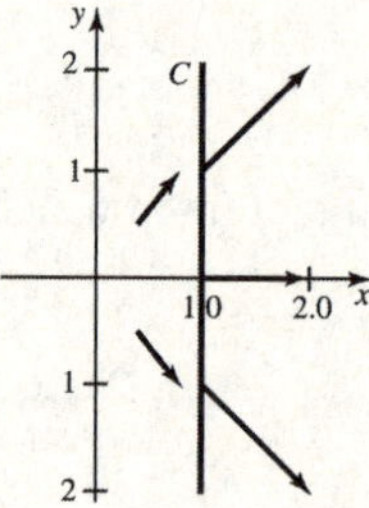

14.1.21

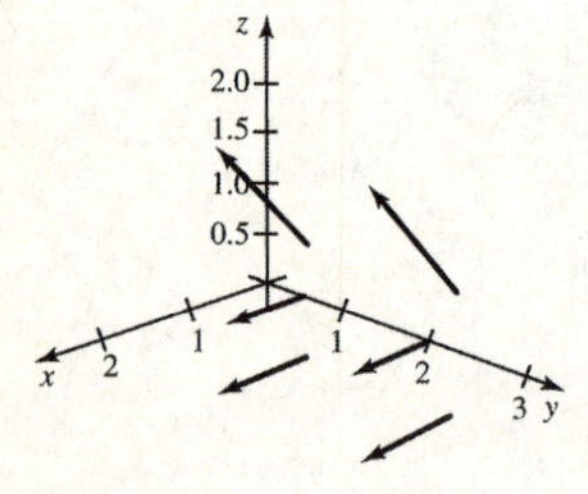

14.1.23

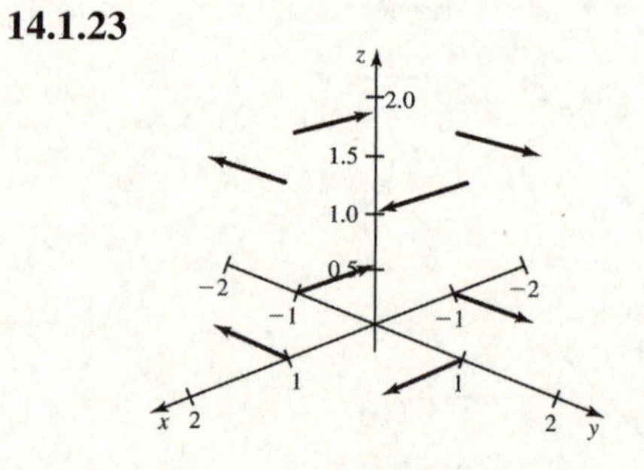

14.1.25 The gradient field is $\langle \varphi_x, \varphi_y \rangle = \langle 2x, 2y \rangle$.

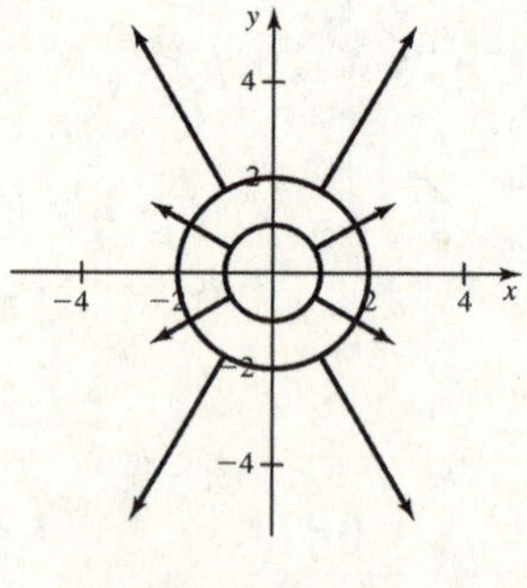

14.1.27 The gradient field is $\langle \varphi_x, \varphi_y \rangle = \langle \cos(x)\sin(y), \sin(x)\cos(y) \rangle$.

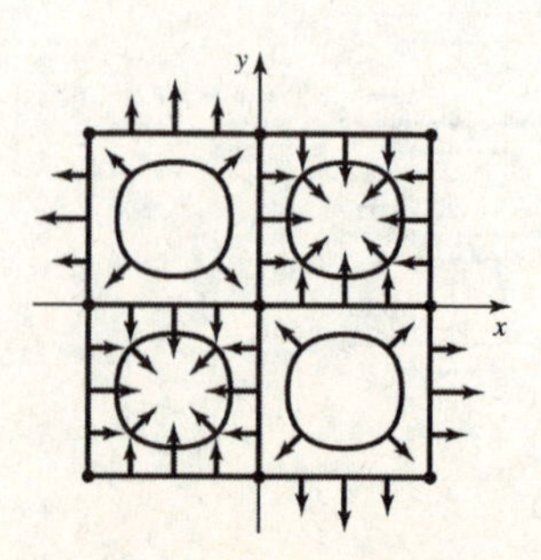

14.1.29 $\nabla\varphi = \langle x,\, y,\, z\rangle = \mathbf{r}$

14.1.31 $\nabla\varphi = -\left(x^2+y^2+z^2\right)^{-3/2}\langle x,\, y,\, z\rangle = -\dfrac{\mathbf{r}}{|\mathbf{r}|^3}$

14.1.33

a. The gradient field is $\langle 2,\, 3\rangle$.

b. The equipotential curve at $(1,\, 1)$ is $2x+3y=5$, which is a line of slope $-\frac{2}{3}$ so has a tangent vector at $(x,\, y)$ parallel to $\left\langle 1,\, -\frac{2}{3}\right\rangle$. But $\langle 2,\, 3\rangle \cdot \left\langle 1,\, -\frac{2}{3}\right\rangle = 0$, so the gradient field is normal to the equipotential line through $(1,\, 1)$.

c. The equipotential curve at any point is a line of slope $-\frac{2}{3}$ and thus has a tangent vector at $(x,\, y)$ parallel to $\left\langle x,\, -\frac{2}{3}y\right\rangle$. The same argument as in part (b) shows that this is normal to the gradient field.

d.

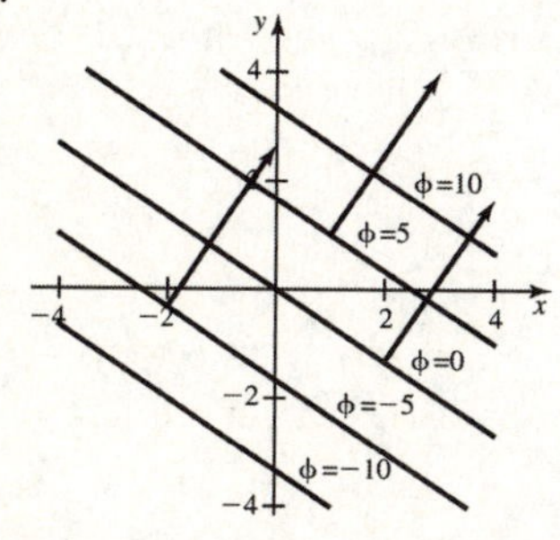

14.1.35

a. The gradient field is $\left\langle e^{x-y},\, -e^{x-y}\right\rangle$.

b. At $(1,\, 1)$, the tangent vector is parallel to $\left\langle -\varphi_y(1,\, 1),\, \varphi_x(1,\, 1)\right\rangle = \langle 1,\, 1\rangle$, which is normal to the gradient $\langle 1,\, -1\rangle$ at $(1,\, 1)$.

c. At $(x,\, y)$, the tangent vector is parallel to $\left\langle e^{x-y},\, e^{x-y}\right\rangle$, and $\left\langle e^{x-y},\, -e^{x-y}\right\rangle \cdot \left\langle e^{x-y},\, e^{x-y}\right\rangle = 0$, so the gradient is everywhere normal to the equipotential curves.

d.

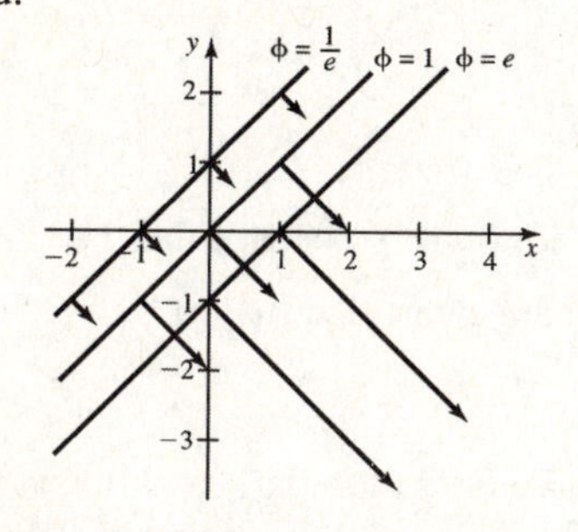

14.1.37

a. True. $\left(\varphi_1\right)_x = \left(\varphi_2\right)_x = 3x^2$, and $\left(\varphi_1\right)_y = \left(\varphi_2\right)_y = 1$.

b. False. It is constant in magnitude (magnitude 1) but not direction.

c. True. For example, it points outwards along the line $y=x$ but horizontally along the line $x=0$.

14.1.39

a. This is a rotational field with magnitude $\sqrt{x^2+y^2}$ at $(x,\, y)$. Thus the answer to this question is the same as for the previous question: on C, the magnitude is 1 on the boundary of C and less than 1 elsewhere. On S, the magnitude is a maximum at the corners of S, where it is $\sqrt{2}$. Finally, on S, the magnitude is a maximum at the corners, where it is 1.

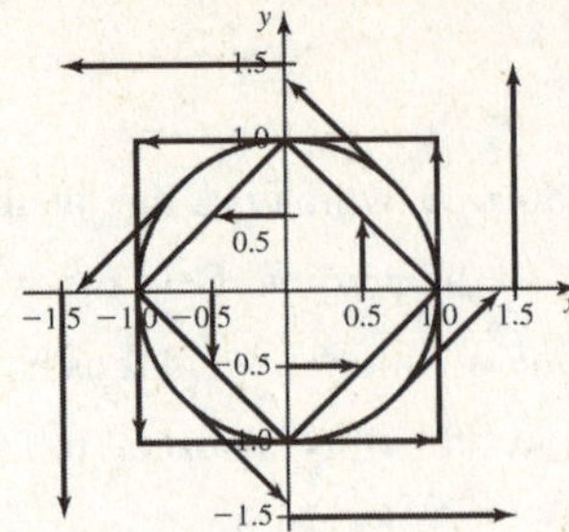

b. Since the vector field is rotational, it is not directed out of C anywhere. For S and for D, it points outwards on half of each edge.

14.1.41 For example, $\mathbf{F} = \langle -y,\, x\rangle$ or $\mathbf{F} = \langle -1,\, 1\rangle$.

14.1.43 For example, $\mathbf{F} = \dfrac{1}{\sqrt{x^2+y^2}}\langle x,\, y\rangle = \dfrac{\mathbf{r}}{|\mathbf{r}|}$, $\mathbf{F}(0,\, 0) = \mathbf{0}$.

14.1.45

a. $V_x = \dfrac{c\sqrt{x^2+y^2}}{r_0}\cdot\left(-\dfrac{1}{2}r_0\left(x^2+y^2\right)^{-3/2}\cdot 2x\right) = \dfrac{-cx}{x^2+y^2}$ and similarly, $V_y = \dfrac{-cy}{x^2+y^2}$, so that

$$\mathbf{E} = -\nabla V = \frac{c}{x^2+y^2}\langle x,\, y\rangle = \frac{c}{|\mathbf{r}|^2}\mathbf{r} = \frac{c}{|\mathbf{r}|}\cdot\frac{\mathbf{r}}{|\mathbf{r}|}.$$

b. From the above formula, the field is a varying multiple of $\langle x,\, y\rangle$, which is a radial field pointing away from the origin. The radial component of $\mathbf{E}$ is thus

$$|\mathbf{E}| = \frac{c}{x^2+y^2}\sqrt{x^2+y^2} = \frac{c}{\sqrt{x^2+y^2}}$$

c. The equipotential curves are curves of the form

$$c\ln\left(\frac{r_0}{\sqrt{x^2+y^2}}\right) = K$$

so are solutions to $\frac{r_0}{\sqrt{x^2+y^2}} = e^{cK} = C$ and thus of the form $\sqrt{x^2+y^2} = K_0$ for some constant K_0. Hence the equipotential curves are circles, so have tangent vectors proportional to $\langle -y,\, x\rangle$; these are clearly normal to $\mathbf{E}$, which is proportional to $\langle x,\, y\rangle$.

14.1.47 The flow curve $y(x)$ of the vector field $\mathbf{F}$ at $(x,\, y)$ is defined to be a continuous curve through $(x,\, y)$ that is aligned with the vector field, i.e. whose tangent at $(x,\, y)$ is given by $\mathbf{F}(x,\, y) = \langle f(x,\, y),\, g(x,\, y)\rangle$. The slope of the tangent line is then $\frac{g(x,y)}{f(x,y)}$, so this is $y'(x)$.

14.1.49 The streamlines satisfy $y'(x) = 1$, so that $y(x) = x + C$.

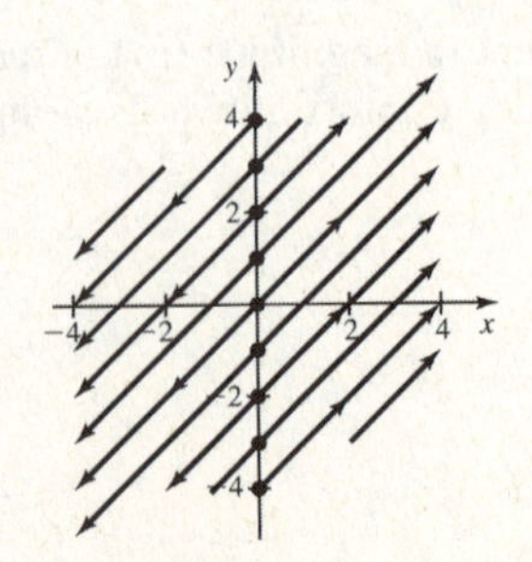

14.1.51 The streamlines satisfy $y'(x) = -\dfrac{x}{y}$. Since $\frac{d}{dx}(y^2) = 2yy'(x)$, we have $y'(x) = \dfrac{\frac{d}{dx}(y^2)}{2y}$ and thus $\frac{d}{dx}(y^2) = -2x$. Thus $y^2 = -x^2 + C$ and the streamlines are the circles $x^2 + y^2 = C$.

14.1.53 For $\theta = 0$, $\mathbf{u}_r$ is coincident with **i**, and $\mathbf{u}_0$ with **j** from the picture; from the formula

$$\mathbf{u}_r = \cos(0)\mathbf{i} + \sin(0)\mathbf{j} = \mathbf{i}$$
$$\mathbf{u}_0 = -\sin(0)\mathbf{i} + \cos(0)\mathbf{j} = \mathbf{j}$$

For $\theta = \frac{\pi}{2}$, the picture implies that we should have $\mathbf{u}_r = \mathbf{j}$, $\mathbf{u}_{\pi/2} = -\mathbf{i}$. From the formulas,

$$\mathbf{u}_r = \cos\left(\frac{\pi}{2}\right)\mathbf{i} + \sin\left(\frac{\pi}{2}\right)\mathbf{j} = \mathbf{j}$$
$$\mathbf{u}_{\pi/2} = -\sin\left(\frac{\pi}{2}\right)\mathbf{i} + \cos\left(\frac{\pi}{2}\right)\mathbf{j} = -\mathbf{i}$$

For $\theta = \pi$, $\mathbf{u}_r$ is coincident with $-\mathbf{i}$, and $\mathbf{u}_\pi$ with $-\mathbf{j}$ from the picture; from the formula

$$\mathbf{u}_r = \cos(\pi)\mathbf{i} + \sin(\pi)\mathbf{j} = -\mathbf{i}$$
$$\mathbf{u}_\pi = -\sin(\pi)\mathbf{i} + \cos(\pi)\mathbf{j} = -\mathbf{j}$$

For $\theta = \frac{3\pi}{2}$, the picture implies that we should have $\mathbf{u}_r = -\mathbf{j}$, $\mathbf{u}_{3\pi/2} = \mathbf{i}$. From the formulas,

$$\mathbf{u}_r = \cos\left(\frac{3\pi}{2}\right)\mathbf{i} + \sin\left(\frac{3\pi}{2}\right)\mathbf{j} = -\mathbf{j}$$
$$\mathbf{u}_{\pi/2} = -\sin\left(\frac{3\pi}{2}\right)\mathbf{i} + \cos\left(\frac{3\pi}{2}\right)\mathbf{j} = \mathbf{i}$$

14.1.55 $\mathbf{F}(r,\theta) = \mathbf{u}_\theta = -\sin(\theta)\mathbf{i} + \cos(\theta)\mathbf{j} = -\dfrac{y}{\sqrt{x^2+y^2}}\mathbf{i} + \dfrac{x}{\sqrt{x^2+y^2}}\mathbf{j} = \dfrac{1}{\sqrt{x^2+y^2}}(-y\mathbf{i} + x\mathbf{j})$.

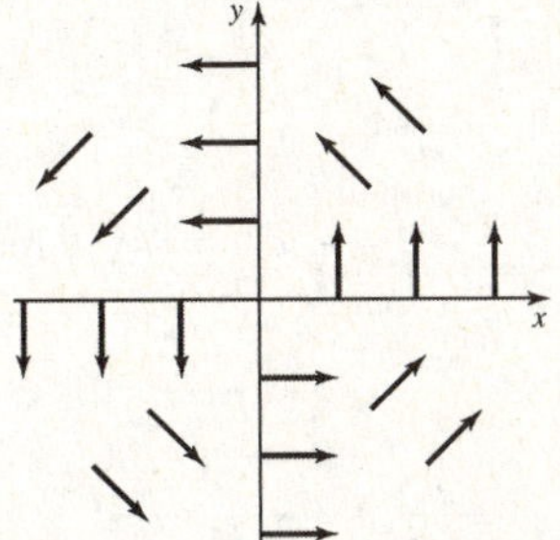

14.1.57
$$\begin{aligned}\mathbf{F}(x,y) &= -y(\mathbf{u}_r\cos(\theta) - \mathbf{u}_\theta\sin(\theta)) + x(\mathbf{u}_r\sin(\theta) + \mathbf{u}_\theta\cos(\theta))\\ &= -r\sin(\theta)(\cos(\theta)\mathbf{u}_r - \sin(\theta)\mathbf{u}_\theta) + r\cos(\theta)(\sin(\theta)\cos(\theta)\mathbf{u}_r + \cos(\theta)\mathbf{u}_\theta)\\ &= r\left(\sin^2(\theta) + \cos^2(\theta)\right)\mathbf{u}_\theta = r\,\mathbf{u}_\theta\end{aligned}$$

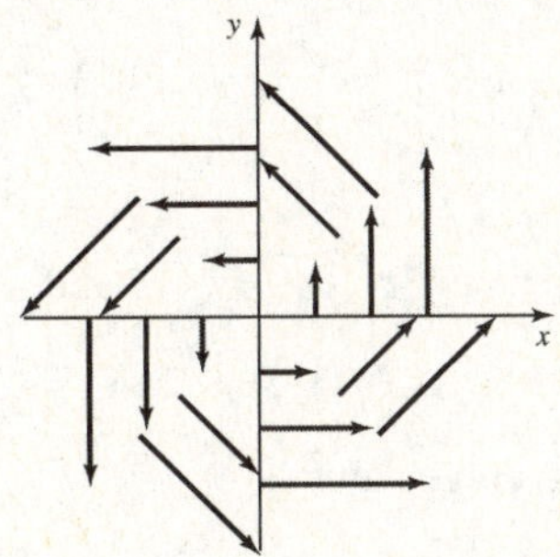

14.2 Line Integrals

14.2.1 A single-variable integral integrates along a segment while a line integral integrates along an arbitrary curve.

14.2.3 $|\mathbf{r}'(t)| = \sqrt{(\mathbf{r}'_x)^2 + (\mathbf{r}'_y)^2} = \sqrt{1+4t^2}$

14.2.5 Since $\mathbf{T} = \langle x'(t),\, y'(t),\, z'(t)\rangle$, $\displaystyle\int_a^b (f\,x'(t) + g\,y'(t) + h\,z'(t))dt$ is simply a rewriting of the dot product.

14.2.7 Take the line integral of $\mathbf{F}\cdot\mathbf{T}$ along the curve using arc length parameterization.

14.2.9 Take the line integral of $\mathbf{F}\cdot\mathbf{n}$ along the curve using arc length parameterization.

14.2.11 $\mathbf{r}(s)$ is an arc length parameterization, so we get $\displaystyle\int_C xy\,ds = \int_0^{2\pi}\cos(s)\sin(s)\,ds = 0$.

14.2.13 With $\mathbf{r}(s) = \left(\dfrac{s}{\sqrt{2}}, \dfrac{s}{\sqrt{2}}\right)$, $|\mathbf{r}'(t)| = 1$, so that we have

$$\int_C (x^2 - 2y^2)\,ds = \int_0^4 \left(\frac{s^2}{2} - 2\frac{s^2}{2}\right) ds = -\int_0^4 \frac{s^2}{2}\,ds = -\frac{s^3}{6}\bigg|_0^4 = -\frac{32}{3}$$

14.2.15

a. $\mathbf{r}(t) = \langle 4\cos(t),\, 4\sin(t)\rangle,\ 0 \le t \le 2\pi$.

b. $|\mathbf{r}'(t)| = \sqrt{(-4\sin(t))^2 + (4\cos(t))^2} = 4$

c. $\displaystyle\int_C (x^2 + y^2)\,ds = \int_0^{2\pi} 4\left(16\cos^2(t) + 16\sin^2(t)\right) dt = \int_0^{2\pi} 64\,dt = 128\pi$

14.2.17

a. $\mathbf{r}(t) = \langle t,\, t\rangle,\ 1 \le t \le 10$.

b. $|\mathbf{r}'(t)| = \sqrt{(1)^2 + (1)^2} = \sqrt{2}$

c. $\displaystyle\int_C \frac{x}{x^2+y^2}\,ds = \int_1^{10} \frac{t}{t^2+t^2}\cdot\sqrt{2}\,dt = \frac{\sqrt{2}}{2}\int_1^{10}\frac{1}{t}\,dt = \frac{1}{2}\sqrt{2}\ln 10$

14.2.19

a. $\mathbf{r}(t) = \langle 2\cos(t),\, 4\sin(t)\rangle,\ 0 \le t \le \pi$.

b. $|\mathbf{r}'(t)| = \sqrt{(-2\sin(t))^2 + (4\cos(t))^2} = 2\sqrt{\sin^2(t) + 4\cos^2(t)}$

c. $\displaystyle\int_C (x - y)\,ds = \int_0^{\pi} (2\cos(t) - 4\sin(t))\cdot 2\sqrt{\sin^2(t) + 4\cos^2(t)}\,dt$

$$= 4\int_0^{\pi}\cos(t)\sqrt{\sin^2(t)+4\cos^2(t)}\,dt - 8\int_0^{\pi}\sin(t)\sqrt{\sin^2(t)+4\cos^2(t)}\,dt$$

$$= 0 - 8\left(2 + \frac{\sqrt{3}}{3}\ln\left(2+\sqrt{3}\right)\right) = -8\left(2 + \frac{\sqrt{3}}{3}\ln\left(2+\sqrt{3}\right)\right)$$

14.2.21 Let $\mathbf{r}(t) = \langle t+1,\, 4t+1\rangle,\ 0 \le t \le 1$; then $|\mathbf{r}'(t)| = \sqrt{17}$ and

$$\int_C (x + 2y)\,ds = \int_0^1 ((t+1) + 2(4t+1))\cdot\sqrt{17}\,dt = \sqrt{17}\int_0^1 (9t+3)\,dt = \frac{15}{2}\sqrt{17}$$

The length of the line segment is $\sqrt{17}$, so the average value is $\frac{15}{2}$.

14.2.23 Let $\mathbf{r}(t) = \langle t^2,\, t\rangle,\ -1 \le t \le 1$; then $|\mathbf{r}'(t)| = \sqrt{4t^2+1}$ and

$$\int_C (4\sqrt{x} - 3y)\,ds = \int_{-1}^1 (4t - 3t)\sqrt{4t^2+1}\,dt = \int_{-1}^1 (t)\sqrt{4t^2+1}\,dt = 0$$

14.2.25 $|\mathbf{r}'(t)| = \sqrt{4\sin^2(t) + 0 + 4\cos^2(t)} = 2$, so

$$\int_C (x + y + z)\,ds = 2\int_0^{2\pi} (2\cos(t) + 2\sin(t))\,dt = 0$$

14.2.27 Let $\mathbf{r}(t) = \langle t,\, 2t,\, 3t\rangle$, $0 \le t \le 1$. Then $|\mathbf{r}'(t)| = \sqrt{14}$, so

$$\int_C (xyz)\,ds = \sqrt{14}\int_0^1 6t^3\,dt = \sqrt{14}\left(\frac{3t^4}{2}\right)\Bigg|_0^1 = \frac{3}{2}\sqrt{14}$$

14.2.29 $|\mathbf{r}'(t)| = \sqrt{10}$, so $\displaystyle\int_C (y-z)\,ds = \sqrt{10}\int_0^{2\pi} (3\sin(t) - t)\,dt = -2\sqrt{10}\,\pi^2$.

14.2.31 The length of the curve is the line integral of 1 along the curve. $\mathbf{r}(t)$ simplifies to $\left\langle 5\sin(t),\, 5\cos(t),\, \frac{t}{2}\right\rangle$, and then

$$|\mathbf{r}'(t)| = \sqrt{50 + \frac{1}{4}} = \frac{1}{2}\sqrt{101}$$

so that the arc length is

$$\frac{1}{2}\int_0^2 \sqrt{101}\,dt = \sqrt{101}$$

14.2.33 $\mathbf{r}'(t) = \langle 4,\, 2t\rangle$, and $\mathbf{F}\cdot\mathbf{r}'(t) = \langle 4t,\, t^2\rangle\cdot\langle 4,\, 2t\rangle = 16t + 2t^3$, so that

$$\int_C \mathbf{F}\cdot\mathbf{T}\,ds = \int_0^1 \mathbf{F}\cdot\mathbf{r}'(t)\,dt = \int_0^1 (16t + 2t^3)\,dt = \frac{17}{2}$$

14.2.35 Let $\mathbf{r}(t) = \langle 4t+1,\, 9t+1\rangle$, $0 \le t \le 1$; then $\mathbf{r}'(t) = \langle 4,\, 9\rangle$ and

$$\mathbf{F}\cdot\mathbf{r}'(t) = \langle 9t+1,\, 4t+1\rangle\cdot\langle 4,\, 9\rangle = 72t + 13\,.$$

Then

$$\int_C \mathbf{F}\cdot\mathbf{T}\,ds = \int_0^1 (72t + 13)\,dt = 49$$

14.2.37 $\mathbf{r}'(t) = \langle 2t,\, 6t\rangle$; then $\mathbf{F}\cdot\mathbf{r}'(t) = \left(10t^4\right)^{-3/2}\langle t^2,\, 3t^2\rangle\cdot\langle 2t,\, 6t\rangle = \dfrac{20t^3}{10\sqrt{10}\,t^6} = \dfrac{2}{\sqrt{10}}t^{-3}$ and

$$\int_C \mathbf{F}\cdot\mathbf{T}\,ds = \frac{2}{\sqrt{10}}\int_1^2 t^{-3}\,dt = \frac{3}{40}\sqrt{10}$$

14.2.39 $\mathbf{r}_1(t) = \langle 1-t,\, 2-2t\rangle$, $\mathbf{r}_2(t) = \langle 0,\, 4t\rangle$, $0 \le t \le 1$. Then $\mathbf{r}_1'(t) = \langle -1,\, -2\rangle$, $\mathbf{r}_2'(t) = \langle 0,\, 4\rangle$, so that

$$\int_C \mathbf{F}\cdot\mathbf{r}'(t)\;dt = \int_0^1 \langle 2-2t,\, 1-t\rangle\cdot\langle -1,\, -2\rangle\,dt + \int_0^1 \langle 4t,\, 0\rangle\cdot\langle 0,\, 4\rangle\,dt = \int_0^1 0\,dt = 0$$

14.2.41 $\mathbf{r}(t) = \langle 2t,\, 8t^2\rangle$, $0 \le t \le 1$. Then $\mathbf{r}'(t) = \langle 2,\, 16t\rangle$, so

$$\int_C \mathbf{F}\cdot\mathbf{r}'\,(t)\,dt = \int_0^1 \langle 8t^2,\, 2t\rangle\cdot\langle 2,\, 16t\rangle\,dt = \int_0^1 48t^2\,dt = 16$$

14.2.43 $\mathbf{r}'(t) = \langle -4\sin(t),\, 4\cos(t),\, -4\sin(t)\rangle$, so

$$\int_C \mathbf{F}\cdot\mathbf{r}'(t)\;dt = \int_0^{2\pi} \langle 4\cos(t),\, 4\sin(t),\, 4\cos(t)\rangle\cdot\langle -4\sin(t),\, 4\cos(t),\, -4\sin(t)\rangle\,dt$$
$$= \int_0^{2\pi} -16\sin(t)\cos(t)\,dt = 0$$

14.2.45 Let $\mathbf{r}(t) = \langle t+1,\, t+1,\, t+1\rangle$, $0 \le t \le 9$, so that $\mathbf{r}'(t) = \langle 1,\, 1,\, 1\rangle$. Then

$$\int_C \mathbf{F}\cdot\mathbf{r}'(t)\;dt = \int_0^9 \frac{1}{\left(3(t+1)^2\right)^{3/2}}\langle t+1,\, t+1,\, t+1\rangle\cdot\langle 1,\, 1,\, 1\rangle\,dt$$
$$= \frac{1}{\sqrt{3}}\int_0^9 \frac{1}{(t+1)^3}(t+1)\,dt = \frac{1}{\sqrt{3}}\int_0^9 \frac{1}{(t+1)^2}\,dt = \frac{3}{10}\sqrt{3}$$

14.2.47

a. Looking at the vector field, it appears that the vector field points counterclockwise just as much as it points clockwise at the boundary of the region, so we would expect the circulation to be zero.

b. $\mathbf{r}'(t) = \langle -2\sin(t), 2\cos(t)\rangle$, so

$$\int_C \mathbf{F}\cdot\mathbf{r}'(t)\,dt = \int_0^{2\pi} \langle 2(\sin(t)-\cos(t)), 2\cos(t)\rangle \cdot \langle -2\sin(t), 2\cos(t)\rangle\,dt$$
$$= -4\int_0^{2\pi} \left(\cos^2(t) + \sin(t)\cos(t) - \sin^2(t)\right)dt = 0$$

14.2.49

a. Looking at the vector field, the inward-pointing vectors (in quadrants II and IV) appear larger than the outward-pointing vectors (in quadrants I and III). Thus we would expect the flux to be negative.

b. $\mathbf{r}'(t) = \langle -2\sin(t), 2\cos(t)\rangle$, so that

$$\int_C \mathbf{F}\cdot\mathbf{n}\,ds = \int_0^{2\pi} (2(\sin(t)-\cos(t))\cdot 2\cos(t) - 2\cos(t)\cdot(-2\sin(t)))\,dt$$
$$= -4\int_0^{2\pi} \left(\cos^2(t) - 2\sin t\cos t\right)dt = -4\pi$$

14.2.51

a. True. This is the definition of an arc length parameterization.

b. True. Let $\mathbf{r}(t) = \langle \cos(t), \sin(t)\rangle$; then $\mathbf{r}'(t) = \langle -\sin(t), \cos(t)\rangle$, and

$$\int_C \mathbf{F}\cdot\mathbf{r}'(t)\,dt = \int_0^{2\pi} \langle \sin(t), \cos(t)\rangle \cdot \langle -\sin(t), \cos(t)\rangle\,dt = \int_0^{2\pi} \left(\cos^2(t) - \sin^2(t)\right)dt = 0$$
$$\int_C \mathbf{F}\cdot\mathbf{n}\,ds = \int_0^{2\pi} (\sin(t)\cdot\cos(t) - \cos(t)(-\sin(t)))\,dt = 2\int_0^{2\pi} \sin(t)\cos(t)\,dt = 0$$

c. True.

d. True. It is the line integral $\int_C \mathbf{F}\cdot\mathbf{n}\,ds$.

14.2.53

a. For the first path, the work done is

$$\int_C \mathbf{F}\cdot\mathbf{r}'(t)\,dt = \int_{-100}^{100} -141\,dt = 28200$$

and for the second path

$$\int_C \mathbf{F}\cdot\mathbf{r}'(t)\,dt = \int_0^{\pi} (-14100\sin(t) - 5000\cos(t))\,dt = 28200$$

so again they are equal.

b. For the first path, the work done is

$$\int_C \mathbf{F}\cdot\mathbf{r}'(t)\,dt = \int_{-100}^{100} -141\,dt = 28200$$

while for the second path,

$$\int_C \mathbf{F}\cdot\mathbf{r}'(t)\,dt = \int_0^{\pi} (-14100\sin(t) - 5000\cos(t))\,dt = 28200$$

so the amount of work is still equal along the two paths.

14.2.55

a. Let $\mathbf{r}(t) = \langle t, t^2\rangle$ for $0 \le t \le 1$ so that $|\mathbf{r}'(t)| = \sqrt{4t^2+1}$, and

$$\int_C f\,ds = \int_0^1 t\sqrt{4t^2+1}\,dt = \frac{5\sqrt{5}-1}{12}$$

b. $\mathbf{r}(t) = \langle 1-t,\,(1-t)^2\rangle$ for $0 \le t \le 1$. Then $|\mathbf{r}'(t)| = \sqrt{(-1)^2 + (-2(1-t))^2} = \sqrt{4t^2 - 8t + 5}$ and

$$\int_C f\,ds = \int_0^1 (1-t)\sqrt{4t^2 - 8t + 5}\,dt = \frac{5\sqrt{5}-1}{12}$$

c. The two integrals are equal.

14.2.57 Let $\mathbf{r}(t) = \langle r\cos(t),\,r\sin(t)\rangle$ for a circle of radius r, so that $\mathbf{r}'(t) = \langle -r\sin(t),\,r\cos(t)\rangle$. Then

$$\begin{aligned}\int_C \mathbf{F}\cdot\mathbf{r}'(t)\,dt &= \int_0^{2\pi} ((ar\cos(t) + br\sin(t))(-r\sin(t)) + (cr\cos(t) + dr\sin(t))(r\cos(t)))\,dt \\ &= r^2\int_0^{2\pi} \left(-a\sin(t)\cos(t) - b\sin^2(t) + c\cos^2(t) + d\sin(t)\cos(t)\right)dt = \pi(c-b)\,r^2\end{aligned}$$

so that the circulation is zero provided $b = c$.

14.2.59 Let $\mathbf{r}(t) = \langle r\cos(t),\,r\sin(t)\rangle$ for a circle of radius r; then the flux is

$$\begin{aligned}\int_C \mathbf{F}\cdot\mathbf{n}\,ds &= \int_0^{2\pi} ((ar\cos(t) + br\sin(t))(r\cos(t)) - (cr\cos(t) + dr\sin(t))(-r\sin(t)))\,dt \\ &= r^2\int_0^{2\pi} \left(a\cos^2(t) + b\sin(t)\cos(t) + c\sin(t)\cos(t) + d\sin^2(t)\right)dt = r^2(a+d)\pi\end{aligned}$$

so that the flux is zero provided that $a = -d$.

14.2.61 Using the same parameterizations as for the previous problem, we have

$$\int_{C_1} \mathbf{F}\cdot\mathbf{T}\,ds = \int_0^1 ((t)(-1) + (1-t)(1))\,dt = \int_0^1 (1-2t)\,dt = 0$$

$$\int_{C_2} \mathbf{F}\cdot\mathbf{T}\,ds = \int_0^{\pi/2} ((\sin(t))(-\sin(t)) + (\cos(t))(\cos(t)))\,dt = \int_0^{\pi/2} \left(\cos^2(t) - \sin^2(t)\right)dt = 0$$

$$\int_{C_3} \mathbf{F}\cdot\mathbf{T}\,ds = \int_0^1 ((0)(-1) + (1-t)(0))\,dt + \int_0^1 (t\cdot 0 + 0\cdot 1)\,dt = 0$$

and all three are equal to zero.

14.2.63 Parameterize C by $\mathbf{r}(t) = \langle t,\,2t^2\rangle$ for $0 \le t \le 3$; then $|\mathbf{r}'(t)| = \sqrt{1 + 16t^2}$ and

$$\int_C \rho\,ds = \int_0^3 \left(1 + 2t^3\right)\sqrt{1 + 16t^2}\,dt \approx 409.5$$

14.2.65 $\mathbf{r}'(t) = \langle 1,\,1,\,1\rangle$ and $|\mathbf{r}(t)| = t\sqrt{3}$, so the work is

$$\int_C \mathbf{F}\cdot\mathbf{T}\,ds = \int_1^a \frac{3t}{\left(t\sqrt{3}\right)^p}dt = 3^{1-p/2}\int_1^a t^{1-p}\,dt$$

a. For $p = 2$, we have $\displaystyle\int_1^a \frac{1}{t}dt = \ln a$.

b. The work is not finite.

c. For $p = 4$, we have $\displaystyle\frac{1}{3}\int_1^a t^{-3}\,dt = -\frac{1}{6t^2}\Big|_1^a = \frac{1}{6} - \frac{1}{6a^2}$.

d. As $a \to \infty$, the work approaches $\frac{1}{6}$.

e. For the general $p > 1$, the analysis above shows that the integral is (for $p \ne 2$)

$$3^{1-p/2}\frac{1}{2-p}t^{2-p}\Big|_1^a = \frac{3^{1-p/2}}{2-p}\left(-1 + a^{2-p}\right)$$

while for $p = 2$ the integral is (from part (a)) $\ln a$.

e. This approaches a limit only for $p > 2$ (when $2 - p < 0$). This limit is $\displaystyle\frac{3^{1-p/2}}{p-2}$.

14.2.67 We use four line segment parameterizations for the rectangle, all for $0 \le t \le 1$:

$$\begin{array}{ll} \mathbf{r}_1(t) = \langle at, 0\rangle & \mathbf{r}_1'(t) = \langle a, 0\rangle \\ \mathbf{r}_2(t) = \langle a, bt\rangle & \mathbf{r}_2'(t) = \langle 0, b\rangle \\ \mathbf{r}_3(t) = \langle a - at, b\rangle & \mathbf{r}_3'(t) = \langle -a, 0\rangle \\ \mathbf{r}_4(t) = \langle 0, b - bt\rangle & \mathbf{r}_4'(t) = \langle 0, -b\rangle \end{array}$$

Then

$$\int_C x\,dy = \int_0^1 (at\cdot 0 + a\cdot b + (a - at)\cdot 0 + 0\cdot(-b))\,dt = \int_0^1 ab\,dt = ab$$

14.3 Conservative Vector Fields

14.3.1 See Figure 14.27.

14.3.3 See Theorem 14.3. (Equality of mixed partial derivatives)

14.3.5 Integrate f with respect to x to get an answer where the "constant" is actually a function of y. Take the partial with respect to y and equate with g to figure out what the constant is.

14.3.7 The integral is zero.

14.3.9 Yes, since $1_y = 1_x = 0 = \nabla(x + y)$.

14.3.11 Yes, since $-y_y = -x_x = -1$.

14.3.13 Yes. $\dfrac{\partial}{\partial y}e^{-x}\cos(y) = -e^{-x}\sin(y) = \dfrac{\partial}{\partial x}e^{-x}\sin(y)$.

14.3.15 Yes. $\dfrac{\partial}{\partial y}(x) = \dfrac{\partial}{\partial x}(y) = 0$. A potential function is $\dfrac{x^2 + y^2}{2}$.

14.3.17 No. $\dfrac{\partial}{\partial y}(x^3 - xy) = -x$, and $\dfrac{\partial}{\partial x}\left(\dfrac{x^2}{2} + y\right) = x$.

14.3.19 Yes. $\dfrac{\partial}{\partial y}\left(\dfrac{x}{\sqrt{x^2 + y^2}}\right) = \dfrac{-xy}{(x^2 + y^2)^{3/2}} = \dfrac{\partial}{\partial x}\left(\dfrac{y}{\sqrt{x^2 + y^2}}\right)$. Integrating $\dfrac{x}{\sqrt{x^2 + y^2}}$ with respect to x, we get a potential function of $\sqrt{x^2 + y^2}$.

14.3.21 Yes, since the mixed partials are pairwise equal. A potential function is $xz + y$.

14.3.23 Yes, since the mixed partials are pairwise equal. To find the potential function, integrate $y + z$ with respect to x to get $x(y + x) + f(y, z)$; differentiating with respect to y gives $x + z = x + f_y(y, z)$ so that $f_y(y, z) = z$ and $f(y, z) = yz$. Thus a potential function is $xy + yz + xz$.

14.3.25 Yes, since the mixed partials are pairwise equal. As in problem 19, a potential function is $\sqrt{x^2 + y^2 + z^2}$.

14.3.27

a. $\nabla\varphi = \langle y, x\rangle$, so

$$\int_C \nabla\varphi\cdot\mathbf{r}'(t)\,dt = \int_0^\pi \langle \sin(t), \cos(t)\rangle\cdot\langle -\sin(t), \cos(t)\rangle\,dt = \int_0^\pi \cos(2t)\,dt = 0$$

b. Since $\nabla\varphi$ is obviously conservative, the integral is simply $\varphi(\cos(\pi), \sin(\pi)) - \varphi(\cos(0), \sin(0)) = 0$.

14.3.29

a. $\nabla\varphi = \langle 1,\ 3\rangle$, so

$$\int_C \nabla\varphi \cdot \mathbf{r}'(t)\,dt = \int_0^2 \langle 1,\ 3\rangle \cdot \langle -1,\ 1\rangle\,dt = \int_0^2 2\,dt = 4$$

b. The integral is $\varphi(0,\ 2) - \varphi(2,\ 0) = 6 - 2 = 4$. (Note that $\varphi(0,\ 2)$ is φ evaluated at the point where $t = 2$).

14.3.31

a. $\nabla\varphi = \langle x,\ y,\ z\rangle$, so

$$\int_C \nabla\varphi \cdot \mathbf{r}'(t)\,dt = \int_0^{2\pi} \left\langle \cos(t),\ \sin(t),\ \frac{t}{\pi}\right\rangle \cdot \left\langle -\sin(t),\ \cos(t),\ \frac{1}{\pi}\right\rangle dt = \int_0^{2\pi} \left(\frac{t}{\pi^2}\right) dt = 2$$

b. The integral is $\varphi(1,\ 0,\ 2) - \varphi(1,\ 0,\ 0) = \frac{5}{2} - \frac{1}{2} = 2$.

14.3.33 Parameterize C by $\mathbf{r}(t) = \langle 4\cos(t),\ 4\sin(t)\rangle$, $0 \le t \le 2\pi$. Then

$$\oint_C \mathbf{F}\cdot d\mathbf{r} = \int_0^{2\pi} \langle 4\cos(t),\ 4\sin(t)\rangle \cdot \langle -4\sin(t),\ 4\cos(t)\rangle\,dt = \int_0^{2\pi} 0\,dt = 0$$

14.3.35 Parameterize C by three paths, all for $0 \le t \le 1$:

$$\begin{aligned} \mathbf{r}_1(t) &= \langle t, t-1\rangle & \mathbf{r}_1'(t) &= \langle 1,\ 1\rangle \\ \mathbf{r}_2(t) &= \langle 1-t,\ t\rangle & \mathbf{r}_2'(t) &= \langle -1,\ 1\rangle \\ \mathbf{r}_3(t) &= \langle 0,\ 1-2t\rangle & \mathbf{r}_3'(t) &= \langle 0,\ -2\rangle \end{aligned}$$

Then

$$\begin{aligned} \oint_C \mathbf{F}\cdot d\mathbf{r} &= \int_0^1 \langle t, t-1\rangle \cdot \langle 1,\ 1\rangle\,dt + \int_0^1 \langle 1-t,\ t\rangle \cdot \langle -1,\ 1\rangle\,dt + \int_0^1 \langle 0,\ 1-2t\rangle \cdot \langle 0,\ -2\rangle\,dt \\ &= \int_0^1 (t + (t-1) + (t-1) + t - 2(1-2t))\,dt = \int_0^1 (8t-4)\,dt = 0 \end{aligned}$$

14.3.37 Using the given parameterization,

$$\oint_C \mathbf{F}\cdot d\mathbf{r} = \int_0^{2\pi} \langle \cos(t),\ \sin(t),\ 2\rangle \cdot \langle -\sin(t),\ \cos(t),\ 0\rangle\,dt = \int_0^{2\pi} 0\,dt = 0$$

14.3.39

a. False, since **F** is not conservative.

b. True, since **F** is conservative.

c. True. If the vector field is $\langle a,\ b,\ c\rangle$, then a potential function is $ax + by + cz$.

d. True, since $\frac{\partial}{\partial y} f(x) = \frac{\partial}{\partial x} g(y) = 0$.

14.3.41 Write $\varphi(x,\ y) = e^{-x}\cos(y)$. Then using the Fundamental Theorem, this integral is equal to $\varphi(\ln 2,\ 2\pi) - \varphi(0,\ 0) = e^{-\ln 2}\cos(2\pi) - e^0\cos(0) = \frac{1}{2} - 1 = -\frac{1}{2}$.

14.3.43 **F** is a conservative vector field; a potential function can be found by integrating x^2 with respect to y to get $x^2y + f(x,\ z)$; differentiate with respect to z to get $f_z(x,\ z) = 2xz$, so that $f(x,\ z) = xz^2 + g(x)$. Thus the potential function is $x^2y + xz^2 + g(x)$; differentiating with respect to x gives $2xy + z^2 + g_x(x) = 2xy + z^2$, so that $g_x(x) = 0$ and we may take $g(x) = 0$. So if $\varphi = x^2y + xz^2$, then $\nabla\varphi = \mathbf{F}$ and thus the integral is zero since both sin and cos, and thus φ, have the same values at the two endpoints of C.

14.3.45 This is a conservative vector field with potential function $\varphi(x,\ y) = \frac{1}{2}x^2 + 2y$, so the work is $\varphi(2,\ 4) - \varphi(0,\ 0) = 10$.

14.3.47 This is a conservative vector field with potential function $\varphi(x,\ y,\ z) = \frac{1}{2}(x^2 + y^2 + z^2)$, so the work is $\varphi(2,\ 4,\ 6) - \varphi(1,\ 2,\ 1) = 28 - 3 = 25$.

14.3.49 For C_1, the vector field points "against" the curve for most of its length, and with larger magnitude, so the integral is negative. For C_2, the vector field points with the curve for its entire length, so the integral is positive.

14.3.51 $\mathbf{F} = \langle a, b, c\rangle$ is a conservative force field with potential function $\varphi(x, y, z) = ax + by + cz$, so the work done is $\varphi(B) - \varphi(A) = \mathbf{F}\cdot B - \mathbf{F}\cdot A = \mathbf{F}\cdot(B - A) = \mathbf{F}\cdot\overrightarrow{AB}$.

14.3.53

a. Away from the origin (where the denominator of the force field equation is undefined), the force field is conservative since, for example,

$$\frac{\partial}{\partial y}GMm\frac{x}{(x^2+y^2+z^2)^{3/2}} = GMm\frac{2xy}{(x^2+y^2+z^2)^{5/2}} = \frac{\partial}{\partial x}GMm\frac{y}{(x^2+y^2+z^2)^{3/2}}$$

b. A potential function for the force field is

$$\varphi(x, y, z) = GMm\left(x^2+y^2+z^2\right)^{-1/2} = GMm\frac{1}{|\mathbf{r}|}.$$

c. The work done in moving the point from A to B, since the force field is conservative, is

$$\varphi(B) - \varphi(A) = GMm\left(\frac{1}{|B|} - \frac{1}{|A|}\right) = GMm\left(\frac{1}{r_2} - \frac{1}{r_1}\right)$$

d. Since the field is conservative, the work done does not depend on the path.

14.3.55

a. This field is $\mathbf{F} = \langle -y, x\rangle\left(x^2+y^2\right)^{-p/2}$, and we have

$$\begin{aligned}\frac{\partial}{\partial y}\left(-y\left(x^2+y^2\right)^{-p/2}\right) &= -\left(x^2+y^2\right)^{-p/2} + py^2\frac{\left(x^2+y^2\right)^{-p/2}}{x^2+y^2}\\ &= -\left(x^2+y^2\right)^{-p/2} + py^2\left(x^2+y^2\right)^{-1-p/2}\\ \frac{\partial}{\partial x}\left(x\left(x^2+y^2\right)^{-p/2}\right) &= \left(x^2+y^2\right)^{-p/2} - px^2\frac{\left(x^2+y^2\right)^{-p/2}}{x^2+y^2}\\ &= \left(x^2+y^2\right)^{-p/2} - px^2\left(x^2+y^2\right)^{-1-p/2}\end{aligned}$$

For the force field to be conservative, these two would have to be equal. However, their difference is

$$\begin{aligned}\frac{\partial N}{\partial x} - \frac{\partial M}{\partial y} &= 2\left(x^2+y^2\right)^{-p/2} - p\left(x^2+y^2\right)\left(x^2+y^2\right)^{-1-p/2}\\ &= 2\left(x^2+y^2\right)^{-p/2} - p\left(x^2+y^2\right)^{-p/2} = (2-p)\left(x^2+y^2\right)^{-p/2}\end{aligned}$$

which is in general nonzero.

b. From the above formula, if $p = 2$, then the mixed partials are equal, so that **F** is conservative.

c. For $p = 2$,

$$\mathbf{F} = \frac{1}{x^2+y^2}\langle -y, x\rangle$$

Integrating the x component of F with respect to x gives $\varphi = \arctan\left(\frac{y}{x}\right)$.

14.3.57

a.

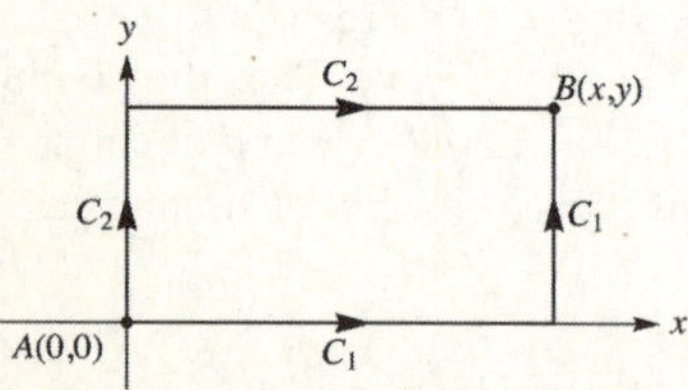

b. Parameterize C_1 by two paths: $\mathbf{r}_1(t) = \langle t, 0\rangle,\ 0 \le t \le x$, and $\mathbf{r}_2(t) = \langle x, t\rangle,\ 0 \le t \le y$. Then

$$\begin{aligned}\int_C \mathbf{F}\cdot d\mathbf{r} &= \int_0^x \langle 2t, -t\rangle\cdot\langle 1, 0\rangle\,dt + \int_0^y \langle 2x - t, -x + 2t\rangle\cdot\langle 0, 1\rangle\,dt\\ &= \int_0^x 2t\,dt + \int_0^y (2t - x)\,dt = x^2 + y^2 - xy\end{aligned}$$

c. Parameterize C_2 by the paths $\mathbf{r}_1(t) = \langle 0,\, t\rangle,\ 0 \le t \le y$, and $\mathbf{r}_2(t) = \langle t,\, y\rangle,\ 0 \le t \le x$. Then

$$\int_C \mathbf{F}\cdot d\mathbf{r} = \int_0^y \langle -t,\, 2t\rangle \cdot \langle 0,\, 1\rangle\, dt + \int_0^x \langle 2t - y,\, -t + 2y\rangle \cdot \langle 1,\, 0\rangle\, dt$$
$$= \int_0^y 2t\, dt + \int_0^x (2t - y)\, dt = x^2 + y^2 - xy$$

14.3.59 Using problem 57 and the same paths $\mathbf{r}_1$, $\mathbf{r}_2$, we have

$$\int_C \mathbf{F}\cdot d\mathbf{r} = \int_0^x \langle t,\, 0\rangle \cdot \langle 1,\, 0\rangle\, dt + \int_0^y \langle x,\, t\rangle \cdot \langle 0,\, 1\rangle\, dt$$
$$= \int_0^x t\, dt + \int_0^y t\, dt = \frac{1}{2}(x^2 + y^2)$$

14.3.61 Using problem 57 and the same paths $\mathbf{r}_1$, $\mathbf{r}_2$, we have

$$\int_C \mathbf{F}\cdot d\mathbf{r} = \int_0^x \langle 2t^3,\, 0\rangle \cdot \langle 1,\, 0\rangle\, dt + \int_0^y \langle 2x^3 + xt^2,\, 2t^3 + x^2 t\rangle \cdot \langle 0,\, 1\rangle\, dt$$
$$= \int_0^x 2t^3\, dt + \int_0^y (2t^3 + x^2 t)\, dt = \frac{1}{2}(x^4 + x^2y^2 + y^4)$$

14.4 Green's Theorem

14.4.1 As with the Fundamental Theorem of Calculus, it allows evaluation of the integral of a derivative by looking at the value of the underlying function on the boundary of a region (or, in the case of the Fundamental Theorem, an interval).

14.4.3 The curl is $\frac{\partial g}{\partial x} - \frac{\partial f}{\partial y} = y^2 + 4x^3 - 4x^3 = y^2$.

14.4.5 The area is $\dfrac{1}{2}\oint_C (x\, dy - y\, dx)$ where C is the boundary of the region.

14.4.7 Because the flux is the integrand in Green's theorem, so the integral vanishes.

14.4.9

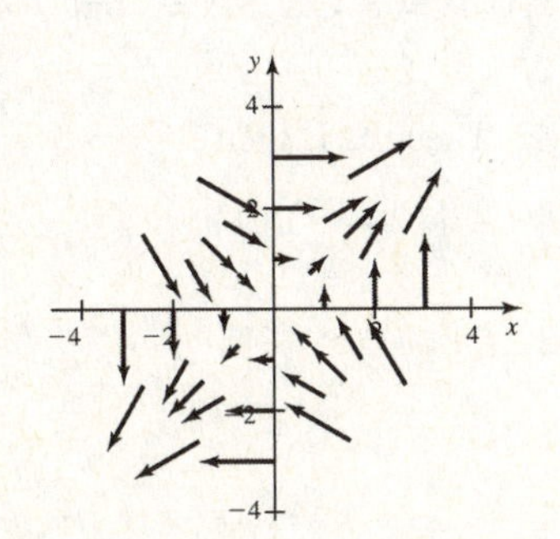

14.4.11

a. The curl is $\dfrac{\partial}{\partial x}g - \dfrac{\partial}{\partial y}f = 0 - 0 = 0$

b. $\displaystyle\int_C \mathbf{F}\cdot d\mathbf{r} = \int_0^{2\pi} \langle \cos(t),\, \sin(t)\rangle \cdot \langle -\sin(t),\, \cos(t)\rangle\, dt = 0$

$$\iint_R \left(\frac{\partial}{\partial x}g - \frac{\partial}{\partial y}f\right) dA = \iint_R 0\, dA = 0$$

c. The vector field is conservative since its curl is zero.

14.4.13

a. The curl is $\dfrac{\partial}{\partial x}g - \dfrac{\partial}{\partial y}f = -2 - 2 = -4$

b. $$\int_C \mathbf{F}\cdot d\mathbf{r} = \int_0^{\pi} \langle 0,\, -2t\rangle \cdot \langle 1,\, 0\rangle\, dt + \int_{\pi}^{0} \langle 2\sin(t),\, -2t\rangle \cdot \langle 1,\, \cos(t)\rangle\, dt$$
$$= \int_{\pi}^{0} (2\sin(t) - 2t\cos(t))\, dt = -8$$

$$\iint_R \left(\frac{\partial}{\partial x}g - \frac{\partial}{\partial y}f\right) dA = \iint_R (-4)\, dA = -4\int_0^{\pi} \sin(x)\, dx = -8$$

c. It is not conservative; the curl is nonzero.

14.4.15

a. The curl is $\dfrac{\partial}{\partial x}g - \dfrac{\partial}{\partial y}f = 2x - 2x = 0$

b. $$\int_C \mathbf{F}\cdot d\mathbf{r} = \int_0^{2} \langle 0,\, t^2\rangle \cdot \langle 1,\, 0\rangle\, dt + \int_2^{0} \langle 2t^2(2-t), t^2 - t^2(2-t)^2\rangle \cdot \langle 1,\, 2-2t\rangle\, dt$$
$$= \int_2^{0} \left(2t^2(2-t) + t^2\left(1-(2-t)^2\right)(2-2t)\right) dt = 0$$

$$\iint_R \left(\frac{\partial}{\partial x}g - \frac{\partial}{\partial y}f\right) dA = \iint_R 0\, dA = 0$$

c. Yes, since the curl is zero.

14.4.17 Parameterize the boundary by $x = 5\cos(t)$, $y = 5\sin(t)$, $0 \le t \le 2\pi$. Then $dx = -5\sin(t)\, dt$, $dy = 5\cos(t)\, dt$, and the area is

$$\frac{1}{2}\oint_C x\, dy - y\, dx = \frac{1}{2}\int_0^{2\pi} ((5\cos(t))\cdot(5\cos(t)) - (5\sin(t))\cdot(-5\sin(t)))\, dt = \frac{25}{2}\int_0^{2\pi} 1\, dt = 25\pi$$

14.4.19 Parameterize the boundary by $x = 4\cos(t)$, $y = 4\sin(t)$, $0 \le t \le 2\pi$. Then $dx = -4\sin(t)\, dt$, $dy = 4\cos(t)\, dt$, and the area is

$$\frac{1}{2}\oint_C x\, dy - y\, dx = \frac{1}{2}\int_0^{2\pi} ((4\cos(t))\cdot(4\cos(t)) - (4\sin(t))\cdot(-4\sin(t)))\, dt = 8\int_0^{2\pi} 1\, dt = 16\pi$$

14.4.21 Traverse the first path from -2 to 2, then the second path back from 2 to -2. The area is then

$$\frac{1}{2}\oint_C x\, dy - y\, dx = \frac{1}{2}\int_{-2}^{2} (t\cdot(4t) - (2t^2)\cdot 1)\, dt + \frac{1}{2}\int_2^{-2} (t\cdot(-2t) - (12 - t^2)\cdot 1)\, dt$$
$$= \frac{1}{2}\int_{-2}^{2} 2t^2\, dt + \frac{1}{2}\int_2^{-2} (-2t^2 - 12 + t^2)\, dt = 32$$

14.4.23

a. The divergence is $\dfrac{\partial}{\partial x}f + \dfrac{\partial}{\partial y}g = 1 + 1 = 2$

b. $$\int_C \mathbf{F}\cdot\mathbf{n}\, ds = 4\int_0^{2\pi} (\cos(t)(\cos(t)) - \sin(t)(-\sin(t)))\, dt = 8\pi$$

$$\iint_R \left(\frac{\partial}{\partial x}f + \frac{\partial}{\partial y}g\right) dA = \iint_R (2)\, dA = 2\cdot 4\pi = 8\pi$$

c. It is not source-free since its divergence is nonzero.

14.4.25

a. The divergence is $\dfrac{\partial}{\partial x}f + \dfrac{\partial}{\partial y}g = 0$

b. $\int_C \mathbf{F}\cdot\mathbf{n}\,ds = \int_{-2}^{2}(0(0) + 3t(1))\,dt + \int_{-2}^{2}((4-t^2)(-2t) + 3t(1))\,dt = 0$

$$\iint_R \left(\frac{\partial}{\partial x}f + \frac{\partial}{\partial y}g\right) dA = \iint_R 0\,dA = 0$$

c. Yes, since its divergence is zero.

14.4.27

a. The divergence is $\frac{\partial}{\partial x}f + \frac{\partial}{\partial y}g = 2y - 2y = 0$

b. Parameterize the region by $\mathbf{r}_1(t) = \langle t,\, 0\rangle$, and $\mathbf{r}_1(t) = \langle t,\, t(2-t)\rangle$ for $0 \le t \le 2$; traverse the second path from $t = 2$ to $t = 0$ to make a counterclockwise closed curve. Then

$$\int_C \mathbf{F}\cdot\mathbf{n}\,ds = \int_0^2 (0(0) - t^2(1))\,dt + \int_2^0 (2t^2(2-t)(2-2t) - t^2(1-(2-t)^2)(1))\,dt = 0$$

$$\iint_R \left(\frac{\partial}{\partial x}f + \frac{\partial}{\partial y}g\right) dA = \iint_R 0\,dA = 0$$

c. Yes, since its divergence is zero.

14.4.29 The line integral, using the flux form of Green's theorem, is equal to

$$\iint_R \left(\frac{\partial}{\partial x}\left(2x + e^{y^2}\right) + \frac{\partial}{\partial y}\left(4y^2 + e^{x^2}\right)\right) dA = \iint_R (2+8y)\,dA = \int_0^1\int_0^1 (2+8y)\,dx\,dy = \int_0^1 (2+8y)\,dy = 6$$

14.4.31 Using the flux form of Green's theorem, the integral is equal to

$$\iint_R \left(\frac{\partial}{\partial x}(0) + \frac{\partial}{\partial y}(xy)\right) dA = \int_0^2\int_0^{4-2x} x\,dy\,dx = \int_0^2 (4x - 2x^2)\,dx = \frac{8}{3}$$

14.4.33 Using the circulation form of Green's theorem, the integral is equal to

$$\iint_R \left(\frac{\partial}{\partial x}(4x^3 + y) - \frac{\partial}{\partial y}(2xy^2 + x)\right) dA = \iint_R (12x^2 - 4xy)\,dA$$

$$= \int_0^{\pi}\int_0^{\sin(x)} (12x^2 - 4xy)\,dy\,dx = \frac{23\pi^2}{2} - 48$$

14.4.35

a. Using Green's theorem, the circulation is

$$\iint_R \left(\frac{\partial}{\partial x}(y) - \frac{\partial}{\partial y}(x)\right) dA = \iint_R 0\,dA = 0$$

b. Using Green's theorem, the flux is

$$\iint_R \left(\frac{\partial}{\partial x}(x) + \frac{\partial}{\partial y}(y)\right) dA = \iint_R 2\,dA = 2\cdot\text{area of } R = 2\cdot\frac{1}{2}(4\pi - \pi) = 3\pi$$

14.4.37

a. Using Green's theorem, the circulation is

$$\iint_R \left(\frac{\partial}{\partial x}(x - 4y) - \frac{\partial}{\partial y}(2x + y)\right) dA = \iint_R (1-1)\,dA = 0$$

b. Using Green's theorem, the flux is

$$\iint_R \left(\frac{\partial}{\partial x}(2x + y) + \frac{\partial}{\partial y}(x - 4y)\right) dA = \iint_R (-2)\,dA = -2\cdot\text{area of } R = -2\cdot\frac{1}{4}(16\pi - \pi) = -\frac{15}{2}\pi$$

14.4.39

a. True. This is the definition of work along a path.

b. False. Divergence corresponds to flux, so if the divergence is zero throughout a region, the flux is zero across the boundary.

c. True, by Green's theorem.

14.4.41

a. Since $\mathbf{F}$ is conservative, the circulation on the boundary of R is zero.

b. $\mathbf{F} = (x^2+y^2)^{-1/2}\langle x, y\rangle$, so the flux is

$$\iint_R \left(\frac{\partial}{\partial x}\left(\frac{x}{\sqrt{x^2+y^2}}\right) + \frac{\partial}{\partial y}\left(\frac{y}{\sqrt{x^2+y^2}}\right)\right) dA = \iint_R (x^2+y^2)^{-1/2}\, dA = \int_0^\pi \int_1^3 \frac{1}{r} r\, dr\, d\theta$$
$$= \int_0^\pi \int_1^3 1\, dr\, d\theta = 2\pi$$

14.4.43 Note that the region is the area between $x = 3y^2$ and $x = 36 - y^2$, which intersect at $y = 3$.

a. The circulation is

$$\iint_R \left(\frac{\partial}{\partial x}(x^2 - y) - \frac{\partial}{\partial y}(x + y^2)\right) dA = \iint_R (2x - 2y)\, dA = \int_{-3}^{3}\int_{3y^2}^{36-y^2} (2x - 2y)\, dx\, dy$$
$$= \int_{-3}^{3} (1296 - 72y - 72y^2 + 8y^3 - 8y^4)\, dy = \frac{28512}{5}$$

b. The flux is

$$\iint_R \left(\frac{\partial}{\partial x}(x + y^2) + \frac{\partial}{\partial y}(x^2 - y)\right) dA = \iint_R (0)\, dA = 0$$

14.4.45 By the circulation form of Green's theorem,

$$\oint_C (f(x)\, dx + g(y)\, dy) = \iint_R \left(\frac{\partial}{\partial x}[g(y)] - \frac{\partial}{\partial y}[f(x)]\right) dA = 0$$

14.4.47 By Green's theorem,

$$\oint_C (x^2y + 2x)\, dy + xy^2\, dx = \oint_C xy^2\, dx + (x^2y + 2x)\, dy = \iint_R \left(\frac{\partial}{\partial x}(x^2y + 2x) - \frac{\partial}{\partial y}(xy^2)\right) dA$$
$$= \iint_R (2xy + 2 - 2xy)\, dA = \iint_R 2\, dA = 2 \cdot \text{area of } A$$

14.4.49

a. The divergence is $\frac{\partial}{\partial x}(4) + \frac{\partial}{\partial y}(2) = 0$.

b. $\psi = 4y - 2x$.

14.4.51

a. The divergence is $\frac{\partial}{\partial x}(-e^{-x}\sin(y)) + \frac{\partial}{\partial y}(e^{-x}\cos(y)) = e^{-x}\sin(y) - e^{-x}\sin(y) = 0$.

b. $\psi = e^{-x}\cos(y)$.

14.4.53

a. The curl and divergence are

$$\textbf{curl F} = \frac{\partial}{\partial x}(-e^x\sin(y)) - \frac{\partial}{\partial y}(e^x\cos(y)) = -e^x\sin(y) + e^x\sin(y) = 0$$
$$\textbf{div F} = \frac{\partial}{\partial x}(e^x\cos(y)) + \frac{\partial}{\partial y}(-e^x\sin(y)) = e^x\cos(y) - e^x\cos(y) = 0$$

b.

$$\varphi(x, y) = e^x \cos(y)$$
$$\psi(x, y) = e^x \sin(y)$$

c. $\varphi_{xx} + \varphi_{yy} = \dfrac{\partial^2}{\partial x^2}(e^x \cos(y)) + \dfrac{\partial^2}{\partial y^2}(e^x \cos(y)) = e^x \cos(y) - e^x \cos(y) = 0$

$$\psi_{xx} + \psi_{yy} = \frac{\partial^2}{\partial x^2}(e^x \sin(y)) + \frac{\partial^2}{\partial y^2}(e^x \sin(y)) = e^x \sin(y) - e^x \sin(y) = 0$$

14.4.55

a. The curl and divergence are

$$\textbf{curl F} = \frac{\partial}{\partial x}\left(\frac{1}{2}\ln(x^2+y^2)\right) - \frac{\partial}{\partial y}\left(\arctan\left(\frac{y}{x}\right)\right) = 0$$
$$\textbf{div F} = \frac{\partial}{\partial x}\left(\arctan\left(\frac{y}{x}\right)\right) + \frac{\partial}{\partial y}\left(\frac{1}{2}\ln(x^2+y^2)\right) = 0$$

b.

$$\varphi(x, y) = x\tan^{-1}\left(\frac{y}{x}\right) - y + \frac{1}{2}y\ \ln(x^2+y^2)$$
$$\psi(x, y) = \int -\frac{1}{2}\ln(x^2+y^2)\,dy = y\tan^{-1}\left(\frac{y}{x}\right) - \frac{x}{2}\ln(x^2+y^2) + x$$

c. $\varphi_{xx} + \varphi_{yy} = \dfrac{\partial^2}{\partial x^2}\left(x\tan^{-1}\left(\dfrac{y}{x}\right) - y + \dfrac{1}{2}y\ln(x^2+y^2)\right) + \dfrac{\partial^2}{\partial y^2}\left(x\tan^{-1}\left(\dfrac{y}{x}\right) - y + \dfrac{1}{2}y\ln(x^2+y^2)\right)$

$$= \frac{\partial}{\partial x}\left(\tan^{-1}\left(\frac{y}{x}\right) - \frac{xy}{x^2+y^2} + \frac{xy}{x^2+y^2}\right) + \frac{\partial}{\partial y}\left(\frac{x^2}{x^2+y^2} - 1 + \frac{1}{2}\ln(x^2+y^2) + \frac{y^2}{x^2+y^2}\right)$$
$$= \frac{\partial}{\partial x}\left(\tan^{-1}\left(\frac{y}{x}\right)\right) + \frac{\partial}{\partial y}\left(\frac{1}{2}\ln(x^2+y^2)\right) = 0$$

$$\psi_{xx} + \psi_{yy} = \frac{\partial^2}{\partial x^2}\left(y\tan^{-1}\left(\frac{y}{x}\right) - \frac{x}{2}\ln(x^2+y^2) + x\right) + \frac{\partial^2}{\partial y^2}\left(y\tan^{-1}\left(\frac{y}{x}\right) - \frac{x}{2}\ln(x^2+y^2) + x\right)$$
$$= \frac{\partial}{\partial x}\left(-\frac{y^2}{x^2+y^2} - \frac{x^2}{x^2+y^2} - \frac{1}{2}\ln(x^2+y^2) + 1\right) + \frac{\partial}{\partial y}\left(\tan^{-1}\left(\frac{y}{x}\right) + \frac{xy}{x^2+y^2} - \frac{xy}{x^2+y^2}\right)$$
$$= \frac{\partial}{\partial x}\left(\frac{1}{2}\ln(x^2+y^2)\right) + \frac{\partial}{\partial y}\left(\tan^{-1}\left(\frac{y}{x}\right)\right) = 0$$

14.4.57

a. The velocity field is $\langle -4\cos(x)\,\sin(y), 4\sin(x)\,\cos(y)\rangle$.

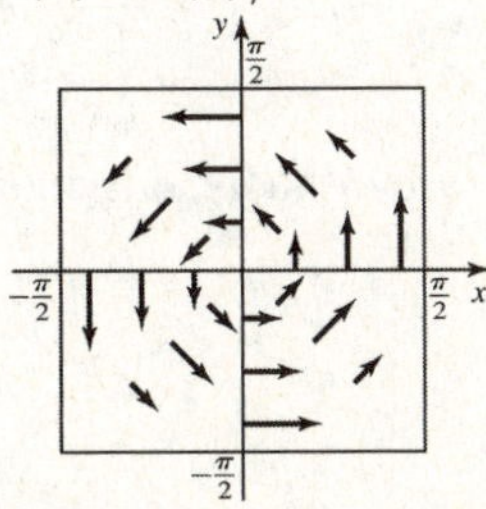

b. The field is source-free if its divergence is zero.

$$\textbf{div F} = \frac{\partial}{\partial x}(-4\cos(x)\,\sin(y)) + \frac{\partial}{\partial y}(\,4\sin(x)\,\cos(y)) = 4\sin(x)\sin(y) - 4\sin(x)\sin(y) = 0,$$

so the field is source-free.

c. The field is irrotational if its curl is zero.

$$\textbf{curl F} = \frac{\partial}{\partial x}(4\sin(x)\,\cos(y)) - \frac{\partial}{\partial y}(-4\cos(x)\,\sin(y)) = 4\cos(x)\cos(y) + 4\cos(x)\cos(y) = 8\cos(x)\cos(y),$$

so the field is not irrotational.

d. Since the field is source-free, it has zero flux across the boundary.

e. The circulation around the boundary of the rectangle is (by Green's theorem) given by

$$\iint_R 8\cos(x)\cos(y)\,dA = \int_{-\pi/2}^{\pi/2}\int_{-\pi/2}^{\pi/2} 8\cos(x)\cos(y)\,dy\,dx = \int_{-\pi/2}^{\pi/2} 16\cos(x)\,dx = 32$$

14.4.59 If $f(x)$ is continuous, then the flux form of Green's theorem says that

$$\oint_C \frac{f(x)}{c}dx = \frac{1}{c}\iint_R \frac{df}{dx}\,dA$$

The right side of this equation evaluates to

$$\frac{1}{c}\iint_R \frac{df}{dx}\,dA = \frac{1}{c}\int_a^b\int_0^c \frac{df}{dx}dy\,dx = \int_a^b \frac{df}{dx}\,dx$$

To evaluate the left side, parameterize the boundary of R with four paths, each for $0 \le t \le 1$:

$$\begin{array}{ll} \mathbf{r}_1(t) = \langle a + (b-a)t, 0\rangle & \mathbf{r}_1'(t) = \langle b-a,\, 0\rangle \\ \mathbf{r}_2(t) = \langle b, ct\rangle & \mathbf{r}_2'(t) = \langle 0,\, c\rangle \\ \mathbf{r}_3(t) = \langle b + (a-b)t,\, c\rangle & \mathbf{r}_3'(t) = \langle a-b,\, 0\rangle \\ \mathbf{r}_4(t) = \langle a,\, c - ct\rangle & \mathbf{r}_4'(t) = \langle 0,\, -c\rangle \end{array}$$

Then we evaluate $\mathbf{F}\cdot\mathbf{r}_i$ for each i and add:

$$\oint_C \frac{f(x)}{c}\,dy = \frac{1}{c}\int_0^1 (0 + f(b)\cdot c + 0 + f(a)\cdot(-c))dt = f(b) - f(a)$$

so that

$$\int_a^b \frac{df}{dx}\,dx = f(b) - f(a)$$

14.4.61

a. The divergence is $\frac{\partial}{\partial x}\left(\frac{x}{x^2+y^2}\right) + \frac{\partial}{\partial y}\left(\frac{y}{x^2+y^2}\right) = 0$

b. Take a line integral around the unit circle, parameterized as $\langle\cos(t),\, \sin(t)\rangle$. The flux is then

$$\oint_C \frac{x}{x^2+y^2}dy + \frac{y}{x^2+y^2}dx = \int_0^{2\pi} (\cos(t)\cos(t) - \sin(t)(-\sin(t)))\,dt = \int_0^{2\pi} 1\,dt = 2\pi$$

c. The vector field is not defined everywhere in R; specifically, it is undefined at the origin.

14.4.63 Since ψ is a stream function, $d\psi = \psi_x\,dx + \psi_y\,dy$, so the flux integral is

$$\int_C \mathbf{F}\cdot\mathbf{n}\,ds = \int_C f\,dy - g\,dx = \int_C \psi_y\,dy - \psi_x\,dx = \int_C d\psi = \psi(B) - \psi(A)$$

so that the integral is independent of the path.

14.4.65 Showing that the level curves of φ and ψ are orthogonal is equivalent to showing that the gradients of φ and ψ are orthogonal. But $\nabla\varphi\cdot\nabla\psi = \langle f,\, g\rangle\cdot\langle -g,\, f\rangle = 0$.

14.5 Divergence and Curl

14.5.1 The divergence is $\frac{\partial f}{\partial x} + \frac{\partial g}{\partial y} + \frac{\partial h}{\partial z}$.

14.5.3 It means that the field has no sources or sinks.

14.5.5 The curl indicates the axis and speed of rotation of a vector field at each point.

14.5.7 $\boldsymbol{\nabla}\cdot(\boldsymbol{\nabla}\times\mathbf{F}) = 0$; see Theorem 14.10.

14.5.9 $\frac{\partial}{\partial x}(2x) + \frac{\partial}{\partial y}(4y) + \frac{\partial}{\partial z}(-3z) = 3\,.$

14.5.11 $\frac{\partial}{\partial x}(12x) + \frac{\partial}{\partial y}(-6y) + \frac{\partial}{\partial z}(-6z) = 0\,.$

14.5.13 $\frac{\partial}{\partial x}(x^2 - y^2) + \frac{\partial}{\partial y}(y^2 - z^2) + \frac{\partial}{\partial z}(z^2 - x^2) = 2x + 2y + 2z\,.$

14.5.15 $\frac{\partial}{\partial x}\left(\frac{x}{1+x^2+y^2}\right) + \frac{\partial}{\partial y}\left(\frac{y}{1+x^2+y^2}\right) + \frac{\partial}{\partial z}\left(\frac{z}{1+x^2+y^2}\right) = \frac{x^2+y^2+3}{(1+x^2+y^2)^2}\,.$

14.5.17

$$\frac{\partial}{\partial x}\left(\frac{x}{x^2+y^2+z^2}\right) + \frac{\partial}{\partial y}\left(\frac{y}{x^2+y^2+z^2}\right) + \frac{\partial}{\partial z}\left(\frac{z}{x^2+y^2+z^2}\right)$$
$$= \frac{1}{(x^2+y^2+z^2)^2}\left((z^2+y^2-x^2) + (x^2+z^2-y^2) + (x^2+y^2-z^2)\right)$$
$$= \frac{1}{(x^2+y^2+z^2)^2}(x^2+y^2+z^2) = \frac{1}{|\mathbf{r}|^2}$$

14.5.19

$$\frac{\partial}{\partial x}\left(\frac{x}{(x^2+y^2+z^2)^2}\right) + \frac{\partial}{\partial y}\left(\frac{y}{(x^2+y^2+z^2)^2}\right) + \frac{\partial}{\partial z}\left(\frac{z}{(x^2+y^2+z^2)^2}\right)$$
$$= \frac{1}{(x^2+y^2+z^2)^3}\left((z^2+y^2-3x^2) + (x^2+z^2-3y^2) + (x^2+y^2-3z^2)\right)$$
$$= \frac{-1}{(x^2+y^2+z^2)^3}(x^2+y^2+z^2) = \frac{-1}{|\mathbf{r}|^4}$$

14.5.21

a. At both P and Q, the arrows owing away from the point are larger in both number and magnitude than those owing in, so we would expect the divergence to be positive at both points.

b. The divergence is $\frac{\partial}{\partial x}(x) + \frac{\partial}{\partial y}(x+y) = 1 + 1 = 2$ so is positive everywhere.

c. The arrows all point roughly away from the origin, so we the flux is outward everywhere.

d. The net flux across C should be positive.

14.5.23

a. The axis of rotation is $\langle 1,\, 0,\, 0\rangle$, the x-axis.

$$\nabla \times \mathbf{F} = \nabla \times \begin{vmatrix} \mathbf{i} & \mathbf{j} & \mathbf{k} \\ 1 & 0 & 0 \\ x & y & z \end{vmatrix} = \nabla \times (-z\mathbf{j} + y\mathbf{k}) = (1+1)\mathbf{i} + (0-0)\mathbf{j} + (0-0)\mathbf{k} = 2\mathbf{i}$$

It is in the same direction as the axis of rotation.

b. The magnitude of the curl is $|2\mathbf{i}| = 2$

14.5.25

a. The axis of rotation is $\langle 1,\, -1,\, 1\rangle$.

b. $$\nabla \times \mathbf{F} = \nabla \times \begin{vmatrix} \mathbf{i} & \mathbf{j} & \mathbf{k} \\ 1 & -1 & 1 \\ x & y & z \end{vmatrix} = \nabla \times (-(y+z)\mathbf{i} + (x-z)\mathbf{j} + (x+y)\mathbf{k})$$
$$= (1+1)\mathbf{i} + (-1-1)\mathbf{j} + (1+1)\mathbf{k} = 2\langle 1,\, -1,\, 1\rangle,$$

and the curl is in the same direction as the axis of rotation.

c. The magnitude of the curl is $2|\langle 1,\, -1,\, 1\rangle| = 2\sqrt{3}$

14.5.27 $\nabla \times \langle x^2 - y^2,\, xy,\, z\rangle = (0-0)\mathbf{i} + (0-0)\mathbf{j} + (y+2y)\mathbf{k} = 3y\mathbf{k}$

14.5.29 $\nabla \times \langle x^2 - z^2, 1, 2xz \rangle = (0-0)\mathbf{i} + (-2z - 2z)\mathbf{j} + (0-0)\mathbf{k} = -4z\mathbf{j}$

14.5.31 $\nabla \times \frac{1}{(x^2+y^2+z^2)^{3/2}}\langle x, y, z \rangle = \frac{1}{(x^2+y^2+z^2)^{5/2}}((-3yz + 3yz)\mathbf{i} + (-3xz + 3xz)\mathbf{j} + (-3xy + 3xy)\mathbf{k})$
$= \mathbf{0}$

14.5.33 $\nabla \times \langle z^2\sin(y), xz^2\cos(y), 2xz\sin(y) \rangle$
$= (2xz\cos(y) - 2xz\cos(y))\mathbf{i} + (2z\sin(y) - 2z\sin(y))\mathbf{j} + (z^2\cos(y) - z^2\cos(y))\mathbf{k} = \mathbf{0}$

14.5.35 Simply compute it:

$$\left\langle \frac{\partial}{\partial x}\left(\frac{1}{(x^2+y^2+z^2)^{3/2}}\right), \frac{\partial}{\partial y}\left(\frac{1}{(x^2+y^2+z^2)^{3/2}}\right), \frac{\partial}{\partial z}\left(\frac{1}{(x^2+y^2+z^2)^{3/2}}\right)\right\rangle$$
$$= \left\langle \frac{-3x}{(x^2+y^2+z^2)^{5/2}}, \frac{-3y}{(x^2+y^2+z^2)^{5/2}}, \frac{-3z}{(x^2+y^2+z^2)^{5/2}} \right\rangle = \frac{-3\mathbf{r}}{|\mathbf{r}|^5}$$

14.5.37 $\nabla\left(\frac{1}{|\mathbf{r}|^2}\right) = \frac{-2\mathbf{r}}{|\mathbf{r}|^4}$ from Problem 36; applying Theorem 14.8 we get

$$\nabla \cdot \nabla\left(\frac{1}{|\mathbf{r}|^2}\right) = -2\nabla \cdot \frac{\mathbf{r}}{|\mathbf{r}|^4} = -2\frac{3-4}{|\mathbf{r}|^4} = \frac{2}{|\mathbf{r}|^4}$$

14.5.39

a. False. For example, $\mathbf{F} = \langle y, z, x \rangle$ has zero divergence yet is not constant.
b. False. For example, $\mathbf{F} = \langle x, y, z \rangle$ is a counterexample.
c. False. For example, consider the vector field $\langle 0, 1 - x^2 \rangle$ from problem 66 in the previous section.
d. False. For example, $\mathbf{F} = \langle x, 0, 0 \rangle$ has divergence 1.
e. False. For example, the curl of $\langle z, -z, y \rangle$ is $\langle 2, 1, 0 \rangle$.

14.5.41

a. No; divergence is a concept that applies to vector fields.
b. No; the gradient applies to functions.
c. Yes; this is the divergence of the gradient and is thus a scalar function.
d. No, since $\nabla \cdot \varphi$ does not make sense (part (a)).
e. No; curl applies to vector fields, so $\nabla \times \varphi$ does not make sense.
f. No, since $\nabla \cdot \mathbf{F}$ is a function, so that applying $\nabla \cdot$ to it does not make sense.
g. Yes, this is the curl of a vector field and is thus a vector field.
h. No, since $\nabla \cdot \mathbf{F}$ is a function, not a vector field.
i. Yes; this is the curl of the curl of a vector field and is thus a vector field.

14.5.43 Let $\mathbf{a} = \langle a_1, a_2, a_3 \rangle$; then $\mathbf{F} = \mathbf{a} \times \mathbf{r} = (a_2 z - a_3 y)\mathbf{i} + (a_3 x - a_1 z)\mathbf{j} + (a_1 y - a_2 x)\mathbf{k}$, so that

$$\nabla \times \mathbf{F} = \frac{\partial}{\partial x}(a_1 + a_1)\mathbf{i} + \frac{\partial}{\partial y}(a_2 + a_2)\mathbf{j} + \frac{\partial}{\partial z}(a_3 + a_3)\mathbf{k} = 2\mathbf{a}$$

14.5.45 **div F** $= 2x + 2xyz + 2x = 2x(yz + 2)$; this function clearly achieves its maximum magnitude at $(-1, 1, 1)$, $(-1, -1, -1)$, $(1, 1, 1)$, and $(1, -1, -1)$, where its magnitude is 6.

14.5.47 **curl F** $= \langle 0 + 2, 0 + 1, 0 - 1 \rangle = \langle 2, 1, -1 \rangle$. If $\mathbf{n} = \langle a, b, c \rangle$, then **curl F** $\cdot\, \mathbf{n} = 0$ when $2a + b - c = 0$ so that $c = 2a + b$; thus all such vectors are of the form $\langle a, b, 2a + b \rangle$, where a, b are real numbers.

14.5.49 $\mathbf{F} = \frac{1}{2}\langle y^2 + z^2, 0, 0 \rangle$ or $\mathbf{F} = \langle 0, -xy, -xz \rangle$, so it is not unique.

14.5.51 The curl of this vector field is $\langle 0,\,1,\,0\rangle$. The component of the curl along some unit vector $\mathbf{n}$ is $(\nabla \times \mathbf{F})\cdot\mathbf{n}$.

a. $\langle 0,\,1,\,0\rangle\cdot\langle 1,\,0,\,0\rangle = 0$, so the wheel does not spin.

b. $\langle 0,\,1,\,0\rangle\cdot\langle 0,\,1,\,0\rangle = 1$, so the wheel spins clockwise (looking towards positive y).

c. $\langle 0,\,1,\,0\rangle\cdot\langle 0,\,0,\,1\rangle = 0$, so the wheel does not spin

14.5.53 The curl of the vector field is $\nabla\times\mathbf{v} = \langle -20,\,0,\,0\rangle$. Since the wheel is placed with its axis normal to the plane $x+y+z=1$, its axis must point in the direction $\langle 1,\,1,\,1\rangle$ (with unit vector $\frac{1}{\sqrt{3}}\langle 1,\,1,\,1\rangle$). Thus, the component of velocity along that direction is $\dfrac{1}{\sqrt{3}}\langle -20,\,0,\,0\rangle\cdot\langle 1,\,1,\,1\rangle = \dfrac{-20}{\sqrt{3}}$ and then ω is the absolute value of one half of that amount, or $\omega = \frac{10}{\sqrt{3}}$ or $\frac{5}{\pi\sqrt{3}} \approx 0.9189$ revolutions per time unit.

14.5.55 $\mathbf{F} = -100k\nabla e^{-x^2+y^2+z^2} = -200k\,e^{-x^2+y^2+z^2}\langle -x,\,y,\,z\rangle$, so the divergence is

$$\begin{aligned}
-200k&\left(\frac{\partial}{\partial x}\left(-x\,e^{-x^2+y^2+z^2}\right)+\frac{\partial}{\partial y}\left(y\,e^{-x^2+y^2+z^2}\right)+\frac{\partial}{\partial z}\left(z\,e^{-x^2+y^2+z^2}\right)\right)\\
&= -200k\Big(-e^{-x^2+y^2+z^2}+2x^2e^{-x^2+y^2+z^2}+e^{-x^2+y^2+z^2}+2y^2e^{-x^2+y^2+z^2}\\
&\qquad\qquad +e^{-x^2+y^2+z^2}+2z^2e^{-x^2+y^2+z^2}\Big)\\
&= -200k\left(e^{-x^2+y^2+z^2}+2(x^2+y^2+z^2)e^{-x^2+y^2+z^2}\right)\\
&= -200k\,(2x^2+2y^2+2z^2+1)e^{-x^2+y^2+z^2}
\end{aligned}$$

14.5.57

a. $$\begin{aligned}
\mathbf{F} = -\nabla\varphi &= -GMm\left\langle \frac{\partial}{\partial x}\left[\frac{1}{\sqrt{x^2+y^2+z^2}}\right],\,\frac{\partial}{\partial y}\left[\frac{1}{\sqrt{x^2+y^2+z^2}}\right],\,\frac{\partial}{\partial z}\left[\frac{1}{\sqrt{x^2+y^2+z^2}}\right]\right\rangle\\
&= -GMm(x^2+y^2+z^2)^{-3/2}\langle -x,\,-y,\,-z\rangle = GMm(x^2+y^2+z^2)^{-3/2}\langle x,\,y,\,z\rangle\\
&= GMm\frac{\mathbf{r}}{|\mathbf{r}|^3}
\end{aligned}$$

b. $\dfrac{\partial}{\partial y}x(x^2+y^2+z^2)^{-3/2} = -3xy(x^2+y^2+z^2)^{-5/2}$; applying this pattern in computing the curl gives

$$\nabla\times\mathbf{F} = GMm(x^2+y^2+z^2)^{-5/2}((-3yz+3yz)\mathbf{i}+(-3xz+3xz)\mathbf{j}+(-3xy+3xy)\mathbf{k}) = \mathbf{0},$$

so the field is irrotational.

14.5.59 Using Exercise 40, we have

$$\rho\left(\left\langle\frac{\partial u}{\partial t},\,\frac{\partial v}{\partial t},\,\frac{\partial w}{\partial t}\right\rangle+\left(u\frac{\partial}{\partial t}+v\frac{\partial}{\partial t}+w\frac{\partial}{\partial t}\right)\langle u,\,v,\,w\rangle\right) = -\left\langle\frac{\partial p}{\partial t},\,\frac{\partial p}{\partial t},\,\frac{\partial p}{\partial t}\right\rangle+\mu\left(\frac{\partial^2}{\partial t^2}+\frac{\partial^2}{\partial t^2}+\frac{\partial^2}{\partial t^2}\right)\langle u,\,v,\,w\rangle$$

so that

$$\begin{aligned}
\rho\left(\frac{\partial u}{\partial t}+u\frac{\partial u}{\partial x}+v\frac{\partial u}{\partial y}+w\frac{\partial u}{\partial z}\right) &= -\frac{\partial p}{\partial x}+u\left(\frac{\partial^2 u}{\partial x^2}+\frac{\partial^2 u}{\partial y^2}+\frac{\partial^2 u}{\partial z^2}\right)\\
\rho\left(\frac{\partial v}{\partial t}+u\frac{\partial v}{\partial x}+v\frac{\partial v}{\partial y}+w\frac{\partial v}{\partial z}\right) &= -\frac{\partial p}{\partial y}+u\left(\frac{\partial^2 v}{\partial x^2}+\frac{\partial^2 v}{\partial y^2}+\frac{\partial^2 v}{\partial z^2}\right)\\
\rho\left(\frac{\partial w}{\partial t}+u\frac{\partial w}{\partial x}+v\frac{\partial w}{\partial y}+w\frac{\partial w}{\partial z}\right) &= -\frac{\partial p}{\partial z}+u\left(\frac{\partial^2 w}{\partial x^2}+\frac{\partial^2 w}{\partial y^2}+\frac{\partial^2 w}{\partial z^2}\right)
\end{aligned}$$

14.5.61

a. We have $\nabla\times\mathbf{B} = -\dfrac{\partial}{\partial z}(A\sin(kz-\omega t))\mathbf{i}+0\mathbf{j}+\dfrac{\partial}{\partial x}(A\sin(kz-\omega t))\mathbf{k} = -Ak\cos(kz-\omega t)\mathbf{i}$

$$C\frac{\partial\mathbf{E}}{\partial t} = C\frac{\partial}{\partial t}(A\sin(kz-\omega t)\mathbf{i}) = -\omega CA\cos(kz-\omega t)\mathbf{i}$$

so that the two are equal when $k=\omega C$, or $\omega = \frac{k}{C}$.

b.

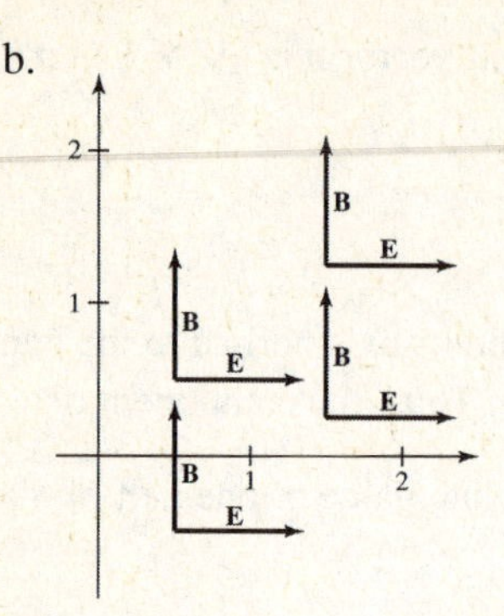

14.5.63 Let $\mathbf{F} = \langle f, g, h\rangle$ and $\mathbf{G} = \langle k, m, n\rangle$. Then

a. $$\nabla\cdot(\mathbf{F}+\mathbf{G}) = \nabla\cdot\langle f+k, g+m, h+n\rangle = \frac{\partial}{\partial x}(f+k)+\frac{\partial}{\partial y}(g+m)+\frac{\partial}{\partial z}(h+n)$$
$$= \frac{\partial f}{\partial x}+\frac{\partial g}{\partial y}+\frac{\partial h}{\partial z}+\frac{\partial k}{\partial x}+\frac{\partial m}{\partial y}+\frac{\partial n}{\partial z} = \nabla\cdot\mathbf{F}+\nabla\cdot\mathbf{G}$$

b. $$\nabla\times(\mathbf{F}+\mathbf{G}) = \left(\frac{\partial}{\partial y}(h+n)-\frac{\partial}{\partial z}(g+m)\right)\mathbf{i}+\left(\frac{\partial}{\partial z}(f+k)-\frac{\partial}{\partial x}(h+n)\right)\mathbf{j}$$
$$+\left(\frac{\partial}{\partial x}(g+m)-\frac{\partial}{\partial y}(f+k)\right)\mathbf{k}$$
$$= \left(\frac{\partial h}{\partial y}-\frac{\partial g}{\partial z}\right)\mathbf{i}+\left(\frac{\partial f}{\partial z}-\frac{\partial h}{\partial x}\right)\mathbf{j}+\left(\frac{\partial g}{\partial x}-\frac{\partial f}{\partial y}\right)\mathbf{k}$$
$$+\left(\frac{\partial n}{\partial y}-\frac{\partial m}{\partial z}\right)\mathbf{i}+\left(\frac{\partial k}{\partial z}-\frac{\partial n}{\partial x}\right)\mathbf{j}+\left(\frac{\partial m}{\partial x}-\frac{\partial k}{\partial y}\right)\mathbf{k}$$
$$= \nabla\times\mathbf{F}+\nabla\times\mathbf{G}$$

c. $$\nabla\cdot(c\mathbf{F}) = \frac{\partial}{\partial x}(cf)+\frac{\partial}{\partial y}(cg)+\frac{\partial}{\partial z}(ch) = c\left(\frac{\partial f}{\partial x}+\frac{\partial g}{\partial y}+\frac{\partial h}{\partial z}\right) = c(\nabla\cdot\mathbf{F})$$

d. $$\nabla\times(c\mathbf{F}) = \left(\frac{\partial}{\partial y}(ch)-\frac{\partial}{\partial z}(cg)\right)\mathbf{i}+\left(\frac{\partial}{\partial z}(cf)-\frac{\partial}{\partial x}(ch)\right)\mathbf{j}+\left(\frac{\partial}{\partial x}(cg)-\frac{\partial}{\partial y}(cf)\right)\mathbf{k}$$
$$= c\left[\left(\frac{\partial h}{\partial y}-\frac{\partial g}{\partial z}\right)\mathbf{i}+\left(\frac{\partial f}{\partial z}-\frac{\partial h}{\partial x}\right)\mathbf{j}+\left(\frac{\partial g}{\partial x}-\frac{\partial f}{\partial y}\right)\mathbf{k}\right] = c(\nabla\times\mathbf{F})$$

14.5.65 $$\nabla\cdot(\varphi\mathbf{F}) = \nabla\cdot\langle\varphi f, \varphi g, \varphi h\rangle = \frac{\partial}{\partial x}(\varphi f)+\frac{\partial}{\partial y}(\varphi g)+\frac{\partial}{\partial z}(\varphi h)$$
$$= \varphi\frac{\partial f}{\partial x}+f\frac{\partial\varphi}{\partial x}+\varphi\frac{\partial g}{\partial y}+g\frac{\partial\varphi}{\partial y}+\varphi\frac{\partial h}{\partial z}+h\frac{\partial\varphi}{\partial z}$$
$$= \varphi\left(\frac{\partial f}{\partial x}+\frac{\partial g}{\partial y}+\frac{\partial h}{\partial z}\right)+\langle f, g, h\rangle\left\langle\frac{\partial\varphi}{\partial x}, \frac{\partial\varphi}{\partial y}, \frac{\partial\varphi}{\partial z}\right\rangle$$
$$= \varphi\nabla\cdot\mathbf{F}+\nabla\varphi\cdot\mathbf{F}$$

14.5.67 If $\mathbf{F} = \langle f, g, h\rangle$ and $\mathbf{G} = \langle k, m, n\rangle$, then
$$\mathbf{G}\cdot(\nabla\times\mathbf{F})-\mathbf{F}\cdot(\nabla\times\mathbf{G}) = \langle k, m, n\rangle\cdot\left\langle\frac{\partial h}{\partial y}-\frac{\partial g}{\partial z}, \frac{\partial f}{\partial z}-\frac{\partial h}{\partial x}, \frac{\partial g}{\partial x}-\frac{\partial f}{\partial y}\right\rangle$$
$$-\langle f, g, h\rangle\cdot\left\langle\frac{\partial n}{\partial y}-\frac{\partial m}{\partial z}, \frac{\partial k}{\partial z}-\frac{\partial n}{\partial x}, \frac{\partial m}{\partial x}-\frac{\partial k}{\partial y}\right\rangle$$
$$= k\frac{\partial h}{\partial y}-k\frac{\partial g}{\partial z}+m\frac{\partial f}{\partial z}-m\frac{\partial h}{\partial x}+n\frac{\partial g}{\partial x}-n\frac{\partial f}{\partial y}$$
$$-f\frac{\partial n}{\partial y}+f\frac{\partial m}{\partial z}-g\frac{\partial k}{\partial z}+g\frac{\partial n}{\partial x}-h\frac{\partial m}{\partial x}+h\frac{\partial k}{\partial y}$$

$$= n\frac{\partial g}{\partial x} + g\frac{\partial n}{\partial x} - h\frac{\partial m}{\partial x} - m\frac{\partial h}{\partial x} + h\frac{\partial k}{\partial y} + k\frac{\partial h}{\partial y} - f\frac{\partial n}{\partial y} - n\frac{\partial f}{\partial y}$$
$$+ f\frac{\partial m}{\partial z} + m\frac{\partial f}{\partial z} - g\frac{\partial k}{\partial z} - k\frac{\partial g}{\partial z}$$
$$= \frac{\partial}{\partial x}(gn - hm) + \frac{\partial}{\partial y}(hk - fn) + \frac{\partial}{\partial z}(fm - gk) = \boldsymbol{\nabla} \cdot (\mathbf{F} \times \mathbf{G})$$

14.5.69 Use the values of $(\mathbf{G} \cdot \boldsymbol{\nabla})\mathbf{F}$ and $(\mathbf{F} \cdot \boldsymbol{\nabla})\mathbf{G}$ from the previous problem. Then

$$\mathbf{G} \times (\boldsymbol{\nabla} \times \mathbf{F}) = \langle k,\, m,\, n\rangle \times \left\langle \frac{\partial h}{\partial y} - \frac{\partial g}{\partial z}, \frac{\partial f}{\partial z} - \frac{\partial h}{\partial x}, \frac{\partial g}{\partial x} - \frac{\partial f}{\partial y} \right\rangle$$
$$= \left\langle m\frac{\partial g}{\partial x} - m\frac{\partial f}{\partial y} - n\frac{\partial f}{\partial z} + n\frac{\partial h}{\partial x}, n\frac{\partial h}{\partial y} - n\frac{\partial g}{\partial z} - k\frac{\partial g}{\partial x} + k\frac{\partial f}{\partial y}, k\frac{\partial f}{\partial z} - k\frac{\partial h}{\partial x} - m\frac{\partial h}{\partial y} + m\frac{\partial g}{\partial z} \right\rangle$$

and similarly for $\mathbf{F} \times (\boldsymbol{\nabla} \times \mathbf{G})$. Thus

$$(\mathbf{G} \cdot \boldsymbol{\nabla})\mathbf{F} + (\mathbf{F} \cdot \boldsymbol{\nabla})\mathbf{G} + \mathbf{G} \times (\boldsymbol{\nabla} \times \mathbf{F}) + \mathbf{F} \times (\boldsymbol{\nabla} \times \mathbf{G})$$
$$= \left\langle k\frac{\partial f}{\partial x} + m\frac{\partial f}{\partial y} + n\frac{\partial f}{\partial z}, k\frac{\partial g}{\partial x} + m\frac{\partial g}{\partial y} + n\frac{\partial g}{\partial z}, k\frac{\partial h}{\partial x} + m\frac{\partial h}{\partial y} + n\frac{\partial h}{\partial z} \right\rangle$$
$$+ \left\langle f\frac{\partial k}{\partial x} + g\frac{\partial k}{\partial y} + h\frac{\partial k}{\partial z}, f\frac{\partial m}{\partial x} + g\frac{\partial m}{\partial y} + h\frac{\partial m}{\partial z}, f\frac{\partial n}{\partial x} + g\frac{\partial n}{\partial y} + h\frac{\partial n}{\partial z} \right\rangle$$
$$+ \left\langle m\frac{\partial g}{\partial x} - m\frac{\partial f}{\partial y} - n\frac{\partial f}{\partial z} + n\frac{\partial h}{\partial x}, n\frac{\partial h}{\partial y} - n\frac{\partial g}{\partial z} - k\frac{\partial g}{\partial x} + k\frac{\partial f}{\partial y}, k\frac{\partial f}{\partial z} - k\frac{\partial h}{\partial x} - m\frac{\partial h}{\partial y} + m\frac{\partial g}{\partial z} \right\rangle$$
$$+ \left\langle g\frac{\partial m}{\partial x} - g\frac{\partial k}{\partial y} - h\frac{\partial k}{\partial z} + h\frac{\partial n}{\partial x}, h\frac{\partial n}{\partial y} - h\frac{\partial m}{\partial z} - f\frac{\partial m}{\partial x} + f\frac{\partial k}{\partial y}, f\frac{\partial k}{\partial z} - f\frac{\partial n}{\partial x} - g\frac{\partial n}{\partial y} + g\frac{\partial m}{\partial z} \right\rangle$$
$$= \left\langle \frac{\partial}{\partial x}(fk + gm + hn), \frac{\partial}{\partial y}(fk + gm + hn), \frac{\partial}{\partial z}(fk + gm + hn) \right\rangle = \nabla(\mathbf{F} \cdot \mathbf{G})$$

14.5.71 $\nabla \cdot \dfrac{\langle x,\, y,\, z\rangle}{(x^2+y^2+z^2)^{p/2}}$

$$= \frac{(1-p)x^2 + y^2 + z^2}{(x^2+y^2+z^2)^{1+p/2}} + \frac{x^2 + (1-p)y^2 + z^2}{(x^2+y^2+z^2)^{1+p/2}} + \frac{x^2 + y^2 + (1-p)z^2}{(x^2+y^2+z^2)^{1+p/2}}$$
$$= \frac{(3-p)(x^2+y^2+z^2)}{(x^2+y^2+z^2)^{1+p/2}} = \frac{3-p}{(x^2+y^2+z^2)^{p/2}} = \frac{3-p}{|\mathbf{r}|^p}$$

14.5.73 $\boldsymbol{\nabla} \cdot \boldsymbol{\nabla}\left(\dfrac{1}{|r|^p}\right) = \boldsymbol{\nabla} \cdot \left(-\dfrac{p\mathbf{r}}{|\mathbf{r}|^{p+2}}\right)$ by Exercise 72, and then by Exercise 71,

$$\boldsymbol{\nabla} \cdot \left(-\frac{p\mathbf{r}}{|\mathbf{r}|^{p+2}}\right) = -p\boldsymbol{\nabla} \cdot \frac{\mathbf{r}}{|\mathbf{r}|^{p+2}} = \frac{-p(3-(p+2))}{|\mathbf{r}|^{p+2}} = \frac{p(p-1)}{|\mathbf{r}|^{p+2}}$$

14.6 Surface Integrals

14.6.1 $\mathbf{r}(u,\, v) = \langle a\cos u,\, a\sin u,\, v\rangle$ where $0 \le u \le 2\pi$; $0 \le v \le h$.

14.6.3 $\mathbf{r}(u,\, v) = \langle a\sin u\cos v,\, a\sin u\sin v,\, a\cos u\rangle$ where $0 \le u \le \pi$; $0 \le v \le 2\pi$.

14.6.5 Use the parametric description from problem 3 and compute

$$\int_0^{\pi}\int_0^{2\pi} a^2 f(a\sin u\cos v,\, a\sin u\sin v,\, a\cos u)\sin u\, du\, dv$$

14.6.7 Using the parameterization from the text, and the fact that for the sphere (see Example 2(b)), $\mathbf{t}_u \times \mathbf{t}_v = \left\langle a^2\sin^2 u\cos v,\ a^2\sin^2 u \sin v,\ a^2 \sin u \cos u\right\rangle$, compute

$$\iint_S \mathbf{F}\cdot(\mathbf{t}_u\times\mathbf{t}_v)\,dS = \iint_R a^2\sin u\,(f\sin u\cos v + g\sin u\sin v + h\cos u)\,dA$$
$$= \int_0^{\pi}\int_0^{2\pi} a^2\sin u\,(f\sin u\cos v + g\sin u\sin v + h\cos u)\,dv\,du$$

14.6.9 The usual orientation of a closed surface is that the normal vectors point outwards.

14.6.11 $\left\langle u,\ v,\ \frac{1}{3}(16-2u+4v)\right\rangle$, $|u|<\infty$, $|v|<\infty$

14.6.13 $\left\langle v\cos u,\ v\sin u,\ v\right\rangle$, $0\le u\le 2\pi$, $2\le v\le 8$.

14.6.15 $\left\langle 3\cos u,\ 3\sin u,\ v\right\rangle$, $0\le u\le \frac{\pi}{2}$, $0\le v\le 3$.

14.6.17 The segment of the plane $z=2x+3y-1$ above $[1,\,3]\times[2,\,4]$.

14.6.19 The portion of the cone $z^2=16x^2+16y^2$ of height 12 and radius 3, where $y\ge 0$.

14.6.21 Using the standard parametric description of the cylinder, we have $\mathbf{r}(u,\,v)=\left\langle 4\cos u,\ 4\sin u,\ v\right\rangle$ for $0\le v\le 7$, $0\le u\le\pi$. Then $|\mathbf{t}_u\times\mathbf{t}_v|=4$ and the area is

$$\iint_S 1\,dS = \iint_R 4\,dA = \int_0^{\pi}\int_0^{7} 4\,dv\,du = 28\pi$$

14.6.23 The plane has parametric description $\mathbf{r}(u,\,v)=\left\langle u,\ v,\ 10-u-v\right\rangle$, for $-2\le u\le 2$, $-2\le v\le 2$. Then

$$\mathbf{t}_u\times\mathbf{t}_v=\left\langle 1,\,0,\,-1\right\rangle\times\left\langle 0,\,1,\,-1\right\rangle=\left\langle 1,\,1,\,1\right\rangle$$

so that

$$\iint_S 1\,dS=\sqrt{3}\iint_R 1\,dA = 16\sqrt{3}$$

14.6.25 Parameterize the cone by $\mathbf{r}(u,\,v)=\left\langle \frac{r}{h}v\cos u,\ \frac{r}{h}v\sin u,\ v\right\rangle$, for $0\le v\le h$, $0\le u\le 2\pi$; then

$$\mathbf{t}_u\times\mathbf{t}_v=\left\langle -\tfrac{r}{h}v\sin u,\ \tfrac{r}{h}v\cos u,\ 0\right\rangle\times\left\langle \tfrac{r}{h}\cos u,\ \tfrac{r}{h}\sin u,\ 1\right\rangle$$

and $|\mathbf{t}_u\times\mathbf{t}_v|=\dfrac{r}{h^2}v\sqrt{h^2+r^2}$. Then

$$\iint_S 1\,dS=\frac{r\sqrt{h^2+r^2}}{h^2}\iint_R v\,dA=\frac{r\sqrt{h^2+r^2}}{h^2}\int_0^{2\pi}\int_0^{h} v\,dv\,du$$
$$=\frac{r\sqrt{h^2+r^2}}{h^2}\int_0^{2\pi}\frac{1}{2}h^2\,du=\frac{2\pi r\sqrt{h^2+r^2}}{2}=\pi r\sqrt{h^2+r^2}$$

14.6.27 Using the standard parameterization of the sphere for $0\le u\le\frac{\pi}{2}$, $0\le v\le 2\pi$, we get

$$\iint_S (x^2+y^2)\,dS=\int_0^{2\pi}\int_0^{\pi/2} 36\sin^2 u\cdot 36\sin u\,du\,dv = 1296\int_0^{2\pi}\int_0^{\pi/2}\sin^3 u\,du\,dv = 1296\int_0^{2\pi}\frac{2}{3}\,dv = 1728\pi$$

14.6.29 Use the standard parameterization, for $0\le u\le 2\pi$, $0\le v\le 3$; then

$$\iint_S x\,dS=\int_0^{2\pi}\int_0^{3}\cos u\cdot 1\,dv\,du = 3\int_0^{2\pi}\cos u\,du = 0$$

14.6.31 $2z\,dz = 8x\,dx$, so $z_x = \dfrac{4x}{z}$; similarly, $z_y = \dfrac{4y}{z}$. Thus

$$\sqrt{z_x^2 + z_y^2 + 1} = \sqrt{\frac{16x^2 + 16y^2 + z^2}{z^2}} = \sqrt{\frac{20(x^2+y^2)}{4(x^2+y^2)}} = \sqrt{5}$$

Further, this cone sits over $x^2 + y^2 = 4$. Then

$$\iint_S 1\,dS = \iint_R \sqrt{5}\,dA = 4\pi\sqrt{5}$$

14.6.33 $z_x = 2x$ and $z_y = 0$, so that $\displaystyle\iint_S 1\,dS = \int_0^4 \int_{-2}^2 \sqrt{4x^2+1}\,dx\,dy = 88\sqrt{17} + 2\ln\left(\sqrt{17}+4\right)$

14.6.35 $z_x = z_y = -1$, so $\displaystyle\iint_S xy\,dS = \sqrt{3}\iint_R xy\,dA = \sqrt{3}\int_0^2 \int_0^{2-x} xy\,dy\,dx = \frac{2\sqrt{3}}{3}$

14.6.37 $x^2 + y^2 + z^2 = 25$, so $z_x = --\dfrac{x}{z}$ and $z_y = -\dfrac{y}{z}$. Then

$$\iint_S (25 - x^2 - y^2)\,dS = \iint_R (25 - x^2 - y^2)\sqrt{\frac{x^2+y^2+z^2}{z^2}}\,dA = 5\iint_R \sqrt{25 - x^2 - y^2}\,dA$$

$$= 5\int_0^{2\pi}\int_0^5 r\sqrt{25 - r^2}\,dr\,d\theta = \frac{1250\pi}{3}$$

14.6.39 $z_x = -3$, $z_y = -4$, so $\sqrt{z_x^2 + z_y^2 + 1} = \sqrt{26}$. The area of the part of the plane is

$$\iint_S 1\,dS = \sqrt{26}\iint_R 1\,dA = 4\sqrt{26}$$

and the surface integral of temperature is

$$\iint_S e^{3x+4y-6}\,dS = \sqrt{26}\int_{-1}^1\int_{-1}^1 e^{3x+4y-6}\,dx\,dy = \frac{\sqrt{26}}{12}\left(e - e^{-5} - e^{-7} + e^{-13}\right)$$

so that the average temperature is the ratio of the two, or $\dfrac{1}{48}\left(e - e^{-5} - e^{-7} + e^{-13}\right)$

14.6.41 $z_x = -\dfrac{x}{z}$ and $z_y = -\dfrac{y}{z}$, so that $\sqrt{z_x^2 + z_y^2 + 1} = \sqrt{\dfrac{x^2+y^2+z^2}{z^2}} = \left(1 - x^2 - y^2\right)^{-1/2}$. The area of the sphere is $\frac{1}{8}\cdot 4\pi = \frac{\pi}{2}$, and the integral of the function is

$$\iint_S xyz\,dS = \iint_R xy\left(1 - x^2 - y^2\right)^{1/2}\left(1 - x^2 - y^2\right)^{-1/2}dA = \int_0^{\pi/2}\int_0^1 r^3\sin(\theta)\cos(\theta)\,dr\,d\theta = \frac{1}{8}$$

so that the average value is $\frac{1}{4\pi}$.

14.6.43 $z_x = z_y = -1$, so the normal vector is $\langle 1, 1, 1\rangle$, which points in the positive z-direction. Then

$$\iint_S \mathbf{F}\cdot\mathbf{n}\,dS = \iint_R (0\cdot 1 + 0\cdot 1 - 1\cdot 1)\,dA = \int_0^4\int_0^{4-x} (-1)\,dy\,dx = -8$$

14.6.45 We have $z = \sqrt{x^2+y^2}$; then $\mathbf{n} = \left\langle -\dfrac{x}{z}, -\dfrac{y}{z}, 1\right\rangle$, which points upwards.

$$\iint_S \mathbf{F}\cdot\mathbf{n}\,dS = \iint_R \left(\left(-\frac{x}{z}\right)\cdot x + \left(-\frac{y}{z}\right)\cdot y + 1\cdot z\right)dA = \iint_R \left(z - \frac{x^2+y^2}{z}\right)dA = \iint_R (z - z)\,dA = 0$$

14.6.47 An outward-pointing normal is $\dfrac{\mathbf{r}}{|\mathbf{r}|}$. The sphere has radius a, so the vector field is in fact $\dfrac{\mathbf{r}}{|\mathbf{r}|^3}$.

$$\iint_S \mathbf{F}\cdot\mathbf{n}\,dS = \iint_S \frac{\mathbf{r}}{|\mathbf{r}|^3}\cdot\frac{\mathbf{r}}{|\mathbf{r}|}\,dS = \iint_S \frac{1}{|\mathbf{r}|^2}dS = \iint_S \frac{1}{a^2}dS = \frac{1}{a^2}\iint_S 1\,dS = \frac{1}{a^2}4\pi a^2 = 4\pi$$

14.6.49

a. True; the formula in Theorem 14.12 gives $\displaystyle\iint_S f(x,\,y,\,z)\,dS = \iint_R f(x,\,y,\,10)\sqrt{0+0+1}\,dA$.

b. False; the formula in Theorem 14.12 gives

$$\iint_S f(x,\,y,\,z)\,dS = \iint_R f(x,\,y,\,x)\sqrt{1+0+1}\,dA = \sqrt{2}\iint_R f(x,\,y,\,x)\,dA.$$

c. True. Substituting $2u$ for u and $\sqrt{v}$ for v in the first parameterization gives $\langle\sqrt{v}\cos 2u,\ \sqrt{v}\sin 2u,\ v\rangle$, $0\le 2u\le\pi$, $0\le\sqrt{v}\le 2$. Simplifying the bounds conditions gives the second parameterization.

d. True. The standard parameterization is $\langle a\sin u\cos v,\ a\sin u\sin v,\ a\cos u\rangle$ for $0\le u\le\pi$, $0\le v\le 2\pi$. Then $\mathbf{t}_u\times\mathbf{t}_v = \langle a^2\sin^2 u\cos v,\ a^2\sin^2 u\sin v,\ a^2\cos u\sin u\rangle$, and it is easily seen that these are outward-pointing vectors, by considering various ranges for u and v.

14.6.51 Parameterize the surface by $\langle 2\cos u,\ 2\sin u,\ v\rangle$ for $0\le u\le 2\pi$, $0\le v\le 8$. Then the normal vector has magnitude 2, and

$$\iint_S |\mathbf{r}|\,dS = \iint_R \sqrt{x^2+y^2+z^2}\,dA = 2\int_0^{2\pi}\int_0^8 \sqrt{4+v^2}\,dv\,du = 8\pi\left(4\sqrt{17}+\ln\left(4+\sqrt{17}\right)\right)$$

14.6.53 The normal vector is $\langle x,\ 0,\ z\rangle$, so

$$\iint_S \frac{1}{\sqrt{x^2+z^2}}\langle x,\,0,\,z\rangle\cdot\langle x,\,0,\,z\rangle\,dS = \iint_R \sqrt{x^2+z^2}\,dA = \int_{-2}^{2}\int_0^{2\pi} a\,dA = 8\pi a$$

14.6.55

a. The surface of the cylinder inside the sphere is defined parametrically by $\langle 1+\cos u,\ \sin u,\ v\rangle$ where $0\le u\le 2\pi$ and (since the z-coordinate must stay inside the sphere of radius 2), $0\le v\le\sqrt{4-(1+\sin u)^2-\cos^2 u}$, or $0\le v\le\sqrt{2-2\sin u}$. The normal is $\langle\cos u,\ \sin u,\ 0\rangle$, which has magnitude 1, so we have

$$\iint_S 1\,dS = \iint_R 1\,dA = \int_0^{2\pi}\int_0^{\sqrt{2-2\sin u}} 1\,dv\,du = \int_0^{2\pi}\sqrt{2-2\sin u}\;du = 8$$

b. To find a parameterization of the portion of the sphere cut by the cylinder above the z-axis, first note that it is sufficient to do this for the portion of the sphere in the first octant, and then double the result. Now, the first octant is determined by $0\le u\le\frac{\pi}{2}$, $0\le v\le\frac{\pi}{2}$ in the standard parameterization $\langle 2\sin u\cos v,\ 2\sin u\sin v,\ 2\cos u\rangle$. For each point on the boundary of the intersection, we must have $(x-1)^2+y^2=1$ or, substituting from the parameterization,

$$1=(2\sin u\cos v-1)^2+(2\sin u\sin v)^2 = 4\sin^2 u\cos^2 v+4\sin^2 u\sin^2 v-4\sin u\cos v+1$$

so that $\sin^2 u = \sin u\cos v$, and we must have $v=\arccos(\sin(u))=\frac{\pi}{2}-u$. Thus, the surface is determined by $0\le u\le\frac{\pi}{2}$, $0\le v\le\frac{\pi}{2}-u$, and the surface area is

$$\iint_S 1\,dS = \iint_R |\mathbf{t}_u\times\mathbf{t}_v|\,dA = \int_0^{\pi/2}\int_0^{\pi/2-u} 4\sin u\,dv\,du = 2\pi-4$$

Doubling this to account for the other quadrant, we get $4\pi-8$.

14.6.57

a. Using the standard parameterization,

$$\iint_S \mathbf{F}\cdot\mathbf{n}\,dS = \iint_R \left(\frac{-x^2}{z}+\frac{-y^2}{z}+z\right)dA = \iint_R 0\,dA = 0$$

since $x^2+y^2=z^2$, so that the flux is zero.

b. $2z\,dz = \left(\frac{2x}{a^2}\right)dx$ so that $z_x = \frac{x}{a^2z}$. Then

$$\iint_S \mathbf{F}\cdot\mathbf{n}\,dS = \iint_R \left(\frac{-x^2}{a^2z}+\frac{-y^2}{a^2z}+z\right)dA = \iint_R \left(\frac{-a^2z^2}{a^2z}+z\right)dA = 0$$

so the flux is again zero. This is because the flow is a radial flow, so is always tangent to this surface.

14.6.59 Since the cap has height h, the circle at the boundary of the cap has radius $\sqrt{a^2-(a-h)^2}=\sqrt{2ah-h^2}$, so that the equation of the base of the region is $x^2+y^2=2ah-h^2$. The outward normals to the sphere are $\left\langle \frac{x}{z}, \frac{y}{z}, 1\right\rangle$. Thus

$$\iint_S 1\,dS = \iint_R \sqrt{\frac{x^2}{z^2}+\frac{y^2}{z^2}+1}\,dA = \iint_R \sqrt{\frac{a^2-z^2}{z^2}+1}\,dA = \iint_R \frac{a}{z}\,dA = \iint_R \frac{a}{\sqrt{a^2-x^2-y^2}}\,dA$$

$$= \int_0^{2\pi}\int_0^{\sqrt{2ah-h^2}} \frac{ar}{\sqrt{a^2-r^2}}\,dr\,d\theta = 2\pi ah$$

(See also problem 54(b)).

14.6.61 $\mathbf{F} = -\nabla T = -\langle T_x, T_y, T_z\rangle = \langle 100e^{-x-y}, 100e^{-x-y}, 0\rangle$. Thus the flow is parallel to the two sides where $z=\pm 1$ so that the flux is zero there. We thus need only compute the flux on the remaining four sides. Parameterize the sides as

$$S_1 : \langle -1, y, z\rangle \quad \mathbf{t}_y\times\mathbf{t}_z = \langle 1, 0, 0\rangle$$
$$S_2 : \langle 1, y, z\rangle \quad \mathbf{t}_y\times\mathbf{t}_z = \langle 1, 0, 0\rangle$$
$$S_3 : \langle x, -1, z\rangle \quad \mathbf{t}_x\times\mathbf{t}_z = \langle 0, -1, 0\rangle$$
$$S_4 : \langle x, 1, z\rangle \quad \mathbf{t}_x\times\mathbf{t}_z = \langle 0, -1, 0\rangle$$

for $-1\le x, y, z\le 1$. We are looking for the outward flux, so we must choose outward normals, which are (respectively) $\langle -1, 0, 0\rangle$, $\langle 1, 0, 0\rangle$, $\langle 0, -1, 0\rangle$, and $\langle 0, 1, 0\rangle$. Then

$$\iint_{S_1} \mathbf{F}\cdot\mathbf{n}\,dS_1 = \iint_R -100e^{-x-y}\,dA = -100\int_{-1}^{1}\int_{-1}^{1} e^{1-y}\,dz\,dy = -200e^2+200$$

$$\iint_{S_2} \mathbf{F}\cdot\mathbf{n}\,dS_2 = \iint_R -100e^{-x-y}\,dA = 100\int_{-1}^{1}\int_{-1}^{1} e^{-1-y}\,dz\,dy = -200e^{-2}+200$$

$$\iint_{S_3} \mathbf{F}\cdot\mathbf{n}\,dS_3 = \iint_R -100e^{-x-y}\,dA = -100\int_{-1}^{1}\int_{-1}^{1} e^{-x+1}dz\,dx = -200e^2+200$$

$$\iint_{S_4} \mathbf{F}\cdot\mathbf{n}\,dS_4 = \iint_R -100e^{-x-y}\,dA = 100\int_{-1}^{1}\int_{-1}^{1} e^{-x-1}dz\,dx = -200e^{-2}+200$$

so that the total flux is $-400(e^2+e^{-2}-2) = -400\left(e-\frac{1}{e}\right)^2$

14.6.63 $\mathbf{F} = -\nabla T = \dfrac{2}{x^2+y^2+z^2}\langle x,\, y,\, z\rangle$. Thus integrating on the top half of the sphere gives

$$\iint_S \mathbf{F}\cdot\mathbf{n}\,dS = \iint_R \left(\frac{x^2}{z}+\frac{y^2}{z}+z\right)dA = 2\iint_R \frac{1}{z}\,dA = 2\int_0^{2\pi}\int_0^a \frac{r}{\sqrt{a^2-r^2}}\,dr\,d\theta = 4\pi a$$

and since the vector field is symmetric, the answer is $2\cdot 4\pi a = 8\pi a$.

14.6.65

a. From problem 60 (which see for the details) the outward flux across a sphere of radius b is $\dfrac{4\pi}{b^{p-3}}$, so the total flux across the concentric spheres when $p=0$ is $4\pi b^3 - 4\pi a^3 = 4\pi(b^3-a^3)$.

b. For $p=3$, the flux across the sphere of radius b is 4π, so the net flux is zero across S.

14.6.67 The cone is rotationally symmetric around the z axis, so $\overline{x}=\overline{y}=0$. Parameterize the cone by $\left\langle \frac{r}{h}v\cos u,\, \frac{r}{h}v\sin u,\, v\right\rangle$. Then from problem 25, the surface area of the cone is $\pi r\sqrt{h^2+r^2}$, so its mass is $\rho\pi r\sqrt{h^2+r^2}$. Using the parameterization from that problem, $|\mathbf{t}_u\times\mathbf{t}_v| = \frac{r}{h^2}v\sqrt{h^2+r^2}$, so that

$$\begin{aligned} M_{xy} &= \rho\iint_S z\,dS = \rho\frac{r\sqrt{h^2+r^2}}{h^2}\iint_R vz\,dA = \rho\frac{r\sqrt{h^2+r^2}}{h^2}\int_0^{2\pi}\int_0^h v^2\,dv\,du \\ &= \rho\frac{r\sqrt{h^2+r^2}}{h^2}\int_0^{2\pi}\frac{1}{3}h^3\,du = \rho\frac{2\pi rh\sqrt{h^2+r^2}}{3} \end{aligned}$$

so that

$$\overline{z} = \frac{M_{xy}}{m} = \rho\frac{2\pi rh\sqrt{h^2+r^2}}{3}\cdot\frac{1}{\rho\pi r\sqrt{h^2+r^2}} = \frac{2h}{3}$$

14.6.69 Using the standard parameterization, the mass of the shell is

$$m = \iint_S (1+z)\,dS = a\iint_R (1+z)\,dA = a\int_0^{2\pi}\int_0^2 (1+v)\,dv\,du = 8\pi a$$

The density does not depend on either x or y, and the cylinder is symmetric about the z axis, so $\overline{x}=\overline{y}=0$. Then

$$\mathrm{M}_{xy} = \iint_S z(1+z)\,dS = a\iint_R z(1+z)\,dA = a\int_0^{2\pi}\int_0^2 v(1+v)\,dv\,du = \frac{28\pi a}{3}$$

and then $\overline{z} = \dfrac{7}{6}$.

14.6.71 The explicit formula $z=g(x,y)$ becomes, on regarding x and y as parameters, the parametric form $\langle x,\, y,\, g(x,y)\rangle$, and now $\mathbf{t}_x = \langle 1, 0,\, z_x\rangle$ and $\mathbf{t}_y = \langle 0,\, 1,\, z_y\rangle$. Then $\mathbf{t}_x\times\mathbf{t}_y = \langle -z_x,\, -z_y,\, 1\rangle$, so that $|\mathbf{t}_x\times\mathbf{t}_y| = \sqrt{z_x^2+z_y^2+1}$. Now the formula

$$\iint_S f(x,\,y,\,z)\,dS = \iint_R f(x,\,y,\,g(x,\,y))\sqrt{z_x^2+z_y^2+1}\,dA$$

follows from the definition of the surface integral for parameterized surfaces.

14.6.73 We have $z=s(x,y)$, so a normal vector is $\langle -z_x,\, -z_y,\, 1\rangle$. Since we are interested in the downward flux, we choose a downward-pointing normal, which is $\langle s_x(x,y),\, s_y(x,y),\, -1\rangle$. Then

$$\iint_S \mathbf{F}\cdot\mathbf{n}\,dS = \iint_R \langle 0,\,0,\,-1\rangle\cdot\langle s_x(x,y),\, s_y(x,y),\, -1\rangle\,dA = \iint_R 1\,dA$$

which is the area of R. Since the vector field is constant and pointed downwards vertically, everything that goes through the surface is matched by something going through R.

14.7 Stokes' Theorem

14.7.1 It measures the circulation of the vector field **F** along the closed curve C.

14.7.3 It says that the circulation of a vector field along a closed curve is equal to the net circulation of the field over a surface whose boundary is that curve, so that either can be calculated from the other.

14.7.5 The line integral is

$$\oint_C \mathbf{F}\cdot d\mathbf{r} = \iint_R \langle \sin t,\ -\cos t,\ 10\rangle \cdot \langle -\sin t,\ \cos t,\ 0\rangle\, dA = \int_0^{2\pi} (-1)\, dt = -2\pi$$

For the surface integral, use the standard parameterization of the sphere; then $\mathbf{n} = \langle \sin^2 u \cos v,\ \sin^2 u \sin v,\ \cos u \sin u\rangle$ and $\nabla \times \mathbf{F} = \langle 0,\ 0,\ -2\rangle$ so that

$$\iint_S (\nabla \times \mathbf{F})\cdot \mathbf{n}\, dS = \int_0^{2\pi}\int_0^{\pi/2} (-2\cos u \sin u)\, du\, dv = -2\pi$$

14.7.7 The line integral is

$$\oint_C \mathbf{F}\cdot d\mathbf{r} = \int_0^{2\pi} \langle 2\sqrt{2}\cos t,\ 2\sqrt{2}\sin t,\ 0\rangle \cdot \langle -2\sqrt{2}\sin t,\ 2\sqrt{2}\cos t,\ 0\rangle\, dt = \int_0^{2\pi} 0\, dt = 0$$

For the surface integral, we have $\nabla \times \mathbf{F}$ so that the surface integral is also zero.

14.7.9 The boundary of the region is the intersection of the sphere with the plane $z = \sqrt{7}$, which has the equation $x^2 + y^2 = 9$ and $z = \sqrt{7}$. Then the line integral is

$$\begin{aligned}\oint_C \mathbf{F}\cdot d\mathbf{r} &= \int_0^{2\pi} \langle 3\sin t - \sqrt{7},\ \sqrt{7} - 3\cos t,\ 3\cos t - 3\sin t\rangle \cdot \langle -3\sin t,\ 3\cos t,\ 0\rangle\, dt \\ &= \int_0^{2\pi} \left(-9\sin^2 t + 3\sqrt{7}\sin t + 3\sqrt{7}\cos t - 9\cos^2 t\right) dt \\ &= \int_0^{2\pi} \left(-9 + 3\sqrt{7}\sin t + 3\sqrt{7}\cos t\right) dt = -18\pi\end{aligned}$$

The surface sits over $x^2 + y^2 = 9$ and the normal to the sphere is $\left\langle \frac{x}{z},\ \frac{y}{z},\ 1\right\rangle$. $\nabla \times \mathbf{F} = \langle -2,\ -2,\ -2\rangle$, so the surface integral is

$$\iint_S (\nabla \times \mathbf{F})\cdot \mathbf{n}\, dS = -2\int_0^{2\pi}\int_0^3 r\left(\frac{r\cos\theta}{\sqrt{16 - r^2}} + \frac{r\sin\theta}{\sqrt{16 - r^2}} + 1\right) dr\, d\theta = -18\pi$$

14.7.11 $\nabla \times \mathbf{F} = \langle 1,\ -1,\ -2\rangle$; for S take the disk $x^2 + y^2 \le 12$ with upward-oriented normal vector $\langle 0,\ 0,\ 1\rangle$. Then

$$\iint_S (\nabla \times \mathbf{F})\cdot \mathbf{n}\, dS = \iint_R (-2)\, dA = -24\pi$$

14.7.13 $\nabla \times \mathbf{F} = \langle 0,\ -4z,\ 0\rangle$. For S take the plane in the first octant, which sits over $0 \le x \le 4, 0 \le y \le 4 - x$. The upward-pointing normal to this plane is $\langle 1,\ 1,\ 1\rangle$. Then

$$\iint_S (\nabla \times \mathbf{F})\cdot \mathbf{n}\, dS = \iint_R (-4z)\, dA = \int_0^4\int_0^{4-x} (-4)(4 - x - y)\, dy\, dx = -\frac{128}{3}$$

14.7.15 $\nabla \times \mathbf{F} = \langle 2z,\ -1,\ -2y\rangle$. Take S to be the disk $\langle 3r\cos t,\ 4r\cos t,\ 5r\sin t\rangle$ for $0 \le r \le 1, 0 \le t \le 2\pi$. $\mathbf{t}_r \times \mathbf{t}_t = \langle 20r,\ -15r,\ 0\rangle$ is a normal vector. Then

$$\iint_S (\nabla \times \mathbf{F})\cdot \mathbf{n}\, dS = \iint_R \langle 2z,\ -1,\ -2y\rangle \cdot \langle 20r,\ -15r,\ 0\rangle\, dA = \int_0^{2\pi}\int_0^1 ((10r\sin t)\cdot 20r + 15r)\, dr\, dt = 15\pi$$

14.7.17 The boundary of the surface is the ellipse $\frac{x^2}{4} + \frac{y^2}{9} = 1$, found by setting $z = 0$. Parameterize the path by $\mathbf{r}(t) = \langle 2\cos t,\ 3\sin t,\ 0\rangle$; then

$$\oint \mathbf{F} \cdot d\mathbf{r} = \int_0^{2\pi} \mathbf{F} \cdot \mathbf{r}'(t)\, dt = \int_0^{2\pi} (-4\cos t \sin t + 9\cos t \sin t)\, dt = \int_0^{2\pi} 5\cos t \sin t\, dt = 0$$

14.7.19 The boundary of the surface is the intersection of the plane $z = 3$ with the sphere $x^2 + y^2 + z^2 = 25$, so is the circle $x^2 + y^2 = 16$ at $z = 3$. With the usual parameterization for the circle, we have $\mathbf{r}'(t) = \langle -4\sin t,\ 4\cos t,\ 0\rangle$, so

$$\oint \mathbf{F} \cdot d\mathbf{r} = \int_0^{2\pi} \langle 8\sin t,\ -3,\ 4\cos t - 4\sin t - 3\rangle \cdot \langle -4\sin t,\ 4\cos t,\ 0\rangle\, dt = \int_0^{2\pi} (-32\sin^2 t - 12\cos t)\, dt = -32\pi$$

14.7.21 $\nabla \times \mathbf{v} = \langle 1,\ 0,\ 0\rangle$. The curl looks like:

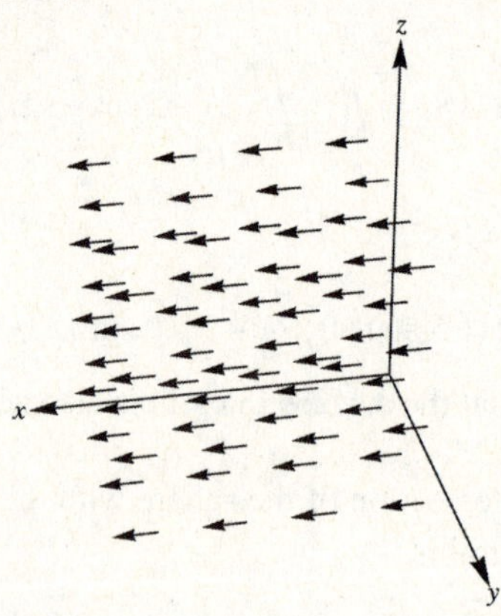

This means that the maximum rotation of the field is in the direction $\langle 1,\ 0,\ 0\rangle$. The rotation is counterclockwise looking in the negative x direction.

14.7.23 $\nabla \times \mathbf{v} = \langle 0,\ -2,\ 0\rangle$. The curl looks like:

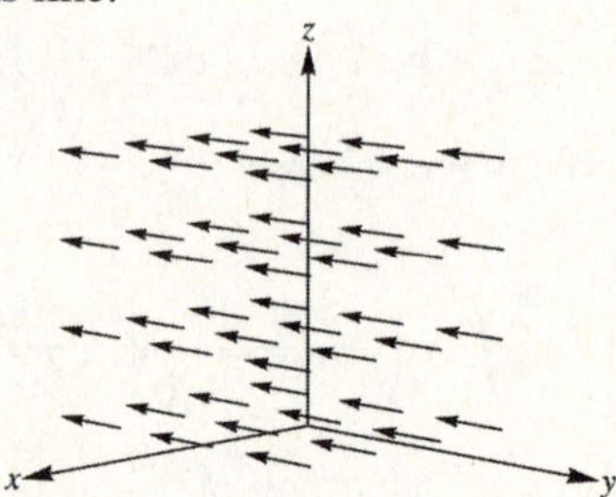

This means that the maximum rotation of the field is in the direction of the y-axis. It is constant at all points, and is clockwise viewed from the positive y-axis.

14.7.25

a. False. This is a rotation field with axis of rotation $\langle 1,\ 1,\ 2\rangle$, but $\langle 0,\ 1,\ -1\rangle \cdot \langle 1,\ 1,\ 2\rangle \neq 0$, so the paddle wheel axis is not perpendicular to the axis of rotation.

b. False. It relates the curl of **F**, not its flux.

c. True, since it is conservative: it is the gradient of $ax + F(x) + by + G(y) + cz + H(z)$, where F, G, H are the antiderivatives of f, g, h respectively.

d. True. See Theorem 14.14.

14.7.27 This is a conservative vector field, so the integral around any closed curve is zero.

14.7.29 This is a conservative vector field with $\varphi = xy^2z^3$, so the integral around any closed curve is zero.

14.7.31 $\mathbf{r}'(t) = \langle -\cos\varphi \sin t,\ \cos t,\ -\sin\varphi \sin t\rangle$, so that $|\mathbf{r}'(t)| = 1$. Then the length of C is

$$\int_C 1\, ds = \int_0^{2\pi} 1\, dt = 2\pi\,,$$

again as expected since C is just an inclined unit circle.

14.7.33 We have

$$\oint \mathbf{F}\cdot d\mathbf{r} = \int_C \langle -y,\ -z,\ x\rangle \cdot \langle -\cos\varphi \sin t,\ \cos t,\ -\sin\varphi \sin t\rangle\, dt$$
$$= \int_0^{2\pi} (\cos\varphi \sin^2 t - \sin\varphi \cos^2 t + \cos\varphi \sin\varphi \cos t \sin t)\, dt = \pi(\cos\varphi - \sin\varphi)$$

This is maximum for $\varphi = 0$, when it is π.

14.7.35 $\nabla \times \mathbf{F} = \langle 3,\ 0,\ 0\rangle$. To evaluate the circulation around C, we instead (using Stokes' theorem) evaluate

$$\iint_S (\nabla \times \mathbf{F})\cdot \mathbf{n}\, dS$$

for the surface of the disk of which C is a boundary. Note that $\mathbf{n} = \langle 1,\ 1,\ 1\rangle$, so that

$$\iint_S (\nabla \times \mathbf{F})\cdot \mathbf{n}\, dS = \iint_R 3\, dA = 3\cdot \text{area of } A = 48\pi$$

From this calculation, it is clear that the result depended on the radius of the circle, since that affects the area of A, but not on the center of the circle.

14.7.37

a. By the right-hand rule, the normal vectors consistent with the orientation of the boundary point inwards.

b. To evaluate the integral, we must add up the integrals on each of the surfaces. $\nabla \times \mathbf{F} = \langle 1,\ 1,\ 1\rangle$. Let S_1 be the surface in the xz-plane, parameterized by $\langle x,\ 0,\ z\rangle$ for $-2 \le x \le 2$, $x^2 \le z \le 4$; then the normal to S_1 is $\mathbf{t}_x \times \mathbf{t}_z = \langle 0,\ 1,\ 0\rangle$, so that the integral over S_1 is

$$\iint_{S_1} \langle 1,\ 1,\ 1\rangle \cdot \langle 0,\ -1,\ 0\rangle\, dS = \int_{-2}^{2}\int_{x^2}^{4} (-1)\, dy\, dx = -\int_{-2}^{2} (4 - x^2)\, dx = -\frac{32}{3}$$

S_2 is the half of the paraboloid for $y \ge 0$, parameterized as $\langle r\cos u,\ r\sin u,\ r^2\rangle$, $0 \le r \le 2$, $-\pi \le u \le 0$. The normal to S_2 is $\mathbf{t}_r \times \mathbf{t}_u = \langle -2r^2 \cos u,\ -2r^2 \sin u,\ r\rangle$. The integral over S_2 is

$$\iint_{S_1} \langle 1,\ 1,\ 1\rangle \cdot \langle -2r^2\cos u,\ -2r^2 \sin u,\ r\rangle\, dS = \int_0^2\int_{-\pi}^{0} (-2r^2(\cos u + \sin u) + r)\, du\, dr = \frac{32}{3} + 2\pi$$

Thus the total is 2π.

c The line integral is the sum of two line integrals:

$$\oint_C \mathbf{F}\cdot d\mathbf{r} = \oint_{C_1} \mathbf{F}\cdot d\mathbf{r}_1 + \oint_{C_2} \mathbf{F}\cdot d\mathbf{r}_2$$

where $C_1 = \langle t,\ 0,\ 4\rangle$ for $-2 \le t \le 2$ and $C_2 = \langle 2\cos t,\ 2\sin t,\ 4\rangle$ for $-\pi \le t \le 0$. Then

$$\oint_{C_1} \mathbf{F}\cdot d\mathbf{r}_1 = \int_{-2}^{2} \langle 2z + y,\ 2x + z,\ 2y + x\rangle \cdot \langle 1,\ 0,\ 0\rangle\, dt = \int_2^{-2} 8\, dt = -32$$
$$\oint_{C_2} \mathbf{F}\cdot d\mathbf{r}_2 = \int_{-\pi}^{0} \langle 2z + y,\ 2x + z,\ 2y + x\rangle \cdot \langle -2\sin t,\ 2\cos t,\ 0\rangle\, dt$$
$$= \int_{-\pi}^{0} (-16\sin t - 4\sin^2 t + 8\cos t + 8\cos^2 t)\, dt = 32 + 2\pi$$

so the total line integral is 2π.

14.7.39 The boundary of the region is the circle C: $x^2 + y^2 = 1$ for $z = 0$. With the usual parameterization, we have

$$\oint_C \mathbf{F} \cdot d\mathbf{r} = \int_0^{2\pi} \langle \cos t - \sin t,\ \sin t,\ -\cos t \rangle \cdot \langle -\sin t,\ \cos t,\ 0 \rangle\, dt$$

$$= \int_0^{2\pi} (-\sin t \cos t + \sin^2 t + \sin t \cos t)\, dt = \int_0^{2\pi} \sin^2 t\, dt = \pi$$

So the integral is independent of a.

14.7.41

a. The boundary of this surface is the circle $x^2 + y^2 = 1$ at $z = 0$, so we choose instead the surface of the disk bounded by that circle. $\nabla \times \mathbf{F} = \langle 2x,\ 0,\ -2z \rangle$, which is $\langle 2x,\ 0,\ 0 \rangle$ at $z = 0$, and the normal to the disk is $\langle 0,\ 0,\ 1 \rangle$. Thus, the integral is equal to

$$\iint_S (\nabla \times \mathbf{F}) \cdot \mathbf{n}\, dS = \iint_S 0\, dS = 0$$

b With the usual parameterization of the boundary circle (and remembering that $z = 0$), we have

$$\oint_C \mathbf{F} \cdot d\mathbf{r} = \int_0^{2\pi} \langle 0,\ 0,\ 1 \rangle \cdot \langle -\sin t,\ \cos t,\ 0 \rangle\, dt = 0$$

14.7.43

a. $\nabla \times \mathbf{F} = \left\langle 0 - 0,\ 0 - 0,\ \dfrac{y^2 - x^2}{(x^2 + y^2)^2} - \dfrac{y^2 - x^2}{(x^2 + y^2)^2} \right\rangle = \mathbf{0}$

b. Let C be the unit circle with the usual parameterization; then

$$\oint_C \mathbf{F} \cdot d\mathbf{r} = \int_0^{2\pi} \langle -\sin t,\ \cos t,\ 0 \rangle \cdot \langle -\sin t,\ \cos t,\ 0 \rangle\, dt = 2\pi$$

c The theorem does not apply since the vector field is not defined at the origin, which is inside the curve C. For example, the limit of the y-coordinate is different depending on the direction.

14.7.45 By the chain rule, $\dfrac{df}{dy} = \dfrac{\partial f}{\partial y} + \dfrac{\partial f}{\partial z}\dfrac{\partial z}{\partial y}$ and similarly for g, h, so

$$M_y = \frac{\partial f}{\partial y} + \frac{\partial f}{\partial z}\frac{\partial z}{\partial y} + h z_{xy} + z_x\left(\frac{\partial h}{\partial y} + \frac{\partial h}{\partial z}\frac{\partial z}{\partial y}\right) = f_y + f_z z_y + h z_{xy} + z_x(h_y + h_z z_y)$$

$$N_x = \frac{\partial g}{\partial x} + \frac{\partial g}{\partial z}\frac{\partial z}{\partial x} + h z_{yx} + z_y\left(\frac{\partial h}{\partial x} + \frac{\partial h}{\partial z}\frac{\partial z}{\partial x}\right) = g_x + g_z z_x + h z_{yx} + z_y(h_x + h_z z_x)$$

14.7.47 Let $\mathbf{F} = \langle 0,\ g(y,\ z),\ h(y,\ z) \rangle$ be a vector field in the yz-plane; for a region R in that plane, with boundary C, the normal is $\langle 1,\ 0,\ 0 \rangle$. Now, $\nabla \times \mathbf{F} = \left\langle \dfrac{\partial h}{\partial y} - \dfrac{\partial g}{\partial z},\ 0,\ 0 \right\rangle$, so by Stokes' theorem,

$$\oint_C \mathbf{F} \cdot d\mathbf{r} = \iint_S (\nabla \times \mathbf{F}) \cdot \mathbf{n}\, dS = \iint_R \left(\frac{\partial h}{\partial y} - \frac{\partial g}{\partial z}\right) dA$$

14.8 The Divergence Theorem

14.8.1 The surface integral measures the flow across the boundary.

14.8.3 The Divergence Theorem says that the flow across the boundary equals the net expansion or contraction of the field within the solid, so that either can be computed from the other.

14.8.5 Since $\nabla \cdot \langle x,\ y,\ z \rangle = 3$, the divergence theorem says that the net outward flux is equal to

$$\iiint_D 3\, dV = 3 \cdot \text{volume of } S = 32\pi$$

14.8.7 The outward fluxes must be equal, since by the divergence theorem the net flux, which is the difference of the two, is zero.

14.8.9 For the volume integral, $\nabla \cdot \mathbf{F} = 9$, so that

$$\iiint_D \nabla \cdot \mathbf{F}\, dV = \iiint_D 9\, dV = 9 \cdot \frac{4}{3} 2^3 \pi = 96\pi$$

For the surface integral, with the usual parameterization of the sphere,

$$\begin{aligned}\mathbf{F} \cdot \mathbf{n} &= \langle 2a \sin u \cos v,\ 3a \sin u \sin v,\ 4a \cos u \rangle \cdot \langle a^2 \sin^2 u \cos v,\ a^2 \sin^2 u \sin v,\ a^2 \cos u \sin u \rangle \\ &= 2a^3 \sin^3 u \cos^2 v + 3a^3 \sin^3 u \sin^2 v + 4a^3 \cos^2 u \sin u\end{aligned}$$

and here $a = 2$, so that

$$\begin{aligned}\iint_S \mathbf{F} \cdot \mathbf{n}\, dS &= 8 \iint_R \left(2 \sin^3 u \cos^2 v + 3 \sin^3 u \sin^2 v + 4 \cos^2 u \sin u\right) dA \\ &= 8 \int_0^{2\pi} \int_0^{\pi} \left(2 \sin^3 u \cos^2 v + 3 \sin^3 u \sin^2 v + 4 \cos^2 u \sin u\right) du\, dv = 96\pi\end{aligned}$$

14.8.11 For the volume integral, $\nabla \cdot \mathbf{F} = 0$, so the volume integral is zero. For the surface integral, the boundary ellipsoid can be parameterized by $\langle 2 \sin u \cos v,\ 2\sqrt{2} \sin u \sin v,\ 2\sqrt{3} \cos u \rangle$, and

$$\mathbf{n} = \mathbf{t}_u \times \mathbf{t}_v = \langle 4\sqrt{6} \sin^2 \mathrm{u} \cos \mathrm{v},\ 4\sqrt{3} \sin^2 u \sin v,\ 4\sqrt{2} \sin u \cos u \rangle$$

so that

$$\begin{aligned}\mathbf{F} \cdot \mathbf{n} &= \langle 2\sqrt{3} \cos u - 2\sqrt{2} \sin u \sin v,\ 2 \sin u \cos v,\ -2 \sin u \cos v \rangle \cdot \\ &\qquad \langle 4\sqrt{6} \sin^2 \mathrm{u} \cos \mathrm{v},\ 4\sqrt{3} \sin^2 u \sin v,\ 4\sqrt{2} \sin u \cos u \rangle \\ &= -8 \sin^2 u \cos v \left(-2\sqrt{2} \cos u + \sqrt{3} \sin u \sin v\right)\end{aligned}$$

and then

$$\iint_S \mathbf{F} \cdot \mathbf{n}\, dS = \int_0^{2\pi} \int_0^{\pi} \left(-8 \sin^2 u \cos v \left(-2\sqrt{2} \cos u + \sqrt{3} \sin u \sin v\right)\right) du\, dv = 0$$

14.8.13 $\nabla \cdot \mathbf{F} = 0$, so by the Divergence Theorem, the net outward flux is zero since the volume integral of $\nabla \cdot \mathbf{F}$ is zero.

14.8.15 $\nabla \cdot \mathbf{F} = 0$, so by the Divergence Theorem, the net outward flux is zero since the volume integral of $\nabla \cdot \mathbf{F}$ is zero.

14.8.17 By the divergence theorem, we can compute the integral of $\nabla \cdot \mathbf{F}$ over the ball of radius $\sqrt{6}$. $\nabla \cdot \mathbf{F} = 2$, so the volume integral is

$$\iiint_D \nabla \cdot \mathbf{F}\, dV = 2 \cdot \text{volume of sphere of radius } \sqrt{6} = 2 \cdot \frac{4}{3}\pi \cdot 6\sqrt{6} = 16\pi\sqrt{6}$$

14.8.19 $\nabla \cdot \mathbf{F} = 4$, so by the divergence theorem, the outward flux is 4 times the volume of the tetrahedron, which is (by the formula for the volume of a pyramid), $\frac{1}{3}$ times the area of the base times the height, or $\frac{1}{6}$. So the outward flux is $\frac{2}{3}$.

14.8.21 $\nabla \cdot \mathbf{F} = -4$, so by the divergence theorem, the outward flux is -4 times the volume of the sphere, so is $-\frac{128}{3}\pi$.

14.8.23 $\nabla \cdot \mathbf{F} = 3$, so the outward flux is 3 times the volume of the paraboloid, which is

$$\int_0^2 \int_0^{2\pi} r(4 - r^2)\, d\theta\, dr = 8\pi,$$

so the outward flux is 24π.

14.8.25 $\nabla \cdot \mathbf{F} = -3$, so the outward flux across the boundary of D is the outward flux across the sphere of radius 4 minus that across the sphere of radius 3, which is $-3 \cdot \frac{4}{3}\pi(4^3 - 2^3) = -224\pi$.

14.8.27 $\nabla \cdot \mathbf{F} = \dfrac{2}{|\mathbf{r}|}$, so the outward flux across a sphere of radius r is

$$\iiint_D \frac{2}{\sqrt{x^2+y^2+z^2}}\,dV = \int_0^{2\pi}\int_0^{\pi}\int_0^{r}\frac{2}{\rho}\rho^2 \sin u\,d\rho\,du\,dv = 4\pi r^2$$

Thus the net outward flux across the boundary of the given region is 12π.

14.8.29 $\nabla \cdot \mathbf{F} = 2(x-y+z)$. The net outward flux is thus the difference in the outward flux across the two planes, so is

$$\iiint_D 2(x-y+z)\,dV = 2\left(\int_0^4\int_0^{4-x}\int_0^{4-x-y}(x-y+z)\,dz\,dy\,dx - \int_0^2\int_0^{2-x}\int_0^{2-x-y}(x-y+z)\,dz\,dy\,dx\right) = 20$$

14.8.31

a. False. For example, $\mathbf{F} = \langle y, 0, 0\rangle$ has $\nabla \cdot \mathbf{F}$ at all points of the unit sphere, but the normal to the unit sphere, $\left\langle \frac{x}{z}, \frac{y}{z}, 1\right\rangle$ is not perpendicular to $\mathbf{F}$ at all points.

b. False. For example, any rotation field has $\nabla \cdot \mathbf{F} = 0$, so that $\iint_S \mathbf{F}\cdot\mathbf{n}\,dS = 0$ by the divergence theorem, but $\mathbf{F}$ is not in general constant.

c. True, since it is bounded by $\iiint_D 1\,dV$.

14.8.33 Since $\nabla \cdot \mathbf{F} = 0$, the outward flux is zero from the Divergence theorem.

14.8.35 $\nabla \cdot \mathbf{F} = 3\sin y$, so the outward flux is

$$\iiint_S 3\sin y\,dV = \int_0^{\pi/2}\int_0^1\int_0^x 3\sin y\,dz\,dx\,dy = \frac{3}{2}$$

14.8.37

a. Either use Ex. 36(a), or compute $\mathbf{F}\cdot\mathbf{n} = \dfrac{\mathbf{r}}{|\mathbf{r}|}\cdot\dfrac{\mathbf{r}}{|\mathbf{r}|} = 1$, so the surface integral is

$$\iint_S \mathbf{F}\cdot\mathbf{n}\,dS = \text{area of sphere} = 4\pi a^2$$

b. $\nabla \cdot \mathbf{F} = 2|\mathbf{r}|^{-1}$, so if D is the shell between the spheres of radius ϵ and a, the volume integral is

$$\iiint_S 2|\mathbf{r}|^{-1}\,dV = 2\int_\epsilon^a\int_0^{2\pi}\int_0^{\pi} r\sin u\,du\,dv\,dr = 4\pi(a^2-\epsilon^2)$$

and $\lim\limits_{\epsilon\to 0} 4\pi(a^2-\epsilon^2) = 4\pi a^2$.

14.8.39

a. By Exercise 36, the flux of $\mathbf{E}$ across a sphere of radius a is $\dfrac{Q}{4\pi\epsilon_0}4\pi a^{3-3} = \dfrac{Q}{\epsilon_0}$.

b. The net outward flux across S is the difference of the fluxes across the inner and outer spheres; but by part (a), these are equal, so the net flux across S is zero.

c. The left-hand side is the flux across the boundary of D, while the right-hand side is the sum of the charge densities at each point of D. The statement says that the flux across the boundary, up to multiplication by a constant, is the sum of the charge densities in the region.

d. By the divergence theorem, and using part (c),

$$\frac{1}{\epsilon_0}\iiint_D q(x,\, y,\, z)\, dV = \iint_S \mathbf{E}\cdot\mathbf{n}\, dS = \iiint_D \nabla\cdot\mathbf{E}\, dV$$

and since this holds for all regions D, we conclude that $\nabla\cdot\mathbf{E} = \dfrac{q(x,\, y,\, z)}{\epsilon_0}$.

e. $\nabla^2\varphi = \nabla\cdot\nabla\varphi = \nabla\cdot\mathbf{E} = \dfrac{q(x,\, y,\, z)}{\epsilon_0}$.

14.8.41 $\mathbf{F} = -\nabla T = \langle -1,\, -2,\, -1\rangle$, so that $\nabla\cdot\mathbf{F} = 0$ and the heat flux is zero.

14.8.43 $\mathbf{F} = -\nabla T = \langle 0,\, 0,\, e^{-z}\rangle$; then $\nabla\cdot\mathbf{F} = -e^{-z}$. The heat flux is then $\displaystyle\int_0^1\int_0^1\int_0^1 -e^{-z}\, dx\, dy\, dz = e^{-1} - 1$.

14.8.45 $\mathbf{F} = -\nabla T = \langle 200xe^{-x^2-y^2-z^2},\, 200ye^{-x^2-y^2-z^2},\, 200ze^{-x^2-y^2-z^2}\rangle$. Then

$$\nabla\cdot\mathbf{F} = 200e^{-x^2-y^2-z^2}(3 - 2x^2 - 2y^2 - 2z^2)$$

so that

$$\iiint_D \nabla\cdot\mathbf{F}\, dV = 200\int_0^{2\pi}\int_0^{\pi}\int_0^{a} e^{-r^2}(3 - 2r^2)r^2\sin u\, dr\, du\, dv = 800\pi a^3 e^{-a^2}$$

14.8.47

a. $\varphi_x(x,\, y,\, z) = G'(\rho)\rho_x = G'(\rho)\cdot\dfrac{x}{\sqrt{x^2+y^2+z^2}} = G'(\rho)\dfrac{x}{\rho}$, so that $\nabla\varphi = \mathbf{F} = G'(\rho)\dfrac{\mathbf{r}}{\rho}$.

b. The normal to the sphere of radius a is $\left\langle \dfrac{x}{z},\, \dfrac{y}{z},\, 1\right\rangle$, so on that sphere (where $\rho = a$)

$$\mathbf{F}\cdot\mathbf{n} = G'(a)\frac{\frac{x^2}{z} + \frac{y^2}{z} + z}{a} = G'(a)\frac{\frac{a^2-z^2}{z} + z}{a} = G'(a)\frac{a}{z},$$

and then the surface integral over the upper hemisphere is

$$\iint_S \mathbf{F}\cdot\mathbf{n}\, dS = aG'(a)\int_0^a\int_0^{2\pi}\frac{r}{\sqrt{a^2-r^2}}\, d\theta\, dr = 2\pi a^2 G'(a),$$

so the total surface integral is twice that, or $4\pi a^2 G'(a)$.

c. By the Chain Rule,

$$\frac{\partial}{\partial x}G'(\rho)\frac{x}{\rho} = G''(\rho)\rho_x\frac{x}{\rho} + G'(\rho)\frac{y^2+z^2}{\rho^3}$$

so that (noting that $\rho_x = \frac{x}{\rho}$)

$$\nabla\cdot\mathbf{F} = G''(\rho)\left(\frac{x^2+y^2+z^2}{\rho^2}\right) + G'(\rho)\frac{2(x^2+y^2+z^2)}{\rho^3} = G''(\rho) + \frac{2G'(\rho)}{\rho}$$

d. By the Divergence theorem, the flux is also given by

$$\begin{aligned}\iiint_D \nabla\cdot\mathbf{F}\, dV &= \int_0^a\int_0^{\pi}\int_0^{2\pi}\rho^2\sin u\left(G''(\rho) + \frac{2G'(\rho)}{\rho}\right) dv\, du\, d\rho\\ &= 2\pi\int_0^a\int_0^{\pi}\sin u(\rho^2 G''(\rho) + 2\rho G'(\rho))\, du\, d\rho\\ &= 2\pi\int_0^a(-\cos u)(\rho^2 G''(\rho) + 2\rho G'(\rho))\Big|_{u=0}^{u=\pi}\, d\rho\\ &= 4\pi\int_0^a(\rho^2 G''(\rho) + 2\rho G'(\rho))\, d\rho\end{aligned}$$

It remains to evaluate this integral. Using integration by parts, we have

$$4\pi\left[\int_0^a(\rho^2 G''(\rho) + 2\rho G'(\rho))\, d\rho\right] = 4\pi\left[\rho^2 G'(\rho)\Big|_{\rho=0}^{\rho=a} - \int_0^a 2\rho G'(\rho)\, d\rho\right] = 4\pi\left[a^2G'(a) - \int_0^a 2\rho G'(\rho)\, d\rho\right]$$

and the remaining integrals cancel, giving $4\pi a^2 G'(a)$ as the final result.

14.8.49 Suppose $\mathbf{F} = \langle f, g\rangle$ where $f = f(x, y)$, $g = g(x, y)$, and suppose $u = u(x, y)$. Then $\nabla \cdot \mathbf{F} = f_x + g_y$, and $\nabla u = \langle u_x, u_y\rangle$. Then we have for this case

$$\iiint_D u\nabla \cdot \mathbf{F}\, dS = \iint_R u(f_x + g_y)\, dA$$

$$\iint_S u\mathbf{F} \cdot \mathbf{n}\, dS = \oint_C u\mathbf{F} \cdot \mathbf{n}\, ds$$

$$\iiint_D \mathbf{F} \cdot \nabla u\, dS = \iint_R (f\, u_x + g\, u_y)\, dA,$$

and the result follows. Setting $u = 1$ then gives

$$\iint_R (f_x + g_y)\, dA = \oint_C \mathbf{F} \cdot \mathbf{n}\, ds,$$

which is the flux form of Green's Theorem.

14.8.51 From Exercise 50, we have both

$$\iiint_D (u\nabla^2 v + \nabla u \cdot \nabla v)\, dV = \iint_S u\nabla v \cdot \mathbf{n}\, dS$$

$$\iiint_D (v\nabla^2 u + \nabla v \cdot \nabla u)\, dV = \iint_S v\nabla u \cdot \mathbf{n}\, dS,$$

where the second formula is obtained by switching u and v in the first formula. Subtracting, and noting that

$$\iiint_D (u\nabla^2 v - v\nabla^2 u)\, dV = \iint_S (u\nabla v - v\nabla u) \cdot \mathbf{n}\, dS$$

14.8.53 The divergence theorem applied to the field $\nabla\varphi$ says that

$$\iiint_D (\nabla^2\varphi)\, dV = \iint_S \nabla\varphi \cdot \mathbf{n}\, dS$$

and if φ is harmonic, the left side is zero.

14.8.55 If $\mathbf{T}$ is a vector field $\langle t, u, v\rangle$, then by $\iint \mathbf{T}$ we mean $\left\langle \iint t, \iint u, \iint v \right\rangle$.

a. Let $\mathbf{F} = \langle f, g, h\rangle$ and suppose $\mathbf{n} = \langle n_1, n_2, n_3\rangle$. Then

$$\mathbf{n} \times \mathbf{F} = \langle n_2 h - n_3 g, n_3 f - n_1 h, n_1 g - n_2 f\rangle$$

$$\nabla \times \mathbf{F} = \left\langle \frac{\partial h}{\partial y} - \frac{\partial g}{\partial z}, \frac{\partial f}{\partial z} - \frac{\partial h}{\partial x}, \frac{\partial g}{\partial x} - \frac{\partial f}{\partial y} \right\rangle$$

Considering first the **i** component of these vectors, note that for the vector field $\mathbf{F}_1 = \langle 0, h, -g\rangle$, the divergence theorem says that

$$\iint_S (n_2 h - n_3 g)\, dS = \iint_S \langle 0, h, -g\rangle \cdot \langle n_1, n_2, n_3\rangle\, dS = \iint_S \mathbf{F}_1 \cdot \mathbf{n}\, dS = \iiint_D (\nabla \cdot \mathbf{F}_1)\, dR$$

$$= \iint_D \left(\frac{\partial h}{\partial y} - \frac{\partial g}{\partial z}\right) dA$$

and similarly for the second and third components.

b. Similarly to part (a), note that

$$\mathbf{n} \times \nabla\varphi = \mathbf{n} \times \langle \varphi_x, \varphi_y, \varphi_z \rangle = \langle n_2\varphi_z - n_3\varphi_y, n_3\varphi_x - n_1\varphi_z, n_1\varphi_y - n_2\varphi_x \rangle$$

Looking first at the x component of this vector, we have

$$\begin{aligned} n_2\varphi_z - n_3\varphi_y &= \langle 0, \varphi_z, -\varphi_y \rangle \cdot \langle n_1, n_2, n_3 \rangle \\ &= (\nabla \times \langle \varphi, 0, 0 \rangle) \cdot \langle n_1, n_2, n_3 \rangle \end{aligned}$$

so that Stokes' theorem says that, writing $\mathbf{F} = \langle \varphi, 0, 0 \rangle$,

$$\iint_S (\nabla \times \langle \varphi, 0, 0 \rangle) \cdot \langle n_1, n_2, n_3 \rangle \, dS = \iint_S (\nabla \times \mathbf{F}) \cdot \mathbf{n} \, dS = \oint_C \mathbf{F} \cdot d\mathbf{r} = \oint_C \langle \varphi, 0, 0 \rangle \cdot d\mathbf{r} = \oint_C \varphi \, d\mathbf{r}$$

and similarly for the second and third components.

14.9 Chapter Fourteen Review

14.9.1

a. False. The curl is $\dfrac{\partial}{\partial x}(x) - \dfrac{\partial}{\partial y}(-y) = 2$.

b. True. The curl of a conservative vector field is zero.

c. False. For example, $\langle -y, x \rangle$ and $\langle 0, 2x \rangle$ both have curl 2.

d. False. For example, $\langle x, 0, 0 \rangle$ and $\langle 0, y, 0 \rangle$ both have divergence 1.

e. True; by the divergence theorem, the integral is equal to $\displaystyle\iiint_D \nabla \cdot \mathbf{F} \, dV = \iiint_D 3 \, dV$

14.9.3 $\nabla\varphi = \langle 2x, 8y \rangle$.

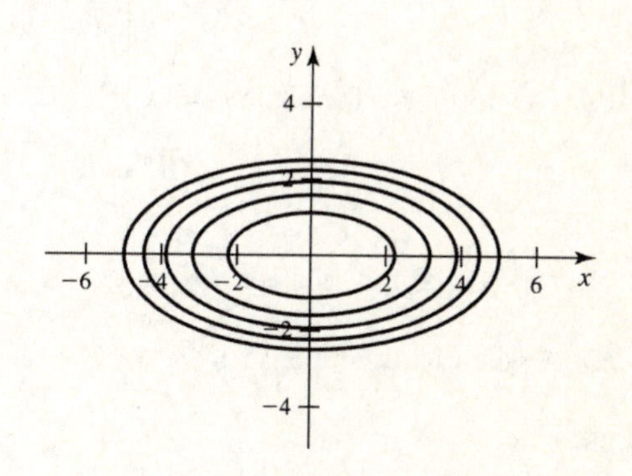

14.9.5 $\nabla\varphi = -\frac{\mathbf{r}}{|\mathbf{r}|^3}$

14.9.7

a. $\langle x, y \rangle$ is an outward normal; for (x, y) on the circle, $|\langle x, y \rangle| = \sqrt{x^2 + y^2} = 2$, so the unit outward normal is $\dfrac{1}{2}\langle x, y \rangle$.

b. The normal component is $2\langle y, -x \rangle \cdot \dfrac{1}{2}\langle x, y \rangle = 0$.

c. The normal component is $\dfrac{1}{x^2 + y^2}\langle x, y \rangle \cdot \dfrac{1}{2}\langle x, y \rangle = \dfrac{1}{x^2 + y^2} \cdot \dfrac{1}{2} \cdot (x^2 + y^2) = \dfrac{1}{2}$

14.9.9 Here $|\mathbf{r}'(t)| = \sqrt{1 + 9 + 36} = \sqrt{46}$, so

$$\int_C y \, e^{-xz} \, ds = \sqrt{46} \int_0^{\ln 8} 3te^{6t^2} \, dt = \frac{\sqrt{46}}{4}\left(e^{54 (\ln 2)^2} - 1\right)$$

14.9.11 For the first parameterization we have $|\mathbf{r}'(t)| = 2$, so

$$\oint_C (x - 2y - 3z) \, ds = 2 \int_0^{2\pi} (2 \cos t - 4 \sin t) \, dt = 0$$

For the second parameterization we have $|\mathbf{r}'(t)| = \sqrt{16t^2\sin^2(t^2) + 16t^2\cos^2(t^2)} = 4t$, so

$$\oint_C (x - 2y - 3z)\,ds = \int_0^{\sqrt{2\pi}} (8t\cos t^2 - 16t\sin t^2)\,dt = 0$$

14.9.13 Parameterize the first path by $\mathbf{r}_1(t) = \langle 0,\, t,\, 0\rangle$, and the second by $\mathbf{r}_1(t) = \langle 0,\, 1,\, 4t\rangle$, both for $0 \le t \le 1$. Then $\mathbf{r}'_1(t) = \langle 0,\, 1,\, 0\rangle$ and $\mathbf{r}'_2(t) = \langle 0,\, 0,\, 4\rangle$

$$\int_C \mathbf{F}\cdot d\mathbf{r} = \int_0^1 (\langle -t,\, 0,\, 0\rangle \cdot \langle 0,\, 1,\, 0\rangle + \langle -1,\, 4t,\, 0\rangle \cdot \langle 0,\, 0,\, 4\rangle)\,dt = 0$$

14.9.15 The circulation is

$$\int_C \mathbf{F}\cdot\mathbf{T}\,ds = \int_0^{2\pi} \langle 2\sin t - 2\cos t,\, 2\sin t\rangle \cdot \langle -2\sin t,\, 2\cos t\rangle\,dt = \int_0^{2\pi} (-4\sin^2 t + 8\sin t\cos t)\,dt = -4\pi$$

The outward flux is

$$\int_C \mathbf{F}\cdot\mathbf{n}\,ds = \int_0^{2\pi} ((2\sin t - 2\cos t)(2\cos t) - (2\sin t)(-2\sin t))\,dt = 4\int_0^{2\pi} (\sin t\cos t - \cos^2 t + \sin^2 t)\,dt = 0$$

14.9.17 The circulation is

$$\int_C \mathbf{F}\cdot\mathbf{T}\,ds = \frac{1}{4}\int_0^{2\pi} \langle 2\cos t,\, 2\sin t\rangle \cdot \langle -2\sin t,\, 2\cos t\rangle\,dt = 0$$

The outward flux is

$$\int_C \mathbf{F}\cdot\mathbf{n}\,ds = \frac{1}{4}\int_0^{2\pi} ((2\cos t)(2\cos t) - (2\sin t)(-2\sin t))\,dt = 2\pi$$

14.9.19 The normal to the plane $x = 0$ is $\langle 1,\, 0,\, 0\rangle$, so the flux is

$$\int_C \mathbf{F}\cdot\mathbf{n}\,ds = \int_{-1/2}^{1/2}\int_{-L}^{L} v_0(L^2 - y^2)\,dy\,dz = \frac{4}{3}v_0 L^3$$

14.9.21 A potential function is $xy + yz^2$.

14.9.23 A potential function is $xy\,e^z$.

14.9.25

a. $\mathbf{F} = \langle yz,\, xz,\, xy\rangle$, so

$$\int_C \mathbf{F}\cdot d\mathbf{r} = \int_0^{\pi} \left\langle \frac{t}{\pi}\sin t,\, \frac{t}{\pi}\cos t,\, \sin t\cos t\right\rangle \cdot \left\langle -\sin t,\, \cos t,\, \frac{1}{\pi}\right\rangle dt = \int_0^{\pi} \left(\frac{t}{\pi}(\cos^2 t - \sin^2 t) + \frac{1}{\pi}\sin t\cos t\right) dt$$
$$= 0$$

b. $\mathbf{F} = \nabla(xyz) = \nabla\varphi$, so

$$\int_C \mathbf{F}\cdot d\mathbf{r} = \varphi(\cos\pi\,\sin\pi) - \varphi\left(\cos 0\,\sin 0\cdot\frac{0}{\pi}\right) = 0 - 0 = 0$$

14.9.27

a.

$$\int_C \mathbf{F}\cdot d\mathbf{r} = \int_0^{2\pi} \left\langle \sin t,\, 4,\, -\cos t\right\rangle \cdot \left\langle -\sin t,\, \cos t,\, 0\right\rangle dt = \int_0^{2\pi} (-\sin^2 t + 4\cos t)\,dt = -\pi$$

b. The vector field is not conservative, since for example $\dfrac{\partial}{\partial y}(y) \neq \dfrac{\partial}{\partial z}(x)$.

14.9.29 By the circulation form of Green's Theorem,

$$\oint_C xy^2\,dx + x^2 y\,dy = \iint_R \left(\frac{\partial}{\partial x}(x^2 y) - \frac{\partial}{\partial y}(xy^2)\right) dA = \iint_R (2xy - 2xy)\,dA = 0$$

14.9.31 By the circulation form of Green's Theorem,

$$\oint_C (x^3+xy)\,dy + (2y^2-2x^2y)\,dx = \iint_R \left(\frac{\partial}{\partial x}(x^3+xy) - \frac{\partial}{\partial y}(2y^2-2x^2y)\right) dA$$

$$= \iint_R (3x^2+y-4y+2x^2)\,dA = \int_{-1}^{1}\int_{-1}^{1}(5x^2-3y)\,dy\,dx = \frac{20}{3}$$

14.9.33 The ellipse is $\frac{x^2}{16}+\frac{y^2}{4}=1$; parameterize it by $\mathbf{r}(t)=\langle x,\,y\rangle=\langle 4\cos t,\,2\sin t\rangle,\ 0\le t\le 2\pi$. Then the area of the region is

$$\frac{1}{2}\oint_C ((4\cos t)(2\cos t)-(2\sin t)(-4\sin t))\,dt = \int_0^{2\pi} 4(\cos^2 t+\sin^2 t)\,dt = 8\pi$$

14.9.35

a. $\mathbf{F}=(x^2+y^2)^{-1/2}\langle x,\,y\rangle$, so the circulation is

$$\oint_C \mathbf{F}\cdot d\mathbf{r} = \iint_R \left(\frac{\partial}{\partial x}\left(\frac{y}{\sqrt{x^2+y^2}}\right)-\frac{\partial}{\partial y}\left(\frac{x}{\sqrt{x^2+y^2}}\right)\right) dA = \iint_R \left(\frac{-xy}{\sqrt{x^2+y^2}}-\frac{-xy}{\sqrt{x^2+y^2}}\right) dA = 0$$

b. The flux is

$$\oint_C \mathbf{F}\cdot\mathbf{n}\,ds = \iint_R \left(\frac{\partial}{\partial x}\left(\frac{x}{\sqrt{x^2+y^2}}\right)+\frac{\partial}{\partial y}\left(\frac{y}{\sqrt{x^2+y^2}}\right)\right) dA = \iint_R \left(\frac{y^2}{(x^2+y^2)^{3/2}}+\frac{x^2}{(x^2+y^2)^{3/2}}\right) dA$$

$$= \iint_R \left(\frac{1}{\sqrt{x^2+y^2}}\right) dA = \int_0^{\pi}\int_1^3 1\,dr\,d\theta = 2\pi$$

14.9.37

a. For $\mathbf{F}$ to be conservative, we must have $\frac{\partial}{\partial y}(ax+by)=\frac{\partial}{\partial x}(cx+dy)$, or $b=c$.

b. For $\mathbf{F}$ to be source-free, we must have $\frac{\partial}{\partial x}(ax+by)=-\frac{\partial}{\partial y}(cx+dy)$, or $a=-d$.

c. $\mathbf{F}$ is both conservative and source-free if $b=c$ and $a=-d$, i.e. if $\mathbf{F}=\langle ax+by,\,bx-ay\rangle$.

14.9.39 The divergence is $4|\mathbf{r}|$. The curl is

$$\left\langle \frac{\partial}{\partial y}(z\,|\mathbf{r}|)-\frac{\partial}{\partial z}(y\,|\mathbf{r}|),\ \frac{\partial}{\partial z}(x\,|\mathbf{r}|)-\frac{\partial}{\partial x}(z\,|\mathbf{r}|),\ \frac{\partial}{\partial x}(y\,|\mathbf{r}|)-\frac{\partial}{\partial y}(x\,|\mathbf{r}|)\right\rangle = \mathbf{0}.$$

The field is irrotational but not source-free.

14.9.41 The divergence is $\frac{\partial}{\partial x}(2xy+z^4)+\frac{\partial}{\partial y}(x^2)+\frac{\partial}{\partial z}(4xz^3)=2y+12xz^2$. The curl is

$$\left\langle \frac{\partial}{\partial y}(4xz^3)-\frac{\partial}{\partial z}(x^2),\ \frac{\partial}{\partial z}(2xy+z^4)-\frac{\partial}{\partial x}(4xz^3),\ \frac{\partial}{\partial x}(x^2)-\frac{\partial}{\partial y}(2xy+z^4)\right\rangle = \mathbf{0},$$

so the field is irrotational but not source-free.

14.9.43

a. The curl is

$$\left\langle \frac{\partial}{\partial y}(-y)-\frac{\partial}{\partial z}(x),\ \frac{\partial}{\partial z}(z)-\frac{\partial}{\partial x}(-y),\ \frac{\partial}{\partial x}(x)-\frac{\partial}{\partial y}(z)\right\rangle = \langle -1,\,1,\,1\rangle$$

So the scalar component in the direction of $\langle 1,\,0,\,0\rangle$ is $\langle -1,\,1,\,1\rangle\cdot\langle 1,\,0,\,0\rangle=-1$, and the scalar component in the direction of $\left\langle 0,\,-\frac{1}{\sqrt{2}},\,\frac{1}{\sqrt{2}}\right\rangle$ is $\left\langle 0,\,-\frac{1}{\sqrt{2}},\,\frac{1}{\sqrt{2}}\right\rangle\cdot\langle 1,\,0,\,0\rangle=0$

b. The scalar component of the curl is a maximum in the direction of the curl, i.e. in the direction $\langle -1,\,1,\,1\rangle$, whose unit vector is $\frac{1}{\sqrt{3}}\langle -1,\,1,\,1\rangle$.

14.9.45 Parameterize the sphere by $\langle 3\sin u\cos v,\ 3\sin u\sin v,\ 3\cos u\rangle$, $0 \le u \le \frac{\pi}{2}$; $0 \le v \le 2\pi$. Then $|\mathbf{n}| = 9\sin u$, so

$$\iint_S 1\,dS = \iint_R 9\sin u\,dA = \int_0^{2\pi}\int_0^{\pi/2} 9\sin u\,du\,dv = 18\pi$$

14.9.47 The volume element is $\sqrt{1^2+1^2+1^2} = \sqrt{3}$, so the area is

$$\iint_S 1\,dS = \sqrt{3}\iint_R 1\,dA = \sqrt{3}\int_{-1}^{1}\int_{-1}^{1} 1\,dx\,dy = 4\sqrt{3}$$

14.9.49 The volume element for $z = 2 - x - y$ is $\sqrt{3}$, so the integral is

$$\iint_S (1+yz)\,dS = \sqrt{3}\iint_R (1+yz)\,dA = \sqrt{3}\int_0^2\int_0^{2-x} (1+y(2-x-y))\,dy\,dx = \frac{8\sqrt{3}}{3}$$

14.9.51 Parameterize the curved surface using spherical coordinates, so that $|\mathbf{n}| = 4\sin u$; then for the curved surface we have

$$\iint_S (x-y+z)\,dS = 8\int_0^{2\pi}\int_0^{\pi/2} (\sin u\cos v - \sin u\sin v + \cos u)\sin u\,du\,dv = 8\pi$$

For the planar surface, $\mathbf{n} = \langle 0,\ -1,\ 0\rangle$ so that $|\mathbf{n}| = 1$ and

$$\iint_S (x-y+z)\,dS = \iint_R (x-y+z)\,dA = \int_0^2\int_0^{2\pi} r(\cos\theta - \sin\theta)\,d\theta\,dr = 0$$

and the total integral is thus 8π.

14.9.53 $\mathbf{F} = (x^2+y^2+z^2)^{-1/2}\langle x,\ y,\ z\rangle$; using spherical coordinates to parameterize the sphere gives

$$\iint_S \mathbf{F}\cdot\mathbf{n}\,dS = \iint_R \frac{1}{a}\langle a\sin u\cos v,\ a\sin u\sin v,\ a\cos u\rangle\cdot\langle a^2\sin^2 u\cos v,\ a^2\sin^2 u\sin v,\ a^2\sin u\cos u\rangle\,dA$$

$$= \iint_R a^2\sin u\,dA = \int_0^{2\pi}\int_0^{\pi} a^2\sin u\,du\,dv = 4\pi a^2$$

14.9.55

a. The base of S is the surface where $z = 0$, or the circle $x^2+y^2 = a^2$. Similarly, the base of the paraboloid is found by setting $z = 0$; simplifying gives again $x^2+y^2 = a^2$. The high point of the hemisphere (maximum z-coordinate) occurs when $x = y = 0$; then $z = a$. Similarly, the high point on the paraboloid also occurs when $x = y = 0$ and again this gives $z = a$.

b. The graph of the paraboloid is inside that of the hemisphere everywhere, so we would expect it to have smaller surface area. We know that the surface area of the hemisphere is $4\pi a^2\cdot\frac{1}{2} = 2\pi a^2$. For the paraboloid, we have

$z_x = -\dfrac{2x}{a}$, $z_y = -\dfrac{2y}{a}$, so that $|\mathbf{n}| = \sqrt{\dfrac{4(x^2+y^2)}{a^2}+1}$, so

$$\iint_S 1\,dS = \frac{1}{a}\iint_R \sqrt{4(x^2+y^2)+a^2}\,dA = \frac{1}{a}\int_0^a\int_0^{2\pi} r\sqrt{4r^2+a^2}\,d\theta\,dr = \frac{\left(5\sqrt{5}-1\right)}{6}\pi a^2$$

which is in fact smaller than the area of the hemisphere.

c. $\mathbf{n} = \langle a^2\sin^2 u\cos v,\ a^2\sin^2 u\sin v,\ a^2\cos u\sin u\rangle$, so

$$\iint_S \mathbf{F}\cdot\mathbf{n}\,dS = \iint_R \langle a\sin u\cos v,\ a\sin u\sin v,\ a\cos u\rangle\cdot\langle a^2\sin^2 u\cos v,\ a^2\sin^2 u\sin v,\ a^2\cos u\sin u\rangle\,dA$$
$$= a^3\int_0^{2\pi}\int_0^{\pi/2}(\sin^3 u\cos^2 v+\sin^3 u\sin^2 v+\cos^2 u\sin u)\,du\,dv$$
$$= a^3\int_0^{2\pi}\int_0^{\pi/2}(\sin^3 u+\cos^2 u\sin u)\,du\,dv = a^3\int_0^{2\pi}\int_0^{\pi/2}\sin u\,du\,dv = 2\pi a^3$$

d. For the paraboloid, the parameterization is $\left\langle v\cos u,\ v\sin u,\ a-\dfrac{v^2}{a}\right\rangle$, $0\le v\le a$, $0\le u\le 2\pi$, and

$$\mathbf{n} = \langle -v\sin u,\ v\cos u,\ 0\rangle\times\left\langle\cos u,\ \sin u,\ -\frac{2v}{a}\right\rangle = \left\langle -\frac{1}{a}2v^2\cos u,\ -\frac{1}{a}2v^2\sin u,\ -v\right\rangle,$$

so that the outward-pointing normal is $\left\langle\dfrac{2v^2}{a}\cos u,\ \dfrac{2v^2}{a}\sin u,\ v\right\rangle$ and

$$\iint_S \mathbf{F}\cdot\mathbf{n}\,dS = \iint_R\left\langle v\cos u,\ v\sin u,\ a-\frac{v^2}{a}\right\rangle\cdot\left\langle\frac{2v^2}{a}\cos u,\ \frac{2v^2}{a}\sin u,\ v\right\rangle dA$$
$$= \int_0^a\int_0^{2\pi}\left(\frac{2}{a}v^3+av-\frac{v^3}{a}\right)du\,dv = \frac{3}{2}\pi a^3$$

14.9.57 Parameterize $x^2+y^2=4$ for $z=0$ using $\mathbf{r}(t)=\langle 2\cos t,\ 2\sin t,\ 0\rangle$ for $0\le t\le 2\pi$.

$$\oint_C \mathbf{F}\cdot d\mathbf{r} = \int_0^{2\pi}\mathbf{F}\cdot\mathbf{r}'(t)\,dt = \int_0^{2\pi}\langle 0,\ 0,\ 4\cos t\sin t\rangle\cdot\langle -2\sin t,\ 2\cos t,\ 0\rangle\,dt = 0$$

14.9.59 The boundary of this region is in the xy-plane, found by setting $z=0$, so it is $x^2+y^2=99$, the circle of radius $\sqrt{99}$ about the origin. Parameterize the circle in the usual way; then

$$\iint_S(\nabla\times\mathbf{F})\cdot\mathbf{n}\,dS = \oint_C\mathbf{F}\cdot d\mathbf{r} = \oint_C\langle 0,\ \sqrt{99}\cos t,\ \sqrt{99}\sin t\rangle\cdot\langle -\sqrt{99}\sin t,\ \sqrt{99}\cos t,\ 0\rangle\,dt$$
$$= \int_0^{2\pi}99\cos^2 t\,dt = 99\pi$$

14.9.61 By Stokes' theorem, the circulation around a closed curve C can be found by choosing a surface S of which C is the boundary; then

$$\oint_C\mathbf{F}\cdot d\mathbf{r} = \iint_S(\nabla\times\mathbf{F})\cdot\mathbf{n}\,dS$$

But for $\mathbf{F}=\nabla(10-x^2+y^2+z^2)$, $\nabla\times\mathbf{F}=\mathbf{0}$, so the right-hand side is zero.

14.9.63 $\nabla\cdot\mathbf{F}=x^2+y^2+z^2$, so

$$\iint_S\mathbf{F}\cdot\mathbf{n}\,dS = \iiint_D(x^2+y^2+z^2)\,dV = \int_0^3\int_0^{2\pi}\int_0^{\pi}r^2\cdot r^2\sin u\,du\,dv\,dr = \frac{972}{5}\pi$$

14.9.65 $\nabla\cdot\mathbf{F}=3(x^2+y^2)$, so the outward flux across the boundary S of a hemisphere D of radius a is

$$\iint_S\mathbf{F}\cdot\mathbf{n}\,dS = \iiint_D 3(x^2+y^2)\,dV = \int_0^a\int_0^{2\pi}\int_0^{\pi/2}3r^2\sin^2 u\cdot r^2\sin u\,du\,dv\,dr = \frac{4}{5}\pi a^5$$

so that the net flux across the region bounded by the hemispheres of radii 1 and 2 is $\dfrac{4}{5}\pi(32-1)=\dfrac{124}{5}\pi$.

14.9.67 Using the Divergence theorem, $\nabla\cdot\mathbf{F}=2x+\sin y+2y-2\sin y+2z+\sin y=2(x+y+z)$, so that

$$\iint_S\mathbf{F}\cdot\mathbf{n}\,dS = \iiint_D 2(x+y+z)\,dV = \int_0^4\int_0^1\int_0^{1-x}2(x+y+z)\,dy\,dx\,dz = \frac{32}{3}$$